Nutrition and Metabolism in Sports, Exercise and Health

A key determinant of successful athletic performance is the high-level energy transformation, which begins with combustion of the food that we eat. By developing a sound understanding of good nutrition we can improve athletic performance, help maintain good health and prevent disease. This clear and comprehensive introduction to nutrition in sport, exercise and health goes further than any other textbook in integrating key nutritional facts, concepts and dietary guidelines with a thorough discussion of the fundamental biological science underpinning our physiological and metabolic processes. By clearly explaining how nutrients function within our biological system, the book helps students to develop a better understanding of the underlying mechanisms, which in turn will help them to apply their knowledge in practice.

The book includes in-depth discussion of key contemporary topics within nutrition, including:

- nutrient bioenergetics
- nutrition and metabolic disease
- nutritional ergogenic aids
- nutrition for special populations
- nutritional assessment.

Each chapter includes useful pedagogical features, such as review questions, case studies, definitions of key terms and practical laboratory exercises, including techniques for assessing nutritional status, body composition and physical activity patterns. A companion website offers additional teaching and learning features, such as PowerPoint slides, multiple-choice question banks and web links (www.routledge.com/cw/kang).

As the most up-to-date introduction to sport and exercise nutrition currently available, this book is essential reading for all students of sport and exercise science, kinesiology, physical therapy, nutrition, dietetics or health sciences.

Jie Kang is a Professor in the Department of Health and Exercise Science, College of New Jersey, USA. His primary research interests are alterations in energy metabolism, substrate utilization, cardiorespiratory function and perceived exertion in response to both aerobic and resistance exercise in healthy individuals and individuals with diseases.

Nutrition and Metabolism in Sports, Exercise and Health

Jie Kang

Routledge
Taylor & Francis Group

LONDON AND NEW YORK

First published 2012
by Routledge
2 Park Square, Milton Park, Abingdon, Oxon OX14 4RN

Simultaneously published in the USA and Canada
by Routledge
711 Third Avenue, New York, NY 10017

Routledge is an imprint of the Taylor & Francis Group, an informa business

British Library Cataloguing in Publication Data
A catalogue record for this book is available from the British Library

Library of Congress Cataloging-in-Publication Data
Nutrition and metabolism in sports, exercise and health / edited by Jie Kang.
   p. cm.
   1. Nutrition. 2. Athletes–Nutrition. I. Kang, Jie.
   QP141.N77 2012
   612.3–dc23   2011038292

ISBN: 978-0-415-57878-3 (hbk)
ISBN: 978-0-415-57879-0 (pbk)
ISBN: 978-0-203-85191-3 (ebk)

Typeset in Garamond
by Wearset Ltd, Boldon, Tyne and Wear

MIX
Paper from
responsible sources
FSC® C004839
www.fsc.org

Printed and bound in Great Britain by
TJ International Ltd, Padstow, Cornwall

Contents

Figures

Tables

1 Introduction

KEY TERMS

- nutrition
- obesity
- morbidity
- risk factor
- sports nutrition
- nutrients
- essential nutrients
- non-essential nutrients
- energy-yielding nutrients
- macronutrients
- micronutrients
- inorganic nutrients
- organic compounds

- malnutrition
- under-nutrition
- over-nutrition
- hunger
- appetite
- satiety
- hypothesis
- control group
- placebo
- single-blind study
- double-blind study
- epidemiological research
- experimental research

GOOD HEALTH AND STRONG PERFORMANCE: NUTRITION CONNECTION

Nutrition and its impact on health and performance are of crucial importance. Nutritional deficiencies were once a major health challenge in most developed countries. However, what we are facing now is the fact that nutritional abundance contributes to many of today's health problems. In order to choose foods that satisfy your personal and cultural preferences, but which also contribute to a healthy diet and prevent diseases, you must have information about what nutrients you require, what role they play in health and performance, and what foods contain them. You must also be able to judge the validity of the nutrition information you encounter. Your body uses the nutrients from foods to make all its components, fuel all its activities, and defend itself against diseases. How successfully your body handles these tasks depends, in part, on your food choices and your understanding of the principles of nutrition. Nutritious food choices support a healthy and strong body.

1

What is nutrition?

Nutrition is a science that links foods to health and diseases. It studies the structure and function of various food groups and the nutrients they contain. It also examines the biological processes by which our body consumes food and utilizes the nutrients. The science of nutrition also concerns the psychological, social, cultural, economic, and technological factors that influence which food we choose to eat.

Why study nutrition?

Nutrition has played a significant role in your life, even from before your birth, although you may not always be aware of it – and it will continue to affect you in major ways, depending on the foods you select. Not meeting nutrient needs in younger years makes us more likely to suffer health consequences in later years. At the same time, taking too much of a nutrient can be harmful. A poor diet and a sedentary life-style are known to be the major risk factors for life-threatening chronic diseases such as heart disease, hypertension, diabetes, and some forms of cancer, which together amount for two-thirds of all deaths in North America (Table 1.1). Such linkage between lifestyle and chronic disease is, in part, mediated through the development of **obesity**, a condition attributable to a positive energy balance (i.e., energy brought in via foods being greater than energy expended via physical activities). Most of these chronic diseases are the co-**morbidity** (a diseased state, disability, or poor health) of obesity. In fact, obesity is considered to be the second highest preventable cause of death in the United States (Figure 1.1)

Needless to say, your food choice today can affect your health tomorrow. Under-standing nutrition will allow you to make wise choices in terms of the foods you

Table 1.1 Leading causes of death in the United States		
RANK	CAUSE OF DEATH	PERCENTAGE OF TOTAL DEATHS
1	Heart diseases (primarily coronary heart disease)[1,2]	29
2	Cancer[1,3]	23
3	Cerebrovascular diseases (stroke)[1,2,3]	7
4	Chronic obstructive pulmonary diseases and allied conditions (lung diseases)[3]	5
5	Accidents and adverse effects[2]	4
6	Diabetes[1]	3
7	Influenza and pneumonia	3
8	Alzheimer's diseases[1]	2
9	Kidney diseases[1,3]	2
10	Blood-borne infections	1

Source: Center for Disease Control and Prevention, *National Vital Statistical Report*, final data.

Notes:
1 Causes of death in which diet plays a part.
2 Causes of death in which excessive alcohol consumption plays a part.
3 Causes of death in which tobacco use plays a part.

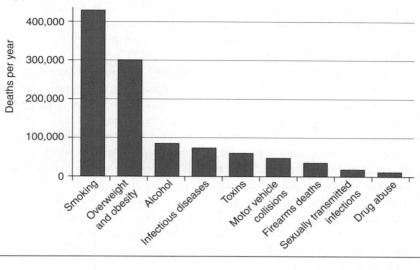

Figure 1.1 Leading preventable causes of death.

consume, thus improving your health and fitness. You must be aware, however, that making appropriate food choices is not an easy task, and can be influenced by many outside factors. A decision should only follow questions such as: Are you active? Are you an athlete? Are you planning a pregnancy? Are you trying to prevent the physical decline that occurs with aging? Has either of your parents died due to a heart attack? Does cancer run in your family? Are you trying to lose weight or eat a vegetarian diet? Is your heritage Asian, African, European, Central or South American? In order to choose foods that satisfy your personal and cultural preferences, but which also contribute to a healthy diet and prevent diseases, you must not only have information about what nutrients you require and what foods contain them, but also understand the role nutrients play in the body and how they may contribute to an enhanced physical performance or a pathological process that leads to a disease. You must also be able to judge the validity of the nutrition information you encounter. Should you be taking antioxidant supplements, eating fat-free foods, or drinking calcium-fortified orange juice? Should you believe the story or testimony you see on the television about a weight-loss diet or protein supplement? Filtering out the worthless requires a solid understanding of principles of nutrition, the nutrient contents of foods, the function of nutrition in the body, and the process by which scientists study nutrition.

Role of nutrition in fitness, health, and performance

The two primary factors that influence one's health status are genetics and lifestyle. Most chronic diseases have a genetic basis. The Human Genome Project, which deciphered the DNA code of our 80,000–100,000 genes, has identified various genes associated with many chronic diseases. Genetically, females whose mothers had breast cancer are at increased risk of breast cancer, while males whose fathers had prostate cancer are at increased risk of prostate cancer. Scientists now have the ability to analyze the genetics underlying various diseases, and such information may be used to evaluate individual susceptibility. For individuals with genetic profiles predisposing them to a specific chronic disease, gene therapy may provide an effective treatment or cure.

Genetic influence may play an important role in the development of chronic diseases, but so too does lifestyle. Recent studies have suggested that lifestyle, particularly one that incorporates a healthy diet and exercise, may provide the best hope for living a healthier and longer life. It is the most proactive and cost-effective approach to address an increasing prevalence of these chronic diseases in our society. Over the years, scientists in the field of epidemiology have identified a number of lifestyle-related **risk factors**. A risk factor is a health behavior or pre-existing condition that is associated with a particular disease, such as cigarette smoking, physical inactivity, stress, insulin resistance, hyperlipidemia, etc. Proper diet and exercise have been found to reduce many of these risk factors, thereby preventing diseases. It is believed that such healthier lifestyles can also intertwine with one's genetic profile – in other words, what you eat and how you exercise may influence your genes.

Proper nutrition is an important component in the total training program of athletes. Consumption of energy-containing nutrients such as carbohydrates provides the fuel necessary for increased biological work. Nutrient deficiencies can seriously impair performance, whereas nutrient supplementation may delay fatigue and improve performance. Nutritional status can be a major factor differentiating athletes of comparable genetic endowment and training. Regular training allows athletes to improve their performance by enhancing biomechanical skills, sharpening psychological focus, and maximizing physiological functions. However, gains in these areas can be directly enhanced or undermined by various dietary factors associated with the athlete. For example, losing excess body fat will enhance biomechanical efficiency; consuming carbohydrates during exercise may prevent hypoglycemia and thus fatigue; and providing adequate dietary iron may ensure optimal oxygen delivery to the working muscles.

Sports nutrition has become a fast-growing area of study within recent years. It is the study and practice of nutrition and diet as it relates to athletic performance. Although scientists have studied the interactions between nutrition and various forms of sports and physical activities for more than a century, it is only within the past few decades that extensive research has been undertaken regarding the specific guidelines and recommendations to athletes. Louise Burke, a prominent sports nutritionist from Australia, defines sports nutrition as the application of eating strategies to promote good health and adaptation to training, to recover quickly after each exercise training session, and to perform optimally during competition. A sound understanding of sports nutrition enables one to appreciate the importance of adequate nutrition, and to critically evaluate the validity of claims concerning specific dietary modifications and nutrient supplements to enhance physique, physical performance, and exercise training responses. Knowledge of the nutrition–metabolism interaction forms the basis for preparation, performance, and recovery phases of intense exercise or training. Many physically active individuals, including some of the world's best athletes, obtain nutritional information from magazine and newspaper articles, advertisements, training partners, and testimonials from successful athletes, rather than from well-informed and well-educated coaches, trainers, physicians, and fitness and sports nutrition professionals. Far too many cases have been reported in which athletes devote considerable time and energy striving for optimum performance, only to fall short due to inadequate, counterproductive, and sometimes harmful nutritional practices.

Nutrition plays a significant role in one's life. "Good nutrition" encompasses more than preventing nutrient deficiencies or inadequacies related to diseases. It also forms the foundation of one's fitness, physical performance, and overall well-being. As you gain an understanding about your nutritional habits and increase your knowledge about optimal nutrition, you will have the opportunity to reduce your risk of many common diseases, to meet the demands placed upon your body, and to stay healthy, fit, and strong.

NUTRIENTS

People eat to receive nourishment. Do you ever think of yourself as a biological being made of carefully arranged atoms, molecules, cells, tissues, and organs? Are you aware of the activity going on within your body even as you sit still? The atoms, molecules, and cells of your body continually move and change, even though the structures of your tissues and organs and your external appearance remain relatively constant. The skin that has covered you since birth is replaced entirely by new cells every seven years. The fat beneath your skin is not the same fat that was there a year ago. Your oldest red blood cell is only 120 days old, and the entire lining of your digestive tract is renewed every three days. To support these on-going changes, you must continually replenish, from foods, the energy and the nutrients you deplete in maintaining the life of your body.

What are nutrients?

Nutrients are substances contained in food that are necessary to support growth, maintenance, and repair of the body tissues. Nutrients can be further assigned to three functional categories: (1) those that provide us with energy; (2) those important for growth, development, and maintenance; and (3) those that regulate biological processes to keep body function running smoothly. Nutrients can also be divided into essential and non-essential. **Essential**, also referred to as indispensable, nutrients are those substances necessary to support life but which must be supplied in the diet because the body cannot either create them at all or create them in a large enough quantity to meet needs. Protein, for example, is an essential nutrient needed for growth and maintenance of the body tissues and the synthesis of regulatory molecules. Food also contains nutrients considered **non-essential**. Some of these are not essential to sustain life, but have health-promoting properties. For example, a phytochemical (e.g., carotenoids) found in orange, red, and yellow fruits and vegetables is not essential but may reduce the risk of cancer. Others are required by the body but can be produced in sufficient amounts to meet demand. For example, lecithin, which is needed for nerve function, is not an essential nutrient because it can be manufactured in enough quantity by the body.

Classes of nutrients

Chemically, there are six classes of nutrients: carbohydrates, lipids, proteins, vitamins, minerals, and water. Carbohydrates, lipids, and proteins provide energy to the body and thus are also referred to as **energy-yielding nutrients**. Along with water, they constitute the major portion of most foods. They are also known as **macronutrients** because they are required in relatively large amounts. Their requirements are measured in kilograms (kg) or grams (g).

Carbohydrates include sugars, such as those in table sugar, fruit, and milk, and starches such as those in vegetables and grains. Sugars are the simplest form of carbohydrate; starches are more complex carbohydrates made of many sugars linked together. Carbohydrates provide a readily available source of energy to the body. Most fiber is also carbohydrate and is important for gastrointestinal health. It cannot be completely broken down, so it provides a little energy. It is found in vegetables, fruits, legumes, and whole grains.

Lipids, commonly referred to as fats and oils, provide a storage form of energy. Lipids in our diets come from foods that naturally contain fats, such as meat and milk, and from processed fats, such as vegetable oils and butter, which we add to food. Most lipids contain fatty acids, some of which are essential in the diet. Lipids contain more energy than carbohydrates, but energy utilization from lipids is limited because it involves a more complex metabolic process. The amount and type of lipid in our diet affects the risk of cardiovascular and metabolic diseases, as well as certain types of cancer.

Protein, such as that found in meat, fish, poultry, milk, grains, and legumes, is needed for growth and maintenance of body structure, and regulation of biological processes. It rarely serves as an energy source. Protein is made up of units called amino acids. Twenty or so amino acids are found in food, and some of them are considered essential – that is, they must be obtained from food. Dietary protein must meet the demand for these essential amino acids. Most North Americans eat about 1.5–2 times as much protein as the body needs to remain healthy. This amount of extra protein in the diet is generally not harmful – it reflects the standard of living and dietary habits, but one should keep in mind that the excess can contribute to storage of fat.

Vitamins and minerals are referred to as **micronutrients** because they are needed in small amounts in the diet. The amounts required are expressed in milligrams (mg) or micrograms (μg). They do not provide energy, but many help regulate the production of energy from macronutrients. They also have unique roles in processes such as bone growth, oxygen transport, fluid regulation, and tissue growth and development. Vitamins and minerals are found in most of the foods we eat. Fresh foods are good natural sources of vitamins and minerals, and many processed foods have micronutrients added to them during manufacture. For example, breakfast cereals are a good source of iron and B vitamins because they are added during processing. While processing can cause nutrient loss due to light, heat, and exposure to oxygen, with addition of certain nutrients, frozen, canned, and otherwise processed foods can still be good sources of vitamins and minerals. In today's diet, vitamin and mineral supplements are also a common source of micronutrients.

Water is the sixth class of nutrient. About 60% of the human body is water. Although sometimes overlooked as a nutrient, water has numerous vital functions in the body. It acts as a solvent and lubricant, as a vehicle for transporting nutrients and wastes, and as a medium for temperature regulation and chemical reactions. Water is considered a macronutrient and is required in a large quantity in the daily diet. The average man should consume about 3000 ml or 13 cups of water or other fluids containing water every day. Women need close to 2200 ml or about 9 cups per day.

Together, macronutrients and micronutrients provide energy, structure, and regulation. These functions are important for growth, maintenance, repair, and reproduction. Each nutrient provides one or more of these functions, but all nutrients together are needed to maintain health (Table 1.2).

Chemical composition of nutrients

The simplest nutrients are the minerals. Each is a chemical element; its atoms are all alike. As a result, its identity never changes. For example, iron may change its form, but it remains as iron when food is cooked, when a person eats the food, when the iron becomes part of a red blood cell, when the cell is broken down, and when the iron is lost from the body by excretion. The next simplest nutrient is water, a compound made of two elements: hydrogen and oxygen. Minerals and water are called **inorganic nutrients** because they do not contain carbon.

The other four classes of nutrients – carbohydrates, lipids, proteins, and vitamins – are more complex. In addition to hydrogen and oxygen, they all contain carbon, an element found in all living species. They are therefore called **organic compounds**. Proteins and vitamins also contain nitrogen and may contain other elements as well (Table 1.3).

Table 1.2 Nutrient functions in the body

NUTRIENTS	MAJOR FUNCTION	EXAMPLE
Carbohydrates	Energy	Muscle glycogen is stored carbohydrate that fuels the body cells
Lipids	Energy	Fat is the most plentiful sources of stored fuel in the body.
	Structure	The membranes that surround each cell are primarily lipids.
	Regulation	Estrogen is a lipid hormone that helps regulate the reproductive cycle in women.
Proteins	Energy	Proteins can be used for energy when consumed in excess or carbohydrate becomes depleted.
	Structure	Proteins form important parts of body tissues, including muscles, tendons, and ligaments.
	Regulation	Insulin is a protein that helps regulate blood glucose concentrations.
Vitamins	Regulation	B vitamins help regulate energy metabolism using macronutrients.
Minerals	Structure	The mineral calcium and phosphorus make bones and teeth solid and hard.
	Regulation	Sodium helps regulate blood volume.
Water	Structure	Water makes up nearly 60% of body weight.
	Regulation	Water evaporated as sweat helps reduce body temperature.

Table 1.3 Chemical elements in the six classes of nutrients				
NUTRIENTS	CARBON	HYDROGEN	OXYGEN	NITROGEN
Carbohydrate	✓	✓	✓	
Lipids	✓	✓	✓	
Proteins[1]	✓	✓	✓	✓
Vitamins[2]	✓	✓	✓	
Minerals[3]				
Water		✓	✓	

Notes:
1 Some proteins also contain the mineral sulfur.
2 Some vitamins also contain nitrogen and other elements.
3 Each mineral is a chemical element.

Energy-yielding nutrients

Carbohydrates, lipids, and proteins provide the fuel or energy required to maintain life, and therefore are considered as energy-yielding nutrients. If less energy is taken into the body than is needed, the body will burn its own fat, as well as carbohydrates and proteins, to meet the energy needs. If more energy is consumed than is needed, the extra is stored as body fat. The energy contained in foods or needed for all body processes and activities is measured in kilocalories (kcal) or kilojoules (kj). The term "calorie" is technically one-thousandth of a kilocalorie, but when spelled with capital "C" it indicates kilocalories. For example, the term "Calories" on food labels actually refers to kilocalories or kcal. When completely broken down in the body, 1 g of carbohydrate or protein provides 4 kcal, whereas 1 g of lipid provides 9 kcal. Therefore lipids have a greater energy density than either carbohydrates or proteins (Table 1.4).

Another substance that contributes energy is alcohol. Alcohol is not considered a nutrient because it interferes with the growth, maintenance, and repair of the body, but it does yield energy. When metabolized in the body, alcohol contributes about 7 kcal per gram (Table 1.4).

Most foods contain all three energy-yielding nutrients, as well as water, vitamins, and minerals. For example, meat contains water, fat, vitamins, and minerals, as well as protein. Bread contains carbohydrate, water, a trace of fat, a little protein, and some vitamins and minerals. Only a few foods are exceptions to this rule, the common ones being table sugar (pure carbohydrate) and cooking oil (pure fat).

Table 1.4 Energy content of macronutrients and alcohol		
	KILOCALORIES/GRAM	KILOJOULES/GRAM
Carbohydrate	4	16.7
Lipids	9	37.6
Proteins	4	16.7
Alcohol	7	29.3

Note: 1 kilocalorie = 4.18 kilojoules.

How much of each nutrient do we need?

In order to support life, an adequate amount of each nutrient must be consumed in the diet. The exact amount that is optimal is different for each individual. It depends on genetic makeup, lifestyle, and overall diet. A person with a genetic predisposition to a disease needs to consume a different amount of certain nutrients to maintain health than does a person with no genetic risk of the disease. Individuals who smoke cigarettes need more vitamin C than non-smokers because vitamin C neutralizes free radicals associated with tobacco. Athletes, or those who are more active, need more carbohydrates and greater total energy intake than their less active counterparts. The amount of each nutrient required is also dependent on the other nutrients and non-nutrient substances present in the diet. For example, adequate consumption of fat is essential for the absorption of vitamin A. The amount of iron absorbed is affected by the presence of vitamin C and calcium. Thus, it is difficult to make generalized recommendations about how much is enough or not without considering both individual needs and overall diet.

Consuming either too much or too little of one or more nutrients or energy forms can cause **malnutrition**. Malnutrition is often interpreted as **under-nutrition** or a deficiency of energy and nutrients. Under-nutrition may occur due to reduced intake of energy and nutrients, increased requirements, or an inability to absorb or use nutrients. It can cause weight loss, poor growth, an inability to reproduce, and if severe enough, death. Iron deficiency is a form of under-nutrition commonly seen in young children, adolescents, and some women because of their increased need for iron. Vitamin B-12 deficiency is a risk for older adults because the ability to absorb B-12 in the stomach decreases with age. When under-nutrition is caused by a specific nutrient deficiency, the symptoms often reflect the body functions that rely on the deficient nutrient. For example, vitamin D is necessary for bone growth and maintenance, so a deficiency of vitamin D can result in osteoporosis. Vitamin A is necessary for vision, so a deficiency of vitamin A can result in blindness.

Over-nutrition is also a form of malnutrition. When food is consumed in excess of energy requirements, the extra is stored as body fat. Some fat is necessary as insulation and protection, and as an energy store, but an excess of body fat increases the risk of high blood pressure, heart disease, diabetes, and other chronic diseases. These conditions can take months and years to manifest. When excesses of specific nutrients are consumed, an adverse or toxic reaction may occur. Because food generally does not contain high enough concentrations of nutrients to be toxic, most nutrient toxicities result from over-use of specific supplements.

FACTORS AFFECTING FOOD INTAKE AND CHOICE

We need nutrients to survive, but we eat food, not nutrients. There are hundreds of food choices to make and hundreds of reasons for making them. Each of these choices contributes to our total nutrient intake. Some foods are rich in protein and minerals, others in vitamins and phytochemicals. Choosing a healthy diet does not mean you have to give up your favorite foods. Most of us understand that nutrition is important to our health, yet people don't want to give up their favorite foods and they don't want to eat foods they don't like. Our food choices are primarily affected by hunger and appetite. They can also be influenced by what is available to us, where we eat, what is within our budget, what is compatible with our lifestyle, what is culturally acceptable, what mood we are in, and what we think we should eat.

Hunger and appetite

Hunger and appetite are the two drives that influence our desire to eat. These two differ dramatically. **Hunger** is considered biological in origin and is controlled by internal body mechanisms. For example, as nutrients are processed by the stomach and small intestine, these organs send signals to the liver and brain to reduce further food intake. **Appetite** is considered psychological in origin and is controlled by external food choice mechanisms. For example, your appetite is intensified as you see a tempting dessert or smell popcorn at the movie theater. We eat in response to hunger. However, what, when, and how much we eat are also affected by appetite, which is not necessarily related to hunger. Fulfilling either or both drives by eating sufficient food normally brings a state of **satiety**, a feeling of fullness and satisfaction, which halts our desire to continue eating.

Role of the hypothalamus

The hypothalamus, a region of the brain, mediates the effect of hunger and appetite and helps regulate satiety (Figure 1.2). Signals to eat or stop eating can be external, originating from the environment, or they can be internal, originating from the gastro-intestinal tract, circulating nutrients, or higher centers in the brain. External factors that stimulate eating include the sight, taste, and smell of food, the time of day, culture and social gathering, the appeal of the foods available, and ethnic and religious rituals. We often eat lunch at noon out of social convention, not because we are hungry. We eat turkey at Thanksgiving because it is a tradition. We eat cookies or cinnamon rolls while walking through the mall because the smell entices us to buy them. Likewise, external factors such as religious dietary obligations or negative experiences associated with certain foods can signal us to stop eating. In addition, our knowledge and beliefs regarding nutrition, body weight, and image can also influence our eating behavior. For example, what we think of as "healthy foods" often direct our food purchase, and some people select certain foods and supplements that they believe will improve their physical appearance or performance.

Internal signals that promote hunger and satiety originate both before and after foods are consumed. The simplest type of signal about food intake comes from local nerves in the walls of the stomach and small intestine, which sense the volume and pressure of the food and send a message to the brain to either start or stop eating. The presence of food in the gastrointestinal tract also triggers the release of gastrointestinal hormones such as cholecystokinin, which causes satiety. Absorbed nutrients may also send information to the brain to modulate food intake. Circulating levels of nutrients, including glucose, fatty acids, amino acids, and ketones are monitored by the brain and may trigger signals to eat or not to eat. Nutrients that are taken up by the brain

Figure 1.2 Process of satiety.

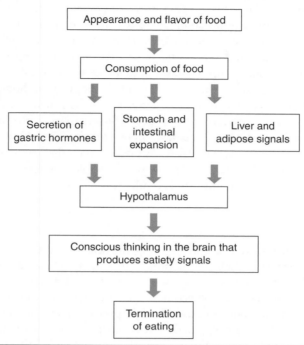

may affect the neurotransmitter concentrations, which then affect the amount and type of nutrients consumed. For example, some studies suggest that when brain serotonin is low, carbohydrates are craved, but when it is high, protein is preferred. The pancreas is also involved in food intake regulation because it releases insulin, which triggers a feeling of fullness and, as a result, decreases the drive to eat.

Psychological factors can also affect eating behavior. Psychological distress may come from events in the external environment, but processing of these events occurs in the brain cortex. The effect that emotions have on appetite depends on the individual. Some people eat for comfort and to relieve stress. Others may lose their appetite when they become emotional or distressed. A depressed person may choose to eat chocolates rather than call a friend. A person who has returned home from an exciting evening out may unwind with a late-night sandwich. These people may find emotional comfort, in part, because foods can influence the brain's chemistry.

Overall, daily food intake is a complicated mix of biological and social influences. It means so much more to us than nourishment and it reflects much of what we think about ourselves.

WHAT IS RELIABLE NUTRITIONAL INFORMATION?

The science of nutrition is young but is growing quickly, especially as our society becomes more health-conscious. We are bombarded with nutrition information, and much of it reaches us through television, the internet, radio, newspapers, and magazines. Although dietitians, nutritionists, and physicians are viewed as the most valuable source of nutrition information, it seems that we get most our food and nutrition information from mass media. Much of the information is reliable, but some can be misleading. One should always be aware that the motivation for news stories is often to sell subscriptions, improve ratings, or make news headlines more enticing, rather than to promote the nutritional health of the population. Some nutrition and health information originates from food companies. It is usually in the form of marketing designed to sell existing or new products. Sifting through the information and distinguishing the useful from the useless can be overwhelming. However, an understanding of the process of science and how it is used to study the relationship between nutrition and health or performance will allow you to develop the knowledge and ability needed to judge the validity of nutritional products.

Scientific methods

Advances in nutrition are made using the scientific method. The scientific method offers a systematic, unbiased approach to evaluating the relationships between food and health or performance. As shown in Figure 1.3, the first step of the scientific method is to make an observation and ask questions about the observation. For example: "What foods or nutrients might protect against the common cold?" In search of an answer, scientists then make a scientific prediction or **hypothesis**, such as "Foods rich in vitamin C reduce the number of common colds." Once a hypothesis has been proposed, experiments can be designed to test it. The experiments must provide objective results that can be measured and repeated. If the results fail to prove the hypothesis to be wrong, a theory or a scientific explanation can be established. Even a theory that has been accepted by the scientific community for years can be proved wrong; this allows the body of knowledge to grow, but it can be confusing as old or more conventional theories give way to new ones.

A well-conducted experiment must collect quantifiable data using proper experimental controls and the right experimental population. For example, body weight and blood pressure are parameters that can be measured reliably. However, feeling and perceptions are more difficult to assess. They can be quantified with standardized questionnaires, but individual testimonies or opinions – referred to as "anecdotal" – cannot be measured objectively, and thus are considered non-quantifiable.

Experimental controls ensure that each factor or variable studied can be compared with a known situation. They are often accomplished by using a **control group** that serves as a standard of comparison for the treatment being tested. A control group is treated the same way as the experimental groups, except no experimental treatment is given. For example, to investigate the effect of creatine supplementation on strength and power performance, the experiment group consists of athletes consuming creatine monohydrate, whereas the control group consists of athletes of similar age, gender, and training backgrounds eating similar diets and following similar workout regimens, but not consuming the creatine product.

A **placebo** can be used to further minimize differences between experimental and control groups. The placebo should be identical in appearance and taste to the actual

Figure 1.3 The scientific method.

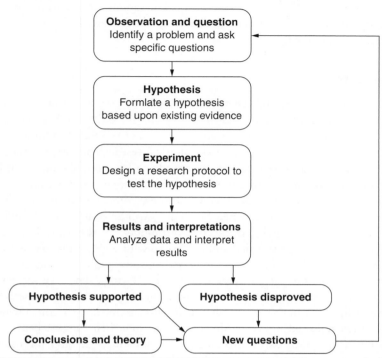

treatment, but should provide no therapeutic value. By using a placebo, participants in the experiment do not know if they are receiving the actual treatment. When subjects do not know which treatment they are receiving, the study is called a **single-blind study**. Using a single-blind study helps prevent the expectations of subjects from biasing the results. For example, if the athletes think they are undergoing the actual treatment, they develop a higher expectation of themselves and as a result work harder during the treatment and/or testing period. Errors can also occur if investigators allow their own desire for a specific result to affect the interpretation of the data. This type of error can be avoided by using a **double-blind study** in which neither the subjects nor the investigators know who is in which group until the results are analyzed.

Another important issue with the scientific method is to determine a sample size. To be successful, an experiment must show that the treatment being tested causes a result to occur more frequently than it would occur by chance alone. To ensure that chance variation between the two groups does not influence the results, the sample size must be large. If a change occurred by chance in one member of a group of five, it can easily alter the whole group's average; but if such change occurs in one member of a group of 500, it will not overly affect the group average. Fewer subjects are needed to demonstrate an effect that rarely occurs by chance, and vice versa. For example, if only one person out of one million can improve muscle power without creatine supplementation, then the experiment to see if creatine supplementation increases muscle power would require fewer subjects to demonstrate the effect. If one in four athletes can improve muscle power without creatine supplementation, then more subjects are needed for the study. The sample size needs to show that the effect of an experimental treatment can be determined using the statistical method before a study is conducted.

Types of research

Several types of research techniques have been used to determine nutrient requirements, to learn more about nutrient metabolism, and to understand the role of nutrition in health, fitness, and performance. These research techniques may be broadly divided into two categories: **epidemiological research** and **experimental research**.

Epidemiological research involves studying large populations in order to suggest a relationship between two or more variables. For example, it is epidemiological research that has helped scientists to observe that those who consume diets high in fat are more likely to develop heart disease. There are various forms of epidemiological research. One general form uses retrospective techniques. In this case, individuals who have certain diseases are identified and compared with a group of peers who don't have the disease. Researchers then trace the lifestyle history and eating habits of both groups to determine whether dietary practices or other factors may have increased the risk for developing the disease. Another general form of epidemiological research uses prospective techniques. In this case, individuals who are free of a specific disease are identified and then followed for years, during which time their lifestyle behaviors – including eating habits – are scrutinized. As some individuals develop the disease and others don't, the researchers are then able to discern whether dietary behaviors may increase the risk of the disease.

Epidemiological research helps scientists identify important relationships between variables, but it does not prove a cause-and-effect relationship. For example, in epidemiological studies that revealed an association between high fat intake and heart disease, the experimental approach typically involves examining the incidence of heart disease and dietary factors either cross-sectionally using multiple groups of subjects, or longitudinally using the same group of subjects, and the conclusion can be drawn based on the observation that those with high fat intake also have high incidences of heart disease. However, a question could be raised as to whether high fat intake directly causes heart disease. To answer that question, one may hypothesize that a high fat intake predisposes individuals to cardiovascular disease, but this hypothesis must then be tested by studies containing tightly controlled experimental approaches.

Experimental research constitutes another common form of research in nutritional science. The observation and hypotheses that come from epidemiological research can be tested in experimental research, which will then allow scientists to establish a cause-and-effect relationship. This type of research actively intervenes in the lives of individuals, and usually involves studying a smaller group of subjects that receive a treatment or placebo under either tightly controlled or free-living conditions. In such studies, often called intervention studies, an independent variable (cause) is manipulated so that changes in dependent variables (effect) can be studied. For example, if it is determined by epidemiological research that individuals who eat a low-fat diet have a lower incidence of heart disease, an intervention trial may be designed with an experimental group that consumes a diet lower in fat than is typical in the population and a control group that consumes the typical, higher-fat diet, while both groups are kept the same in terms of other aspects of lifestyle, e.g., total caloric intake and participation in physical activities. The two groups can be monitored and compared to see if the dietary intervention affects the incidence of heart disease.

The experimental approach appears to be a common choice in studies that examine the effect of nutrition on sports performance, though they are of shorter timeframes compared to those studies that investigate the relationship between nutrition and health. Additionally, most sports nutrition studies are conducted in a laboratory with tight control of extraneous variables. In order to make research findings more relevant to actual sports, many of these studies also attempt to use laboratory protocols designed to mimic the physiological demands of the sport. Although research that possesses both the rigorous control in its experimental design and the ability for its findings to be readily applied is always preferable, achieving both simultaneously has often been difficult.

Judging nutritional information

Knowledge relative to all facets of life, the science of nutrition included, has increased phenomenally in recent years. As knowledge advances, new nutritional principles are developed. Sometimes, established beliefs and concepts must give way to new ideas, which then result in a change in recommendations. It has long been suggested that margarine is better for you than butter. However, current research indicates that it is just as bad. Indeed, nutrition is a fast-growing discipline in part because consumers are now taking greater responsibility for self-care and are eager to receive food and nutrition information. Such an increase in societal interest creates opportunities for nutritional misinformation to flourish. According to the American Dietetic Association, the media are consumers' leading source of nutrition information, but news reports of nutrition research often provide inadequate depth for consumers to make wise decisions. Consumers must also be aware that certain individuals may capitalize on research findings for personal financial gain. For example, isolated nutritional facts may be distorted or results of a single study may be used to market a specific nutritional product.

Just as scientists use the scientific method to expand their understanding of the world around us, each of us can use an understanding of how science is done to evaluate nutritional claims. Claims associated with nutrition products are always appealing. It is up to us as consumers to decide whether and how we should accept them. In judging the validity of a nutritional product, one should always question whether product claims make sense and, if they do, where they come from. If it is claimed that a product can change body composition in only one or two weeks without changing diet and exercise habits, common sense should tell you it is too good to be true. If the claim seems to be reasonable, then the question is about where the claim comes from. Was it a personal testimony, a government recommendation, or advice from a health professional? Was it the result of a research study? Was such a research study found in a peer-reviewed journal?

Claims that come from individual testimonies have not been tested by experimentation, and therefore it cannot be assumed that similar results will occur in other people. On the other hand, government recommendations regarding healthy dietary practices are developed by a panel of scientists who use the results of well-controlled research studies to develop recommendations for the population as a whole. The government provides information about food safety and recommendations on food choices and the quantities of specific nutrients needed to avoid nutrient deficiencies and excesses and to prevent chronic diseases. These recommendations are used to develop food labeling regulations and are the basis for public health policies and programs.

Results from research studies published in peer-reviewed journals are generally considered to be accurate because these studies have been scrutinized by the scientific community to determine their validity and reliability. On the other hand, results presented at conferences or published in popular magazines, although they may be legitimate, should be considered with caution as they are usually not subject to the scrutiny of others who are experts in the same field. Even well-designed, carefully executed, peer-reviewed experiments can be a source of misinformation if the experimental results are interpreted incorrectly or if the implications of the results are exaggerated. For example, a mineral called boron has been considered to be an ergogenic aid because a study showed that consuming boron enhances blood testosterone levels in those with boron deficiency. However, just because supplementing boron increases testosterone levels in those with boron deficiency, it doesn't necessarily mean that to increase boron consumption in those with a normal boron level will have the same effect. It is usually true that once an adequate intake of a nutrient is attained, consuming it in excess is ineffective. A study that shows that rats fed a diet high in vitamin E live longer than those consuming less vitamin E could lead people to conclude that vitamin E supplementation can prolong one's life. However, can results from this animal study be extrapolated to humans? Just because rats consuming diets high in vitamin E live longer does not mean that the same is true for humans.

The best means to evaluate claims of enhanced health and sports performance made by nutritional products or practices is to possess a good background in nutrition and a familiarity with the experimental process of high-quality research. However, this may not be possible for all individuals who are seeking nutritional products. For those who have a minimal background in nutrition, it is recommended that the following be used as basic guidelines in evaluating the claims made for a nutritional product or practice. If the answer to any of the following questions is "yes," then one should be skeptical of such a product and investigate their real efficacy before investing any money.

- Is its claim "too good to be true"?
- Does the product promise quick improvement in health and physical performance?
- Is it advertised mainly by use of anecdotes, case histories, or individual testimonies?
- Are currently popular personalities or star athletes featured on its advertisement?
- Does the person or organization who recommends it also sell the product?

- Is it expensive, especially when compared to the cost of equivalent nutrients that may be obtained from ordinary foods?
- Does it use the results of a single study or poorly controlled research to support its claims?

SUMMARY

- Nutrition is a science that links foods to health and diseases. It studies the structure and function of various food groups and the nutrients they contain. It also includes the biological processes by which our bodies consume food and utilize the nutrients.
- Sports nutrition represents a fast-growing area of study in recent years. It is the study and practice of nutrition and diet as it relates to athletic performance.
- Your food choice today can affect your health tomorrow. Understanding nutrition will allow you to make wise choices on foods you consume, thus improving health and fitness.
- Proper nutrition is an important component in the total training program of the athlete. Nutrient deficiencies can seriously impair performance, whereas nutrient supplementation may delay fatigue and improve performance. Nutritional status can be a major factor differentiating athletes of comparable genetic endowment and training.
- Nutrients are substances contained in food that are necessary to support growth, maintenance, and repair of the body tissues. The six classes of nutrients include: carbohydrates, lipids, proteins, vitamins, minerals, and water.
- Carbohydrates, lipids, and proteins provide the fuel or energy that is required to maintain life, and therefore are considered as energy-yielding nutrients. The energy contained in foods or needed for all body processes and activities is measured in kilocalories (kcal) or kilojoules (kj).
- Hunger and appetite are the two drives that influence our desire to eat. The hypothalamus, a region of the brain, mediates the effect of hunger and appetite and helps regulate satiety.
- Signals to eat or stop eating can be external, originating from the environment, including sight, taste, and smell, or they can be internal, originating from the gastrointestinal tract, circulating nutrients, or higher centers in the brain.
- Epidemiological research and experimental research are the two types of research frequently used in the study of nutrition. The former involves studying large populations in order to suggest relationships between two or more variables, whereas the latter involves studying a smaller group of subjects that receive a treatment or placebo under either tightly controlled or free-living conditions.
- In judging the validity of a nutritional product, one should always question whether product claims make sense and, if they do, where they come from. Claims that come from individual testimonies have not been tested by experimentation. However, claims supported by research studies published in peer-reviewed journals are generally considered accurate.

CASE STUDY: JUDGING THE SCIENTIFIC MERIT OF A NUTRITIONAL PRODUCT

The picture on a nutritional product was eye-catching. A good-looking couple were running along the beach. He was shirtless and had impressive upper-body muscle mass. She was in a tight-fitting sundress and had a perfect figure with long, blonde hair. The product was advertised as a weight-loss supplement. This advertisement also included a statement saying that "Research studies show that this supplement helps people lose weight and feel less fatigued." After seeing this advertisement, Jill wrote to the company and asked about the research carried out to develop this product. She received the company's newsletter, which discussed two research studies.

In the first study, 12 obese subjects were divided into two groups. One was given the supplement and the other was not. The 12 subjects were all consuming a liquid diet of 800 kcal per day. These subjects were living in an experimental research ward where they had no access to food other than the liquid diet. It was found that over a four-week study period, subjects receiving the supplement lost more weight than those not receiving it.

In the second study, eight healthy, male college students were studied in two groups. They were asked to follow their regular physical activities, but one group took the supplement, while the other received a placebo. It was a double-blind study. After three weeks, subjects were asked how energetic they felt during a workout and throughout the day. It was found that the subjects who received the supplement reported high energy levels and less fatigue than those who did not.

Questions
- What were the strengths of these experiments?
- What was wrong with each of these experiments?
- Should consumers be encouraged to use this supplement as a weight-loss aid? Why?

REVIEW QUESTIONS

1 What is nutrition? What is sports nutrition?
2 What is a nutrient? Name the six classes of nutrients found in foods. What is an essential nutrient?
3 Which nutrients yield energy and how much energy do they yield per gram? How is energy measured?
4 Wendy's Big Bacon Classic contains 44 g of carbohydrate, 36 g of fat, and 37 g of protein. Calculate (1) the total calories contained in this meal; and (2) the percentage of calories derived from fat?
5 Define the terms: hunger, appetite, satiety. Explain the role the hypothalamus plays in regulating hunger and satiety.
6 What are the internal and external signals that promote hunger and satiety?
7 Describe the types of research methods often used in acquiring nutrition information.
8 List the steps involved in the scientific method.
9 What is a control group? What is a placebo? What is a double-blind study?
10 What factors should be considered in judging nutrition claims?

SUGGESTED READING

American Heart Association Nutrition Committee, Lichtenstein, A.H., Appel, L.J., Brands, M., Carnethon, M., Daniels, S., Franch, H.A., Franklin, B., Kris-Etherton, P., Harris, W.S., Howard, B., Karanja, N., Lefevre, M., Rudel, L., Sacks, F., Van Horn, L., Winston, M., and Wylie-Rosett, J. (2006) Diet and lifestyle recommendations revision 2006: a scientific statement from the American Heart Association Nutrition Committee. *Circulation*, 114: 82–96.

Improving diet and lifestyle is a critical component of the American Heart Association's strategy for cardiovascular disease risk reduction in the general population. This document presents recommendations designed to meet this objective.

Cordain, L., Eaton, S.B., Sebastian, A., Mann, N., Lindeberg, S., Watkins, B.A., O'Keefe, J.H., and Brand-Miller, J. (2005) Origins and evolution of the Western diet: health implications for the 21st century. *American Journal of Clinical Nutrition*, 81: 341–354.

This article discusses an evolutionary discord between our ancient, genetically determined biology and the nutritional, cultural, and activity patterns of contemporary Western populations, which may explain some of the lifestyle-related diseases we are experiencing today.

Hoffman, D.J., Policastro, P., Quick, V., and Lee, S.K. (2006) Changes in body weight and fat mass of men and women in the first year of college: a study of the "freshman 15." *Journal of American College Health*, 55: 41–45.

This original investigation was designed to measure changes in body weight and percentage of body fat among first-year college students and to address a common but often undocumented myth among college students that there is a high risk of gaining 15 pounds of weight during freshman year.

2 Macronutrients: carbohydrates, lipids, and proteins

KEY TERMS

- glucose
- glycogen
- monosaccharides
- disaccharides
- oligosaccharides
- polysaccharides
- fructose
- galactose
- disaccharide
- sucrose
- maltose
- lactose
- amylase
- fermentation
- lactase
- starch
- amylose
- amylopectin
- cellulose
- fiber
- soluble fiber
- insoluble fiber
- gluconeogenesis
- lipids

- fatty acids
- triglycerides
- phospholipids
- sterols
- saturated fatty acid
- unsaturated fatty acids
- monounsaturated fatty acids
- polyunsaturated fatty acids
- trans fatty acid
- lipolysis
- lipogenesis
- atherosclerosis
- essential or indispensable amino acids
- non-essential or dispensable amino acids
- transamination
- peptide bond
- denaturation
- complete proteins
- incomplete proteins
- limiting amino acids
- complementary proteins
- osmosis
- edema
- deamination

CARBOHYDRATES

A student, quietly reading a textbook, is seldom aware that within his brain cells, billions of **glucose** molecules are splitting to provide the energy that permits him to learn. Similarly, a marathon runner, bursting across the finish line, seldom gives thanks to the glycogen his muscles have devoured to help him finish the race. Your brain needs carbohydrates to power its activities. Your muscles need carbohydrates to fuel their work. Together, glucose and its stored form, **glycogen**, often provide more than half of all the energy the brain, muscles, and other body tissues use. The rest of the body's energy comes mostly from fat.

People don't eat glucose and glycogen directly; they eat food rich in carbohydrates. Then their bodies convert the carbohydrates mostly into glucose for immediate energy and into glycogen for reserve energy. Except for lactose from milk and a small amount of glycogen from animals, plants provide the major source of carbohydrates in the human diet. All plant foods, e.g., whole grains, vegetables, legumes, and fruits, provide ample carbohydrates. Although everyone eats carbohydrates, the amount and type consumed often depend on the wealth and prosperity of the society. In more affluent countries, animal foods become more affordable, so the intake of fat and protein increases. For example, the typical intake of carbohydrates accounts for nearly two-thirds of the energy in the diet in developing countries, while this accounts for only about half of the energy intake in more economically developed countries.

Not all carbohydrates are created equal. Some of them are referred to as simple and well-refined carbohydrates, e.g., candies, cookies, and cakes, while others are considered complex and less processed, e.g., whole grains, vegetables, and legumes. Many people mistakenly think of carbohydrates as "fattening" and believe that diets high in carbohydrates contribute to the epidemic of obesity. They avoid them when trying to lose weight. Such a strategy can be counterproductive if the carbohydrates being consumed are mostly complex. In fact, most dietitians and nutritional scientists consider consuming foods rich in complex carbohydrates and fibers to be one of the most important components of a healthy diet, not only for its potential to prevent certain chronic diseases but also as an integral part of a proper diet to lose excess body fat over the long term.

Chemical basis of carbohydrates

Chemically, carbohydrates contain carbon (carbo), and hydrogen and oxygen in the same proportion as in water (hydrate). Combining atoms of carbon, hydrogen, and oxygen forms a simple carbohydrate or sugar molecule with the general formula $C_6H_{12}O_6$, although the number of carbon can vary from three to seven. Six-carbon sugars are also referred to as hexoses. Accordingly, three-carbon sugars are trioses, four-carbon sugars are tetroses, five-carbon sugars are pentoses, and seven-carbon sugars are heptoses. Of these varieties, the hexose sugars interest nutritionists the most. In these molecules, atoms of carbon (C), oxygen (O), nitrogen (N), and hydrogen (H) are linked by chemical bonds. Atoms form molecules in ways that satisfy the bonding requirement of each atom. For example, as shown in Figure 2.1, each carbon atom has four binding sites that link to other atoms, including carbons. Carbon bonds not linked to other carbon atoms accept hydrogen (with one binding site), oxygen (with two binding sites), or an hydrogen–oxygen combination (OH) referred to as hydroxyl.

Classification of carbohydrate

Carbohydrates have been typically classified as simple carbohydrates that include **monosaccharides** and **disaccharides** and complex carbohydrates that include **oligosaccharides** and **polysaccharides**. "Saccharide" means an organic compound

Figure 2.1 Chemical structure of the three monosaccharides depicted in both the linear and ring configurations.

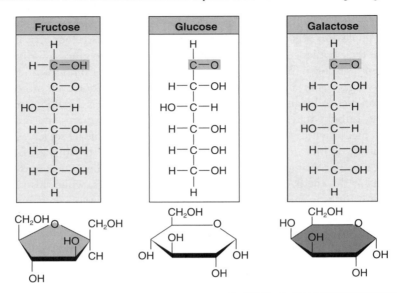

containing a sugar or sugars. It is the number of saccharides or sugars linked within the molecule that distinguishes each carbohydrate category.

Monosaccharides

The basic unit of carbohydrate is a single sugar molecule or a monosaccharide ("mono" means "one"). When two monosaccharides combine, they form a disaccharide ("di" means "two"). Monosaccharides and disaccharides are known as simple sugars, or simple carbohydrates. The three most common monosaccharides in the diet are glucose, **fructose**, and **galactose**. Each contains 6 carbon, 12 hydrogen, and 6 oxygen atoms, but differ in their arrangement (Figure 2.1). Glucose, also called dextrose or blood sugar, is produced in plants by the process of photosynthesis, which uses energy from the sun to combine carbon dioxide and water. Glucose rarely occurs as a monosaccharide in food; it is most often found as part of a disaccharide or starch. The digestion of more complex carbohydrates also produces glucose, which is then absorbed across the wall of the small intestine so it can be (1) used directly by cells for energy, (2) stored as glycogen in muscle and liver, or (3) converted to fat. Inside the body, glucose can be formed from the breakdown of stored carbohydrate, i.e., glycogen, or synthesized from carbon skeletons of specific amino acids, glycerol, pyruvate, and lactate.

Fructose, also called fruit sugar, is another common monosaccharide. It tastes sweeter than glucose. It is found naturally in fruits and vegetables, mostly as a part of sucrose, a disaccharide, and makes up more than half of the sugar in honey. It accounts for about 10% of the average energy intake in the United States. The small intestine absorbs some fructose directly into the blood. It is then transported to the liver, where it is quickly metabolized. Much is converted to glucose, but if fructose is consumed in very high amounts the rest goes on to form other compounds, such as fat. Most of the free fructose in our diets comes from the use of high-fructose corn syrup in soft drinks, candies, jams, jelly, and many other fruit products and desserts.

Galactose occurs most often as a part of lactose, the disaccharide in milk, and is rarely present as a monosaccharide in the food supply. After lactose is digested and absorbed, galactose arrives in the liver. There it is either transformed into glucose or further metabolized into glycogen.

Disaccharides

The combination of two monosaccharides yields a **disaccharide**. The disaccharides (double sugars) include **sucrose** (cane sugar or table sugar), **maltose** (malt sugar), and **lactose** (milk sugar). Sucrose forms when the two sugars glucose and fructose bond together (Figure 2.2). Sucrose is found naturally in sugar cane, sugar beets, honey, and maple sugar. These products are processed to varying degrees to make brown, white, and powdered sugars. Animals do not produce sucrose or much of any carbohydrate except glycogen. Sucrose is considered the most common dietary disaccharide and constitutes up to 25% of the total caloric intake in the United States.

Figure 2.2 Chemical structure of the three disaccharides depicted in the ring configurations.

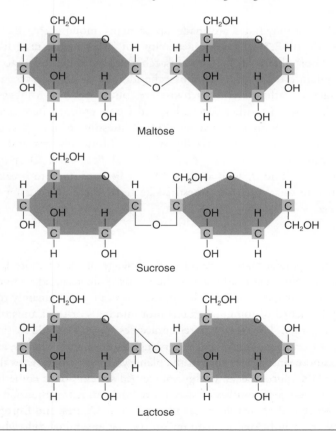

Maltose is a disaccharide consisting of two molecules of glucose (Figure 2.2). This sugar is made whenever starch breaks down, as happens in plants when seeds germinate and in human beings during carbohydrate digestion. For example, this sugar is responsible for the slightly sweet taste experienced when bread is held in the mouth for a few minutes. As salivary **amylase** begins digesting the starch, some sweeter-tasting maltose is formed. Maltose plays an important role in the beer and liquor industry. In the production of alcoholic beverages, starches in various cereal grains are first converted to simpler carbohydrates. The products of this step – maltose, glucose, and other sugars – are then mixed with yeast cells in the absence of oxygen. The yeast cells convert most of the sugars to alcohol or ethanol and carbon dioxide, a process called **fermentation**.

Lactose forms when glucose bonds with galactose during the synthesis of milk (Figure 2.2). Lactose is the only sugar found naturally in animal foods. Depending on milk's fat content, lactose contributes 30–50% of the energy in milks. As the least sweet of the disaccharides, lactose can be artificially processed and is often present in carbohydrate-rich, high-calorie liquid meals. A substantial segment of the world's population is lactose intolerant; these individuals lack adequate quantities of the enzyme **lactase**, which splits lactose into glucose and galactose during digestion.

Complex carbohydrates

Complex carbohydrates are made up of many monosaccharides linked together in chains (Figure 2.3). They are generally not sweet to the taste like simple carbohydrates. Short chains of 3–10 monosaccharides are called oligosaccharides and chains that contain more than ten monosaccharides are called polysaccharides. Oligosaccharides such as raffinose and stachyose are found in beans, cabbage, brussels sprouts, broccoli, asparagus, other vegetables, and whole grains. These cannot be digested by enzymes in the human stomach and small intestine, so they pass undigested into the large intestine. Here, bacteria digest them, producing gas and other by-products, which can cause abdominal discomfort and flatulence. Over-the-counter enzyme tablets and solutions, such as Bean-O, can be consumed to breakdown oligosaccharides before they reach the intestinal bacteria, thereby reducing the amount of gas produced.

Starch

The term "polysaccharide" refers to the linkage of ten or more (into the thousands) monosaccharide residues by glycosidic bonds. Polysaccharides are classified into plant and animal categories. Polysaccharides stored in plants are mainly referred to as **starch**, a long, branched or unbranched chain of hundreds or thousands of glucose molecules linked together. These giant starch molecules are packed side by side in grains such as wheat or rice, in root crops and tubers such as yams and potatoes, and in legumes such as peas and beans. When you eat the plant, your body hydrolyzes the starch to glucose and uses the glucose as an energy source. All starchy foods come from plants. Grains are the richest food sources of starch, providing much of the food energy all over the world – rice in Asia; wheat in Canada, the United States, and Europe; corn in much of Central and South America; and millet, rye, barley, and oats elsewhere.

Figure 2.3 Comparison of some common starches and glycogen.

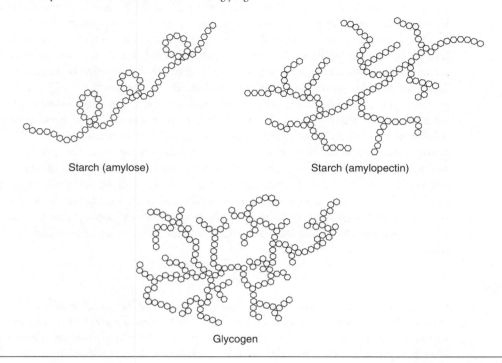

Starch (amylose)   Starch (amylopectin)

Glycogen

There are two forms of starch digestible by humans: **amylose** and **amylopectin** (Figure 2.3). Amylose, a long, straight chain of glucose units, comprises about 20% of the digestible starches found in vegetables, beans, breads, pasta, and rice. Amylopectin is a highly branched chain and makes up the remaining 80% of digestible starches in the diet. The relative proportion of each starch form determines the digestibility of a food containing starch. The enzymes that break down starches to glucose and other related sugars act only at the end of a glucose chain. Amylopectin, because it is branched, provides many more ends for enzyme action. Therefore, amylopectin is digested more rapidly and raises blood glucose much more readily than amylose. **Cellulose** is another form of polysaccharide found in plants. Although similar to amylose, it cannot be digested by human enzymes.

Fiber

Fiber as a class is mostly made up of polysaccharides, but they differ from starches in that the chemical links that join the individual sugar units cannot be digested by human enzymes in the gastrointestinal tract. This prevents the small intestine from absorbing the sugars that make up the fibers. Consequently, fibers contribute little or no energy to the body. Fibers exist exclusively in plants; they make up the structure of leaves, stems, roots, seeds, and fruit covering. Fiber is not a single substance, but a group of substances including cellulose, hemicelluloses, pectins, gums, and mucilages, as well as the non-carbohydrate, lignin. In total, these constitute all the non-starch polysaccharides in foods. Nutrition facts labels generally do not list these individual forms of fiber, but instead lump them together under the term "dietary fiber."

These various forms of fiber differ in many aspects, but generally can be divided into two categories: **soluble fiber** and **insoluble fiber** (Table 2.1). Some forms of fiber are considered soluble because they can be digested by bacteria in the large intestine. These soluble fibers are found around and inside of the plant cells. They include pectins, gums, mucilages, and some hemicelluloses. They can form viscous solutions when placed in water and are therefore referred to as soluble fibers. Food sources of soluble fibers include oats, apples, beans, and seaweed. Fibers that cannot be broken down by bacteria in the large intestine and do not dissolve in water are called insoluble fibers. Insoluble fibers are primarily derived from the structural parts of plants, such as cell walls, and include cellulose, some hemicelluloses, and lignin. Food sources of insoluble fiber include wheat bran and rye bran, which are mostly hemicelluloses and celluloses, and vegetables such as broccoli, which contain woody fibers comprised partly of lignin. Most foods of plant origin contain mixtures of soluble and insoluble fibers.

Glycogen

Glycogen is found to only a limited extent in meats and not at all in plants. For this reason, glycogen is not a significant food source of carbohydrate, but it does perform an important role in the body. Glycogen consists of a chain of glucose units with many branches, providing even more sites for enzyme action than amylopectin (Figure 2.3). Such a highly branched arrangement permits rapid hydrolysis or breakdown. When the hormone message "release energy" arrives at the storage sites in the liver or muscle cell, enzymes respond by attacking the many branches of each glycogen simultaneously, thereby making a surge of glucose possible.

Glycogen is the only stored form of carbohydrate in humans. However, the amount of glycogen in the body is relatively small – about 400–500 g. Glycogen is mainly stored in muscle and liver. A well-nourished 80-kg person can store up to approximately 500 g with 80% or ~400 g of it existing as muscle glycogen. Because each gram of carbohydrate contains about 4 kcal of energy, a typical person stores 1500–2000 kcal of carbohydrate energy – enough total energy to power a high-intensity, 20-mile run. The amount of glycogen stored in muscle can be temporarily increased by a diet and exercise regimen called carbohydrate loading or glycogen

Table 2.1 Classification of fibers			
TYPE	CHEMICAL COMPONENTS	PHYSIOLOGICAL EFFECTS	MAJOR FOOD SOURCES
Insoluble fibers	Celluloses, hemicelluloses, lignin	Increases fecal bulk, decreases intestinal transit time	Wheat bran, rye bran, whole grains, broccoli
Soluble fibers	Pectins, gums, mucilages, some hemicelluloses	Delay stomach emptying, slows glucose absorption	Oats, apples, beans, barley, carrots, citrus fruits, seaweed

supercompensation. This regimen is often used by endurance athletes to build up glycogen stores before an event. Extra glycogen can mean the difference between running only 20 miles and finishing a 26-mile marathon before exhaustion takes over. Details of how glycogen supercompensation is carried out are provided in Chapter 8.

Food sources of carbohydrate

The foods that yield the highest percentage of calories from carbohydrates are table sugar, honey, jam, jelly, and fruits. Table 2.2 lists common food sources of carbohydrates, among which corn flakes, rice, bread, and noodles all contain at least 75% of calories as carbohydrate. Foods with moderate amounts of carbohydrate calories are peas, broccoli, oatmeal, dry beans and other legumes, cream pies, French fries, and fat-free milk. In these foods, carbohydrate content is diluted either by protein or by fat. Foods with essentially no carbohydrates include beef, eggs, poultry, fish, vegetable oils, butter, and margarine.

It is recommended that adults consume 14 g of dietary fiber per 1000 kcal. However, the average daily intake of fiber in the United States is about half this amount (Lang and Jebb 2003). There are many ways to assure adequate fiber intake. Whole grains and cereals, legumes, fruits, and vegetables are probably the best options. It is important to select foods made with whole grains. Nutritionally, white bread, white rice, and white pasta are no match for their whole grain counterparts. This is because the

Table 2.2 Selected food sources of carbohydrate

FOODS	SERVING SIZE	CARBOHYDRATE (G)
Baked potato	1	51
Spaghetti noodles	1 cup	40
Cola drink	12 fluid ounces	39
M&M candies	½ ounce	30
Banana	1	28
Rice (cooked)	½ cup	22
Corn (cooked)	½ cup	21
Low-fat yogurt	1 cup	19
Kidney beans	½ cup	19
Orange	1	16
Carrot (cooked)	1 cup	16
Whole-wheat bread	1 slice	16
Oatmeal	½ cup	13
1% milk	1 cup	13
Kiwi	1	11
Pineapple chunks	½ cup	10
Broccoli	1 cup	7
Peanut butter	2 tablespoon	7
Peanuts	1 oz	6
Tofu	1 cup	4

nutritional value of whole grain is greatest when all three layers of wheat kernel – bran, germ, and endosperm – are intact. Generally speaking, coats of grains and legumes and the skins and peels of fruits and vegetables contain relatively high fiber content. Milling removes the bran and germ layers, resulting in finely ground, refined flour. To restore some of the lost nutrients, food manufacturers fortify their products with a variety of vitamins and minerals. However, many other important nutrients lost during processing are not added back. The fiber contents of selected foods are shown in Table 2.3.

Both monosaccharides and disaccharides are collectively regarded as sugars. Sugars can also be divided into added sugars and naturally occurring sugars. Added sugars are not nutritionally or chemically different to sugars occurring naturally in foods. The only difference is that they have been refined and thus separated from their plant sources, such as sugar cane and sugar beets. Foods in which naturally occurring sugar predominate, such as milk, fruits, and vegetables, provide not only energy but also fiber and micronutrients. In contrast, foods with large amounts of added sugars, such as soft drink, cakes, cookies, and candy, often have little nutritional value beyond the calories they contain. For example, a tablespoon of sugar contains 50 kcal, but almost no nutrients other than sugar. A small orange also has about 50 kcal, but contributes vitamin C, folate, potassium, and some calcium, as well as fiber. Foods that contain most of the added sugars in American diets are:

- regular soft drinks
- candy
- cakes
- cookies
- pies
- fruit drinks, such as fruit juice and fruit punch
- milk-based products, such as ice cream, sweetened yogurt, and sweetened milk
- grain products such as sweet rolls and cinnamon toast.

Major roles of carbohydrate in the body

The major function of carbohydrate in the body is to provide energy, especially during high-intensity exercise. Energy derived from blood-borne glucose and the breakdown of muscle and liver glycogen ultimately powers contractile processes of skeletal, cardiac, and smooth muscle tissues. Some of the energy is also used for many other biological processes such as digestion and absorption, glandular secretion, metabolic reactions, and homeostatic regulation.

Storing glucose as glycogen

After a meal, monosaccharides are absorbed and travel via the hepatic portal vein to the liver, where much of the fructose and galactose is metabolized for energy. The fate of the absorbed glucose depends on the energy needs of the body. If glucose is needed at the tissue, it is transported in the blood, reaching cells throughout the body. The amount of glucose in the blood is regulated at about 70–100 mg per 100 ml of blood. This ensures adequate glucose delivery to body cells, which is particularly important for brain cells and red blood cells, which rely almost exclusively on glucose as an energy source. If blood glucose levels rise too high, the secretion of insulin from the pancreas is increased. This will then cause liver cells to link the excess glucose molecules by condensation reaction into long, branched chains of glycogen. When blood glucose levels fall below the normal range, secretion of glucagon from the pancreas increased, and as a result the liver cells

Table 2.3 The dietary fiber content in selected common foods		
FOODS	SERVING SIZE	FIBER (G)
Fruits		
Raspberries	1 cup	8.0
Pear, with skin	1 medium	5.5
Apple, with skin	1 medium	4.4
Strawberries (halves)	1¼ cup	3.8
Banana	1 medium	3.1
Orange	1 medium	3.1
Raisins	2 tablespoons	1.0
Grains, cereals, and pasta		
Spaghetti, whole-wheat, cooked	1 cup	6.2
Barley, pearled, cooked	1 cup	6.0
Bran flake	3/4 cup	5.3
Oat bran muffin	1 medium	5.2
Oatmeal, quick, regular or instant, cooked	1 cup	4.0
Popcorn, air-popped	3 cups	3.5
Brown rice, cooked	1 cup	3.5
Bread, rye	1 slice	1.9
Bread, whole-wheat or multigrain	1 slice	1.9
Legumes, nuts, and seeds		
Split peas, cooked	1 cup	16.3
Lentils, cooked	1 cup	15.6
Black beans, cooked	1 cup	15.0
Lima beans, cooked	1 cup	13.2
Baked beans, vegetarian, canned, cooked	1 cup	10.4
Almonds	1 oz (23 nuts)	3.5
Pistachio nuts	1 oz (49 nuts)	2.9
Pecans	1 oz (19 halves)	2.7
Vegetables		
Peas, cooked	1 cup	8.8
Broccoli, boiled	1 cup	5.1
Turnip greens, boiled	1 cup	5.0
Sweet corn, cooked	1 cup	4.2
Brussels sprouts, cooked	1 cup	4.1
Potato, with skin, baked	1 medium	2.9
Tomato paste	¼ cup	2.7
Carrot, raw	1 medium	1.7

Source: Adapted from the data provided by Mayo Clinics.

dismantle the glycogen by hydrolysis reactions into single molecules of glucose and release them into the bloodstream. Thus, glucose becomes available to supply energy to the brain and other tissues regardless of whether the person has eaten recently. Muscle cells can also store glucose as glycogen, as mentioned earlier, but they keep most of their supply, using it just for themselves during exercise. Glycogen holds water and thus is rather bulky. This is why the body can only store enough glycogen to provide energy for relatively short periods of time – less than a day during rest and a few hours at most during exercise. For its long-term energy reserves or for use over days or weeks of food deprivation, the body uses its abundant, water-free fuel – fat.

Using carbohydrate as energy

The main function of carbohydrate is to supply calories for use by the body. Certain tissues in the body, such as red blood cells, can use only glucose as fuel. Most parts of the brain and central nervous system also derive energy only from glucose unless the diet contains almost none. In that case, much of the brain can use partial breakdown products of fat, called ketone bodies, for energy needs. Other body cells, including muscle cells, can use carbohydrates as a fuel but they can also use fat or protein for energy needs. Glucose fuels the work of most body cells. Inside a cell, glucose is metabolized through cellular respiration to produce carbon dioxide, water, and energy in the form of adenosine triphosphate (ATP). Providing energy through cellular respiration involves several interconnected chemical pathways that take place primarily in mitochondria. As mentioned earlier, liver cells store glucose in the form of glycogen as a reserve. However, the total glycogen stores in the liver last only for a few hours. To keep providing glucose to meet the body's energy needs, a person has to consume dietary carbohydrate frequently. Those who fail to meet their carbohydrate requirements may draw energy from the other two energy-yielding nutrients, fat and protein. Nevertheless, their level of performance during vigorous exercise can reduce significantly.

Sparing protein from use as an energy source

A diet that supplies enough digestible carbohydrate to prevent breakdown of proteins for energy needs is considered protein-sparing. Under normal circumstances, digestible carbohydrate in the diet mostly ends up as blood glucose, and protein is reserved for functions such as building and maintaining muscles and vital organs. However, if you don't eat enough carbohydrates, your body is forced to make glucose from body protein. This process of producing new glucose using non-glucose molecules such as amino acids is called **gluconeogenesis**. Gluconeogenesis occurs in liver and kidney cells and is stimulated by the hormone glucagon secreted from the pancreas in response to a decreased blood glucose concentration. Newly synthesized glucose is released into the blood to prevent blood glucose from dropping below the normal range. However, such a process of gluconeogenesis, if continued, can drain the pool of amino acids available in cells for other crucial functions. During long-term starvation, the continuous withdrawal of protein from the muscle, heart, liver, kidney, and other vital organs can result in weakness, poor function, and even failure of body systems. Gluconeogenesis can also be stimulated by the hormone cortisol. This hormone responds to dangerous or stressful situations by causing a rapid release of glucose into the blood to meet the energy needs.

Carbohydrate is needed to break down fat

In addition to the loss of protein, insufficient carbohydrate intake can also affect fat metabolism. To metabolize fat completely, a small amount of carbohydrate must be available. This is because acetyl coenzyme A (acetyl-CoA), produced from fat break down, can be used to produce energy via the Krebs cycle only if it can combine with a four-carbon oxaloacetate molecule derived from carbohydrate metabolism. When carbohydrate is in short supply, oxaloacetate is limited and acetyl-CoA cannot be metabolized to carbon dioxide and water. Instead, liver cells convert acetyl-CoA into compounds known as ketones or ketone bodies (Figure 2.4). Ketone production is a normal response to limitations of glucose transport into the cell such as in diabetes or glycogen depletion through starvation, a very low carbohydrate diet, or prolonged exercise. Ketones can be used for energy by tissues, such as those in the heart, muscle, and kidney. Even the brain, which requires glucose, can adapt to obtain a portion of its energy from ketones. Excess ketones are excreted by the kidney in urine. However, if excretion is outpaced by production, or fluid intake is too low to produce enough urine to excrete ketones, ketones can build up in the blood, causing ketosis. Mild ketosis, which occurs with moderate caloric restriction such as occurs during a weight-loss diet, produces symptoms including headaches, dry mouth, foul-smelling breath, and in some cases, a reduction in appetite. High ketone levels, if left untreated, will increase the acidity of the blood and can result in coma and death.

Glycemic response

Our bodies react differently to different sources of carbohydrates. For example, a serving of a high-fiber food, such as baked beans, results in lower blood glucose levels compared to the same serving size of mashed potatoes. How quickly blood glucose levels rise after

Figure 2.4 The availability of carbohydrate determines how fatty acids are metabolized.

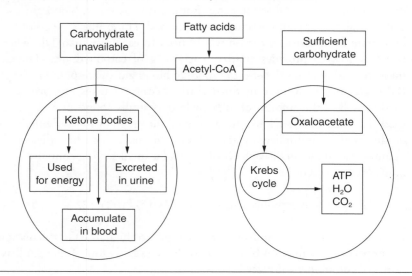

a meal is affected by the composition of the meal. Fat and protein consumed with high-carbohydrate foods cause the stomach to empty more slowly and therefore delay the rate at which glucose enters the small intestine, where it is absorbed. This will then cause a slower rise in blood glucose. Fiber also slows the rise in blood glucose because fiber – due to its unique structure – takes longer to be digested. In contrast, drinking a sugar-sweetened soft drink on an empty stomach will cause blood glucose to increase rapidly. For a meal we consume daily, the glycemic responses can vary depending upon its nutrient composition. For example, a meal of chicken, rice, and green beans, which contains starch, fat, protein, and fiber, will take at least 30 minutes to cause a rise in blood glucose. However, blood glucose will rise much more quickly if we consume a meal consisting primarily of carbohydrate, such as spaghetti or white bread.

Two food measurements have been developed to predict the blood glucose response to various foods and to plan a diet to avoid hyperglycemia. The first of these tools is the glycemic index (GI). GI is the ratio of the blood glucose response to a given food compared to a standard or reference food (typically, glucose or white bread). It is a numerical system of measuring how fast and high ingesting a carbohydrate food triggers a rise in circulating blood glucose. Keep in mind, however, that one person's glycemic response to a given food may be very different to someone else's response. Moreover, people do not normally consume carbohydrate foods by themselves, but often with other foods containing fat and protein, such as a hamburger in a bun. Indeed, GI can be influenced by many factors, including rate of ingestion, fiber content, fat and protein content, starch characteristics, food form, gastric empting, and gastrointestinal digestion and absorption. For example, mashed potatoes are associated with a higher GI due to higher amylopectin content and greater surface area exposed. On the other hand, fructose has a low glycemic response mainly because of its slower absorption rate. In fact, this is one of the reasons why fructose has been recommended for use as a carbohydrate supplement prior to an athletic event. A fast rise in blood glucose can create an insulin response and the potential reactive hypoglycemia, which is detrimental to performance.

Another shortcoming of GI is that it does not account for the amount of carbohydrate found in a typical serving size. Rather, it is based on blood glucose response to consuming 50 g of carbohydrate in a given food. For example, to consume 50 g of carbohydrate from carrots a person would need to eat more than 1 lb, whereas a cup of rice (about 8 oz) provides approximately 42 g of carbohydrates. Therefore, another measure used to assess the effect of food on blood glucose response is the glycemic load (GL). GL may be a more useful measure, because it takes into account the GI of a food as well as the amount of carbohydrate typically found in a single serving of the food. To calculate the GL of a food the amount (in grams) of carbohydrate in a serving of the food is multiplied by the GI of that food, and then divided by 100. For example, vanilla wafers have a GI of 77, and a small serving contains 15 g of carbohydrate. Hence, its GL is:

$$(GI \times \text{grams of carbohydrate}) / 100 = (77 \times 15) / 100 = 12.$$

Even though the GI of vanilla wafers is considered high, the GL calculation shows that the impact of this food on blood glucose levels is fairly low. Table 2.4 lists GI and GL values of commonly consumed foods.

Table 2.4 Glycemic index (GI) and glycemic load (GL) values of common foods

FOOD	GLYCEMIC INDEX[1]	CARBOHYDRATE/SERVING (G)	GLYCEMIC LOAD[2]
Pasta/Grains			
White, long grain	56	45	25
White, short grain	72	53	38
Brown rice	45	33	16
Spaghetti	41	40	16
Bread and Muffins [1]			
Bagel	72	30	12
Pancake	67	58	39
Waffle	76	13	10
Oat bran bread	44	18	8
White bread	70	10	7
Whole wheat bread	69	13	9
Vegetables			
Boiled carrot	49	16	8
Baked potato	85	57	48
Yam	37	36	13
Corn	55	39	21
Fruits			
Apple	38	22	8
Banana	55	29	16
Orange	44	15	7
Grape	43	17	7
Plums	24	14	3
Cherries	22	12	3
Raisin	64	44	28
Raisin bran	61	19	12
Legumes			
Baked beans	48	54	26
Kidney beans	27	38	10
Navy bean	38	54	21
Dairy Foods			
Milk, skim	32	12	4
Yogurt, low fat	33	17	6
Ice cream	61	31	19
Snack Foods			
Potato chips	54	15	8
French fries	75	29	22
Popcorn	54	11	6

Source: Adapted from Foster-Powell *et al.* 2002.

Notes:
1 Low GI foods – below 55; medium GI foods – between 55 and 70; high GI foods – more than 70.
2 Low GL foods – below 15; medium GL foods – between 15 and 20; high GL foods – more than 20.

LIPIDS

Lipids are required for every physiological system in the body, and are thus essential nutrients. For many people, the thought of fatty foods invokes images of unhealthy living. We often shop for "fat-free" foods and try to avoid fats entirely. Food manufacturers have even developed "fat substitutes" to replace the fats normally found in food. However, although diets high in fat can lead to health complications such as obesity and heart disease, getting enough of the right types of fat is just as essential for optimal health.

What are the right types of fat? Should we put butter or margarine on our toast? Should we use canola or corn oil in cooking? There are hundreds of oils, butters, and margarines from which to choose. Some are solid, some are liquid, some come from plants, and some come from animals. Some are said to increase your risk of heart disease while others claim to do the opposite. Recommendations for a healthy diet suggest that we consume a diet moderate in fat and low in saturated fat, trans fat, and cholesterol. In order to follow these guidelines, we must know how much and what types of fats are in the foods we choose.

Common properties and specific types

"Lipid" is the chemical term for what are commonly known as fats and oils. Lipids are a diverse group of chemical compounds. They share one main characteristic: they do not readily dissolve in water. For example, think of an oil and vinegar salad dressing. The oil is not soluble in the water-based vinegar; the two separate into distinct layers, with oil on the top and vinegar on the bottom. Lipids in the diet and in our bodies provide a concentrated source of energy. As mentioned in Chapter 1, each gram of fat provides 9 kcal compared with only 4 kcal per gram from carbohydrate and protein. The major lipid classes include **fatty acids**, **triglycerides**, **phospholipids**, and **sterols**. The triglycerides predominate both in foods and in the body.

Fatty acids

In the body and in foods, fatty acids are found in the main form of lipids – triglycerides. A fatty acid is basically a long chain of carbons bonded together and flanked by hydrogen (Figure 2.5). At one end of the molecule is an acid group (COOH). At the other end, which is often referred to as the omega end, is a methyl group (CH_3). Most naturally occurring fatty acids contain even numbers of carbon in their chains, usually 12 to 22, although some may be as short as 4 or as long as 26 carbons. Fatty acids with fewer than 8 carbons are called short-chain fatty acids; those with 8–12 carbons are medium-chain fatty acids; and those with more than 12 carbons are long-chain fatty acids. The long-chain (12–24 carbons) fatty acids are most common in the diet and are found primarily in meats, fish, and vegetable oils, while short- or medium-chain (6–10 carbons) fatty acids occur mainly in dairy products. The chain length of a fatty acid affects its chemical properties and physiological functions. In general, fatty acids with a shorter chain length tend to be liquid at room temperature, less stable, and more water soluble.

Figure 2.5 Chemical structure of saturated, monounsaturated, and polyunsaturated fatty acids. Each contains 18 carbons, but they differ from each other in the number and location of double bonds.

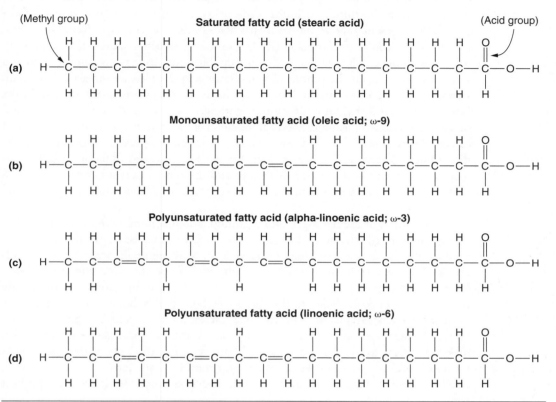

Another way that fatty acids differ is by the types of chemical bonds between the carbon atoms (Figure 2.5). These carbon–carbon bonds can either be single bonds or double bonds. If a fatty acid contains all single carbon–carbon bonds, it is a **saturated fatty acid**. The most common saturated fatty acids are palmitic acid, which has 16 carbons, and stearic acid, which has 18 carbons. These are found most often in animal foods such as meat and dairy products. Vegetable sources of saturated fatty acids include palm oil, palm kernel oil, and coconut oil. These are often called tropical oils because they are found in plants common in tropical climates.

Fatty acids containing one or more double bonds are **unsaturated fatty acids** (Figure 2.5). In other words, an unsaturated fatty acid contains some carbons that are not saturated with hydrogen. More specifically, fatty acids with one double bond are **monounsaturated fatty acids**; those with two or more double bonds are **polyunsaturated fatty acids**. In our diet, the most common monounsaturated fatty acid is oleic acid, which is prevalent in olive and canola oils. The most common polyunsaturated fatty acid is linoleic acid, found in corn, safflower, and soybean oils. Unsaturated fatty acids melt at cooler temperatures than saturated fatty acids of the same chain length. Therefore, the more unsaturated bonds a fatty acid contains, the more likely it is to be liquid at room temperature.

There are different categories of unsaturated fatty acids, depending on the location of the first double bond in the chain. As shown in Figure 2.5, if the first double bond occurs between the third and fourth carbons, counting from the omega end of the chain, the fat is said to be an omega-3 fatty acid. Alpha-linolenic acid, found in vegetable oils, and eicosapentaenoic acid (EPA) and docosahexaenoic acid (DHA), found in fish oil, are omega-3 fatty acids. If the first double bond occurs between the sixth and seventh carbons from the omega end, the fatty acid is called an omega-6 fatty acid. Linoleic acid, found in corn and safflower oils, is the major omega-6 fatty acid in the North American diet. Our body cannot synthesize double bonds in the omega-3 and omega-6 positions. Therefore, both alpha-linolenic acid (omega-3) and linoleic acid (omega-6) are also referred to as essential fatty acids and so must be obtained from the diet. Omega-3 fatty acids are important for the structure and function of cell membranes, particularly in the retina of the eye and the central nervous system. Omega-6 fatty acids are important for growth, skin integrity, fertility, and maintaining red blood cell structure.

The position of the hydrogen atoms around a double bond is another way of classifying unsaturated fatty acids. Unsaturated fatty acids can exist in two different structural forms, the *cis* and *trans* forms (Figure 2.6). Most naturally occurring fatty acids are usually in the *cis* form in which the hydrogens are on the same side of a carbon–carbon double bond. During certain types of food processing, some hydrogens are transferred to opposite sides of the carbon–carbon double bond, creating the *trans* form, or a **trans fatty acid**. The *cis* bond causes the fatty acid backbone to bend. However, the *trans* bond allows the fatty acid backbone to remain straight, which makes it similar to the shape of saturated fatty acid. For this reason, trans fatty acids are also more likely to be solid at room temperature. Trans fatty acids are found in small amounts in nature and are formed during food processing involving high heat and high pressure.

Figure 2.6 *Cis* versus *trans* fatty acids.

A *cis* fatty acid

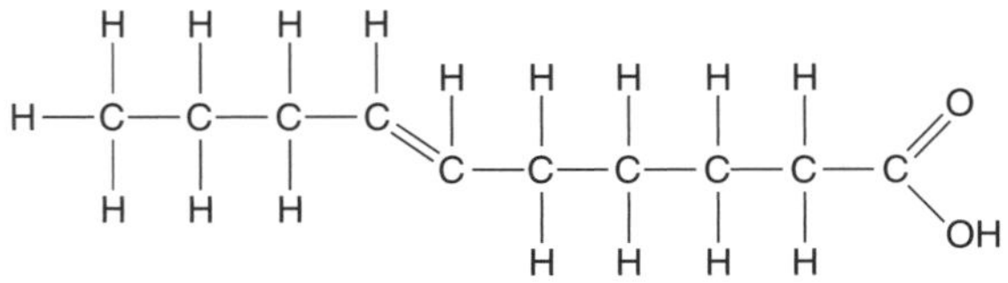

A *trans* fatty acid

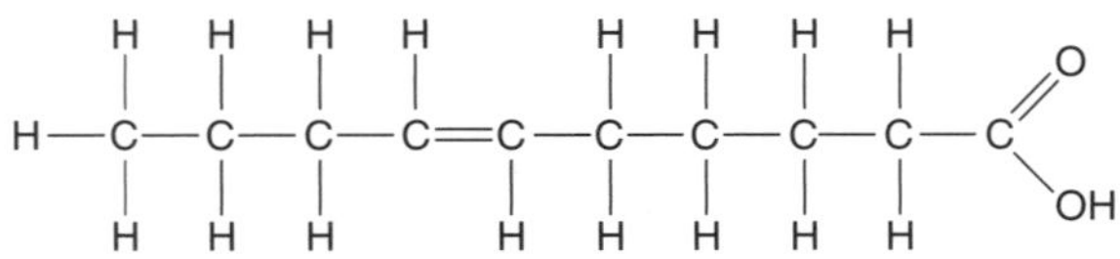

Triglycerides

Most fatty acids do not exist in their free or unbound form in foods or in the body. Instead, they are part of larger, more complex molecules called triglycerides or smaller molecules called diglycerides and monoglycerides. When three fatty acids are attached to a backbone of the three-carbon molecule glycerol, the molecule is called a triglyceride (Figure 2.7a). When one fatty acid is attached, the molecule is called a monoglyceride, and when two fatty acids are attached, it is a diglyceride. Before most dietary fats are absorbed in the small intestine, the two outer fatty acids are typically removed from triglycerides. This produces a mixture of fatty acids and monoglycerides that can be absorbed into intestine cells. After absorption, the fatty acids and monoglycerides are mostly re-joined to form triglycerides. Triglycerides may contain any combination of

Figure 2.7 Chemical forms of common lipids: (a) triglyceride; (b) phospholipids (e.g., lecithin); and (c) sterol (e.g., cholesterol).

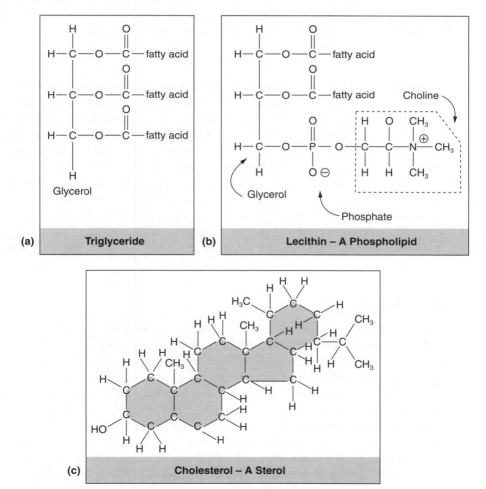

fatty acids: long, medium, short, saturated, or unsaturated. Triglycerides make up most of the lipids in foods and in the body, and usually are what are being referred to when the term "fat" is used.

Phospholipids

Phospholipids are another class of lipid. They are important parts of cell membranes. Like triglycerides, they are built on a backbone of glycerol. However, at least one fatty acid is replaced with a compound containing phosphorus and often other elements such as nitrogen and choline (Figure 2.7b). Lecithin is a common example of a phospholipid that is attached with a molecule of choline. The fatty acid end of phospholipids is soluble in fat and is hydrophobic, whereas the phosphate end is water soluble and hydrophilic. Phospholipids are amphipathic, meaning they contain both polar (hydrophilic) and non-polar (hydrophobic) properties. The structure allows phospholipids to be major components of cell membranes because they are able to mix with both water and fat. Having such a polarized configuration makes phospholipids important in carrying out the digestion, absorption, and transport of lipids. Phospholipids are also found in food sources such as eggs, liver, soybeans, wheat germ, and peanuts.

Sterols

In addition to triglycerides and phospholipids, lipids include sterols, compounds with a multiple-ring structure (Figure 2.7c). A sterol can be attached to a fatty acid via an ester bond, forming a sterol ester. The most famous sterol is cholesterol. Cholesterol is a weakly polar compound. Although some free or unbound cholesterol is found in the body, most is bonded to a fatty acid. This cholesterol–fatty acid is called a cholesteryl ester. Cholesteryl esters are more hydrophobic than free cholesterol. Cholesterol can be manufactured by almost every tissue in the body, especially the liver. Therefore, cholesterol is regarded as a non-essential nutrient. More than 90% of the cholesterol in the body is found in cell membranes. It is also part of myelin, the coating on many nerve cells. Cholesterol is found only in foods from animal sources. Plant foods do not contain cholesterol unless animal products are combined with them in cooking or processing.

Food sources of lipids

The fat content in foods can vary from 100%, as found in most cooking oils and spreads such as butter, margarine, and mayonnaise, to trace amounts – less than 5% – as found in most fruits and vegetables. Some foods obviously have a high fat content. For example, foods high in fat include nuts, bologna, avocados, and bacon, which have about 80% of calories as fat. These are followed by peanut butter, cheddar cheese, steak, hamburgers, ice cream, doughnuts, and whole milk (Table 2.5). However, in other foods the fat content may be high but not as obvious. This is known as hidden fat. For example, some baked goods such as cakes, muffins, croissants, cookies, crackers, and chips contain considerable amounts of fat that we are often unaware of. A 5-oz baked potato contains 145 kcal with about 3% fat, but people often ignore the fact that the same size serving of potato chips contains 795 kcal, over 60% of which are from fat.

FOODS	SERVING SIZE	FAT (G)	CALORIES FROM FAT (%)
			Table 2.5 Fat content of commonly selected foods
Canola oil	1 tablespoon	14	100
Margarine	1 tablespoon	12	100
Butter	1 tablespoon	12	100
Avocado	½ cup	11	86
Mixed nuts	1 oz	16	78
Peanut butter	1 tablespoon	8	76
Cheddar cheese	1 oz	10	74
T-bone steak	3 oz	17	66
Flex seeds	1 tablespoon	3	62
Whole milk	1 cup	8	49
Snack crackers	1 oz	7	45
Doughnut	1	5	45
Hamburger	1	12	39
Chocolate candies	1 oz	6	39
Chicken breast with skin	3 oz	7	36
2% milk	1 cup	5	36
Chicken breast without skin	3 oz	6	32
Baked beans	½ cup	7	31
Yogurt	8 oz	7	28
Low-fat yogurt	8 oz	4	18

The type of fat in food is important to consider along with the total amount of fat. Animal fats are the chief contributors of saturated fatty acids. About 40–60% of the total fat in dairy and meat products is in the form of saturated fatty acids (Figure 2.8). In contrast, plant oils contain mostly unsaturated fatty acids, covering about 70–95% of total fat. Some of the plant oils are good sources of monounsaturated fatty acids, such as canola, olive, and peanut oils. Corn, sunflower, soybean, and safflower oils contain mostly polyunsaturated fatty acids. These plant oils supply the majority of the alpha-linoleic (omega-3) and linoleic (omega-6) in the North American diet. These fatty acids are considered as essential fatty acids, meaning they must be obtained through the diet because human cells lack the enzymes needed to produce these fatty acids. Both omega-3 and omega-6 fatty acids perform important roles in immune function and vision, help form cell membranes, and produce hormone-like compounds. Table 2.6 shows the amounts of omega-3 fatty acid of commonly chosen fish and seafood products.

As mentioned earlier, wheat germ, peanuts, egg yolk, soybeans, and organ meats are rich sources of phospholipids. Phospholipids such as lecithin, a component of egg yolks, are often added to salad dressing. Lecithin is used as an emulsifier because of its ability to keep mixtures of lipids and water from separating. Emulsifiers are added to salad dressing to keep the vegetable oil suspended in water. Eggs being added to cake batter is another example of phospholipids being used to emulsify fat with water.

FOOD	OMEGA-3 FATTY ACID (G)
Salmon	1.15
Swordfish	1.15
Trout	1.15
Shark	0.83
Flounder	0.48
Sole	0.44
Cod	0.44
Squid	0.40
Crab	0.35
Oysters	0.30
Shrimp	0.27
Scallop	0.27
Mussel	0.26
Clam	0.26
Tuna	0.23
Lobster	0.07

Table 2.6 Omega-3 fatty acid content of fish and seafood

Source: Adapted from USDA Nutrient Data Laboratory.

Note: All values represent estimated amounts in a 3-ounce cooked portion and these values can vary markedly with species, season, diet, packaging, and cooking methods.

Figure 2.8 Saturated, monounsaturated, and polyunsaturated fatty acid content of various sources of dietary lipid.

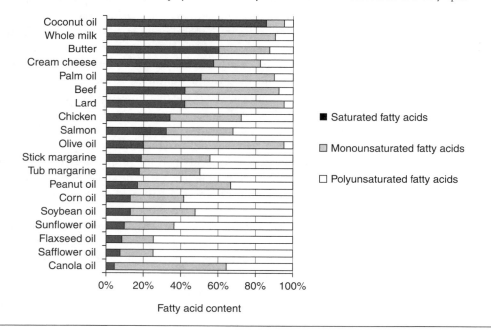

Cholesterol, a common example of sterol and widespread in the plasma membrane of all cells, is obtained either through the diet or through cellular synthesis. Cholesterol that is obtained from the diet is referred to as exogenous cholesterol, while cholesterol that is produced within the body is referred to as endogenous cholesterol. Even if an individual maintains a "cholesterol-free" diet, endogenous cholesterol synthesis varies at 500–2000 mg per day. More endogenous cholesterol forms with a diet high in saturated fatty acids. Exogenous cholesterol is found only in animal foods (Table 2.7). Eggs are our main source of cholesterol, along with meats and whole milk. One egg yolk contains about 200 mg of cholesterol. Organ meats contain about 300 mg per 3-oz serving. Lean red meat and chicken contains 100 mg, whereas fish contains 50 mg in 3 oz. The production of endogenous cholesterol is usually sufficient to meet the body's needs; hence, severely reducing cholesterol intake may cause little harm except in pregnant women and infants.

Table 2.7 Cholesterol content of commonly selected foods

FOODS	SERVING SIZE	CHOLESTEROL CONTENT (MG)
Skimmed milk	1 cup	4
Mayonnaise	1 tablespoon	10
Butter	1 pat	11
Lard	1 tablespoon	12
Cottage cheese	½ cup	15
Low-fat milk (2%)	1 cup	22
Half-and-half	¼ cup	23
Hot dog	1	29
Ice cream	½ cup	30
Cheddar cheese	1 oz	30
Whole milk	1 cup	34
Oysters	3 oz	40
Salmon	3 oz	40
Clams	3 oz	55
Tuna	3 oz	55
Chicken	3 oz	70
Turkey	3 oz	70
Beef	3 oz	75
Pork	3 oz	75
Lamb	3 oz	85
Crab	3 oz	85
Shrimp	3 oz	110
Lobster	3 oz	110
Heart	3 oz	165
Egg yolk	1	210
Beef liver	3 oz	410
Kidney	3 oz	540

Source: USDA National Nutrient Database for Standard Reference, Release 22, 2009.

Major roles of lipids in the body

The blood carries lipids to various sites around the body. Once they arrive at their destinations, the lipids can get to work providing energy, insulating against temperature extremes, protecting against shock, and maintaining cellular integrity. The following sections describe each of these roles in more detail.

Energy source and reserve

Triglycerides provide an important source of energy. For this to happen, they must first be broken down into glycerol and fatty acids. This process, called **lipolysis**, is catalyzed by the hormone-sensitive enzyme lipase, whose activity increases when secretion of the pancreatic hormone insulin is low. Lipolysis is also stimulated by exercise and physiological stress. Compared to other energy-yielding nutrients, triglycerides represent the body's richest source of energy. As noted earlier, the complete breakdown of 1 g of triglyceride yields approximately 9 kcal of energy, which is more than twice the yield from 1 g of carbohydrate or protein. Therefore, gram for gram, high-fat foods contain more calories than do other foods.

The pancreatic hormone insulin stimulates the storage of triglycerides, a process that is the opposite of lipolysis. This occurs during times of energy excess. Insulin causes adipocytes, and to a lesser extent skeletal muscle cells, to take up glucose and fatty acids and convert glucose into fatty acids. Fatty acids are then incorporated into triglycerides. The synthesis of fatty acids and triglycerides is called **lipogenesis**. Triglycerides are stored in adipose tissue and, to a lesser extent, skeletal muscle. Adipose tissue consists of specialized cells called adipocytes, which can accumulate large amounts of lipids. Adipose tissue is found in many parts of the body, including beneath the skin (subcutaneous adipose tissue) and around the vital organs in the abdomen (visceral adipose tissue). Considerable adipose tissue is also associated with many of the body's organs, such as the kidneys and breasts, making it possible for these organs to have ready access to fatty acids for their energy needs. Because lipids are not stored with water as glycogen and protein are, the body can store a large amount of triglycerides in a small space. This will result in our ability to store an almost unlimited amount of energy. An average individual is capable of storing 100,000–150,000 kcal of fat energy, which is equivalent to 75–100 times the carbohydrate energy we normally store.

Insulation and protection

Triglycerides stored in adipose tissue also insulate the body and protect internal organs from injury. Although most of us do not rely on adipose tissue to keep warm, people with very little body fat can have difficulty regulating body temperature. Early research has demonstrated that fats stored just below the skin determine the ability to tolerate extremes of cold exposure. For example, it was found that swimmers who excelled in swimming the English Channel showed only a slight fall in body temperature while resting in cold water and essentially no slowing effect while swimming. In contrast, body temperature of leaner, non-Channel swimmers decreased markedly under rest and exercise conditions. In fact, one common physiological response to becoming excessively lean is to develop very fine hair covering the body. This hair, often referred to as lanugo, partially makes up for the absence of subcutaneous adipose tissue by providing a layer of external insulation for the body. The presence of lanugo is common in very lean individuals, such as those with eating disorders. For large football linemen or athletes involved in contact sports, excess fat storage may provide additional cushioning to protect from high impacts experienced in the sports. However, this protective benefit should be interpreted with caution as such excess fat can have negative consequences on energy expenditure, thermoregulation, and exercise performance.

Components of cell membranes

Phospholipids make up the major structural component of all cell membranes. More specifically, cell membranes consist of two layers of phospholipids with the hydrophilic polar head group pointing to the extra- and intra-cellular spaces. Remember, these compartments are predominantly water. To function effectively, a cell membrane must be able to provide stable barriers between these spaces. If the cell membrane were completely hydrophilic, it would dissolve and not create a barrier. On the other hand, if the cell membrane were complete hydrophobic, there would be no communication between extra- and intra-cellular compartments. The incorporation of phospholipids that are amphipathic and have both the hydrophobic and hydrophilic portions allows cell membranes to effectively carry out their functions.

Many important body compounds are sterols. Among them are bile acids, the sex hormones such as testosterone, adrenal hormones such as cortisol, vitamin D, and cholesterol. Cholesterol in the body can serve as the starting material for the synthesis of these compounds or as a structural component of cell membranes; more than 90% of all the body's cholesterol resides in the cells. Despite popular impressions to the contrary, cholesterol is a necessary compound the body makes and uses, although cholesterol in excess can be harmful. As noted earlier, the liver is the main site at which cholesterol is produced. In fact, the liver makes about 800–1500 mg of cholesterol per day, contributing much more to the body's total than does the diet. Cholesterol's harmful effects in the body occur when it forms deposits in the artery walls. These deposits lead to **atherosclerosis**, a condition in which an artery wall thickens as the result of a build-up of fatty materials such as cholesterol. If left untreated, atherosclerosis can cause heart attacks and strokes.

PROTEINS

Protein is a macronutrient that is distinguished from carbohydrate and lipid by the fact that it contains the element nitrogen. It is made from amino acids that are joined together by peptide bonds. Plants combine nitrogen from the soil with carbon and other elements to form amino acids. They then link these amino acids together to make proteins. Some proteins are very simple, containing only a few amino acids, whereas others contain thousands. However, most proteins are of intermediate size, having 250–300 amino acids. Protein in the diet provides the raw material to make all the various types of proteins the body needs. Thousands of substances in the body are made of proteins. Aside from water, proteins form the major part of lean body mass, totaling about 15–20% of body weight. These body proteins provide important structural and regulatory functions. In some circumstances protein can be used for energy, providing 4 kcal per gram.

Amino acids: the building blocks of protein

The numerous proteins in the body are very chemically diverse due to which amino acids they contain and the ways these are linked together. Each different protein contains a specific number of amino acids in specific proportions that are bound together in a specific order. Although at least 100 amino acids are found in nature, the body uses only about 20 different amino acids to make its own proteins. Each amino acid consists of four common components: (1) a central carbon bonded to hydrogen; (2) an amino group (–NH$_2$) containing nitrogen; (3) a carboxylic acid group (–COOH); and (4) a unique side-chain group that varies in length and structure. Different side chains give specific properties to individual amino acids. Figure 2.9 shows a "generic" amino acid.

The side-chain groups on amino acids vary from one amino acid to the next, making proteins more complex than either carbohydrates or lipids. A polysaccharide such as starch may be several thousand units long, but every unit is a glucose molecule, just like all the others. A protein, on the other hand, is made up of about 20 different amino acids, each with a different side-chain group. Each amino acid is defined by its side-chain group, which can be as simple as a single hydrogen atom or as complex as an organic ring structure. Appendix B presents the chemical structure for each of the 20 amino acids. Some side-chain groups also contain sulfur atoms. These subtle differences in the side-chain groups give each amino acid a unique chemical and physical feature. For example, some of the side-chain groups are negatively charged, some are positively charged, and some don't have charge at all. The charges associated with side-chain groups help determine the final shape and function of the protein.

Classification of amino acids

Of the 20 amino acids commonly found in protein, nine cannot be made by the adult human body. These amino acids are called **essential or indispensable amino acids,**

Figure 2.9 The main components of an amino acid.

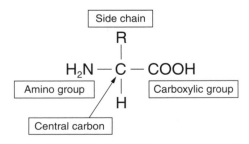

and they must be consumed in the diet (Table 2.8). If the diet is deficient in one or more of these amino acids, new proteins containing them cannot be made without breaking down other body proteins to provide them. The 11 **non-essential** or **dispensable amino acids** can be made by the human body and are not required in the diet. When a non-essential amino acid needed for protein synthesis is absent from the diet, it can be made in the body. Most of the non-essential amino acids can be made by the process of **transamination** in which the amino group from one amino acid is transported to a carbon-containing molecule to form a different amino acid.

Some amino acids are conditionally essential, meaning that they are only essential under certain conditions. For example, the conditionally essential amino acids tyrosine can be made in the body from the essential amino acid phenylalanine. If phenylalanine is in short supply, tyrosine cannot be made and becomes essential in the diet. Likewise, the amino acid cysteine is only essential when the essential amino acid methionine is in short supply. There are other factors that can influence the essentiality of amino acids. For example, some infants, especially those born prematurely, cannot make several of the non-essential amino acids such as cystine and glutamine. Thus, these amino acids must be obtained from the diet during this period. Also, certain diseases can cause a non-essential amino acid to become essential. For example, with a genetic disorder called phenylketonuria (PKU), the body loses its ability to convert phenylalanine to tyrosine due to a lack of an enzyme. Therefore, tyrosine must be supplemented via diet in patients with PKU.

Protein structure

Condensation reactions connect amino acids, just as they combine monosaccharides to form disaccharides, and fatty acids with glycerol to form triglycerides. Amino acids are linked together to form proteins by a unique type of chemical bond called a **peptide bond** (Figure 2.10). The bond is formed between the acid group of one amino acid and the nitrogen atom of the next amino acid. Two amino acids bond together to form a

Table 2.8 Essential and non-essential amino acids	
ESSENTIAL AMINO ACIDS	NON-ESSENTIAL AMINO ACIDS
Histidine	Alanine
Isoleucine	Arginine*
Leucine	Asparagine
Lysine	Aspartic acid
Methionine	Cysteine*
Phenylalanine	Glutamic acid
Threonine	Glutamine*
Tryptophan	Glycine*
Valine	Proline*
	Serine
	Tyrosine*

Note: *These amino acids are also classified as conditionally essential.

Figure 2.10 Condensation of two amino acids to form a dipeptide that contains a peptide bond.

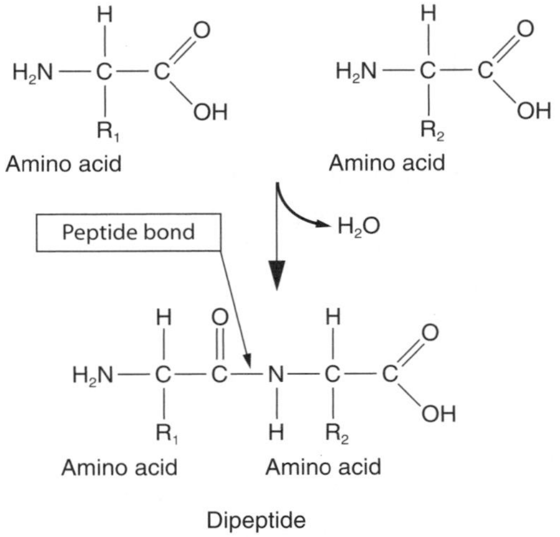

dipeptide. By another such reaction, a third amino acid can be added to the chain to form a tripeptide. As additional amino acids join the chain, a polypeptide is formed. Most proteins are a few dozen to several hundred amino acids long. A protein is made of one or more polypeptide chains folded into a complex three-dimensional structure.

In light of the level of folding complexity of polypeptide chains, proteins can be further divided into four distinct aspects: (1) primary structure; (2) secondary structure; (3) tertiary structure; and (4) quaternary structure. The primary structure concerns only the amino acid sequence of the peptide chains. The primary structure represents the basic identity of the protein. Alterations in the primary structure can be caused by inherited genetic variations. A disease called sickle-cell anemia is such an example in which the shape of hemoglobin is altered because of genetic "error." The secondary structure of a peptide chain results from weak chemical bonds, called hydrogen bonds, that twist and fold the primary structure. Such chemical interaction is due to the fact that the backbone of the peptide chain is made of amino and carboxylic acid groups with positive and negative charges. A normally functional protein always exists in a tertiary structure that is three-dimensional and contains additional folding of the peptide chain. Such additional folding is brought about by interactions between the side-chains. The quaternary structure refers to a protein that is made from more than one polypeptide chain. Hemoglobin is an example of a protein with a quaternary structure and is made from four separate polypeptide chains, each of which combines with an iron-containing unit called a heme. Heme is the portion of the hemoglobin molecule that actually holds the oxygen and carbon dioxide gases as they are transported in the blood.

A protein's final shape determines its ability to carry out its function. However, there are many conditions that can alter a protein's shape. One example is **denaturation**. Denaturation occurs when a protein unfolds in unusual ways. Compounds and conditions that cause denaturation include heat, acid, detergents, bases, salts, alcohol, and heavy metals such as mercury. A familiar example of protein denaturation occurs when an egg white is heated: proteins unfold and the egg white changes from a thin and clear liquid to a cloudy solid. Another example is mercury, which can disrupt bonds between side-chains and thus tertiary structure. Such denaturating action explains why mercury exposure can cause numbness, hearing loss, visual problems, difficulty walking, and severe emotional and cognitive impairments.

Quality of proteins

In a typical day, most people consume about 100 g of protein. This is almost twice their requirement given that the RDA for protein should be 56 g for a 70-kg man. Most of this protein comes from animal sources such as meat, milk, cheese, and eggs, which represent the most concentrated sources of protein. Nuts, seeds, and plants such as legumes also provide good sources of protein. Legumes are special in that they are associated with bacteria that can take nitrogen from the air and incorporate it into protein. Generally speaking, foods of animal origin tend to have larger amounts of certain essential amino acids than do plant-derived foods, and therefore are considered more efficient in terms of being used to make body proteins.

Complete and incomplete proteins

As you might expect, human tissue composition resembles animal tissue more than it does plant tissue. The similarities enable us to use proteins from any single animal source more efficiently to support human growth and maintenance than those from any single plant source. For this reason, animal proteins are generally considered high-quality or **complete proteins**, which contain the nine essential amino acids we need in sufficient amounts. Plant sources of protein, except for soybean, are considered low-quality or **incomplete proteins** because they lack adequate amounts of one or more essential amino acids. Proteins from plants have more diverse amino acid patterns that are quite different from what the body has. Hence, a single plant protein source, such as corn or wheat, cannot easily support body growth and maintenance. Corn protein has low amounts of lysine and tryptophan, whereas wheat protein lacks lysine. The amino acids that are missing or in a low quantity are called **limiting amino acids**.

When only low-quality protein foods are consumed, consumption of essential amino acids may be insufficient. Therefore, when compared to high-quality proteins, a greater amount of low-quality protein is needed to meet the needs of protein synthesis. Moreover, once any of the nine essential amino acids in the plant protein we have consumed is used up, further protein synthesis becomes impossible. Because the depletion of just one of the essential amino acids prevents protein synthesis, the process illustrates the all-or-nothing principle: either all essential amino acids are available or none can be used. The remaining or unused amino acids would then be used for energy needs, or converted to carbohydrate or fat.

Protein complementation

In general, plant proteins are of lower quality than animal proteins, and plants also offer less protein per unit of weight. For this reason, many vegetarians improve the quality of proteins in their diets by combining plant protein foods that are different but have complementary amino acid patterns. When two or more proteins combine to compensate for deficiencies in essential amino acid content in each protein, the proteins are called **complementary proteins**. Examples of commonly consumed meals representing protein complementation are rice and beans, or corns and beans. Both rice and corn have the limiting amino acids such as lysine, but provide adequate methionine. Beans and other legumes are limited in methionine but provide adequate lysine. Protein complementation allows diets containing a variety of plant protein sources to provide all the essential amino acids. The mixed diets that we normally consume generally provide high-quality protein because of protein complementation. Therefore, healthy adults should have little concern about balancing foods to yield the proteins needed to obtain enough of all nine essential amino acids. Even in plant-based diets, complementary proteins need not be consumed at the same meal by adults. Meeting amino acid needs over the course of a day is a reasonable goal because there is a ready supply of amino acids from those present in the body cells and in the blood (Craig *et al.* 2009).

Food sources of proteins

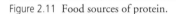

In a typical day, most people consume about 100 g of protein – as mentioned earlier, this is almost twice the required amount. Most of this protein comes from animal sources such as meat, milk, cheese, and eggs, which represent the most concentrated sources of protein. One egg or 1 oz of meat contains about 7 g of protein, and a cup

Figure 2.11 Food sources of protein.

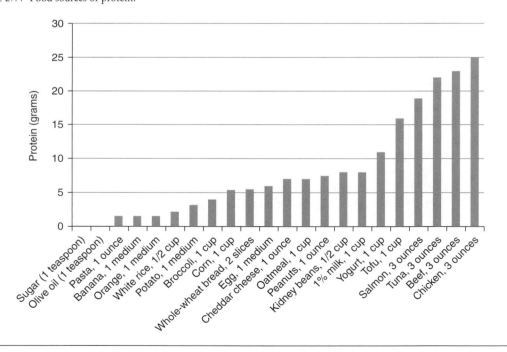

of milk contains 8 g. Plants also provide a good source of protein (Figure 2.11). Legumes, such as lentils, soybeans, peanuts, peas, kidney beans, and black beans provide 6–10 g of protein per half-cup serving. Nuts and seeds are also good sources of proteins, providing about 5–10 g per quarter-cup. As noted earlier, foods of animal origin tend to have larger amounts of essential amino acids than do plant-derived foods. However, a diet including plant proteins from a variety of sources can easily meet most people's needs.

Major roles of protein in the body

The body uses amino acids to synthesize the hundreds of thousands of proteins it needs. Whenever the body is growing, repairing, or replacing tissues, proteins are involved. Sometimes their role is to become part of the structure, other times it is to facilitate or to regulate. We rely on foods to supply the amino acids needed to form these proteins. However, only when we also eat enough carbohydrates and fats can food proteins be used efficiently. If we fail to consume enough calories to meet needs, some amino acids from proteins are broken down to produce energy instead of being available to replenish and build body proteins.

Structure

Proteins provide most of the structural materials in the body. For example, they are important constituents of muscle, skin, bone, hair, and finger nails. An example of a structural protein is collagen, which forms a supporting matrix in bones, teeth, ligaments, and tendons. Proteins are also an integral part of the cell membrane, the cytoplasm, and the organelles. The synthesis of structural proteins such as those in skeletal muscle is especially important during periods of active growth and development, such as infancy and adolescence. If a person's diet is low in protein for a long period, the processes of protein synthesis will slow down. Over time, skeletal muscles and vital organs such as the heart and liver will decrease in size or volume.

Most vital body proteins are in a constant state of breaking down, rebuilding, and repairing. For example, the intestinal tract lining is constantly sloughed off. The digestive tract treats sloughed cells just like food particles, digesting them and absorbing their amino acids. In fact, most of the amino acids released throughout the body can be recycled to become part of the pool of amino acids available for synthesis of future proteins. Overall, protein turnover is a process by which a cell can respond to its changing environment by producing proteins that are needed and degrading proteins that are not needed. It is estimated that an adult makes and degrades an average of 250 g of protein each day. Relative to the 70–100 g of protein typically consumed, recycled amino acids make an important contribution to total protein metabolism.

Enzymes

Enzymes are protein molecules that speed up the metabolic reactions of the body but are not used up or destroyed in the reactions. All the reactions involved in the production of energy and the synthesis and break down of carbohydrates, lipids, proteins, and other molecules are expedited by enzymes. Each reaction requires a specific enzyme with a specific structure. If the structure of the enzyme molecule is altered, it can no longer function in the reaction it is designed to accelerate.

Hormones

Hormones are chemical messengers that are secreted into the blood by one tissue or organ and act on target cells in other parts of the body. Their primary function is to respond to changes that challenge the body by eliciting the appropriate responses to restore the body's homeostasis or normal conditions. Some hormones are made of lipids; others are made of amino acids and so are classified as peptide or protein hormones. For example, insulin and glucagon are protein hormones.

Movement

Some proteins give cells and organisms the ability to move, contract, and change shape. Actin and myosin are proteins that function in the contraction of muscles. The two proteins slide past each other to shorten the muscle and thus cause contraction. A similar process causes contraction in the heart muscle and in the muscles that cause constriction in the digestive tract, blood vessels, and body glands. Nearly half of the body's protein is present in skeletal muscle, and adequate protein intake is required to form and maintain muscle mass throughout life.

Transport

Proteins transport substances throughout the body and into and out of the individual cells. Transport proteins in the blood carry substances from one organ to another. For example, hemoglobin, the protein in red blood cells, binds to oxygen in the lungs and transports it to other organs of the body. The proteins in lipoproteins are needed to transport lipids from the intestine and liver to body cells. Some vitamins, such as vitamin A, must be bound to a specific protein in order to be transported in the blood. When protein is deficient, the nutrients that require protein for transport cannot travel to the cells. For this reason, a protein deficiency can cause a vitamin A deficiency, even if consumption of vitamin A in the diet is adequate. At the cellular level, transport proteins present in cell membranes help move substances such as glucose and amino acids across the cell membrane. For example, transport proteins in the intestinal mucosa are necessary to absorb glucose and amino acids from intestinal lumen into the mucosal cells.

Regulation of fluid balance

Most of the body is made of water. This important fluid is found both inside cells (intracellular space) and outside cells (extracellular space). In addition, the extracellular space can be further divided into that found in blood and lymph vessels (intravascular fluid) and between cells (interstitial fluid). The amount of fluid in these spaces is highly regulated by a variety of means, some of which involve proteins. For example, a protein called albumin is present in the blood in relatively high concentrations. As the blood circulates through the capillaries, fluid and nutrients in the blood get pushed out into the interstitial space, in part because of blood pressure and the narrowness of the capillaries. However, albumin remains in the blood vessels, gradually increasing in concentration as more fluid is lost. When the albumin concentration reaches a certain level in the blood, albumin draws some of the interstitial fluid back into blood vessels via **osmosis**, partially counteracting the force of blood pressure. Osmosis is the movement of water molecules across a partially permeable membrane from an area of high water potential (low solute concentration) to an area of low water potential (high solute concentration).

With an inadequate consumption of protein, the concentration of proteins in the blood drops below normal. Excessive fluid then builds up in the surrounding tissues because the counteracting force produced by blood proteins is too weak to pull enough of the fluid back from the tissues into the blood vessels. As fluids accumulate in the tissues, the tissues swell, causing **edema**. Edema is associated with a variety of medical problems, so its cause must be identified. An important step in diagnosing the cause is to measure the concentration of blood proteins such as albumin.

Regulation of acid–base balance

The chemical reactions of metabolism require a specific level of acidity, or pH, to function properly. In the gastrointestinal tract, acidity levels vary widely. The digestive enzyme pepsin works best in the acid environment of the stomach, whereas the pancreatic enzymes operate most effectively in the more neutral environment of the small intestine. Inside the body, large fluctuations in pH can prevent metabolic reactions from proceeding. Proteins both within cells and in the blood help prevent large changes in acidity. For example, the protein hemoglobin in red blood cells helps neutralize acid produced from cellular respiration, so that the pH of blood can always be maintained as relatively neutral. Recall that components of amino acids, including the side chains, carry charges. In other words, they can accept and donate charged hydrogen ions easily. When hydrogen ion concentration in the blood is too high, proteins can bind excess hydrogen ions. Conversely, proteins can release hydrogen ions into the blood when the hydrogen concentration is too low.

Protection as antibodies

Protein can also defend the body against disease. A virus, whether it is one that causes 'flu, smallpox, measles, or the common cold, enters the cells and multiplies there. One virus may produce over 100 replicas of itself within an hour or so. Each replica can then burst out and invade different cells. Such a process of virus multiplication will ultimately cause disease. Fortunately, when the body detects these invading agents, it manufactures antibodies, giant protein molecules designed specifically to combat them. Each antibody has a unique structure that allows it to attach to a specific invader. When an antibody binds to an invading substance, the production of more antibodies is stimulated, and other parts of the immune system are activated to help destroy the invaders. In a normal, healthy individual, most diseases never have a chance to get started because of antibodies. Without sufficient protein, however, the body cannot maintain an adequate level of antibodies to resist diseases.

A source of energy in times of need

Some amino acids can be used for glucose synthesis and energy production as well as energy storage as fats. Together, these processes allow the body to (1) maintain an appropriate level of blood glucose, and (2) store excess energy for later use when dietary energy intake is more than adequate. When the body's available supply of energy is low, it first turns to glycogen and fatty acids. However, when glycogen is depleted and fatty acid reserves reduce, the body then dismantles its tissue proteins and converts some amino acids to glucose via gluconeogenesis. In addition, many cells can harvest the energy stored in amino acids by oxidizing them directly. Thus, over time energy deprivation always causes wasting of lean body tissue in addition to fat loss. An adequate supply of carbohydrates and fats spares amino acids from being used for energy and allows them to perform their unique roles.

During times of glucose and energy excess, the body redirects the flow of amino acids away from gluconeogenesis and ATP-producing pathways. To do this, the nitrogen-containing group of each amino acid is removed and converted to ammonia in the liver via a process called **deamination**. The remaining carbon skeleton is then converted to lipids and stored in adipose tissue. Thus, eating extra protein during times of glucose and energy sufficiency contributes to fat stores, not to muscle growth.

SUMMARY

- Carbohydrates are chemical compounds that contain carbon, hydrogen, and oxygen, with hydrogen and oxygen in the ratio of $2:1$. Simple carbohydrates include monosaccharides and disaccharides, while complex carbohydrates include oligosaccharides and polysaccharides.
- The common monosaccharides in foods are glucose, fructose, and galactose. Once they are absorbed from the small intestine and delivered to the liver, much of the fructose and galactose is converted to glucose.
- The major disaccharides are sucrose (glucose + fructose), maltose (glucose + glucose), and lactose (glucose + galactose). When digested, they yield their component monosaccharides.
- Polysaccharides include glycogen in animals and starch and fiber in plants. Glycogen and starch can be broken down by digestive enzymes, releasing the glucose units. Fiber cannot be digested by enzymes and therefore is not absorbed by the body. Fiber benefits gastrointestinal function by increasing the ease and rate at which materials move through the gastrointestinal tract.
- The foods that yield the highest percentage of calories from carbohydrates are: table sugar, honey, jam, jelly, and fruits. Other foods rich in carbohydrates include corn

flakes, rice, bread, and noodles. Foods with moderate amounts of carbohydrate calories are peas, broccoli, oatmeal, dry beans and other legumes, cream pies, French fries, and fat-free milk.

- Carbohydrates are a major source of energy, but are stored in limited quantity in the liver and muscles. They are the sole source of energy for most parts of the brain and central nervous system. Carbohydrates are also needed for burning fat as well as to protect against break down of body protein.

- Lipids, like carbohydrates, contain carbon, hydrogen, and oxygen, but with a higher ratio of hydrogen to oxygen. The major lipid classes include fatty acids, triglycerides, phospholipids, and sterols, with the triglycerides predominating both in foods and in the body.

- Fatty acids consist of a carbon chain with an acid group at one end. The length of the carbon chain and the number and position of the carbon–carbon double bonds determine the characteristics of the fat. Some fatty acids, such as alpha-linolenic acid (omega-3) and linoleic acid (omega-6), are considered essential because they cannot be synthesized by the body. Most fatty acids are found as part of triglycerides.

- Foods rich in fat include cooking oils and spreads such as butter, margarine, and mayonnaise. Nuts, bologna, avocados, and bacon are also high in fat, followed by peanut butter, cheddar cheese, steak, hamburgers, ice cream, doughnuts, and whole milk.

- Diets high in total fat, saturated fat, trans fat, and cholesterol increase the risk for developing cardiovascular diseases, metabolic disorders, and certain types of cancer. However, diets high in omega-3 and omega-6 polyunsaturated fatty acids along with plant foods containing fiber, antioxidants, and phytochemicals protect against these chronic conditions.

- Lipids provide the largest nutrient store of potential energy for biological work. Other major function of lipids include insulating against temperature extremes, protecting against shock, maintaining cellular integrity, and transporting the fat-soluble vitamins A, D, E, and K.

- Proteins differ chemically from carbohydrates and lipids because they contain nitrogen in addition to sulfur, phosphorus, and iron. Body proteins are made from individual amino acids that are bonded together. The sequential order of amino acids determines the protein's ultimate shape and function.

- Of the 20 amino acids used by the body, nine must be consumed from foods (essential) and the rest can be synthesized in the body (non-essential).

- High-quality (complete) protein foods contain ample amounts of all nine essential amino acids. They are mainly obtained from animal sources. Low-quality (incomplete) protein foods lack sufficient amounts of one or more essential amino acids. This is typical of plant foods, but different types of plant foods eaten together often complement each other's amino acid deficits.

- Proteins form important body components, such as muscle, connective tissue, transport proteins in the blood, enzymes, and some hormones. The carbon chains of proteins may be used to produce glucose or fat if necessary.

CASE STUDY: BUILDING A HEALTHY BASE AND REDUCING RISK FACTORS

Melissa's mother died of a heart attack at age 55. Melissa is worried about her own health and heart disease risk. She wants to eat a healthy diet and tries to follow the dietary guidelines. She made an appointment with her physician. She filled out a questionnaire about her medical history and lifestyle, met with a dietician to evaluate her diet, and had blood drawn for blood glucose and lipid analysis.

Melissa's diet analysis indicates that she consumes about 2000 kcal, 20% of which come from protein, 41% from fat, and 39% from carbohydrate. The percentage energy from saturated fat and unsaturated fat are 17% and 7%, respectively. Her fiber intake is 19 g per day and her cholesterol intake is 380 mg per day.

The following table provides the results of her medical history and blood analysis.

TABLE A	
Sex	Female
Age	35
Family history	Mother had heart attack at age 55
Height/weight	64 inches/175 lb
Blood pressure	120/70 mmHg
Smoking	No
Activity level	Sedentary
Blood glucose (fasting)	97 mg/100 ml
Blood triglycerides	185 mg/100 ml
Total cholesterol	210 mg/100 ml
LDL cholesterol	160 mg/100 ml
HDL cholesterol	34 mg/100 ml

Questions
- What is your overall impression of Melissa's diet?
- How many more grams of carbohydrate would Melissa need to meet the recommendation of 45–65% of energy from carbohydrates?
- What risk factors does Melissa have for developing cardiovascular disease?
- What dietary and lifestyle changes would you recommend to reduce her risks?

REVIEW QUESTIONS

1 Describe the structure of a monosaccharide and name the three monosaccharides that are important in nutrition.

2 Name the three disaccharides commonly found in foods and their component monosaccharides. In what foods are these sugars found?

3 How does the body maintain its blood glucose concentration? What happens when blood glucose concentration rises too high or falls too low?

4 Describe the structure of polysaccharides and name the ones important in nutrition. How are starch and glycogen similar, and how do they differ?

5 How is fiber different from other polysaccharides? Why do we say that fiber does not provide energy?

6 What health benefits are associated with a diet high in unrefined carbohydrates?

7 What is a lipid? Name three classes of lipids found in the body. What are some of their functions in the body?

8 Describe the chemical structures of saturated, monounsaturated, and polyunsaturated fatty acids and their different effects in food and in the body.

9 What does hydrogenation do to fats? What are trans fatty acids, and how do they influence heart disease?

10 How do phospholipids differ from triglycerides in structure? What role does cholesterol play in the body?

11 Which of the fatty acids are essential? Name their chief dietary sources.

12 How does the chemical structure of proteins differ from the structure of carbohydrates and lipids?

13 Describe the structure of amino acids. What are essential amino acids? Why is it important for essential amino acids lost from the body to be replaced in the diet?

14 Briefly describe the organization of proteins. How can this organization be altered or damaged? What might be a consequence of damaged protein organization?

15 Describe the concept of complementary proteins. How can vegetarians meet their protein needs without eating meat?

16 Describe some of the roles proteins play in the body.

SUGGESTED READING

Burke, L.M., Collier, G.R., and Hargreaves, M. (1998) Glycemic index: a new tool in sport nutrition? *International Journal of Sports Nutrition*, 8: 401–415.
   The GI provides a way to rank foods rich in carbohydrate according to the glucose response following their intake. This review article discusses specifically how the concept of GI may be applied to training and sports competition.

Coyle, E.F. (2000) Physical activity as a metabolic stressor. *American Journal of Clinical Nutrition*, 72: S512–S520.
   Physical activity provides stimuli that promote specific and varied adaptations according to the type, intensity, and duration of exercise performed. This article talks about how diet or supplementation can further enhance the body's responses and adaptations to these positive stimuli.

Jenkins, D.J., Kendall, C.W., Augustin, L.S., Franceschi, S., Hamidi, M., Marchie, A., Jenkins, A.L., and Axelsen, M. (2002) Glycemic index: overview of implications in health and disease. *American Journal of Clinical Nutrition*, 76: S266–S273.
   This article provides a solid review of literature on the GI and its relevance to those chronic Western diseases associated with central obesity and insulin resistance. The authors believe that the GI concept is an extension of the fiber hypothesis, suggesting that fiber consumption reduces the rate of nutrient influx from the gut.

3 Micronutrients: vitamins

KEY TERMS

- vitamins
- fortification
- enrichment
- bioavailability
- provitamins
- chylomicrons
- retinoids
- osteoclasts

- rickets
- osteomalacia
- osteoporosis
- free radical
- microcytic hypochromic anemia
- macrocytic anemia
- pernicious anemia
- collagen

OVERVIEW

Effective regulation of all metabolic processes requires a delicate blending of food nutrients in the watery medium of the cell. Of special significance in this regard are micronutrients – the small quantities of vitamins and minerals that facilitate energy transfer and tissue synthesis. The term "vitamin" was coined in 1912 by Polish biochemist Casimir Funk, who originally used the word "*vitamine*" to refer to substances that are amines (that contains an amino group) and are vital to life. Today, we know that vitamins are vital to life, but they are not all amines, so the "e" has been dropped, and the term "vitamin" refers to all these substances. Initially, the vitamins were named alphabetically in approximately the order in which they were identified: A, B, C, D, and E. The B vitamins were first thought to be one chemical form but were later found to be divided into many different subgroups, so the alphabetical naming system was broken down further by numbers that reflect the chronological order of discovery. For example, thiamin was the first B vitamin identified in 1937, and vitamin B-12 was the last, characterized in 1948. Currently, vitamins B-6 and B-12 are the only ones that are still commonly referred to by their numbers. Thiamin, riboflavin and niacin were originally referred to as vitamins B-1, B-2, and B-3, respectively, but they are now usually referred to by their own names.

What are vitamins?

Vitamins are organic compounds that are essential in the diet in small amounts for the purpose of promoting and regulating body functions necessary for growth, reproduction, and maintenance. Vitamins are generally essential in human diets because they cannot be synthesized in the body or because their synthesis can be decreased by environmental factors. Notable exceptions to having a strict dietary need for a vitamin are vitamin A, which we can synthesize from certain pigments in plants, vitamin D, synthesized in the body if the skin is exposed to adequate sunlight, niacin, synthesized from the amino acid tryptophan, and vitamin K and biotin, synthesized to some extent by bacteria in the intestinal tract.

To qualify as a vitamin, a compound must meet the following two criteria of an essential nutrient: (1) the body is unable to synthesize enough of the compound to maintain health; and (2) absence of the compound from the diet for a certain period produces deficiency symptoms that, if caught in time, are quickly cured when the substance is resupplied. A substance does not qualify as a vitamin merely because the body cannot synthesize it. Evidence must suggest that health declines when the substance is not consumed.

Vitamins differ from carbohydrates, lipids, and proteins in the following ways:

- Structure – vitamins are individual units; they are not linked together as are the molecules glucose, fatty acids, and amino acids.
- Function – vitamins do not yield energy when broken down. They assist enzymes that catalyze energy-yielding pathways involving carbohydrates, lipids, and proteins.
- Food contents – the amounts of vitamins we ingest daily from foods and the amounts we require are measured in micrograms (μg) or milligrams (mg), rather than grams (g).

Classification of vitamins

Vitamins have traditionally been grouped based on their solubility in water or fat. This chemical characteristic allows generalizations to be made about how they are absorbed, transported, excreted, and stored in the body. The water-soluble vitamins include the B vitamins and vitamin C. The fat-soluble vitamins include vitamins A, D, E, and K.

Vitamins in the diet

Almost all foods contain some vitamins. Generally speaking, grains are good sources of thiamin, niacin, riboflavin, pantothenic acid, and biotin. Meat and fish are good sources of all of the B vitamins. Milk provides riboflavin and vitamins A and D; leafy greens provide folate and vitamins A, E, and K; citrus fruit provides vitamin C; and vegetable oils are high in vitamin E. The vitamin content, however, can be affected by cooking, storage, and processing. The vitamins naturally found in foods can be washed away during preparation or destroyed by cooking. Exposure to light and oxygen can also cause vitamin losses. Food processing can both cause nutrient losses and add nutrients to food. The addition of nutrients to foods is called **fortification**. The added nutrients may or may not have been present in the original food. **Enrichment** is a type of fortification in which nutrients are added for the purpose of restoring those lost in processing to the same or higher level than was originally present. For example, the milling of whole grain wheat to make white flour results in the loss of the nutrients contained in the bran and germ. Enrichment adds back the vitamins thiamin, niacin, and riboflavin, and the mineral iron. Foods that are staples of a diet are often fortified to prevent vitamin or mineral deficiencies and to promote health in the population. For example, milk is fortified with vitamin D to promote bone health, and grains are fortified with folic acid to reduce the incidence of birth defects. Some foods are fortified because they are used in place of other foods that are good sources of an essential nutrient. For example, margarine is fortified with vitamin A because it is often used instead of butter, which naturally contains vitamin A.

Preserving vitamins in foods

Substantial amounts of vitamins can be lost from the time a fruit or vegetable is picked until it is eaten. The water-soluble vitamins, particularly thiamin, vitamin C, and folate, can be destroyed with improper storage and excessive cooking. Heat, light, exposure to the air, cooking in water, and alkalinity are all factors that can destroy vitamins. The sooner a food is eaten after harvest, the less chance there is of nutrient loss.

In general, if the food is not eaten within a few days, freezing is the best preservation method to retain nutrients. Fruits and vegetables are often frozen immediately after harvesting, so frozen vegetables and fruits are often as nutrient-rich as freshly picked ones. As part of the freezing process, vegetables are quickly blanched in boiling water. This destroys the enzymes that would otherwise degrade the vitamins. Table 3.1 provides some tips to help prevent vitamin loss.

Table 3.1 Tips for preventing nutrient loss	
WHAT TO DO	WHY
Keep fruits and vegetables cool	Chilling will reduce the degradation of vitamins by enzymes.
Refrigerate fruits and vegetables that have been cut in moisture-proof, air-tight containers	This will help to keep all nutrients and minimize oxidation of vitamins.
Trim, peel, and cut fruits and vegetables minimally	Oxygen breaks down vitamins faster when more surface is exposed, and outer leaves of most vegetables have higher values of vitamins and minerals than inner tender leaves and/or stems.
Rinse fruits and vegetables before cutting	This will prevent nutrients from being washed away.
Use a microwave oven or steam vegetables in a small amount of water	More nutrients are retained when there is less contact with water.
Add vegetables after water has come to a boil	More nutrients are retained with a shorter cooking time.
Minimize reheating of food	Prolonged reheating reduces vitamin content.
Do not add baking soda to vegetables to enhance the green color	Alkalinity destroys most vitamins, especially vitamin D and thiamin.
Do not add fats to vegetables during cooking if you plan to discard the liquid	Fat-soluble vitamins will be lost in discarded fat.

Vitamins in the digestive tract

About 40–90% of the vitamins in foods are absorbed into the body, primarily via the small intestine. The composition of the diet and conditions in the body, however, may influence **bioavailability**, a general term that refers to how well a nutrient can be absorbed and used by the body. The bioavailability of a specific nutrient may also be affected by other foods and nutrients in the diet. For example, the amount of fat in the diet affects the bioavailability of fat-soluble vitamins because they are absorbed along with dietary fat. In other words, fat-soluble vitamins are poorly absorbed when the diet is very low in fat. The transport mechanism by which vitamins are absorbed also determines the amount that enters the body. Fat-soluble vitamins are easily absorbed by simple diffusion. Many of the water-soluble vitamins, however, depend on energy-requiring transport systems or binding molecules in the gastrointestinal tract in order to be absorbed. For example, thiamin and vitamin C are absorbed by energy-requiring transport systems, riboflavin and niacin require carrier proteins for absorption, and vitamin B-12 must be bound to a protein produced in the stomach before it can be absorbed in the intestine. The quantity of vitamins in foods can be easily determined using an analytical approach. However, determining the bioavailability of a vitamin is a more complex task because it depends on many factors, including: (1) efficiency of digestion and time of transit through the GI tract; (2) previous nutrient intake and nutritional status; (3) other foods consumed at the same time; (4) methods of preparation (e.g., raw, cooked, or processed); and (5) source of the nutrients (e.g., synthetic, fortified, or naturally occurring).

Some of the vitamins are available from foods in inactive forms known as vitamin precursors, or **provitamins**. Once inside the body, the precursor is converted to an active form of the vitamin. Thus, in measuring a person's vitamin intake, it is important to count both the amount of the active vitamin and the potential amount available from its precursors.

Vitamins in the body

Once absorbed into the blood, vitamins must be transported to the cells. Despite their solubility in water, most of the water-soluble vitamins are bound to blood proteins for transport. Fat-soluble vitamins must be incorporated into lipoproteins or bound to transport proteins in order to be transported in the aqueous environment of the blood. For example, vitamins A, D, E, and K are all incorporated into chylomicrons for transport from the intestine. The amount of vitamin delivered to the tissues depends on the availability of the transport protein.

The body has the ability to store and excrete vitamins. This helps to regulate and maintain an adequate amount of vitamins present in the body. Except for vitamin K, the fat-soluble vitamins are not readily excreted from the body. In contrast, with the exception of vitamin B-12, excess amounts of water-soluble vitamins are generally lost from the body rapidly, partly because the water in cells dissolves these vitamins and they are excreted from the body via the kidneys. Because of the limited storage of many vitamins, they should be consumed in the diet regularly, although an occasional lapse in the intake of even water-soluble vitamins generally causes no harm. Symptoms of a vitamin deficiency occur only when that vitamin is lacking in the diet for an extended period and the body's stores are essentially exhausted. For example, an average person must consume no vitamin C for about 30 days before developing the first symptoms of vitamin C deficiency.

FAT-SOLUBLE VITAMINS

Fat-soluble vitamins are typically absorbed in the small intestine. This requires the presence of other lipids as well as the action of bile. Fat-soluble vitamins are circulated away from the small intestine in the lymph via **chylomicrons** – large lipoprotein particles that consist primarily of triglycerides – and eventually enter the blood. In the blood, fat-soluble vitamins are circulated as components of very low-density lipoproteins (VLDLs) or bound to transport proteins. Because most of the fat-soluble vitamins are stored in the body, people can eat less than their daily need for days, weeks, or even months or years without ill effects. In fact, consuming large amounts of them, especially in supplement form, can result in toxicities, sometimes with serious consequences. Most fat-soluble vitamins are involved in processes such as regulation of gene expression, cell maturation, and stabilization of free radicals. Table 3.2 gives an overview of the functions and sources, as well as deficiency diseases and toxicity symptoms associated with each of the four fat-soluble vitamins.

Table 3.2 Functions, sources, deficiency diseases, and toxicity symptoms for fat-soluble vitamins

VITAMIN	MAJOR FUNCTION	DEFICIENCY	TOXICITY	FOOD SOURCES
Vitamin A	• Growth • Reproduction • Vision • Cell differentiation • Immune function • Bone health	• Night blindness • Xerophthalmia • Hyperkeratosis	• Hypercarotenemia • Blurred vision • Birth defects • Liver damage • Osteoporosis	• Liver • Pumpkin • Sweet potato • Carrot
Vitamin D	• Calcium homeostasis • Bone health • Cell differentiation	• Rickets • Osteomalacia • Osteoporosis	• Hypercalcemia	• Fish • Mushrooms • Fortified milk • Fortified cereals
Vitamin E	• Antioxidant • Cell membranes • Eye health • Heart health	• Neuromuscular problems • Hemolytic anemia	• Hemorrhage	• Tomatoes • Nuts and seeds • Spinach • Fortified cereals
Vitamin K	• Coenzyme • Blood clotting • Bone health • Tooth health	• Bleeding	• No known effects	• Kale • Spinach • Broccoli • Brussels sprouts

Vitamin A

Vitamin A is found pre-formed and in precursor or provitamin forms in our diet. Pre-formed vitamin A compounds are known as **retinoids**. These include retinol, retinoic acid, and retinal. They are found in animal foods such as liver, fish, egg yolks, and dairy products (Table 3.3). Margarine and non-fat or reduced-fat milk are fortified with vitamin A because they are often consumed in place of butter and whole milk, which are good sources of this vitamin. Plant sources of vitamin A include carrots, cantaloupe, apricots, mangoes, and sweet potatoes, which contain yellow-orange pigments called carotenoids. Beta-carotene, the most potent precursor, is found in carrots, squash, and other red and yellow vegetables and fruits, as well as in leafy greens where the yellow pigment is masked by green chlorophyll. Other carotenoids that provide some provitamin A activity include alpha-carotene, found in leafy green vegetables, carrots, and squash, and beta-cryptoxanthin found in corn, green peppers, and lemons. Lutein, lycopene, and zeaxanthin are carotenoids with no vitamin A activity. To help consumers identify food sources of vitamin A, labels on packaged foods must include the vitamin A content as a percentage of the daily value. All forms of vitamin A in the diet are fairly stable when heated, but may be destroyed by exposure to light and oxygen.

Vitamin A has many roles, including aiding vision, growth, and reproduction. In addition, it is needed for maintaining a healthy immune system and building strong bones. Vitamin A is involved in the perception of light. In the eye, the retinal form of the vitamin combines with the protein opsin to form the visual pigment rhodopsin. Rhodopsin helps transform the energy from light into a nerve impulse that is sent to

Table 3.3 Food sources of vitamin A		
FOOD ITEM	AMOUNT	VITAMIN A CONTENT (μG)
Beef liver, fried	1 oz	3025
Sweet potato	½ cup	958
Carrots, cooked	½ cup	885
Spinach	⅔ cup	494
Mango	1 medium	402
Squash	⅔ cup	244
Eggs	2 large	185
2% milk	1 cup	175
Broccoli	1 cup	138
Apricots	3 medium	137
Cheddar cheese	1 oz	78
Margarine	1 teaspoon	52
Salmon	3 oz	45
Butter	1 teaspoon	45
Tomato, raw	½ cup	40
Orange	1 medium	25
Chicken	3 oz	10

Note: RDA: 900 μg/day for men and 700 μg/day for women.

the brain. This nerve impulse allows us to see. The visual cycle begins when light passes into the eye and strikes rhodopsin. Each time this cycle occurs, some retinal is lost and must be replaced by retinol from the blood. The retinol is converted to retinal in the eye. When vitamin A is deficient, there is a delay in regeneration of rhodopsin, which causes difficulty in adapting to dim light after experiencing a bright light, a condition known as night blindness. Night blindness is one of the first and more easily reversible symptoms of vitamin A deficiency.

Vitamin A affects cell differentiation through its effect on gene expression. In order to affect gene expression, the retinoic acid form of vitamin A enters specific target cells. Inside the nucleus of these target cells, retinoic acid binds to protein receptors to form a retinoic acid–protein receptor complex. This complex then binds to regulatory regions of DNA, which then changes the amount of messenger RNA that is made by the gene. This increases protein synthesis, thereby affecting various cellular functions. For example, vitamin A turns on a gene that makes an enzyme in liver cells, which enables the liver to make glucose by gluconeogenesis.

The ability of vitamin A to regulate the growth and differentiation of cells makes it essential throughout life for normal reproduction, growth, and immune function. In reproduction, vitamin A is believed to play a role during early embryonic development by directing cells to form the shapes and patterns needed for a completely formed organism. Poor overall growth is an early sign of vitamin A deficiency in children. Vitamin A affects the activity of cells that form and break down bone, and a deficiency early in life can cause abnormal jawbone growth, resulting in crooked teeth and poor dental health. Via its role in regulating cell differentiation, vitamin A is also important for producing the different types of immune cells and for stimulating the activity of specific immune cells.

The recommended daily amount (RDA) for vitamin A is set at 900 μg per day for men and 700 μg per day for women. These RDA values are based on the amount needed to maintain normal body stores. There is no recommendation to increase intake above this level for older adults. The RDA is increased in pregnancy to account for the vitamin A that is transferred to the fetus and during lactation to account for the vitamin A lost in milk. Consumption of vitamin A should not exceed 3000 μg per day. Above this upper limit, other possible side effects include an increased risk of hip fracture and poor pregnancy outcomes. The consumption of large amounts of vitamin A-yielding carotenoids does not cause toxic effects. This is because (1) they are less well absorbed, and (2) their rate of conversion into vitamin A is relatively slow and regulated.

Vitamin D

Vitamin D has an interesting and unique place among the nutrients. Although this vitamin is found in food, significant amounts of vitamin D can also be produced in the skin by exposure to ultraviolet light. For this reason, vitamin D is also known as the sunshine vitamin, and many nutritional scientists consider it a conditionally essential nutrient. Egg yolks, butter, whole milk, fatty fish, fish oil, and mushrooms are some of the foods that naturally contain vitamin D (Table 3.4). However, most liquid and dried milk products, as well as breakfast cereals, are fortified with vitamin D, and most dietary vitamin D comes from these foods. Vitamin D is quite stable and is not destroyed during food preparation, processing, or storage.

Table 3.4 Food sources of vitamin D		
FOOD ITEM	AMOUNT	VITAMIN D CONTENT (μG)
Baked herring	1 oz	44.4
Smoked eel	1 oz	25.5
Salmon	3 oz	6
Sardines	1 oz	3.4
2% milk	1 cup	2.5
1% milk	1 cup	2.5
Eggs	2 large	1.3
Total cereal	¾ cup	1.0
Soy milk	1 cup	1.0
Margarine	1 teaspoon	0.65
Chicken	3 oz	0.4
Beef liver	2 oz	0.4
Cheddar cheese	1.5 oz	0.25
Butter	1 teaspoon	0.15

Note: The adequate intake (AI) is 5 μg per day for people under age 50 and increases to 2–3 times for older adults. An RDA could not be set for vitamin D because the amount produced by sunlight exposure is too variable between individuals.

Vitamin D plays an important role in regulating calcium concentration in the blood. This requires several organs, including the small intestine, kidneys, and bone. Vitamin D, or more precisely, the calcitriol-a form of vitamin D, increases calcium absorption in the intestine, decreases calcium excretion in urine, and facilitates the release of calcium from bone. In this context, vitamin D acts like a hormone because it is produced in one organ, the skin, and affects other organs such as the intestine, kidneys, and bone. In the small intestine, vitamin D up-regulates several genes that code for proteins required for the transport of dietary calcium into the cells. In other words, vitamin D is involved in cell signaling. Without vitamin D, these proteins are not made, and calcium absorption is severely limited. In kidneys, vitamin D, along with parathyroid hormone, causes the kidneys to reduce their excretion of calcium via urine. As a result, more calcium remains in the blood. Vitamin D also acts with parathyroid hormone to stimulate bone break down by **osteoclasts** and therefore the release of calcium into the blood. While calcium in bones is important for their structure, calcium in the blood has additional physiological functions. For example, it is needed for muscle contraction, blood pressure regulation, and conduction of neural impulses. Without vitamin D to help maintain adequate levels of calcium in the blood, these vital functions would be impaired. Because of the close relationship between vitamin D and calcium in the body, the US Food and Drug Administration (FDA) encourages vitamin D fortification of milk. It must be noted that regardless of whether consumed in the diet or produced in the skin, vitamin D must be activated before the body can use it. Such activation processes occurs in the liver and kidneys. The role vitamin D plays in regulating calcium homeostasis is illustrated in Figure 3.1.

Figure 3.1 Vitamin D and calcium homeostasis.

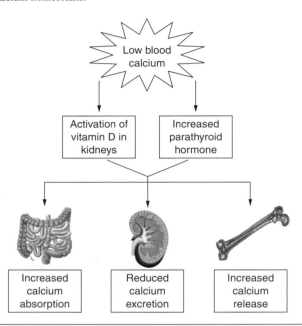

Vitamin D is also involved in a wide variety of other functions, such as regulation of gene expression and cell differentiation. As with vitamin A, vitamin D moves into the nucleus of the cell for the subsequent stimulation of the genes coding for specific proteins. For example, vitamin D causes immature bone cells to become mature bone marrow cells and causes certain intestinal epithelial cells to differentiate into mature enterocytes. As such, vitamin D plays a role in maintaining bone health and gastrointestinal function. Recently, it has also been suggested that vitamin D may help prevent certain types of cancers, such as those of the colon, breast, skin, and prostate (Bikle 2004; Gross 2005; Harris and Go 2005; Holick 2004).

When vitamin D is deficient, dietary calcium cannot be absorbed efficiently. As a result, calcium is not available for proper bone mineralization and abnormalities in bone structure occur. Vitamin D deficiency in infants and children who are in active stages of growth can result in inadequate bone mineralization – a disease called **rickets**. Although fortifying food with vitamin D has almost eliminated rickets in the United States, a significant number of cases has still been reported, especially in inner-city children who have a poor diet and whose exposure to sunlight is limited. Rickets is also a significant public health concern in other parts of the world (Calvo *et al.* 2005). Children with rickets have slow growth and characteristically bowed legs or knocked knees caused by the bending of long, weak bones that cannot support the stress of weight-bearing activities, such as walking.

In adults, the vitamin D deficiency disease comparable to rickets is called **osteomalacia**. Because bone growth is complete in adults, osteomalacia does not cause deformities, but bones are weakened because not enough calcium is available to form the mineral deposits needed to maintain healthy bone. Symptoms of osteomalacia include diffuse bone pain and muscle weakness. People with osteomalacia are at increased risk of bone fracture. It is common in adults with kidney failure because the conversion of vitamin D from inactive to active forms is reduced. The elderly are at risk for vitamin D deficiency because the ability to produce vitamin D in the skin decreases with age, mainly because older adults typically cover more of their skin with clothing and spend less time in the sun than their younger counterparts. In addition, the elderly tend to have a lower intake of dairy products. Vitamin D deficiency can also result in demineralization of bone, ultimately leading to a disease called **osteoporosis**, a condition characterized by a decrease in bone density and strength, resulting in fragile bones that can be frequently fractured. Osteoporosis is a serious chronic disease, and researchers estimate that more than 28 million Americans (1 in 10) suffer from it. To help prevent both osteomalacia and osteoporosis, people over 50 years of age are advised to get at least 15 minutes of sun exposure each day when possible and to increase their vitamin D intake. In some cases, vitamin D supplements may be necessary.

An RDA cannot be set for vitamin D because the amount produced by sunlight exposure is too variable between individuals. Consequently, a notation of "adequate intake" or AI is used to provide a guideline for vitamin D intake. AI for adult males and females is set at $5\,\mu g$ per day, which can be achieved by drinking two cups of vitamin D-fortified milk. This AI value was given based upon the assumption that no vitamin D is synthesized in the skin. If there is sufficient sun exposure, dietary vitamin D is not needed. This assumption is made because of the variation in the extent to which synthesis from sunlight meets the requirement. The amount synthesized in the skin is affected by skin pigmentation, climate, season, clothing, pollution, tall buildings that block sunlight, and the use of sunscreens. The AI of vitamin D for infants and children is the same as for adults. This is to allow sufficient vitamin D for bone development during the periods of rapid growth. Infants and children who are exposed to sunlight for about half an hour per day do not require supplemental vitamin D. The AI for adults 50–70 years of age is $10\,\mu g$ per day to prevent bone loss during periods of low sun exposure. In adults 70 or older, the AI is $15\,\mu g$ per day to maintain blood values of vitamin D and prevent skeletal fracture. The consumption of vitamin D should not exceed $50\,\mu g$ per day. Too much vitamin D can result in the overabsorption of calcium, which eventually leads to calcium deposits in the kidneys and other organs and causes metabolic disturbances and cell death.

Vitamin E

Vitamin E refers to eight different naturally occurring compounds that all have somewhat similar chemical structures. Of these, α-tocopherol is the most biologically active. Vitamin E was initially identified as a fat-soluble component of grains that was necessary for fertility in laboratory rats. In fact, the name tocopherol was derived from the Greek tokos (childbirth) and phero (to bear) and means "to bring forth offspring." However, vitamin E is now known to have many other functions. For example, its potential for decreasing risk of chronic diseases such as heart disease has attracted much public interest.

Vitamin E is widespread in foods. Much of the vitamin E in the diet comes from vegetable oils and products made from them, such as margarine and salad dressing (Table 3.5). Wheat germ oil is especially rich in vitamin E. Some dark green vegetables such as broccoli and spinach contain vitamin E as well. Vitamin E can be easily destroyed during food preparation, processing, and storage. Therefore, fresh or lightly processed foods are preferable sources. Most processed and convenience foods do not contribute enough vitamin E to ensure an adequate intake. Absorption of vitamin E occurs in the small intestine and requires the presence of bile and the synthesis of micelles. Vitamin E is circulated in chylomicrons – via lymph – and in the blood, eventually reaching the liver. In the liver, vitamin E is repackaged into VLDLs for further delivery throughout the body. Excess vitamin E is stored mainly in adipose tissue.

Like the carotenoids, vitamin E acts as an antioxidant preventing oxidation and **free radical** damage. Much of the body's vitamin E is associated with various membranes. Remember, cell membranes consist of a bilayer of phospholipid. In addition, many cell organelles, such as mitochondria and endoplasmic reticula, are enclosed in

Table 3.5 Food sources of vitamin E

FOOD ITEM	AMOUNT	VITAMIN E CONTENT (MG)
Total cereal	¾ cup	22.5
Sunflower oil	2 tablespoons	16.3
Sunflower seeds	1 oz	14.3
Safflower oil	1 tablespoon	5.9
Canola oil	1 tablespoon	5.7
Almonds	1 oz	4.5
Italian dressing	2 tablespoons	3.1
Mayonnaise	1 tablespoon	3.0
Avocado	1 medium	2.7
Peanut butter	2 tablespoons	2.4
Peanuts	1 oz	2.1
Kiwi	2 medium	1.8
Eggs	2 large	1.6
Salmon	3 oz	1.2
Margarine	1 teaspoon	1.2
Apricots	2 medium	0.8
Chicken	3 oz	0.7
Carrots, cooked	½ cup	0.6
Whole-wheat bread	2 slices	0.5
Orange	1 medium	0.4
Tomato, raw	½ cup	0.3
2% milk	1 cup	0.2
Cheddar cheese	1.5 oz	0.2
Oatmeal	1 cup	0.2

Note: RDA: 15 mg/day for men and women.

the phospholipid bilayer membrane. Maintaining these membranes is vital to the stability and function of cells and their organelles, and vitamin E plays a major role. Specifically, it protects the fatty acids in the membrane from free-radical-induced oxidative damage (Figure 3.2). This occurs because vitamin E can donate electrons to free radicals, making them more stable. This protection is especially important in cells that are exposed to oxygen, such as those in the lungs and red blood cells. Vitamin E can also defend cells from damage by heavy metals, such as lead and mercury, and toxins, such as carbon tetrachloride, benzene, and a variety of drugs. It also protects against some environmental pollutants such as ozone. The ability of vitamin E to act as an antioxidant is enhanced in the presence of other antioxidant micronutrients, such as vitamin C and selenium. Because polyunsaturated fats are particularly susceptible to oxidative damage, the vitamin E requirement increases as polyunsaturated fat intake increases.

Antioxidant nutrients protect DNA from cancer-causing free-radical damage, so people are very interested in the possibility that vitamin E might prevent or reduce cancer risk. However, although diets high in vitamin E are associated with decreased cancer risk, there is little experimental evidence that vitamin E by itself decreases the risk of this disease (Bostick *et al.* 1993; Graham *et al.* 1992; Kline *et al.* 2004). As an antioxidant, vitamin E also helps protect low-density lipoprotein (LDL) cholesterol from oxidation, which can lead to atherosclerosis. It may also inhibit an enzyme that allows the build-up of atherosclerotic plaque and increases the synthesis of an enzyme needed to produce eicosanoids, which help lower blood pressure and reduce blood clot formation (Steiner 1999; Emmert and Kirchner 1999). At present, experts do not know whether mega-dose vitamin E supplements taken by otherwise healthy people

Figure 3.2 Vitamin E functions as an antioxidant that protects the unsaturated fatty acids in cell membranes by neutralizing free radicals.

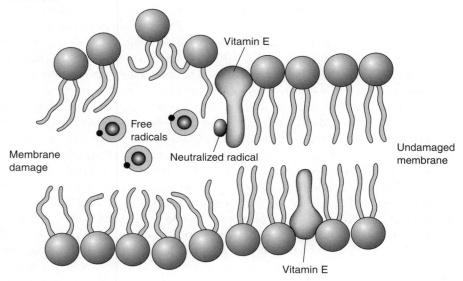

Source: Smolin and Grovenor 2003. Used with permission.

confer any more protection against cardiovascular disease and cancer than that achieved by improving diet, performing regular physical activity, not smoking, and maintaining a healthy body weight. In fact, the American Health Association considers that it is premature to recommend vitamin E supplements to the general public based on current knowledge and the failure of major clinical trials to show any benefit. In addition, the FDA has denied the request of the supplement industry to make a health claim that vitamin E supplements reduce the risk of cardiovascular disease and cancer.

Vitamin E deficiency is uncommon, and cases have been reported only in infants fed with formula that contains inadequate vitamin E, people with genetic abnormalities, and in diseases causing malabsorption of fat. Vitamin E deficiency is characterized by a variety of symptoms, including neuromuscular problems, loss of coordination, and muscular pain. Vitamin E deficiency also causes membranes of red blood cells to weaken and rupture, a condition referred to as hemolytic anemia. This is because vitamin E is especially important in protecting red blood cells from oxidative damage. Hemolytic anemia reduces the blood's ability to transport oxygen, resulting in weakness and fatigue.

The RDA for vitamin E for adult males and females is set at 15 mg per day. This value is based upon the amount needed to maintain plasma concentrations of α-tocopherol that protect red blood cells from breaking. The RDA for vitamin E does not change regardless of age or pregnancy status, although a slight increase is recommended for women who are lactating. The upper level for vitamin E for a healthy population is 1000 mg per day of supplemental α-tocopherol. This upper level was established because excessive amounts of vitamin E can reduce blood clotting by interfering with the action of vitamin K.

Vitamin K

Vitamin K was discovered and named for its role in coagulation by Henrik Dam, a Danish physiologist (it is spelt as "koagulation" in Danish), who found that vitamin K deficiency in chickens caused excessive bleeding. Dam received the Nobel Prize in Physiology or Medicine in 1943 for this discovery. As with all the fat-soluble vitamins, vitamin K is found in several forms. Phylloquinone is the form found in plants and the primary form in the diet. A group of vitamin K compounds, called menaquinones, are found in fish oils and meats and are synthesized by bacteria, including those in the human intestine. Menaquinones are also the form found in vitamin K supplements. Only a small number of foods provide significant amounts of vitamin K; liver, fish, legumes, and leafy green vegetables such as spinach, broccoli, brussels sprouts, kales, and turnip greens provide about half of the vitamin K in a typical North American diet (Table 3.6). Some vegetable oils are also good sources. Some of the vitamin K produced by bacteria in the human gastrointestinal tract is also absorbed. Dietary vitamin K is absorbed, along with other fat-soluble vitamins in the small intestine via micelle. Vitamin K is then incorporated into chylomicrons and put into lymph, eventually entering the blood. Vitamin K produced by bacteria in the large intestine is transported into epithelial cells by simple diffusion and then circulated to the liver via blood. The liver packages both dietary and bacterially produced forms of vitamin K into lipoproteins for delivery to the rest of the body.

FOOD ITEM	AMOUNT	VITAMIN K CONTENT (μG)
Kale, cooked	½ cup	530
Turnip, green, cooked	1 cup	520
Spinach, cooked	1 cup	480
Brussels sprouts, cooked	½ cup	150
Spinach, raw	1 cup	144
Asparagus, cooked	1 cup	144
Broccoli, cooked	½ cup	110
Lettuce	1 cup	97
Green beans, cooked	½ cup	49
Cabbage, raw	1 cup	42
Kiwi	2 medium	38
Green peas	½ cup	26
Soybean oil	1 tablespoon	25
Cauliflower, cooked	1 cup	20
Carrots, cooked	½ cup	18
Canola oil	1 tablespoon	17
Tomato, raw	½ cup	3
Whole-wheat bread	2 slices	2

Table 3.6 Food sources of vitamin K

Note: RDA: 120 μg/day for men and 90 μg/day for women.

Vitamin K is needed for the production of the blood-clotting protein prothrombin and other specific blood-clotting factors. These proteins are needed to produce fibrin, the protein that forms the structure of the blood clot (Figure 3.3). Injuries, as well as the normal wear and tear of daily living, produce micro-tears in blood vessels. To prevent blood loss, these tears must be repaired with blood clots. Other roles for vitamin K are less well understood. It has been suggested that vitamin K also catalyzes the carboxylation of other proteins needed for bone and tooth formation. Only after they have been carboxylated can these proteins bind calcium. Some studies have shown that consuming foods high in vitamin K is associated with decreased risk for hip fracture (Booth *et al.* 2004; Radecki 2005; Sasaki *et al.* 2005). However, further studies are needed to determine whether increased vitamin K intake results in increased bone strength.

Although rare in healthy adults, vitamin K deficiency appears in some infants and people with diseases that cause lipid malabsorption. In addition, prolonged use of antibiotics can kill the bacteria that normally live in the large intestine, resulting in vitamin K deficiency. The main sign of vitamin K deficiency is excessive bleeding. In infants, there is little transfer of this vitamin from mother to fetus, and because the infant gut is free of bacteria, none is made there. Further, breast milk is low in vitamin K. Therefore, to prevent uncontrolled bleeding, infants are typically given a vitamin K injection within six hours of birth.

Unlike other fat-soluble vitamins, vitamin K is used rapidly by the body, so a constant supply is needed. The RDA for vitamin K has been set at about 120 μg per day for men and 90 μg per day for women. Additional vitamin K is provided by bacteria in

Figure 3.3 The role of vitamin K in the blood-clotting process.

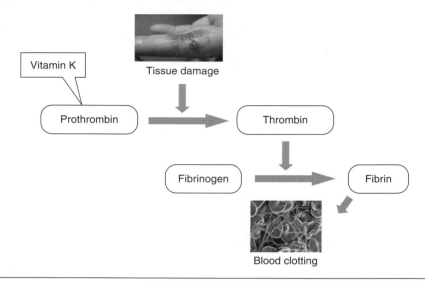

the gastrointestinal tract. The RDA is not increased for pregnancy or lactation, and remains unchanged with advancing age. Oral vitamin K supplementation generally poses no risk of toxicity, so no upper level has been established. Because vitamin K functions in blood clotting, high doses can interfere with anticoagulant drugs used to lessen blood clotting. Therefore, those who are prescribed these medications should consult their physicians before taking supplements containing vitamin K.

WATER-SOLUBLE VITAMINS

Water-soluble vitamins include vitamin C and the B vitamins such as thiamin, riboflavin, niacin, pantothenic acid, biotin, vitamin B-6, folate, and vitamin B-12. They dissolve in water, so large amounts of these vitamins can be lost during food processing and preparation. Vitamin content is best preserved by light cooking methods, such as stir-frying, steaming, and microwaving. Water-soluble vitamins are absorbed mostly in the small intestine, and to a lesser extent, the stomach. The extent to which vitamins are absorbed and used in the body, or bioavailability, is influenced by many factors including nutritional status, other nutrients and substances in foods, medications, age, and illness. Once absorbed, the water-soluble vitamins are circulated to the liver in the blood. Because the body does not store large quantities of most water-soluble vitamins, they generally do not have toxic effects when consumed in large amounts. Most water-soluble vitamins are readily excreted from the body with an excess generally ending up in the urine and stool and very little being stored. Most B vitamins function as coenzymes that help regulate energy metabolism, as illustrated in Figure 3.4, whereas vitamin C may be best known for its role in the synthesis and maintenance of connective tissues, as well as in preventing scurvy. The following is more detailed discussion for each of the water-soluble vitamins.

Figure 3.4 Various roles water-soluble vitamins play in metabolic pathways.

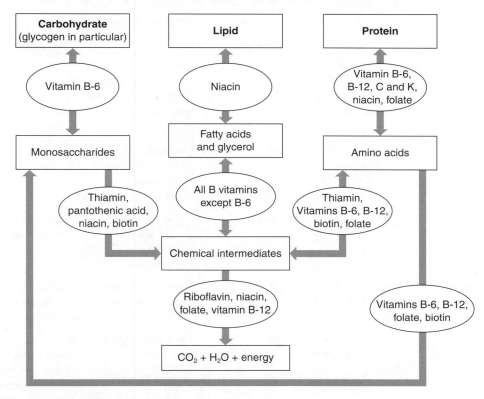

Source: Wardlaw and Smith 2006. Used with permission.

Thiamin (vitamin B-1)

Thiamin – vitamin B-1 – is widely distributed in foods. A large proportion of the thiamin consumed in the United States comes from enriched grains used in foods such as breakfast cereals and baked goods (Table 3.7). Pork, whole grains, legumes, nuts, seeds, and organ meats (e.g., liver, kidney, heart) are also good sources. The adult RDA for thiamin is 1.1 mg and 1.2 mg per day for females and males, respectively. The RDA is based on the amount of thiamin needed to achieve and maintain normal activity of a thiamin-dependent enzyme found in red blood cells and normal urinary thiamin secretion. For an average adult, half of the thiamin can be obtained from 4 oz of pork or three cups of soy milk.

Thiamin does not provide energy, but it is important in the energy-producing reactions in the body. Thiamin functions as a coenzyme in reactions in which carbon dioxide is lost from large molecules. For example, the reaction that forms acetyl-CoA from pyruvate requires thiamin pyrophosphate, an active form of thiamin. Thiamin is therefore essential to the production of energy from glucose. Thiamin is also needed for the synthesis of the neurotransmitter acetylcholine and production of sugar ribose, which is needed to synthesize ribonucleic acid (RNA).

Table 3.7 A summary of water-soluble vitamins

VITAMIN	SOURCES	RDA FOR ADULTS	MAJOR FUNCTIONS	DEFICIENCY DISEASES AND SYMPTOMS
Thiamin (B-1)	Enriched grains, pork, whole grains, legumes, nuts, seeds, organ meats	1.1–1.2 mg	Coenzyme in acetyl-CoA formation, Krebs cycle, nerve function	Beriberi: tingling, poor coordination, weakness, heart changes
Riboflavin (B-2)	Milk, leafy greens, enriched grains, poultry, fish	1.1–1.3 mg	Coenzyme in Krebs cycle, fat metabolism, electron transport chain	Inflammation of the mouth and tongue, dermatitis
Niacin (B-3)	Enriched grains, peanuts, poultry, beef, tuna	14–16 mg	Coenzyme in glycolysis, Krebs cycle, electron transport chain	Pellagra: dementia, diarrhea; dermatitis
Pantothenic acid (B-5)	Meat, seeds, mushrooms, peanuts, eggs	5 mg*	Coenzymes in Krebs cycle, fat metabolism	Tingling in the feet and legs, fatigue, weakness, and nausea
Biotin (B-7)	Cauliflower, egg yolks, peanuts, liver, cheese	30 µg	Coenzyme in glucose production, fat synthesis	Depression, hallucinations, skin irritation, infections, poor muscle control
Vitamin B-6	Meat, legumes, seeds, leafy greens, whole grains	1.3–1.7 mg	Coenzyme in protein metabolism, neurotransmitter, and hemoglobin synthesis	Headache, nausea, poor growth, microcytic hypochromic anemia
Folate (folic acid)	Leafy greens, organ meats, legumes, orange juice, milk	400 µg DFE†	Coenzyme in DNA synthesis and amino acid metabolism	Macrocytic anemia, diarrhea, poor growth, neural tube defects
Vitamin B-12	Meat, milk, poultry, seafood, eggs, organ meats	2.4 µg	Coenzyme in folate metabolism, nerve function	Pernicious anemia, poor nerve function
Vitamin C	Citrus fruits, green peppers, cauliflower, broccoli, strawberries	75–90 g	Collagen synthesis, hormone, and neurotransmitter synthesis, antioxidant	Scurvy: poor wound healing, bleeding gums, bruising, depression, and hysteria

Notes:
* Adequate intake (AI).
† Dietary folate equivalent.

The thiamin deficiency disease is known as beriberi, a word that means "I can't" in Sinhalese, the language of Sri Lanka. The symptoms include weakness, loss of appetite, irritability, nervous tingling throughout the body, poor arm and leg coordination, and deep muscle pain in the calves. A person with beriberi often develops an enlarged heart and sometimes severe edema.

Riboflavin (vitamin B-2)

Riboflavin – vitamin B-2 – consists of a multi-ring structure attached to the simple sugar ribose. Riboflavin in the body is typically found as one of its coenzymes, flavin adenine dinucleotide (FAD), and is important for energy metabolism. Milk is the best source of riboflavin in the North American diet (Table 3.7). Other major sources include liver, red meat, poultry, fish, and whole grains and enriched breads and cereals. Vegetables sources include asparagus, broccoli, mushrooms, and leafy green vegetables such as spinach. The RDAs for riboflavin for adult men and women are 1.3 mg per day and 1.1 mg per day, respectively. Additional riboflavin is recommended during pregnancy to support growth and increased energy utilization. Two cups of milk provide about half the amount of riboflavin recommended for a typical adult. Although riboflavin is relatively stable during cooking, it is easily destroyed by exposure to light. For this reason, milk is packaged in cardboard or cloudy plastic containers, and it is recommended that food be stored in dark containers or covered with paper or foil.

The coenzyme forms of riboflavin participate in many energy-yielding metabolic pathways. When cells generate energy using oxygen-requiring pathways, such as when fatty acids are broken down and burned for energy, the coenzymes of riboflavin are used. Riboflavin is also required for the synthesis of other compounds. For example, it is needed to convert vitamin A and folate (a B vitamin) to their active forms, convert tryptophan (an amino acid) to niacin (a B vitamin), and form vitamins B-6 and K. Riboflavin is also involved in the metabolism of some important neurotransmitters, such as dopamine, and in several important reactions that protect biological membranes from oxidative damage.

The symptoms associated with riboflavin deficiency include inflammation of the mouth and tongue, dermatitis, cracking of tissue around the corners of the mouth, various eye disorders, and sensitivity to the sun. These symptoms usually develop after approximately two months of a riboflavin-poor diet. A deficiency of riboflavin is rarely seen alone. Rather, it often occurs with deficiencies of other B vitamins such as niacin, thiamin, and vitamin B-6, because these nutrients often occur in the same foods.

Niacin (vitamin B-3)

Niacin – vitamin B-3 – takes two forms: nicotinic acid and nicotinamide. The body uses both forms to make the coenzymes nicotinamide adenine dinucleotide (NAD) and nicotinamide adenine dinucleotide phosphate (NADP). NAD and NADP are involved in numerous reactions in the body, many of which are required for energy metabolism. Major sources of niacin are poultry, ready-to-eat breakfast cereals, beef, wheat bran, tuna and other fish, asparagus, and peanuts (Table 3.7). Coffee and tea also contribute some niacin to the diet. Niacin is heat stable, and little is lost in cooking. Niacin can be synthesized from the essential amino acid tryptophan. In a diet that contains high-protein foods such as milk and eggs, which are poor sources of niacin but good sources of tryptophan, much of the need for niacin can be met by tryptophan. However, this happens only if enough tryptophan is available to meet the needs of protein synthesis. The adult RDA of niacin is 14–16 mg per day. The RDA is expressed as niacin equivalents to account for niacin received from the diet as well as that made from tryptophan.

Niacin is important in the production of energy from energy-yielding nutrients as well as in reactions that synthesize other molecules. As mentioned earlier, the body uses niacin to make the two active coenzymes nicotinamide adenine dinucleotide (NAD) and nicotinamide adenine dinucleotide phosphate (NADP). NAD functions in glycolysis and the Krebs cycle, accepting released electrons and passing them to the electron transport chain where ATP is formed. NADP acts as an electron carrier in reactions that synthesize compounds including fatty acids, cholesterol, steroid hormones, and DNA. Niacin has additional functions unrelated to its role as a coenzyme. For example, it is important for maintaining, replicating, and repairing DNA, and may play a role in protein synthesis, glucose homeostasis, and cholesterol metabolism. It has been shown that consuming large amounts of niacin (2–4 g per day) lowers LDL cholesterol and increases high-density lipoprotein (HDL) cholesterol (Ganji *et al.* 2003; Krauss 2004).

Almost every cellular metabolic pathway uses niacin as a coenzyme, so a deficiency causes widespread changes in the body. The group of niacin-deficiency symptoms is known as pellagra, which means "rough and painful skin." The early symptoms of the disease include poor appetite, weight loss, and weakness. If left untreated it can then result in dementia, diarrhea, and dermatitis (especially on areas of skin exposed to the sun). Pellagra is the only dietary deficiency disease ever to reach epidemic proportions in the United States. It became a major problem in the southeastern United States in the late 1800s and persisted until the late 1930s, when standards of living and diets improved. Today, pellagra is rare in Western societies, but can be seen in the developing world.

Pantothenic acid (vitamin B-5)

Pantothenic acid – vitamin B-5 – is a nitrogen-containing vitamin named for the Greek word *pantos*, meaning "everywhere." This is because pantothenic acid is found in almost every plant and animal tissue. Pantothenic acid functions as a component of coenzyme A (CoA) in a variety of metabolic reactions. Rich sources of pantothenic acid are sunflower seeds, mushrooms, peanuts, and eggs (Table 3.7). Other sources are meat, milk, and many vegetables. There is not enough information to establish RDAs for pantothenic acid, so an AI has been set for pantothenic acid at 5 mg per day for adults. Because no evidence exists of toxicity, no upper limits are set for this vitamin.

The primary function of pantothenic acid as CoA is in the metabolism of glucose, amino acids, and fatty acids for ATP production via glycolysis and the Krebs cycle. For example, one of the pivotal steps in energy metabolism involves converting pyruvate to acetyl-CoA. This reaction requires pantothenic acid. Ability to produce acetyl-CoA is essential for the body to metabolize energy-yielding nutrients for ATP production. Pentothenic acid is also required for synthesizing many other critical compounds in the body, including heme (a portion of hemoglobin), cholesterol, bile salts, phospholipids, fatty acids, and steroid hormones.

Because it is found in almost all foods, pantothenic acid deficiency is rare. Nonetheless, a condition called "burning feet syndrome" is thought to be due to severe pantothenic acid deficiency. Burning feet syndrome causes a tingling in the feet and legs, as well as fatigue, weakness, and nausea. A deficiency in pantothenic acid might also occur when alcoholism is accompanied by a nutrient-deficient diet. However, the symptoms would probably be hidden among deficiencies of thiamin, riboflavin, vitamin B-6, and folate, so the pantothenic acid deficiency might be unrecognizable.

Biotin (vitamin B-7)

Biotin – vitamin B-7 – is a sulfur-containing molecule with two connected ring structures and a side chain. The body obtains biotin from both the diet and via biotin-producing bacteria in the large intestine. Cauliflower, egg yolks, peanuts, and cheese are good sources of biotin (Table 3.7). Food containing raw egg whites should be avoided not only because a protein in egg white, called avidin, binds biotin and prevents its absorption, but because raw eggs also may be contaminated with bacteria that can cause blood-borne illness. Thoroughly cooking eggs destroys bacteria and denatures avidin, which prevents it from binding to biotin. No RDA is available for this vitamin. However, the AI for biotin has been set to be 30 μm per day for adults. Biotin is relatively non-toxic, so no upper limit for biotin has been set.

Biotin acts as a coenzyme for several enzymes, all of which catalyze carboxylation reaction. In other words, each biotin-requiring enzyme causes the acid group COOH to be added to a molecule. In general, these enzymes are involved in energy metabolism pathways. For example, a biotin-requiring enzyme converts pyruvate to oxaloacetate, a key step in gluconeogenesis. Biotin is also a coenzyme for reactions that allow the body to use some amino acids in the Krebs cycle for the synthesis of fatty acids, and for the breakdown of the amino acid leucine. In addition to biotin's role as a coenzyme, it has non-coenzyme functions related to gene expression, in particular influencing cell growth and development.

Although biotin deficiency is uncommon, it occurs in small portions of the population, such as people who routinely consume large quantities of raw egg whites. However, in theory it would take daily consumption of at least 12 raw egg whites for a prolonged period of time to cause biotin deficiency. Biotin can also be caused by conditions impairing intestinal absorption such as inflammatory bowel disease. Signs and symptoms of biotin deficiency include depression, hallucinations, skin irritations, inflections, hair loss, poor muscle control, seizures, and developmental delay in infants.

Vitamin B-6

Almost all of the B vitamins discussed so far have a common role of functioning as a coenzyme involved in energy production. Vitamin B-6, however, is somewhat unique in that it is mainly involved in protein and amino acid metabolism. There are three forms of vitamin B-6: pyridoxine, pyridoxal, and pyridoxamine, all made of a modified, nitrogen-containing ring structure. All three forms can be changed to the active vitamin B-6 coenzyme involved in numerous chemical reactions. Major sources of vitamin B-6 are animal products, ready-to-eat breakfast cereals, potatoes, and milk (Table 3.7). Other sources are fruits and vegetables such as bananas, cantaloupes, broccoli, and spinach. Overall, animal sources and fortified products are the most reliable because the vitamin B-6 they contain is more absorbable than that in plant foods. The adult RDA of vitamin B-6 is 1.3–1.7 mg per day. Vitamin B-6 is easily destroyed in processing, such as heating and freezing. It is not one of the vitamins added to "enrich" products, but fortified breakfast cereals make an important contribution to vitamin B-6 intake.

Vitamin B-6 is comprised of a group of compounds including pyridoxine, pyridoxal, and pyridoxamine, as mentioned above. All three forms can be converted into the active coenzyme form, pyridoxal phosphate. Pyridoxal phosphate is needed for the activation of more than 100 enzymes involved mainly in protein and amino acid metabolism. As shown in Figure 3.5, vitamin B-6 is used to synthesize non-essential amino acids by transamination and to remove the amino group so amino acids can be used to produce energy or to synthesize glucose. This vitamin is also needed to remove the carboxyl group (COOH) from amino acids for the synthesis of neurotransmitters such as serotonin and dopamine, as well as hemoglobin. Without pyridoxal phosphate, the non-essential amino acids cannot be synthesized. Recall that only nine essential amino acids must be obtained from foods, whereas 20 amino acids are needed for life. Without vitamin B-6, all 20 amino acids would be essential. Pyridoxal phosphate is important for the immune system because it is needed to form white blood cells. It is also needed for the synthesis of the lipids that are part of the myelin coating on nerves.

Vitamin B-6 deficiency results in inadequate heme production, and thus lower concentrations of hemoglobin in red blood cells. This condition, called **microcytic hypochromic anemia**, results from the fact that red blood cells are small in size and

Figure 3.5 The role of vitamin B-6 in protein metabolism, synthesis of neurotransmitters and energy production.

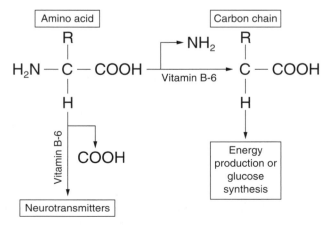

light in color. It decreases oxygen availability in tissues and impairs the ability to produce ATP via aerobic metabolism. Vitamin B-6 deficiency also causes neurological symptoms, including depression, headaches, confusion, numbness and tingling in the extremities, and seizures. These symptoms may be related to the role of vitamin B-6 in neurotransmitter synthesis and myelin formation. Other deficiency symptoms such as poor growth, skin lesions, and decreased antibody formation may occur because vitamin B-6 is important in protein and energy metabolism. Since vitamin B-6 is needed for amino acid metabolism, the onset of deficiency can be hastened by a diet that is low in vitamin B-6 but high in protein.

Folate

Folate consists of three parts: (1) a nitrogen-containing double-ring structure; (2) a nitrogen-containing single-ring structure; and (3) a glutamic acid (also called glutamate). Folate typically has additional glutamic acids attached to it. The interconversion of these "polyglutamate" forms of folate is important for functions of folate. Folic acid, which is the oxidized and stable form of folate, is rarely found in foods but is used in vitamin supplements and food fortification. Folate is derived from the Latin word *folium*, which means foliage or leaves. Green, leafy vegetables, as well as organ meats, sprouts, legumes, and orange juice, are the richest sources of folate. In addition, ready-to-eat cereals, milk, and bread are also important sources of folate (Table 3.7). Folate is susceptible to destruction by heat. Food processing and preparation can destroy at least half of the folate in food. This underscores the importance of eating fresh fruits and raw or lightly cooked vegetables regularly.

Recommendations concerning folate intake have received substantial attention since its relationship with neural tube defects was determined in the late 1980s. The adult RDA for folate is set at $400\,\mu g$ of dietary folate equivalents (DFEs) per day. One DFE is equal to $1\,\mu g$ of food folate or $0.5\,\mu g$ of synthetic folic acid consumed on an empty stomach. In order to reduce the risk of neural tube defects, a special recommendation is made for women capable of becoming pregnant. A daily intake of $400\,\mu g$ of synthetic folic acid from fortified foods and/or supplements is recommended in addition to the food folate consumed in a diet. The RDA for folate during pregnancy is increased to $600\,\mu g$ per day due to the increase in cell division. Although this level can be met by a carefully selected diet, folate is typically supplemented during pregnancy.

Folate acts as a coenzyme for many reactions, all involving the transfer of single-carbon or methyl group such as $-CH_3$. These reactions shift carbons from one molecule to another to form the many organic substances the body needs. An example of folate's single-carbon role is the conversion of homocysteine to the amino acid methionine. In this reaction, 5-methyltetrahydrofolate (an inactive form of folate) loses a methyl group $(-CH_3)$ to homocysteine, which produces methionine and tetrahydrofolate. This important reaction provides the body with the amino acid methionine as well as tetrahydrofolate, an active form of folate. However, this reaction does not happen by itself. It occurs in synchronization with another reaction involving vitamin B-12, so the production of methionine from homocysteine requires both folate and vitamin B-12. Folate is also involved in single-carbon transfer reactions required to make purines and pyrimidines, the molecules that comprise DNA and RNA. Because DNA must be synthesized each time a new cell is made, folate is essential for the growth, maintenance, and repair of all tissues in the body.

Folate deficiency symptoms include poor growth, problems in nerve development and function, gastrointestinal deterioration, and anemia. Anemia results when folate is deficient because the bone marrow cells that develop into blood cells cannot divide. Instead, they just grow bigger. These large red blood cells, called macrocytes, are immature and have limited oxygen-carrying capacity. This type of anemia is also known as **macrocytic anemia**. A lack of ability for cells to divide due to folate deficiency also contributes to the deterioration of the gastrointestinal tract. This is because the cells that form the inner lining of the intestinal wall cannot successfully grow or be repaired. Folate supplementation has been considered necessary for preventing neural tube defects. Women in early pregnancy are advised to take up to 800 μg per day of synthetic folic acid in addition to food folate in order to reduce the incidence of neural tube defects. It must be noted that neural tube defects are not true folate deficiency symptoms because not every pregnant woman with inadequate folate levels will give birth to a child with a neural tube defect. Instead, neural tube defects are probably due to a combination of factors that include low folate levels and a genetic predisposition.

Vitamin B-12

Vitamin B-12 was the last of the B vitamins to be discovered. It is also referred to as cobalamin due to the fact that it contains the trace element cobalt (Co) and several nitrogen atoms. Major sources of vitamin B-12 include meat, milk, ready-to-eat breakfast cereals, poultry, seafood, and eggs (Table 3.7). Organ meats, especially liver, kidneys, and heart, are especially rich sources of vitamin B-12. Vitamin B-12 can also be made by bacteria, fungi, and algae, but not by plants and animals. Microorganisms in the human colon produce vitamin B-12, but it cannot be absorbed. Vitamin B-12 is not supplied by plant products unless they have been contaminated with bacteria, soil, insects, or other sources of vitamin B-12, or have been fortified with vitamin B-12. Diets that do not include animal products must include supplements or foods fortified with vitamin B-12 in order to meet needs. The RDA of vitamin B-12 for adults is 2.4 μg per day. On average, adults consume two or more times the RDA. Such over-consumption seems to be necessary because it is assumed that only 50% of the vitamin B-12 ingested is absorbed.

Vitamin B-12 participates as coenzyme in only two reactions. One reaction catalyzes the production of succinyl-CoA, an intermediate in the Krebs cycle. This reaction ultimately allows the body to use some amino acids and fatty acids for energy production. The other reaction catalyzes the conversion of homocysteine to the amino acid methionine, as mentioned earlier. This reaction also regenerates the active form of folate that functions in DNA synthesis. Without adequate vitamin B-12, homocysteine levels build up in the blood, and folate becomes "trapped" as its inactive 5-methyltetrahydrofolate form. Thus, folate deficiency symptoms appear. In this context, a deficiency of vitamin B-12 can cause a secondary folate deficiency and its related symptoms such as macrocytic anemia.

Symptoms of vitamin B-12 deficiency include an increase in blood homocysteine levels and a macrocytic anemia that is indistinguishable from that seen in folate deficiency. The symptoms also include numbness and tingling, abnormalities in gait, memory loss, and disorientation due to degeneration of the myelin that coats the nerves, spinal cord, and brain. If not treated, these neurological symptoms can eventually lead to paralysis and death. A severe deficiency in vitamin B-12 can be caused by **pernicious anemia**, a condition in which the parietal cells of the stomach that produce intrinsic factor are damaged. Intrinsic factor is a protein that binds to vitamin B-12 and allows the vitamin to be absorbed in the ileum of the small intestine. Only small portions of vitamin B-12 can be absorbed without intrinsic factor. Vitamin B-12 deficiency is especially common in the elderly population. This is due to a variety of factors including inadequate vitamin B-12 intake, decreased synthesis of intrinsic factor, and reduced stomach acid secretion.

Vitamin C (ascorbic acid)

Vitamin C appears to play a role in almost every physiological system. For example, it is important for immune, cardiovascular, neurological, and endocrine systems. This relatively simple compound can be made from glucose in all plants and most animals, but not in humans. Thus, for humans vitamin C is considered an essential nutrient. Vitamin C is also referred to as ascorbic acid. Major sources of vitamin C are citrus fruits, green peppers, cauliflower, broccoli, cabbage, strawberries, and romaine lettuce (Table 3.7). Potatoes, breakfast cereals, and fortified fruit drinks are also good sources of vitamin C. The adult RDA of vitamin C is 75–90 mg per day. Cigarette smokers need to add an extra 35 mg per day to the RDA because of the great stress on their lungs from oxygen and toxic by-products of cigarette smoke. Nearly all North Americans likely meet their daily needs for vitamin C via their regular diet. Nevertheless, nutrition experts who advocate increased use of vitamin C often recommend intake of about 200 mg per day, but this amount can still be obtained by sufficient fruit and vegetable intake. Vitamin C is rapidly lost in processing and cooking as it is water soluble and is unstable in the presence of heat, iron, copper, or oxygen.

The most notable function of vitamin C is its role in synthesizing the protein **collagen**. This protein is highly concentrated in connective tissue, bone, teeth, tendons, and blood vessels. It is important for wound healing. Vitamin C increases the cross-connections between amino acids in collagen, greatly strengthening the structural tissues it helps form. Vitamin C also functions as an antioxidant. Antioxidants are substances that protect against oxidative damage, which is damage caused by reactive oxygen molecules. Reactive oxygen molecules such as free radicals can be generated by normal oxidation reactions inside the body or can come from environmental sources such as air pollution or cigarette smoke. Free radicals cause damage by removing electrons from DNA, proteins, carbohydrates, or unsaturated fatty acids. This results in unstable structure and function in these molecules. DNA damage is considered a major reason for the increase in cancer incidence that occurs with age. Damage to lipoproteins and lipids in membranes is also implicated in the development of atherosclerosis. Vitamin C is vital for the function of the immune system, especially for the activity of certain immune cells. Thus, disease states that increase the need for immune function can increase the need for vitamin C, possibly above the RDA. Due to such associations between vitamin C and immunity, most North Americans supplement their diets with vitamin C in order to combat the common cold. However, it remains questionable whether vitamin C effectively works against colds and other infections. Numerous well-designed, double-blind studies have failed to show mega-doses of vitamin C to reliably prevent colds, though it seems to reduce the duration of the symptoms.

Vitamin C deficiency can causes a sometimes deadly condition called scurvy. Scurvy results in a multitude of signs and symptoms, including bleeding gums, skin irritation, bruising, and poor wound healing, many of which are due to inadequate collagen production. Without vitamin C, the bonds holding adjacent collagen molecules together cannot be formed and maintained. The psychological manifestations of scurvy include depression and hysteria. Although it used to be very common, the increased availability of fruits and vegetables has made this disease rare. Scurvy can develop in infants fed diets consisting exclusively of cow's milk, but this condition can be reversed by adding fruit juice to the infant diet. Scurvy may also occur in alcoholics and elderly individuals consuming nutrient-poor diets.

SUMMARY

- Vitamins are carbon-containing compounds that are required by the body daily in small quantities.
- Vitamins do not provide energy directly, but many contribute to energy-yielding chemical reactions in the body and promote growth and development.
- We consume vitamins that are naturally present in foods, added to foods by fortification and enrichment, and contained in supplements.
- Vitamins are classified as either fat soluble or water soluble. Fat-soluble vitamins are vitamins A, D, E, and K; water-soluble vitamins are B vitamins and vitamin C.
- Absorption of fat-soluble vitamins requires an adequate consumption of fat in the diets.
- Vitamins A, C, and E, and carotenoids, precursors to vitamin A, serve important protective functions as antioxidants; diets containing these micronutrients can reduce the potential for tissue damage and protect against heart disease and cancer.
- Excess fat-soluble vitamins accumulate in body tissues and can become toxic. However, excess water-soluble vitamins generally remain non-toxic as they can be excreted via urine.
- Most B vitamins, such as B-1 (thiamin), B-2 (riboflavin), B-3 (niacin), B-5 (pantothenic acid), B-6, and B-7 (biotin), function as coenzymes in reactions involved in the metabolism of carbohydrate, fat, and protein.
- The B vitamin folate and vitamin B-12 share the similar role of regulating the synthesis of DNA, which is required for cells to divide. Therefore, deficiency in these vitamins can result in problems such as anemia and neural tube defects.
- Vitamin C may be best known for its role in the synthesis and maintenance of connective tissues, as well as in preventing scurvy.

CASE STUDY: CHOOSE A VITAMIN SUPPLEMENT WISELY

John works the late shift in a local warehouse four days per week. John is also a full-time student, and a combination of taking a full course load at college and working late hours has created a lot of stress for him. John also plays intramural soccer and writes regularly for the college newspaper. His many commitments make it important that he not become ill. Recently, one of his roommates suggested that he take vitamin supplements to help prevent colds, 'flu, and other illnesses. The special multivitamin product he is interested in recommends taking three tablets daily for health maintenance and two tablets every three hours at the first sign of feeling ill. John looks at the supplement facts label on the bottle and finds that each tablet contains (as a percentage of the RDA): 35% for vitamin A (three-quarters is pre-formed vitamin A); 500% for vitamin C; 50% for zinc; and 20% for selenium. A monthly supply costs about $60.

Questions

• How many tablets would be taken per day if John was feeling ill?
• Given that the RDA for vitamin A is 1000 µg, what is the total amount of vitamin A in this larger dose?
• How does this amount compare with the upper level of 3000 µg for pre-formed vitamin A?
• How does the cost of this supplement compare to that of a typical multivitamin/mineral supplement available at your local drug store?

REVIEW QUESTIONS

1 What is a vitamin?
2 Why is the risk of toxicity generally greater with fat-soluble vitamins than with water-soluble vitamins?
3 What do enrichment and fortification mean?
4 How would you determine which fruits and vegetables displayed in your super-market are likely to provide plenty of carotenoids?
5 Define the terms: rickets, osteomalacia, and osteoporosis.
6 The need for certain vitamins increases as energy expenditure increases. Name two such vitamins and explain why this is the case.
7 Why should milk be packaged in opaque containers?
8 While is a low folate intake of particular concern for women in pregnancy or of child-bearing age?
9 Why does vitamin C deficiency cause poor wound healing?
10 Describe the relationship between folate and vitamin B-12.

SUGGESTED READING

American Dietetic Association (2005) Position of the American Dietetic Association: fortification and nutritional supplements. *Journal of the American Dietetic Association*, 105: 1300–1311.

This position statement from the American Dietetic Association emphasizes that the best nutritional strategy for promoting optimal health and reducing the risk of chronic disease is to wisely choose a wide variety of foods, although they acknowledge that additional nutrients from fortified foods and/or supplements can help some people meet their nutritional needs.

Fairfield, K.M. and Fletcher, R.H. (2002) Vitamins for chronic disease prevention in adults: scientific review. *Journal of the American Medical Association*, 287: 3116–3126.

This article reviews the clinically important vitamins with regard to their biological effects, food sources, deficiency syndromes, potential for toxicity, and relationship to chronic disease.

Voutilainen, S., Nurmi, T., Mursu, J., and Rissanen, T.H. (2006) Carotenoids and cardiovascular health. *American Journal of Clinical Nutrition*, 83: 1265–1271.

Nutrition has a significant role in the prevention of many chronic diseases such as cardiovascular disease, cancers, and degenerative brain diseases. In this article the role of the main dietary carotenoids, i.e., lycopene, beta-carotene, alpha-carotene, beta-cryptoxanthin, lutein, and zeaxanthin, in the prevention of heart diseases is discussed.

4 Micronutrients: minerals and water

KEY TERMS

- major minerals
- trace minerals
- hypertension
- hypokalemia
- osteoblasts
- osteoclasts
- osteopenia
- tetany
- hemoglobin
- myoglobin
- ferric iron
- ferrous iron
- cytochromes
- microcytic hypochromic anemia
- goiter

- cretinism
- cupric
- cuprous
- fluorosis
- intracellular fluid
- extracellular fluid
- interstitial fluid
- osmosis
- specific heat
- thirst
- dehydration
- anti-diuretic hormone
- aldosterone
- hyponatremia

MINERALS

In addition to organic molecules such as protein, carbohydrates, lipids, and vitamins, our bodies are also made of inorganic matter. These inorganic substances include minerals and water, which together constitute over 60% of the body's weight. Minerals are needed by the body as structural components and regulators of various biological processes. They make up the structure of our bones and teeth, and participate in hundreds of chemical reactions. The metabolic roles of minerals vary considerably. Some minerals, such as copper and selenium, work as cofactors, enabling various proteins, such as enzymes, to function. Minerals also contribute to many body compounds. For example, iron is a component of red blood cells. Sodium, potassium, and calcium aid in the transfer of nerve impulses throughout the body. Body growth and development also depend on certain minerals, such as calcium and phosphorous. Minerals may combine with other elements in the body, but they retain their chemical identity. Unlike vitamins, they are not destroyed by heat, oxygen, or acid. Minerals are divided into **major minerals** – those needed in the diet in amounts greater than 100 mg per day or present in the body in amounts greater than 0.01% of the body weight – and **trace minerals** – those required by the body in an amount of 100 mg or less per day or present in the body in an amount of 0.01% or less of body weight.

Dietary sources and bioavailability

Minerals in the diet come from both plant and animal sources. For example, iron is a component of muscle tissue, so it is found in meat, while magnesium is a component of chlorophyll, so it is found in leafy greens. In general, the quantities of most minerals in foods are quite predictable because minerals are regular components of the plant or animal. However, the amounts of some trace minerals in food can vary depending on the mineral concentration in the soil and water at the food's source. For example, the soil content of iodine is high near the ocean but usually quite low in inland areas. Therefore, foods grown near the ocean are better sources of iodine than those grown inland. The mineral content of foods can also be affected by food processing and refining. For example, when the skins of produce and bran and germ of grains are removed, many trace elements, such as iron, selenium, zinc, and copper, are lost. Such food processing will also decrease the potassium content of foods. Some minerals are added inadvertently through contamination. For example, the iodine content of dairy products is increased by contamination by the cleaning solutions used in milking machines. Minerals can also be added intentionally. For example, the fortification of breakfast cereals can add calcium and other minerals. Choosing a variety of nutrient-dense foods including those that are unprocessed or less processed can help maximize the mineral content of the diet.

Foods offer a plentiful supply of many minerals, but the ability of our body to absorb and use them varies. The bioavailability of minerals depends on many factors. Mineral ions that carry the same charge compete for absorption in the gastrointestinal tract. For example, calcium, magnesium, zinc, copper, and iron all carry 2^+ charge, and a high intake of one may reduce the absorption of others. Mineral bioavailability is also affected by the binding of minerals to other substances in the gastrointestinal tract. For example, spinach contains plenty of calcium, but only about 5% of it can be absorbed because of the vegetable's high concentration of oxalate, a calcium binder. Components found in fibers, such as phytate, can also limit absorption of some minerals – such as calcium, magnesium, zinc, and iron – by binding to them. On the other hand, absorption of minerals can be facilitated by consuming several vitamins. For example, the active vitamin D improves calcium absorption. In addition, when consumed in conjunction with vitamin C, absorption of iron improves.

General functions of minerals

Minerals are transported in the blood bound to transport proteins. The binding of minerals to transport proteins helps regulate their absorption and prevent reactive minerals from forming free radicals that could cause oxidative damages to various tissues. Minerals function in a variety of ways in the body. For example, calcium and phosphorous are vitally important for the structure and strength of bones and teeth; iodine is a component of the thyroid hormones, which regulate metabolic rate; and chromium plays a role in regulating blood glucose levels. Many minerals participate in chemical reactions by serving as cofactors. They are also required for energy metabolism, nerve function, and muscle contraction. In addition, electrolytes such as sodium, chloride, and potassium are essential for maintaining a proper fluid distribution across the body's different compartments.

MAJOR MINERALS

Some of the general characteristics of minerals and how they function in the body were discussed in the previous section. It appears that minerals serve three broad roles in the body: (1) providing structure in forming bones and teeth; (2) maintaining normal heart rhythm, muscle contractibility, and neural conductivity; and (3) regulating metabolism by becoming constituents of enzymes and hormones. The following sections provide more detailed coverage of each of the major minerals, including sodium, chloride, potassium, calcium, phosphorus, magnesium, and sulfur.

Sodium and chloride

Sodium and chloride are almost always found together in foods, and in many ways have similar functions in the body. This is because they join together via ionic bonds to form salt or sodium chloride (NaCl). These minerals are essential part of our diets and add flavor to our foods. Table salt is 40% sodium and 60% chloride, so a dietary intake of 10 g of salt translates into about 4 g of sodium and 6 g of chloride. About 80% of the sodium and chloride we consume is added to foods during food processing and cooking. A teaspoon of salt contains about 2 g (2000 mg) of sodium. Other food additives, such as monosodium glutamate, also contain sodium. In general, unprocessed foods such as fresh fruits and vegetables contain small amounts of sodium and chloride, whereas manufactured and highly processed foods like fast foods and frozen entrées contains large amounts. Some meats, dairy products, poultry, and seafood naturally contain moderate amounts of both sodium and chloride. We obtain these minerals also from salt-containing condiments, such as soy source and ketchup. The more processed and restaurant food one consumes, generally the higher one's sodium intake. Conversely, the more home cooking one does, the more sodium control one has. Other foods that are especially high in sodium include salted snack foods, French fries and potato chips, and sauces and gravies (Table 4.1).

Sodium and chloride, the most abundant ions in the blood, are the body's principal electrolytes. When the ionic bond of an NaCl molecule dissociates in water, sodium is released as a cation (Na^+), whereas chloride is released as an anion (Cl^-). Both ions play major roles in fluid balance. Because water naturally moves to areas that have high sodium and/or chloride concentrations, the body can maintain fluid balance by selectively moving these electrolytes where more water is needed. Diets high in salt can increase extracellular volume, include plasma volume. This can in turn cause high blood pressure or **hypertension**. Blood pressure refers to the systemic arterial pressure measured at a person's upper arm and is usually expressed in terms of the systolic pressure over the diastolic pressure (mmHg). A healthy blood pressure is 120/80 mmHg or

Table 4.1 A summary of major minerals				
VITAMIN	SOURCES	RDA OR AI*	MAJOR FUNCTIONS	DEFICIENCY DISEASES AND SYMPTOMS
Sodium	Table salts, processed foods, condiments, sauces, soups, chips	Age 19–50 years: 1500 mg Age 51–70 years: 1300 mg Age >70 years: 1200 mg	Major positive ions of the extracellular fluid, aids nerve impulse transmission, water balance	Muscle cramps, diarrhea, vomiting
Chloride	Table salts, processed foods, some vegetables	2300 mg	Major negative ions of the extracellular fluid, used for acid production in the stomach, aids nerve impulse transmission, water balance	Muscle cramps, diarrhea, vomiting
Potassium	Spinach, squash, bananas, orange juice, milk, meat, legumes, whole grains	4700 mg	Major positive ions of the intracellular fluid, aids nerve impulse transmission, water balance	Irregular heartbeat, muscle cramps, loss of appetite, confusion
Calcium	Dairy products, canned fish, leafy vegetables, tofu, fortified beverages, and foods	Age 9–18 years: 1300 mg Age >18 years: 1000–1200 mg	Maintenance of bones and teeth, aids in nerve impulse transmission, muscle contraction, blood clotting	Stunted growth in children, increased risk of osteoporosis in adults, muscle cramps and, if extreme, muscle pain or tetany
Phosphorus	Dairy products, processed foods, fish, soft drinks, bakery goods, meats	Age 9–18 years: 1250 mg Age >18 years: 700 mg	Bone and tooth structure and strength, part of metabolic compounds, acid–base balance	Possibility of poor bone maintenance, muscle weakness
Magnesium	Wheat bran, green vegetables, nuts, chocolate, legumes	Men: 400–420 mg Women: 310–320 mg	Bone structure, aids enzyme function and energy metabolism, aids nerve and heart function	Weakness, muscle pain, poor heart function, confusion and, if extreme, convulsions
Sulfur	Protein foods	None	Parts of vitamins and amino acids, acid–base balance, aids in drug detoxification	None observed

Note. * Adequate intake (AI).

less. However, hypertension is generally defined as a blood pressure of 140/90 mmHg or greater. Sodium is also important for nerve function and muscle contraction, both of which also involve potassium (K^+). In addition, chloride is needed for production of hydrochloric acid (HCl) in the stomach, removal of carbon dioxide (CO_2) by the lungs, and for optimal immune function.

The Adequate Intake (AI) for sodium for adults under age 51 is 1500 mg, and this number should be reduced by 100–200 mg for older adults. Under FDA food and supplement labeling rules, the daily value for sodium is 2400 mg or 2.4 grams. However, the amount typically eaten in North America ranges from 2300 mg to 4700 mg. If we ate only unprocessed foods and added no salt, we would consume about 500 mg of sodium per day. Nevertheless, the body really needs only about 200 mg per day to maintain physiological functions.

Deficiencies of sodium and chloride are rare in healthy individuals. However, they can occur in infants and small children with diarrhea and/or vomiting. These conditions result in loss of sodium and chloride through the gastrointestinal tract or loss of nutrients before they even enter the intestine. Diarrhea and vomiting can be life threatening because of the rapid loss of both electrolytes and accompanying water. Less severe sodium and chloride deficiencies due to excessive sweating can occur in athletes, especially those involved in endurance sports such as marathon running. Symptoms of electrolytes deficiency include nausea, dizziness, muscle cramps, and, in severe cases, coma.

Potassium

Whereas sodium is the most abundant cation in the extracellular fluids, potassium (K) is the most abundant cation in the intracellular fluids. Potassium performs many of the same functions as sodium, such as fluid balance and nerve impulse transmission. However, it operates inside, rather than outside, the cell. Intracellular fluids, those inside cells, contain 95% of the potassium in the body. Also unlike sodium, increasing potassium intake is associated with lower rather than high blood pressure.

Generally, unprocessed foods are rich sources of potassium. This includes fruits, vegetables, milk, whole grains, dried beans, and meats. Major sources of potassium in the adult diet include milk, potatoes, beef, coffee, tomatoes, and orange juice (Table 4.1). Diets are more likely to be lower in potassium than sodium because we generally do not add potassium to foods. Some diuretics used to treat high blood pressure can also deplete the body's potassium stores. Thus, people who take diuretics need to monitor their potassium intake carefully. For them, high-potassium foods are necessary additions to the diet, as are potassium chloride supplements if prescribed by a physician. The bioavailability of potassium from these foods is high and is not influenced by other factors.

The potassium cation (K^+) is an important electrolyte, working with sodium and chloride to maintain proper fluid balance in the body. In addition, potassium is critical for muscle function (especially in heart tissue), nerve function, and energy metabolism. Unlike sodium, which causes a rise in blood pressure, consuming high amounts of potassium can decrease blood pressure in some people. As with sodium and chloride, regulation of the blood potassium level is achieved mostly by the kidneys. In other words, when blood potassium is elevated, the kidneys excrete more potassium. The opposite is true when blood potassium is low. The hormone aldosterone, which is released by the adrenal glands and works on the kidneys, causes blood levels of sodium and chloride to increase while simultaneously causing blood levels of potassium to decrease.

The AI for potassium for adults is 4700 mg per day. The daily value used to express potassium content on food and supplement labels is 3500 mg. On average, North Americans consume 2000–3000 mg per day. Thus, many of us need to increase our potassium intake, preferably by increasing fruit and vegetable consumption.

Potassium deficiency is rarely seen, because of the abundance of the mineral in the diet, although it can result from diarrhea and vomiting. Heavy use of certain diuretics can also result in excessive potassium loss in the urine. Diuretics are drugs used to lower blood pressure by helping the body eliminate water. This reduces blood volume and helps decrease blood pressure. However, when the body excretes excessive amounts of water, it also loses electrolytes. This can lead to low blood potassium, a condition called **hypokalemia**. People with eating disorders involving vomiting, such as bulimia nervosa, are at increased risk for hypokalemia. Potassium deficiency causes muscle weakness, irritability, and confusion. Recent studies also suggest that it may cause insulin resistance (Stumvoll *et al.* 2005). In severe cases, potassium deficiency can cause irregular heartbeat, muscle paralysis, decreased blood pressure, and difficulty breathing.

Calcium

Calcium is the most abundant mineral in the body. It represents 40% of all the minerals present in the body, which is equal to about 1.2 kg or 2.5 lb. Calcium accounts for 1–2% of adult body weight. All cells need calcium, but more than 99% of the calcium in the body is used to strengthen bones and teeth. The remaining 1% is present in intracellular fluid, blood, and other extracellular compartments, where it plays vital roles in nerve transmission, muscle contraction, blood pressure regulation, and release of hormones.

Dairy products, such as milk and cheese, provide about 75% of the calcium in our diets. Cottage cheese is an exception because most calcium is lost during production. Bread, rolls, crackers, and other foods made with milk products are secondary contributors. Other calcium sources are leafy greens such as spinach, broccoli, sardines, and canned salmon (Table 4.1). It is important to note that much of the calcium in some leafy green vegetables, notably spinach, is not absorbed because of the presence of oxalate. Oxalate will bind with calcium, thereby impeding its absorption. Fat-free milk is the most nutrient-dense source of calcium because of its high bioavailability and low caloric content. Other common sources of calcium in our diets include calcium-fortified orange juice and other beverages such as soy milk, as well as calcium-fortified cottage cheese, breakfast cereals, breakfast bars, snacks, and chewable candies.

Calcium is absorbed by both active transport and passive diffusion. Active transport, a process that requires energy, depends on the active form of vitamin D and accounts for most absorption when intakes are low to moderate. When vitamin D is deficient, absorption decreases dramatically. At high intakes, passive transport becomes more important. Unlike sodium, chloride, and potassium, the amount of calcium in the body depends greatly on its absorption from the diet. Calcium requires an acidic environment in the gastrointestinal tract to be absorbed efficiently. Absorption of calcium is also affected by a number of other dietary factors, including the presence of lactose, which enhances calcium absorption, and tannins, fibers, phytates, and oxalates, which decrease calcium absorption. For example, as mentioned earlier, spinach is a high-calcium vegetable but only about 5% of its calcium is absorbed; the rest is bound by oxalates and excreted in the feces. Other factors that enhance calcium absorption include blood levels of parathyroid hormone and the gradual flow of digestive contents through the intestine.

Calcium is important in the maintenance of bones and teeth, where it is primarily found with phosphorous as solid mineral crystals. Bones are made of two different kinds of cells: **osteoblasts** and **osteoclasts**. Together, these cells help to maintain healthy, strong bones. In general, osteoblasts are involved in bone formation, whereas osteoclasts facilitate the breakdown of older bone. Calcium also plays important roles in cell communication and the regulation of various biological processes. Calcium helps regulate enzymes and is necessary in blood clotting. It is involved in transmitting chemical and electrical signals in nerves and muscles. It is necessary for the release of neurotransmitters, which allow nerve impulses to pass from one nerve to another and from nerves to other tissues. Inside the muscle cells, calcium allows the two contractile proteins, actin and myosin, to interact to cause muscle contraction. Calcium is also involved in blood pressure regulation by modulating the contraction of smooth muscle in the blood vessel walls.

There is insufficient data available to generate an RDA for calcium. The AI for calcium for adults aged 19–50 years is set at 1000 mg per day, which is based on the amount of calcium needed each day to offset calcium losses in urine, feces, and other routes. Since absorption decreases with age, the AI for men and women aged 51 and older is increased to 1200 mg per day. For adolescents the AI is higher than for adults, at 1300 mg per day for boys and girls aged 9–18. This greater amount supports bone growth. The AI for calcium during pregnancy is not increased above non-pregnant levels. This is because there will be an increase in maternal calcium absorption during pregnancy, which helps to supply the calcium needed for the fetal skeleton.

In children, calcium deficiency results in rickets, a disease that can also be caused by vitamin D deficiency. As mentioned in Chapter 3, sufferers of rickets have poor bone mineralization and characteristically "bowed" bones, especially in legs. In adults, calcium deficiency can cause **osteopenia**, the moderate loss of bone mass. In older adults, calcium deficiency can cause osteoporosis, a more serious chronic disease that can lead to an increase in risk of bone fracture. A low calcium intake is the most significant dietary factor contributing to osteoporosis, but intake alone does not predict the risk of osteoporosis. Genetics, as well as other dietary and lifestyle factors, also affect calcium status and bone mass. As mentioned earlier, diets high in phytates, oxalate, and tannins reduce calcium absorption, as does low vitamin D status. Adequate protein is necessary for bone health, but increasing protein intake increases urinary calcium losses. Despite this, high protein intakes are generally associated with a lower risk of osteoporosis. This is because diets higher in protein are typically higher in calcium, and bone mass depends more on the ratio of calcium to protein than the amounts of protein alone.

Calcium deficiency also affects other tissues. Because of calcium's role in muscle contraction and nerve function, low blood calcium can cause muscle pain, muscle cramps, and tingling of hands and feet. More serious calcium deficiency causes muscles to tighten and become unable to relax, a condition called **tetany**.

Phosphorus

Phosphorus makes up about 1% of the adult body by weight; 85% of this is found as a structural component of bones. Phosphorus is also a component of enzymes, genetic material such as DNA, and cell membranes. In nature, phosphorus is most often found in combination with oxygen as phosphate. Although no disease is currently associated with an inadequate phosphorus intake, a deficiency may contribute to bone loss in older women. The body can efficiently absorb phosphorus at about 70% of dietary intake. This high absorption rate, plus the wide availability of phosphorus in foods, makes this mineral less important than calcium in dietary planning. The active form of vitamin D enhances phosphorus absorption, as it does for calcium.

Like calcium, phosphorus is found in dairy products such as milk, yogurt, and cheese, but meat and bread are also common sources of phosphorus in the adult diet (Table 4.1). Breakfast cereals, bran, eggs, nuts, and fish are also good sources. About 25% of dietary phosphorus comes from food additives, especially in baked goods, cheeses, processed meats, and many soft drinks. In a 12-oz (~340 ml) serving of soft drink, there is about 50–75 mg of phosphorus in the form of phosphoric acid. Reliance on soft drinks to supply dietary phosphorus is not recommended because they typically do not contain any other essential nutrients. In other words, they have low nutrient densities.

Cell membranes are made from phospholipids, which consist of a phosphorus-containing polar head group. Therefore, a primary role of phosphorus in the body is its function as a component of cell membranes. Phosphorus, along with calcium, is also required to form hydroxyapatite, which contains a calcium to phosphate ratio of 2:1. This crystal compound is believed to contribute to the rigidity of bones. Phosphorus is a component of the high-energy compound adenosine triphosphate (ATP), as well as our genetic materials such as DNA and RNA. In addition, phosphorus-containing compounds help maintain blood pH by acting as buffers that accept and donate hydrogen ions. Phosphorus is also involved in hundreds of metabolic reactions in the body. In these reactions, phosphate groups are transferred from one molecule to another, producing "phosphorylated" molecules. In fact, some molecules remain inactive until they are phosphorylated. For example, the enzyme needed to breakdown glycogen into its glucose subunits must be phosphorylated before it can work.

The RDA for phosphorus is set at 700 mg for adults 19+ years of age. Because neither absorption nor urinary losses change significantly with age, the RDA is the same for adults over 50 as those under 50. For growing children and adolescents, the RDA is based on the phosphorus intake necessary to meet the needs for bone and soft-tissue growth. There is no evidence that phosphorus requirements are increased during pregnancy. This is because during the period of pregnancy, intestinal absorption increases by 10%, which is sufficient to provide the additional phosphorus needed by the mother and fetus.

Phosphorus deficiency results in loss of appetite, anemia, muscle weakness, poor bone development and, in extreme cases, death. However, because phosphorus is so widely distributed in food, dietary deficiency of this particular mineral is rare. Marginal phosphorus deficiencies can be found in pre-term infants, vegans, people with alcoholism, older people on nutrient-poor diets, and people with long-term bouts of diarrhea.

Magnesium

There are approximately 25 g of magnesium in the adult human body. Over half of the body's magnesium is in the bones. Most of the rest is in the muscle and soft tissues, with only 1% in the extracellular fluid. Magnesium is important for nerve and heart function and aids many enzyme reactions. Over 300 enzymes use magnesium, and many energy-yielding compounds in cells require magnesium to function properly.

Rich sources for magnesium are plant products, such as whole grains like wheat bran, broccoli, potatoes, squash, beans, nuts, and seeds (Table 4.1). Magnesium is also found in leafy greens such as spinach and kale. Animal products, such as milk, fish, and meats, supply some magnesium. Two other sources of magnesium are hard tap water, which has a high mineral content, and coffee. We normally absorb about 40–60% of the magnesium in our diets, but absorption efficiency can increase up to about 80% if intakes are low. The active form of vitamin D can enhance magnesium absorption, whereas the presence of phytate reduces absorption. As calcium in the diet increases, the absorption of magnesium decreases, so the use of calcium supplements can reduce the absorption of magnesium.

The majority of magnesium in the body is associated with bone, in which it is essential for the maintenance of structure. Magnesium is a cofactor for over 300 enzymes. It is necessary for the production of energy from carbohydrates, lipids, and proteins. In these reactions, magnesium functions as either a stabilizer of ATP or an enzyme activator. Magnesium is also involved in regulating calcium homeostasis and is needed for the action of vitamin D and many hormones, including parathyroid hormone.

The adult RDA for magnesium is about 400 mg per day for men and 310 mg per day for women. The daily value used to express magnesium content on food and supplement labels is 400 mg. Adult men consume an average of 320 mg daily, whereas women consume closer to 220 mg. This suggests that many of us should improve our consumption of magnesium-rich foods, such as whole grain breads and cereals. If dietary means are not enough, a balanced multivitamin and mineral supplement containing approximately 100 mg of magnesium can help to close the gap between intake and needs. As with most other supplements, the typical form used in supplements is not as well absorbed as the forms of magnesium found in foods, but still contributes to meeting magnesium needs.

Magnesium deficiency is rare in the general population, but is sometimes seen in those with alcoholism, malnutrition, kidney disease, and gastrointestinal disease, as well as those who use diuretics that increase magnesium loss via urine. Deficiency symptoms include nausea, muscle cramping, irritability, heart palpitations, and an increase in blood pressure. There is much interest in the possibility that mild magnesium deficiency may increase risk for cardiovascular disease (Alghamdi *et al.* 2005; Bobkowski *et al.* 2005; Weglicki *et al.* 2005). Some research also suggests that magnesium deficiency may predispose people to type-2 diabetes (Guerrero-Romero *et al.* 2005).

Sulfur

The body does not use sulfur by itself as a nutrient. Sulfur is mentioned here because it is a major mineral that occurs in essential nutrients such as the vitamins biotin and thiamin and the amino acids methionine and cysteine. Being part of these amino acids, sulfur is important for protein synthesis. Sulfur plays a role in determining the contour or structure of protein molecules. The sulfur-containing side chains in cysteine molecules can link to each other, forming disulfide bonds, which stabilize the protein structure. Sulfur also helps in the balance of acids and bases in the body.

There is no recommended intake for sulfur. Proteins supply the sulfur we need. As such, no deficiencies are known when protein needs are met (Table 4.1).

TRACE MINERALS

As mentioned earlier, essential minerals are classified as major minerals or trace minerals, depending upon how much we need. Major minerals are required in amounts greater than 100 mg per day, whereas less than 100 mg of each trace mineral is required daily. Information about trace minerals is one of the most rapidly expanding areas of knowledge in nutrition. With the exceptions of iron and iodine, the importance of trace minerals to humans has been recognized only within the last 50 years. Although we need 100 mg or less of each trace mineral daily, they are just as essential to good health as major minerals. Further details on some of the major trace minerals including iron, zinc, selenium, iodine, chromium, copper, and fluoride are provided here.

Iron

Of all the trace minerals, iron (Fe) is likely the most studied. Its role as a major constituent of blood was identified in the eighteenth century, when iron tablets were available for treating young women in whom "coloring matter" was lacking in the blood. Today, we know that the red color in blood is due to the iron-containing protein called **hemoglobin** and that a deficiency of iron decreases hemoglobin production. Although the importance of dietary iron has long been recognized, iron deficiency is still one of the most common nutritional deficiencies worldwide today. Iron is found in every living cell, adding up to about 5 g (1 teaspoon) for the entire body.

Iron in the diet comes from both plant and animal sources. Much of the iron in animal products is heme iron, which is part of a chemical complex found in animal protein such as hemoglobin in blood and **myoglobin** in muscle. Meat, poultry, and fish are good sources of heme iron. Heme iron accounts for 10–15% of dietary iron. Leafy green vegetables, legumes, and whole and enriched grains are good sources of non-heme iron, which may not be absorbed as well as heme iron. Most of the iron in bakery items has been added to refined flour in the enrichment process. Another source of non-heme iron in the diet is iron cooking utensils from which iron leaches into food. Such leaching is enhanced by acidic foods – for example, spaghetti sauce cooked in a glass pan contains about 3 mg of iron, but the same sauce cooked in an iron skillet may contain as much as 50 mg. Milk is a poor source of iron. A common cause of iron-deficiency anemia in children is an over-reliance on milk, coupled with an insufficient meat intake. Vegetarians who omit all animal products are particularly susceptible to iron-deficiency anemia because of their lack of dietary heme iron.

The bioavailability of iron is complex and is influenced by many factors, including its form, a person's iron status, and the presence or absence of other dietary components. For example, the bioavailability of heme iron is two to three times greater than of non-heme iron. Absorption of heme iron is high and most affected by iron status. However, many factors can influence absorption of non-heme iron. One of the most important factors affecting non-heme iron absorption is its ionic state. Non-heme iron is found in two ionic forms in foods: the more oxidized **ferric iron** (Fe^{3+}) and more reduced **ferrous iron** (Fe^{2+}). The more reduced ferrous form is found to be more readily absorbed. One of the best-known enhancers of iron absorption is vitamin C, which converts ferric iron to ferrous iron in the intestinal lumen. Thus, consuming vitamin C in a meal that contains non-heme iron enhances the bioavailability of the iron. Stomach acid also helps reduce ferric iron to ferrous iron, and some studies suggest that chronic use of antacids to neutralize stomach acidity can decrease non-heme iron absorption.

Dietary factors that interfere with the absorption of non-heme iron include fiber. Phytate found in cereals, tannins found in tea, and oxalates found in leafy greens such as spinach can prevent absorption by binding to iron in the gastrointestinal tract. The presence of other minerals may also decrease iron absorption. For example, calcium supplements decrease iron absorption, particularly when both are consumed at the same meal.

Iron is part of the hemoglobin in red blood cells and myoglobin in muscle cells. Hemoglobin molecules transport oxygen from lungs to cells and assist in the return of some carbon dioxide from cells to the lungs for excretion. Without sufficient hemoglobin, oxygen availability to tissues decreases. This can result in lack of energy and fatigue. Myoglobin is another oxygen-carrying molecule. It acts as a reservoir of oxygen, releasing oxygen to muscle cells when needed for energy production. In addition to transporting and delivering oxygen to cells, iron is also needed for other aspects of energy metabolism. For example, it is a basic component of the **cytochromes**, which are heme-containing protein complexes that function in the electron transport chain. Cytochromes serve as electron carriers, allowing the conversion of adenosine diphosphate (ADP) to adenosine triphosphate (ATP). Other functions of iron include helping metabolize drugs and remove toxin from the body and serving as a cofactor for antioxidant enzymes that stabilize free radicals and for enzymes needed for DNA synthesis.

The daily adult RDA for iron for men aged 19–50 years and for women over 50 years is 8 mg. For women aged 19–50 years the RDA is 18 mg. The higher RDA for young and middle-age women is primarily because of menstrual blood loss. Women who menstruate more heavily and longer than average may need even more dietary iron than those who have lighter and shorter flows. The daily value used to express iron content on food and supplement labels is 18 mg, but it increases to 27 mg for pregnant women. Most women do not consume 18 mg of iron daily. The average daily intake is closer to 13 mg, while in men it is about 18 mg per day. Therefore, women should seek out iron-fortified foods such as ready-to-eat breakfast cereals that contain at least 50% of the daily value. Use of a balanced multivitamin and mineral supplement containing up to 100% of daily value for iron is another option.

Iron deficiency is the most common nutritional deficiency in the United States, as well as the rest of the world. Because iron requirements increase during growth and development, iron deficiency is typically seen in infants, growing children, and pregnant women. Iron is lost in the blood each month during the menstrual cycle. Therefore, women of child-bearing age are also at increased risk of iron deficiency. Although iron deficiency was once thought to cause only anemia, scientists now know that it can influence many aspects of health. Mild iron deficiency is associated with fatigue and impaired physical work performance. In addition, it can cause behavioral abnormalities and impaired cognitive function in children (Black 2003; Bryan *et al.* 2004). Mild iron deficiency also impairs body temperature regulation, especially in cold conditions (Rosenzweig and Volpe 1999) and may negatively influence the immune function (Cunningham-Rundles and McNeeley 2005; Failla 2003). Some studies also suggest that mild iron deficiency during pregnancy increases the risk of premature delivery, low birth weight, and maternal mortality (Gambling *et al.* 2003).

Severe iron deficiency causes **microcytic hypochromic anemia**, a condition characterized by small, pale red blood cells. Microcytic hypochromic anemia due to iron deficiency is caused by the inability to produce enough heme, and thus hemoglobin. The term anemia refers to a decreased ability of the blood to carry oxygen. Signs and symptoms of anemia include fatigue, difficulty maintaining concentration, and compromised immune function.

Zinc

The essentiality of zinc in the human diet was only recognized in the 1960s in Egypt and Iran, when a syndrome of growth retardation and poor sexual development, seen in Egyptian and Iranian men consuming a diet based on vegetable protein, was alleviated by supplemental zinc. Although the diet was not low in zinc, it was found that the absorption of zinc was reduced because of a lack of animal protein and almost exclusive use of unleavened bread. Unleavened bread is very high in phytate, which can interfere with zinc bioavailability. The zinc deficiencies were first observed in the United States in the early 1970s in hospitalized patients who were fed with an intravenous injection of certain amino acids. Such amino acid formulas are low in trace minerals compared to whole protein.

In general, protein-rich diets are also rich in zinc. Animal foods supply almost half of an individual's zinc intake. Major sources of zinc are beef, poultry, eggs, milk, seafood, bread, and fortified breakfast cereals. Animal foods are our primary sources of zinc because zinc from animal sources is not bound to phytate. Whole grains are also a good source, but refined grains are not because zinc is lost in milling and not added back in enrichment. Grain products leavened with yeast provide more zinc than unleavened products because the yeast leavening of breads reduces the phytate content. Because zinc, iron, and calcium share the same transport proteins in the intestinal cell, high intake of iron and calcium can decrease zinc absorption.

Zinc is the most abundant intracellular trace mineral. It is found in the cytosol, organelles, and nucleus. Zinc is involved in the functioning of nearly 100 different enzymes, including superoxide dismutase, which is vital for protecting cells from free radical damage. Zinc is also needed by enzymes that function in the synthesis of DNA and RNA, in carbohydrate metabolism, in acid–base balance, and in a reaction that is necessary for folate absorption. Zinc also plays a role in the storage and release of insulin, the mobilization of vitamin A from the liver, and the stabilization of cell membranes. It influences hormonal regulation of cell division and is therefore needed for the growth and repair of tissues. Although zinc supplements are often touted as helping "cure" the common cold, most studies do not support this claim (Jackson *et al.* 2000).

The adult RDA for zinc is 11 mg for men and 8 mg for women. These values are based on the amount needed to cover daily losses of zinc. During pregnancy, the recommendation for zinc is increased to account for the zinc that accumulates in maternal and fetal tissues. The daily value used to express zinc content on food and supplement labels is 15 mg. There are no indications of moderate or severe zinc deficiencies in an otherwise healthy adult population. It is likely, however, that some people such as women, poor children, vegans, the elderly, and people with alcoholism can have a borderline zinc deficiency. People who show deterioration in taste sensation, recurring infections, poor growth, or slow wound healing should have their zinc status checked.

The symptoms of zinc deficiency include poor growth and development, skin rashes, impaired immune function, and delayed sexual maturation. These symptoms reflect the fact that zinc is important in protein synthesis and gene expression. The risk of zinc deficiency is greater in areas of the world where the diet is high in phytate, fiber, tannins, and oxalates. Such a risk is also higher in elderly, low-income children, and vegetarians. Zinc supplements have been shown to reduce the incidence of diarrhea and infections in children in developing countries (Fraker *et al.* 2000).

Selenium

Although discovery of selenium can be traced back more than 150 years, the essentiality of this trace mineral in human nutrition was not recognized until the 1960s. Since that time, much has been learned about how the body uses selenium for carrying out various vital functions. Selenium exists in many readily absorbed forms. Like zinc, selenium has indirect antioxidant function. Selenium's best understood role is as a part of enzymes, such as glutathione peroxidase, that works to reduce damage to cell membranes from electron-seeking, free radical compounds. Food contains several forms of selenium, but typically it is associated with the amino acid methionine. Usually, methionine contains sulfur. However, selenium often substitutes for sulfur due to the similarity in chemical characteristics between sulfur and selenium. When methionine contains selenium it is called selenomethionine.

The best animal sources of selenium are nuts, seafood, and meats. Fruits, vegetables, and drinking water are generally poor sources. Grains can be a good plant source depending on the selenium content of the soil where they were grown. For example, in some areas of China, where the soil selenium content is very low, grains contain negligible amounts. However, in some parts of the western United States, where the soil selenium content is very high, grains can contain toxic levels. Soil selenium content can have a significant impact on the selenium intake of populations consuming primarily locally grown food. However, as we normally eat a variety of foods supplied from many geographic areas, it is unlikely that low soil selenium in a few locations will mean inadequate selenium in our diets. Major selenium contributors to the adult diet are animal and grain products.

The bioavailability of selenium in foods is high, and absorption of this mineral in the intestine is not regulated. Therefore, almost all selenium that is consumed enters the blood. Once selenium is absorbed, homeostasis is maintained by regulating its excretion in urine. As mentioned above, selenium is an essential part of the enzyme glutathione peroxidase. Glutathione peroxidase neutralizes peroxides so they no longer form free radicals that cause oxidative damage. By reducing free radical formation, selenium can spare some of the requirement for vitamin E, because vitamin E is used to stop the action of free radicals once they are produced (Figure 4.1). Recall from Chapter 3 that vitamin E helps prevents attacks on cell membranes by donating electrons to electron-seeking compounds. In this regard, selenium and vitamin E work together toward the same goal. Selenium is also needed for synthesis of the thyroid hormones, which regulate basal metabolic rate.

The RDA for selenium is $55\,\mu g$ per day for adults. This intake maximizes the activity of the selenium-dependent enzyme glutathione peroxidase in the blood. The daily value used to express selenium content on food and supplement labels is $70\,\mu g$. In general, adults meet the RDA, consuming on average $105\,\mu g$ of selenium each day. Although selenium could prove to have a role in immune function and prevention of cancer, at this point it seems premature to recommend selenium supplementation for this purpose.

Figure 4.1 The action of the selenium-containing enzyme glutathione peroxidase. The enzyme neutralizes peroxides before they form free radicals, which can then spare some of the need for vitamin E.

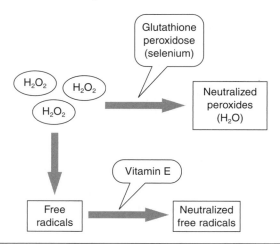

Symptoms of selenium deficiency include muscle pain, discomfort, and weakness. A form of heart disease called Keshan disease may also occur with selenium deficiency. Keshan disease is a potentially fatal form of cardiomyopathy (disease of the heart muscle). The incidence of such muscular and heart disorders has been found to be high in some areas of China, New Zealand, and Finland, where the soil selenium content is low and people consume primarily locally grown foods. The treatment for these disorders is selenium supplementation. It is considered that selenium supplements may relieve most of the symptoms of Keshan disease and reduce its incidence.

Iodine

Iodine is needed for synthesis of thyroid hormones, which regulates growth, reproduction, and energy metabolism. In the early 1900s, iodine deficiency was common in the central United States and Canada, and a condition associated with deficiency of iodine is known as **goiter**, an enlarged thyroid gland. The soils in these areas have low iodine contents. In 1920s, researchers in Ohio found that low doses of iodine given to children over a four-year period could prevent goiter. This finding led to the addition of iodine to salt, beginning in the 1920s. Today, many nations such as Canada require iodine fortification of salt. In the United States, salt can be purchased either iodized or plain. Iodine deficiency still remains a world health problem. About two billion people worldwide are at risk of iodine deficiency, and approximately 800 million of these people have suffered the various effects of the deficiency.

Saltwater fish, seafood, and iodized salt contain various forms of iodine. Dairy products may contain iodine because of the iodine-containing additives used in cattle feed and the use of iodine-containing disinfectants on cows, milking machines, and storage tanks. Sea salt found in health-food stores, however, is generally not a good source because the iodine is lost during processing.

Iodine is highly bioavailable, being almost completely absorbed in the small intestine and, to a lesser extent, the stomach. Once in the blood, iodine is rapidly taken up by the thyroid gland and used for the production of thyroid hormones. Thyroid hormones are synthesized using iodine and the amino acid tyrosine. If a person's iodine intake is insufficient, the thyroid gland enlarges as it attempts to take up more iodine from the bloodstream. This eventually leads to the development of goiter. Simple goiter is a painless condition, but if uncorrected can lead to pressure on the trachea, which may cause difficulty in breathing. Although iodine can prevent goiter formation, it does not significantly shrink goiter once it has formed. Surgical removal may be required in severe cases.

The RDA for iodine for adults is 150 μg to support thyroid gland function. This is the same as the daily value used to express iodine content on food and supplement labels. A half teaspoon of iodine-fortified salt supplies that amount. Most adults consume more iodine than the RDA. The iodine in our diets adds up because dairies and fast-food restaurants use it as a sterilizing agent, bakeries use it as a dough conditioner, food producers use it as part of food colorants, and it is added to salt. There is concern, however, that vegans may not consume enough unless iodized salt is used.

Iodine deficiency reduces the production of thyroid hormones. As a result, the metabolic rate slows, causing fatigue and weight gain. As mentioned earlier, the most obvious sign of deficiency is goiter, an enlarged thyroid gland. If iodine is deficient during pregnancy, it increases the risk of stillbirth or fetal death and spontaneous abortion. Deficiency can also cause a condition called **cretinism** in the offspring. Cretinism is characterized by symptoms such as mental retardation, deaf mutism, and growth failure. Iodine deficiency during childhood and adolescence can also result in goiter and impaired mental function.

Chromium

The importance of chromium in human diets has been recognized only in the past 40 years. The most-studied function of chromium is the maintenance of glucose uptake into cells. Our current understanding is that chromium enters the cell and acts to enhance the transport of glucose and amino acids across the cell membrane by aiding insulin function. In the sports-supplement market, chromium is recognized by many as the popular supplement chromium picolinate, which is promoted to increase lean body mass, although this claim has yet to be validated.

Specific data regarding the chromium content of various foods are limited, and most food composition tables do not include values for this trace mineral. Egg yolk, whole grain, organ meats, mushrooms, nuts, and beer are good sources. Milk, vegetables, fruits, and refined carbohydrates such as white breads, pasta, and white rice are poor sources. The amount of chromium in foods is closely tied to the local soil content of chromium. To provide yourself with a good chromium intake, regularly choose whole grains in place of mostly refined grains.

Chromium is needed for the hormone insulin to function properly in the body and appears to be especially important in regulating its function in people with type-2 diabetes. When carbohydrate is consumed, insulin is released and binds to receptors in cell membranes. This binding triggers the uptake of glucose by cells and an increase in protein and lipid synthesis. Chromium is part of a small peptide that stabilizes the bound insulin, thereby augmenting insulin action. When chromium is deficient, it takes more insulin to produce the same effect. Chromium is also required for normal growth and development in children. In addition, it increases lean mass and decreases fat mass – at least in laboratory animals (McNamara and Valdez 2005; Page *et al.* 1993). This is what has resulted in chromium picolinate being widely marketed as an ergogenic aid for athletes (Lukaski 1999). However, most controlled studies investigating the effect of this supplement on athletic performance and blood glucose regulation have shown no beneficial outcomes (Pittler *et al.* 2003; Vincent 2003).

The AI for chromium is 35 μg per day for men and 25 μg per day for women based on the amount present in a balanced diet. The AI is increased during pregnancy and lactation. The AI for older adults is slightly lower because energy intake decreases with age. Average adult intakes are estimated at 30 μg per day, but could be somewhat higher.

Symptoms of chromium deficiency include impaired blood glucose tolerance with diabetes-like symptoms such as elevated blood glucose and insulin levels. Chromium deficiency may also cause elevated blood cholesterol and triglyceride levels. The mechanism by which chromium influences cholesterol metabolism is not known, but may involve enzymes that control cholesterol synthesis.

Copper

Copper is present in two forms: its oxidized **cupric** (Cu^{2+}) and its reduced **cuprous** form (Cu^+). Note that, as with iron, the ending "-ous" represents the more reduced form, whereas "-ic" represents the more oxidized form. Copper is a cofactor for several enzymes involved in a wide variety of processes, such as ATP production and protection from free radicals. Copper and iron share many similarities in terms of food sources, absorption, and functions in the body.

Copper is found primarily in liver, seafood, cocoa, legumes, nuts, seeds, and whole grain breads and cereals. As with many other trace elements, soil content affects the amount of copper in plant foods.

About 30–40% of the copper in a typical diet is absorbed. The absorption of copper is affected by the presence of other minerals and vitamins in the diet. The zinc content of the diet can have a major impact on copper absorption. There is antagonism in absorption between zinc and copper. When zinc intake is high, it stimulates the synthesis of the protein metallothionein in mucosal cells. Metallothionein preferentially binds copper rather than zinc, thereby preventing copper from being moved out of mucosal cells into the blood. Copper absorption is also reduced by high intakes of iron and manganese. Other factors affecting copper absorption include vitamin C and large doses of antacid, which inhibit copper absorption and, over the long term, can cause copper deficiency.

Once absorbed, copper binds to albumin, a protein in the blood, and travels to the liver, where it binds to the protein ceruloplasmin for delivery to other tissues. Copper can be removed from the body and subsequently eliminated in the feces.

Copper is a cofactor for many enzymes involved in reduction-oxidation reactions important in ATP production, iron metabolism, neural function, antioxidant function, and connective tissue synthesis. For example, copper serves as a cofactor for the enzyme cytochrome c oxidase, which combines electrons, hydrogen ions, and oxygen to form water in the electron transport chain. Copper is also a cofactor for the enzyme superoxide dismutase, which converts the superoxide free radical (O_2^-) to the less harmful hydrogen peroxide molecule (H_2O_2). The synthesis of nor-epinephrine, a neurotransmitter, and collagen needed for connective tissue also requires copper.

The RDA for copper is 900 μg daily for adults, based on the amount needed for activity of copper-containing proteins and enzymes in the body. The average adult intake is about 1–1.6 mg per day. The form of copper typically found in multivitamin and mineral supplements is not readily absorbed. It is best to rely on food sources to meet copper needs. The copper status of adults appears to be good, though we lack sensitive measures to determine copper status.

Severe copper deficiency is relatively rare, occurring most often in pre-term infants. The most common manifestation of copper deficiency is anemia. This is due primarily to the fact that the copper-containing protein ceruloplasmin is needed for iron transport. In copper deficiency, even if iron is sufficient in the diet, iron cannot be transported out of the intestinal mucosa. Copper deficiency causes skeletal muscle abnormality similar to those seen in vitamin C deficiency. This is because the enzyme needed for the cross-lining of connective tissue requires copper as well as vitamin C. Because of copper's role in the development and maintenance of the immune system, a diet low in copper can decrease the immune response and thus increase the incidence of infection.

Fluoride

Fluoride has its greatest effect on dental decay prevention early in life. This link was found when dentists in the early 1900s noticed a lower rate of dental decay in the southwestern United States. These areas contain high amounts of fluoride in the water. After experiments showed that fluoride in the water did indeed decrease the rate of dental decay, controlled fluoridation of water in parts of the United States began in 1945. It has been evidenced that those who grew up drinking fluoridated water generally have 40–60% less dental decay than people who did not drink fluoridated water as children.

Tea, seafood, seaweed, and some natural water sources are the only good food sources of fluoride. Most of our fluoride intake comes from fluoride added to drinking water and toothpaste and from fluoride treatments performed by dentists. Fluoride is not added to bottled water. Therefore, if children consume bottled water, they won't receive fluoride from such water intake. Cooking utensils also affect food fluoride content. Foods cooked with Teflon utensils can pick up fluoride from the Teflon, whereas aluminum cookware can decrease fluoride content.

The gastrointestinal tract provides very little fluoride regulation. In fact, almost all fluoride consumed is absorbed in the small intestine and then circulates in the blood to the liver and then bones and teeth. Fluoride has a high affinity for calcium. In teeth, fluoride is incorporated into the enamel crystals, where it forms the compound fluorohydroxyalatite, which is more resistant to acid than the hydroxyalatite crystal it replaces. As a result, fewer cavities are formed. Fluoride also appears to stimulate maturation of osteoblasts, the cells that build new bone, and has therefore been suggested to strengthen bones in adults with osteoporosis.

The AI for fluoride for adults is 3.1–3.8 mg per day. This range of intake provides the benefits of resistance to dental decay without causing ill effects. Typical fluoridated water contains about 1 mg per liter, which works out to about 0.25 mg per cup.

No deficiency symptoms are known. However, fluoride toxicity is well documented. Signs and symptoms include gastrointestinal upset, excessive production of saliva, watering eyes, heart problems and, in severe cases, coma. In addition, very high fluoride intake causes pitting and mottling of teeth, often referred to as dental **fluorosis**, and a weakening of the skeleton called skeletal fluorosis.

WATER

Water (H_2O), the most abundant molecule in the human body, is truly the essence of life. The body needs more water each day than any other nutrient. An average individual can survive only a few days without water, whereas a deficiency of other nutrients may take weeks, months, or even years to develop. Water is found both inside and outside of cells, and although some water is made in the body during metabolism, it is an essential nutrient. Acting as a solvent, it dissolves many body compounds such as sodium chloride (table salt). Water is the perfect medium for body processes because it enables chemical reactions to occur. Water even participates directly in many of these reactions, such as hydrolysis. Water also helps regulate body temperature. Water balance within different compartments of the body is vital for health and is regulated by the movement of electrolytes such as sodium and potassium.

Fluid balance and electrolytes

Water forms the greatest component of the human body, making up 50–70% of the body's weight (about 40 liters or 10 gallons). Water is found in varying proportions in all tissues of the body. Blood is about 90% water, muscle about 75%, bone about 25%, and adipose tissue about 20%. About two-thirds of the body's water is found inside cells – this is known as **intracellular fluid** (Figure 4.2). The remaining one-third is outside cells – this is **extracellular fluid**. Extracellular fluid includes primarily blood plasma, lymph, and the fluid between cells called **interstitial fluid**. The concentration of substances dissolved in body water, or solutes, varies among these body compartments. The concentration of protein is highest in intracellular fluid, lower in extracellular fluid, and even lower in interstitial fluid. Extracellular fluid has a higher concentration of sodium and chloride and lower concentration of potassium, and intracellular fluid is higher in potassium and lower in sodium and chloride.

The concentration of these electrolytes must be maintained within certain ranges for cells to function properly. Cell membranes control the movement of most substances into and out of cells. For example, sodium and chloride cannot cross cell membranes passively, but instead need help from membrane-bound transport proteins – "pumps" – and the input of energy (ATP). Thus, movement of electrolytes into and out of cells is an active transport process. However, water is unique in that it passes freely across cell membranes, making this a passive transport process. The body can couple the active pumping of ions across cell membranes with the passive movement of water. In doing so, fluid movement and balance are maintained in the various compartments at appropriate levels.

Figure 4.2 Fluid compartments and their relative proportions to the total fluid volume for an average individual.

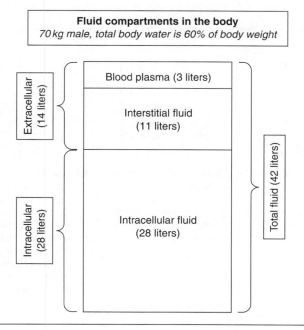

Movement of water across membranes via osmosis

Just as the body maintains the concentrations of other substances such as glucose within specific ranges, it also tightly regulates the amounts of water within the various fluid compartments. The movement of water from one compartment to another depends on fluid pressure and on the concentration of solutes in each compartment. The fluid pressure of blood against blood vessel walls, or blood pressure, causes water to move from the blood into interstitial space. The difference in the concentration of solutes between capillaries and interstitial space causes much of this water to re-enter the capillaries. When the concentration of solutes in one compartment is higher than in another, water will move to equalize the solute concentration. This diffusion of water across a membrane from an area with a lower solute concentration to an area with a higher solute concentration is called **osmosis**. Osmosis occurs when there is a selectively permeable membrane, such as a cell membrane, which allows water to pass freely but regulates the passage of other substances (Figure 4.3). For example, when sugar is sprinkled on fresh strawberries, the water inside the strawberries moves across the skin of the fruit to try to equalize the sugar concentration on each side, causing the fruits to shrink.

The body can regulate the amount of water in each compartment by adjusting the concentration of solutes and relying on osmosis to move water. One such example is the active transport of sodium into the cells that line the colon, causing water to be absorbed as well. Without this absorption, excessive amounts of water would be lost in the feces. An example of how osmosis may have negative health consequences is the regulation of blood volume and blood pressure. Recall that high salt (sodium chloride)

Figure 4.3 Water flows in the direction of the more highly concentrated solutions due to osmosis.

Water flows in the direction of the more highly concentrated solution.

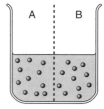 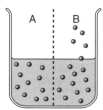 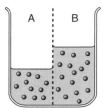

❶ With equal numbers of solute particles on both sides, the concentrations are equal, and the tendency of water to move in either direction is about the same.

❷ Now additional solute is added to side B. Solute cannot flow across the divider (in the case of a cell, its membrane).

❸ Water can flow both ways across the divider, but has a greater tendency to move from side A to side B, where there is a greater concentration of solute. The volume of water becomes greater on side B, and the concentrations in side A and B become equal.

Source: Whitney and Rolfes 2005. Used with permission.

intake can lead to increased blood volume and blood pressure in some people. This is because salt-sensitive people are unable to excrete excess sodium, resulting in high levels of sodium in the blood. This in turn causes water to move into the intravascular space, increasing blood volume and thus blood pressure.

Function of water

Water plays an active role in many processes, including hundreds of chemical reactions. Water also helps keep the body at a constant temperature, even when the environment is very cold or hot. In addition, water provides protection, helps remove waste products, and serves as an important solvent and lubricant.

Water as a solvent, transport medium, and lubricant

Water is an ideal solvent for some substances because it is polar – that is, the two poles of a water molecule have different electrical charges. The polar nature of water allows it to surround other charged molecules and disperse them. Table salt consists of a positively charged sodium ion bound to a negatively charged chloride ion. When placed in water, the sodium and chloride ions move apart or dissociate because the positively charged sodium ion is attracted to the negative pole of the water molecule and the negatively charged chloride ion is attracted to the positive pole. Because of this polar nature, water is the primary solvent in blood, saliva, and gastrointestinal secretions. For example, blood is a solution consisting of water and a variety of dissolved solutes, including nutrients and metabolic waste products such as carbon dioxide and urea. Substances dissolved in blood can move from inside the blood vessel out into the watery environment within and around tissue and cells, delivering important nutrients and allowing removal of waste products.

Water is also a lubricant. This is especially true in the gastrointestinal tract, respiratory tract, skin, and reproductive system, which produce important secretions such as digestive juices, mucus, sweat, and reproductive fluids, respectively. The ability of the body to incorporate water into these secretions is vital for health. For example, water is needed for producing functional mucus in the lungs. Mucus both protects the lungs from environmental toxins and pathogens and lubricates lung tissue so it can remain moist and supple. Water also helps form the lubricants found in knees and other joints of the body. It is the basis for saliva, bile, and amniotic fluid. Amniotic fluid acts as a shock absorber surrounding the growing fetus in the mother's womb.

Water regulates body temperature

When energy-yielding nutrients such as amino acids, glucose, and fatty acids are metabolized, energy is released. Some of this energy becomes heat and helps maintain the internal body temperature at a comfortable 98.6°F (37°C). However, excess heat generated by metabolism must be removed from the body so the body's temperature does not rise. Hot environments can also raise the body's internal temperature.

The fact that water changes temperature slowly in response to changes in the external environment helps the human body resist temperature change when the outside temperature fluctuates. The term **specific heat** (also called specific heat capacity) refers to the amount of energy it takes to increase the temperature of 1 g of a substance by 1°C. Water has a high specific heat, meaning that changing its temperature takes a large amount of energy. Because water can handle so much energy without heating up, our bodies can maintain a relatively stable internal temperature even when metabolic rates are high or the environment is hot. The water in blood actively regulates body temperature. When body temperature starts to rise, the blood vessels in the skin dilate, causing blood to flow close to the surface of the body and to release some of the heat to the environment. This occurs with fevers as well as when the environmental temperature rises. The most obvious way water helps regulate body temperature is through the evaporation of sweat. When body temperature increases, the sweat glands in the skin increase their secretions. As the sweat evaporates from the skin, heat is lost. Each liter of perspiration evaporated represents approximately 600 kcal of energy lost from the skin and surrounding tissues. Similarly, the heat lost when we have a fever increases one's need for calories. The role water plays in regulating body temperature is further discussed in Chapter 14.

Water helps remove waste products

Water is an important vehicle for transporting substances throughout the body and for removing waste products from the body. Most unusable substances in the body can dissolve in water and exit the body through the urine. Urea is a major waste product. This by-product of protein metabolism contains nitrogen. The more protein we eat in excess of our need, the more nitrogen we remove from amino acids and excrete in the form of urea in the urine. Likewise, the more sodium we consume, the more sodium we excrete in the urine. Overall, the amount of urine a person needs to produce is determined primarily by excess protein and salt intake.

A typical volume of urine produced per day is about 1 liter or more, depending mostly on the intake of fluid, protein, and sodium. A somewhat greater urine output than that is fine, but less, especially less than 500 ml (~2 cups), forces the kidneys to form concentrated urine. The easy way to determine if water intake is adequate is to observe the color of urine – it should be clear or pale yellow, but concentrated urine is very dark yellow.

Regulating water balance

A constant supply of water without excess or deficiency is needed in the body. Water intake must be equal to water loss in order to maintain water balance. The desire to drink, or **thirst**, is triggered by the thirst center in the brain, particularly the hypothalamus, when it senses a decrease in blood volume and an increase in the concentration of dissolved substances in the blood. A decrease in the amount of water in the blood also decreases saliva secretion, resulting in a dry mouth. Together, signals from the brain and a dry mouth motivate the consumption of fluid.

Thirst drives a person to seek water, but it often lags behind the body's need. For example, it has been found that athletes exercising in hot weather lose water rapidly, but do not experience intense thirst until they have lost so much body water that their performance is compromised. For this reason, athletes should carefully monitor fluid status – they should weigh themselves before and after training sessions to determine their rate of water loss and, thus, their water needs. A person with fever, vomiting, or diarrhea may also be losing water rapidly, and the thirst mechanism may not be adequate to replace the fluid. In children and the elderly, the thirst mechanism can also become less sensitive or unreliable, so an individual may not be thirsty even though body water is depleted. Even if a person does respond to thirst, the amount of fluid consumed can be insufficient to replace the water loss because thirst is quenched almost as soon as fluid is consumed and long before water balance is restored.

When too much water is lost from the body and not replaced, **dehydration** begins. Dehydration occurs when the drop in body water is great enough for blood volume to decrease, thereby reducing the ability to deliver oxygen and nutrients to cells and to remove waste products. A first sign of dehydration is thirst, the signal that the body has already lost some fluid. If a person is unable to obtain fluid or, as in many elderly people, fails to perceive the thirst message, the symptoms of dehydration may progress rapidly from thirst to weakness, exhaustion, and delirium, ending in death if not corrected. Early symptoms of dehydration (a body water loss of 1–2% of body weight) include headache, fatigue, loss of appetite, dry eyes and mouth, and dark-colored urine. A loss of 5–6% of body water can cause nausea and difficulty concentrating. Confusion and disorientation can occur when water loss approaches 7–8%. A loss of about 9–10% can result in exhaustion, collapse, and death (Table 4.2).

Once the body experiences a shortage of available water, it increases fluid conservation. Two hormones that participate in this process are **anti-diuretic hormone** and **aldosterone**. The pituitary gland releases antidiuretic hormone to force the kidneys to conserve water. The kidneys respond by reducing urine flow. At the same time as fluid volume decreases in the bloodstream, blood pressure falls. This eventually triggers the release of the hormone aldosterone, which signals the kidneys to retain more sodium and, in turn, more water. Alcohol inhibits the action of anti-diuretic hormone. One

Table 4.2 Signs and symptoms of dehydration	
BODY WEIGHT LOSS (%)	SYMPTOMS
1–2	Thirst, fatigue, weakness, discomfort, loss of appetite, impaired physical and cognitive performance
3–4	Decreasing blood volume and urine output, dry mouth, flushed skin, impatience, nausea, apathy, declining physical performance
5–6	Headache, difficulty concentrating, irritability, impaired temperature regulation, increased pulse and breathing rate, sleepiness
7–8	Dizziness, labored breathing with exertion, loss of balance, disorientation, mental confusion, unclear speech
9–10	Muscle spasms, impaired circulation, hypotension, delirium, renal failure, exhaustion, collapse

Note: The onset and severity of symptoms at various percentages of body weight lost depend on the intensity of activity, fitness level, degree of acclimatization, environmental temperature, and humidity.

reason people feel so weak the day after heavy drinking is that they are dehydrated. Even though they may have consumed a lot of liquid in their drinks, they have lost even more liquid because alcohol has inhibited the anti-diuretic hormone. Despite mechanisms that work to reduce water loss via the kidneys, fluid continues to be lost via the feces, skin, and lungs. Those losses must be replaced. In addition, there is a limit to how concentrated urine can become. Eventually, if fluid is not consumed, the body can still suffer ill effects of dehydration despite the action of these two regulatory hormones.

Water intoxication, on the other hand, is rare but can occur with excessive water ingestion and kidney disorders that reduce urine production. The symptoms may include confusion, convulsions, and even death in extreme cases. Excessive water ingestion contributing to the dangerous condition is known as **hyponatremia**. This water intoxication is mainly seen in endurance athletes.

Estimating water needs

The recommended total intake of water per day is 2.7 liters (~11 cups) for adult women and 3.7 liters (~15 cups) for adult men. This is based primarily on our typical total water intake from a combination of fluids and foods. For fluid alone this corresponds to about 2.2 liters (~9 cups) for women and about 3 liters (~13 cups) for men. Water needs can also be calculated based on energy requirements – the greater the energy requirement, the greater the water need. Adults need about 1 ml of water per kilocalorie of energy requirement, or about 2–3 liters per day.

We consume water in various liquids, such as fruit juice, coffee, tea, soft drinks, and water. Coffee, tea, and soft drinks often contain caffeine, which increases urine output. However, the fluid consumed from these beverages is not completely lost in urine, so these fluids still help to meet water needs. Foods also supply water, and many fruits and vegetables are more than 80% water (Table 4.3).

This amount is sufficient under normal conditions, but water needs can be increased by variations in activity, environment, and diet. For example, a person exercising in a hot environment can require an additional 1–2 liters or more per day to replace water losses through sweating. Water needs can also be affected by the composition and adequacy of the diet. A low-energy, high-protein diet increases water needs because water losses increase due to the need to excrete waste such as ketone and/or ammonia. A high-sodium diet increases water needs because the excess salt must be excreted in the urine. A high-fiber diet also increases water needs because more fluid is retained in the gastrointestinal tract.

SUMMARY

- Minerals occur freely in nature, in the water of rivers, lakes, and oceans, and in the soil. The root systems of plants absorb minerals and they eventually become incorporated into the tissues of animals that consume plants.
- Minerals come from plant and animal sources, and their bioavailability is affected by interactions with other minerals, vitamins, and other dietary components, such as fibers, phytates, oxylates, and tannins.

FRUITS	WATER (%)	VEGETABLES	WATER (%)	OTHERS	WATER (%)
Apple	84	Broccoli	91	Beer	90
Apricot	86	Cabbage (green)	93	Bread	38
Banana	74	Cabbage (red)	92	Butter	16
Blueberry	85	Carrot	87	Chicken	64
Cantaloupe	90	Cauliflower	92	Crackers	4
Cherry	81	Celery	95	Honey	20
Cranberry	87	Cucumber	96	Jam	28
Grape	81	Eggplant	92	Milk	89
Grapefruit	91	Lettuce	96	Shortening	0
Orange	87	Pea	79	Steak	50
Peach	88	Pepper	92		
Pear	84	Potato	79		
Pineapple	87	Radish	95		
Plum	85	Spinach	92		
Raspberry	87	Tomato (red)	95		
Strawberry	92	Tomato (green)	94		
Watermelon	92	Zucchini	93		

Table 4.3 Water content of various foods

Note: Values are expressed as percentages by weight.

- Minerals are divided into major minerals – those needed in the diet in amounts greater than 100 mg per day – and trace minerals –those required by the body in an amount of 100 mg or less per day.
- A balanced diet generally provides adequate mineral intake, except in some geographic locations lacking specific minerals (e.g., iodine and selenium).
- Minerals function in a variety of ways in the body. For example, calcium and phosphorous are vitally important for the structure and strength of bones and teeth; iodine is a component of the thyroid hormones, which regulate metabolic rate; and chromium plays a role in regulating blood glucose levels.
- Many minerals participate in chemical reactions by serving as cofactors. They are also required for energy metabolism, nerve function, and muscle contraction. In addition, the electrolytes, such as sodium, chloride, and potassium are essential for maintaining a proper fluid distribution across different body compartments.
- Over- or under-consumption of certain minerals has been linked to the development of some chronic conditions. For example, a diet low in iron results in anemia, which reduces oxygen-carrying capacity; a diet high in sodium increases blood pressure and thus leads to hypertension; and an inadequate consumption of calcium contributes to osteoporosis.
- Water forms the greatest component of the human body, making up 50–70% of the body's weight (about 40 liters or 10 gallons). It is consumed in beverages and food, and a small amount is produced by metabolism.

- Body water is distributed between intracellular and extracellular compartments. The body regulates the distribution of water by adjusting the concentration of solutes in each compartment. This is because of osmosis, by which water always moves from a region of lower solute concentration to a region of higher solute concentration.
- Water plays an active role in many processes, including hundreds of chemical reactions. Water also helps keep the body at a constant temperature, even when the environment is very cold or hot. In addition, water provides protection, helps remove waste products, and serves as an important solvent and lubricant.
- The recommended total water intake per day is 2.7 liters (~11 cups) for adult women and 3.7 liters (~15 cups) for adult men. Water needs can also be calculated based on energy requirements – the greater the energy requirement, the greater the water need. Adults need about 1 ml of water per kilocalorie of energy requirement, or about 2–3 liters per day. Water needs can be increased by variations in activity, environment, and diet.

CASE STUDY: NUTRITION AND BONE HEALTH

Michelle is a 19-year-old sophomore in college. She is a strict vegetarian. She became a near-vegan (but consumes some fish) at the age of 12 when she stopped eating meat and most dairy products after urging from her mother, who had been a vegan her whole adult life. Michelle is now concerned that her diet may be inadequate in vitamins and minerals. Michelle also started smoking, and her only activity is practice for the Frisbee club, which occurs once per week. Michelle's typical diet consists of oatmeal, made with water, an apple, and a cup of fruit juice for breakfast. At lunch, she eats pasta with vegetables, bread with olives, and a soft drink. In the afternoon, she buys a snack cake or candy bar from the vending machine. For dinner she has pasta or breads along with a mixed-vegetable salad, soy-based milkshake, one ounce of mixed nuts, and another soft drink. For the evening snack, she often eats cookies along with hot tea or water.

Questions

- What nutrients are low in Michelle's typical diet?
- Which of Michelle's lifestyle factors contribute to the increased risk of osteoporosis?
- What changes to her current diet could reduce the risk of osteoporosis?
- Which bone assessment test would you recommend Michelle take?

REVIEW QUESTIONS

1 Describe the function of water in the body.
2 What is the recommended water intake for adults? List three factors that can increase water needs.
3 Identify three factors that influence the bioavailability of minerals from food.

4 How do sodium, potassium, and chloride function in the body? What types of foods contribute the most sodium to our diet? What types of foods are good sources of potassium?

5 What is the relationship between dietary sodium and blood pressure?

6 Calcium and phosphorus are the first and second most abundant minerals, respectively. What function do these minerals have in common?

7 What is the function of magnesium in the body?

8 What is goiter and what causes it?

9 What role does iron play in the body? Describe the symptoms of iron deficiency-related anemia.

10 What is the function of zinc in the body? What are the best food sources for zinc?

11 Why does selenium decrease the need for vitamin E?

12 Explain why a deficiency of copper can contribute to anemia.

SUGGESTED READING

Appel, L.J., Brands, M.W., Daniels, S.R., Karanja, N., Elmer, P.J., Sacks, F.M., and the American Heart Association (2006) Dietary approaches to prevent and treat hypertension: a scientific statement from the American Heart Association. *Hypertension*, 47: 296–308.

This article represents an official view of the American Heart Association toward the dietary approaches to prevent and treat hypertension. It discusses multiple dietary factors affecting blood pressure and provides recommendations as to how one's diet may be modified to lower blood pressure.

Kohrt, W.M., Bloomfield, S.A., Little, K.D., Nelson, M.E., Yingling, V.R. and the American College of Sports Medicine (2004) American College of Sports Medicine position stand: physical activity and bone health. *Medicine & Science in Sports & Exercise*, 36: 1985–1996.

This official document released by the American College of Sports Medicine summarizes the current literature supporting the role physical activity plays in maximizing bone mass in various groups of individuals, including children, adults, and the elderly.

Shirreffs, S.M. and Maughan, R.J. (2000) Rehydration and recovery of fluid balance after exercise. *Exercise and Sport Sciences Reviews*, 28: 27–32.

Restoration of fluid balance after exercise-induced dehydration avoids the detrimental effects of a body water deficit on subsequent exercise performance and physiological function. This article discusses various key issues in restoring fluid balance, including consumption of a volume of fluid greater than that lost in sweat and replacement of electrolyte losses, particularly sodium.

5 Digestion and absorption

KEY TERMS

- hydrolysis
- condensation
- enzymes
- energy of activation
- substrates
- coenzymes
- gastrointestinal tract
- lumen
- segmentation
- peristalsis
- enteric nervous system
- chemoreceptors
- mechanoreceptors
- enteric endocrine system
- bolus
- epiglottis
- sphincter
- lower esophageal sphincter
- chyme
- pyloric sphincter
- cephalic phase
- gastric phase
- gastrin

- pepsin
- peptic ulcers
- intestinal phase
- gastric inhibitory protein
- duodenum
- jejunum
- ileum
- secretin
- cholecystokinin (CCK)
- villi
- microvilli
- lacteals
- cecum
- colon
- rectum
- appendix
- probiotic
- veins
- arteries
- arterioles
- venules
- hepatic portal circulation

CHEMICAL BASIS OF DIGESTION AND ABSORPTION

Proper food intake provides an uninterrupted supply of energy and tissue-building chemicals to sustain life. For exercise and sports participants, the ready availability of specific nutrients takes on added importance because physical activity increases energy expenditure and the need for tissue repair and synthesis. Nutrient uptake by the body involves complex physiological and metabolic processes that usually progress unnoticed for a lifetime. Hormones and enzymes work in concert throughout the digestive tract, at proper pH levels, to facilitate the breakdown of complex nutrients into simpler absorbable subunits. Substances produced during digestion are absorbed through the thin lining of small intestine and pass into blood and lymph. Self-regulating processes within the digestive tract usually move food along at a slow rate to allow its complete absorption, yet rapid enough to ensure timely delivery of its nutrients.

Hydrolysis and condensation

Hydrolysis reactions digest or breakdown complex molecules such as carbohydrates, lipids, and proteins into simpler forms that the body absorbs and assimilates. During the decomposition process chemical bonds split by the addition of hydrogen ions (H^+) and hydroxyl ions (OH^-), the constituents of water, to form the reaction by-products. Examples of hydrolysis reactions include the digestion of starches and disaccharides to monosaccharides, protein to amino acids, and lipids to glycerol and fatty acids. A specific enzyme catalyzes each step in the breakdown process. For breaking down disaccharides, the enzymes are lactase, sucrase, and maltase for lactose, sucrose, and maltose, respectively. The enzyme lipase degrades the triglyceride molecule by adding water, thereby cleaving the fatty acids from their glycerol backbone. During protein degradation, the enzyme protease accelerate amino acid release when the addition of water splits the peptide bonds. All of these examples represent catabolism, which in some cases can result in a release of energy. Figure 5.1a illustrates the hydrolysis reaction for the disaccharide sucrose to its end product molecules of glucose and fructose.

The reactions illustrated for hydrolysis also occur in the opposite direction – this is known as **condensation**. In this reverse process, as shown in Figure 5.1b, a hydrogen atom is cleaved from one molecule and a hydroxyl group is removed from another. As

Figure 5.1 (a) Hydrolysis reaction of the disaccharide sucrose to the end product molecules glucose and fructose and (b) condensation reaction of two glucose molecules forming maltose.

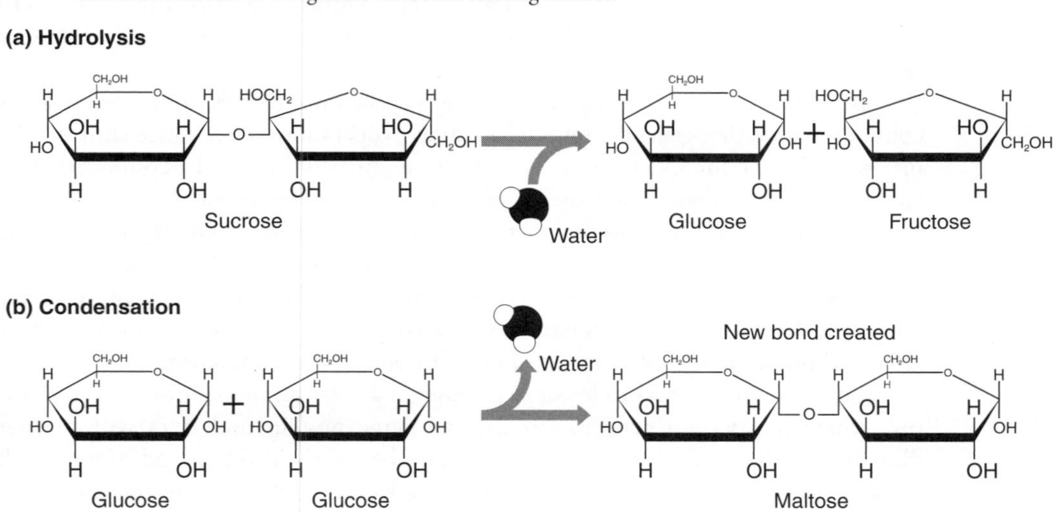

(a) Hydrolysis

(b) Condensation

a result, a compound of maltose is synthesized and a water molecule is formed. Condensation reactions are also referred to as anabolic processes during which individual components of the nutrients bind together in condensation reactions to form more complex molecules. Condensation reactions also apply to protein synthesis. In this process, as a peptide bond is formed from two amino acids; a water molecule is created from a hydroxyl ion cleaved from one amino acid and a hydrogen ion from the other amino acid. For lipids, water molecules form when a glycerol binds with three fatty acids to form a triglyceride molecule.

Enzymes: the biological catalysts

The speed of cellular chemical reactions is regulated by catalysts called **enzymes**. Enzymes are proteins that play a major role in digestion as well as the regulation of metabolic pathways in the cell. Enzymes do not cause a reaction to occur, but simply regulate the rate or speed at which the reaction takes place. The great diversity of protein structures enables enzymes to perform highly specific functions. Enzymes only affect reactions that would normally take place but at a much slower rate. Enzymes do not change the nature of the reaction nor its final result.

Chemical reactions occur when the reactants have sufficient energy to proceed. The energy required to initiate chemical reactions is called the **energy of activation**. Enzymes work as catalysts by lowering the energy of activation. By reducing the energy of activation, enzymes increase the speed of chemical reactions and therefore increase the rate of product formation.

Enzymes possess the unique property of not being altered by the reactions they affect. Consequently, the turnover of enzymes in the body remains relatively slow and they are continually re-used. A typical mitochondrion may contain up to ten billion enzyme molecules, each carrying out millions of operations within a brief time. During exercise, enzyme activity increases tremendously within the cell because of an increase in energy demand. Enzymes make contact at precise locations on the surfaces of cell structures, e.g. mitochondria; they also operate within the structure itself. Many enzymes function outside the cell – in the bloodstream, digestive mixture, or fluids of the small intestine.

Although there is a standardized naming system for enzymes, most textbooks use common names that generally reflect the mode of operation or substance with which it interacts. Except for older enzymes such as rennin, trypsin, and pepsin, almost all enzyme names end with the "-ase" suffix. For example, hydrolase adds water during hydrolysis reactions, protease interacts with protein, oxidase adds oxygen to a substance. In addition, kinases are a group of enzymes that add phosphate groups to the reactants or substances with which they react. Dehydrogenases are enzymes that remove hydrogens from substances they catalyze. In the realm of human biology, we refer to those substances that are acted upon by enzymes as **substrates**.

The ability of enzymes to lower the energy of activation results from unique structural characteristics. In general, enzymes are large protein molecules with a three-dimensional shape. Each type of enzyme has characteristic ridges and grooves. The pockets that are formed from the ridges or grooves located on the enzyme are called active sites. These active sites are important because it is the unique shape of the active site that causes a specific enzyme to adhere to a particular reactant molecule or substrate. The concept of how enzymes fit with a particular substrate molecule is analogous to the idea of a lock and key (Figure 5.2). The shape of the enzyme's activity site is specific for the shape of a particular substrate, which allows the two molecules, enzyme and substrate, to form a complex known as the enzyme–substrate complex. After the formation of the enzyme–substrate complex, the energy of activation needed for the reaction to occur is lowered, and the reaction is more easily brought to completion. This is followed by the dissociation of the enzyme and product.

Figure 5.2 Sequences and steps in the "lock and key" mechanism of enzyme action.

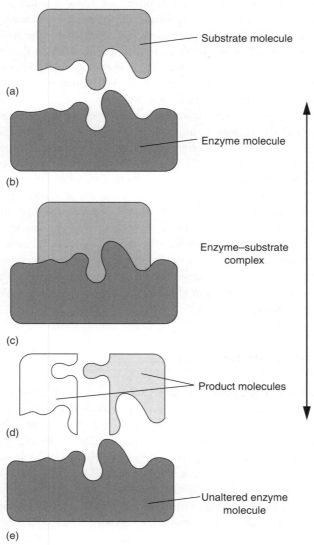

The lock-and-key mechanism offers a protective function so only the correct enzyme activates the targeted substrate. Consider the enzyme hexokinase, which accelerates a chemical reaction by linking with a glucose molecule. As a result of the action of this enzyme, a phosphate group transfers from adenosine triphosphate (ATP) to a specific binding site on one of the glucose's carbon atoms. Once the two binding sites join to form a glucose–hexokinase complex, the substrate begins its stepwise degradation (controlled by other specific enzymes) to form less complex molecules during energy metabolism.

The temperature and hydrogen ion concentrations of the reactive medium dramatically affect enzyme activity. Each enzyme performs its maximum activity at a specific pH. The optimum pH of an enzyme usually reflects the pH of the body fluids in which it bathes. For some enzymes, optimal activity requires a relatively high acidity level. For example, the protein-splitting enzyme pepsin released by the stomach is most active in hydrochloric acid, whereas trypsin released by the pancreas functions more effectively on the alkaline side of neutrality. Increases in temperature generally accelerate enzyme reactivity. As temperature rises above 40–50°C, enzymes can become denatured (a process in which proteins lose their tertiary or secondary structure) and therefore lose their function permanently.

Some enzymes require activation by additional ions and/or smaller organic molecules termed **coenzymes**. These complex non-protein substances facilitate enzyme action by binding the substrate with its specific enzyme. The metallic ions iron and zinc function as coenzymes, as do the B vitamins or their derivatives. Oxidation-reduction reactions use the B vitamins riboflavin and niacin, while other vitamins serve as transfer agents for groups of compounds in metabolic processes. A coenzyme requires less specificity in its action than an enzyme because the coenzyme affects a number of different reactions. It can serve as a temporary carrier of intermediary products in the reaction. For example, the coenzyme nicotinamide adenine dinucleotide (NAD) forms NADH to transport hydrogen atoms and electrons that split from food fragments during energy metabolism. The electrons then pass to special transporter molecules in another series of chemical reactions that ultimately deliver the electrons to molecules of oxygen.

THE DIGESTIVE SYSTEM: AN OVERVIEW

The foods and beverages we consume, for the most part, must undergo extensive alteration by the digestive system to provide us with usable nutrients. The digestive system provides two major functions: (1) digestion, the physical and chemical breakdown of food, and (2) absorption, the transfer of nutrients from the digestive tract into the blood or lymphatic circulatory systems. Carbohydrates, lipids, and proteins are digested and absorbed as sugars, fatty acids, and amino acids, respectively. Some substances, such as water, can be absorbed without digestion, whereas others, such as dietary fibers, cannot be digested by humans and therefore cannot be absorbed. These unabsorbed substances pass through the digestive tract and are excreted in the feces.

The digestive system is made up of the digestive tract and accessory organs. The digestive tract, more commonly known as the **gastrointestinal tract** or alimentary tract, can be thought of as a hollow tube that runs from the mouth to the anus (Figure 5.3). Organs that make up the gastrointestinal tract include the mouth, pharynx, esophagus, stomach, small intestine, and large intestine. The inside of the tube that these organs form is called the **lumen**. Food within the lumen of the gastrointestinal tract has not been absorbed and is therefore technically still outside of the body. Only

Figure 5.3 The gastrointestinal tract and accessory organs of the digestive system.

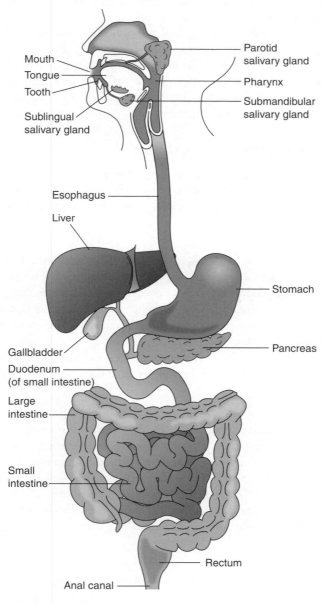

Source: Shier *et al.* 2010. Used with permission.

after food is transferred into the cells of the intestine by the process of absorption is it actually considered to be inside the body. The accessory organs participate in digestion but are not part of the gastrointestinal tract, and include the salivary glands, pancreas, liver, and gallbladder (Figure 5.3). The accessory organs release their secretions needed for the process of digestion into ducts, which empty into the lumen of the gastrointestinal tract.

The amount of time between the consumption of food and its elimination as solid waste is called transit time. It takes approximately 24 to 72 hours for food to pass from the mouth to the anus. Many factors affect transit time, such as composition of diet, illness, certain medications, physical activity, and emotions. Bands of smooth muscle called sphincters act like one-way valves, regulating the flow of the luminal contents from one organ to the next. The gastrointestinal tract has several sphincters, which are often named according to their anatomical locations. For example, the ileocecal sphincter is between the ileum, the last segment of the small intestine, and the cecum, the first portion of the large intestine.

Organization of the gastrointestinal tract

The digestive tract contains our major tissue layers – the mucosa, submucosa, muscular layer, and serosa. Each tissue layer contributes to the overall function of the gastrointestinal tract by providing secretions, movement, communication, and protection.

Mucosa

The innermost lining of the digestive tract, called the mucosa, consists mainly of epithelial cells. It carries out a variety of digestive functions. The mucosa, often called the mucosal lining, produces secretions needed for digestion, such as enzymes, hormones, and mucus. The digestive system produces and releases a variety of substances and secretions, collectively referred to as digestive juices, some of which are quite acidic. Because mucosal cells are continuously exposed to harsh digestive secretions within the gastrointestinal tract, their life span is a mere 2–5 days. Once the mucosal epithelial cells wear out, they slough off and are replaced by new cells.

Submucosa

A layer of connective tissue called the submucosa surrounds the mucosal layer. The mucosal layer contains a rich supply of blood vessels, which nourish the inner mucosal layer and the next outer muscular layer. In addition to blood vessels, the submucosa contains lymphatic vessels, which are filled with fluid called lymph. Lymph transports fluid away from body tissues and aids in the circulation of fat. The submucosa also contains a network of nerves called the submucosal plexus, which regulates the release of gastrointestinal secretions from cells making up the mucosal lining.

Muscular layer

Moving outward from the submucosa, the next layer in the gastrointestinal tract is the two layers of smooth muscle organized as an outer longitudinal and an inner circular layer. Located between these two muscle layers is the myenteric plexus, a network of nerves that control the contraction and relaxation of the muscle. Such contraction and relaxation promotes mixing of the food mass with digestive secretions and keeps food moving through the entire length of the gastrointestinal tract.

Serosa

The serosa is the outer-most layer that encloses the gastrointestinal tract. It consists of connective tissue and provides overall support and protection. In particular, the serosa secrets a fluid that lubricates the digestive organs, preventing them from adhering to one another. In addition, much of the gastrointestinal tract is anchored within the abdominal cavity by the mesentery, a membrane that is continuous with the serosa.

Gastrointestinal motility and secretions

The term "motility" refers to the mixing and propulsion of material by muscular contractions in the gastrointestinal tract. These movements result from the contraction and relaxation of circular and longitudinal muscle in the muscular layer. There are two types of movement in the gastrointestinal tract: **segmentation** and **peristalsis**. Segmentation occurs when circular muscles in the small intestine move the food mass back and forth, thereby increasing the contact between food particles and digestive secretions. Peristalsis involves rhythmic, wave-like muscle contractions that propel food along the entire length of the gastrointestinal tract. The contraction of circular muscles behind the food mass causes the longitudinal muscle to shorten. When the longitudinal muscles lengthen, the food is propelled forward. Peristalsis is similar to the motion exhibited when an earthworm moves.

Gastrointestinal secretions are important for digestion and protection of the gastrointestinal tract, and include water, acid, electrolytes, mucus, salts, enzymes, bicarbonate, and other substances (Table 5.1). For example, mucus forms a protective coating that lubricates the mucosal lining. Digestive enzymes are biological catalysts that facilitate chemical reactions that breakdown complex food particles. More specifically, digestive enzymes catalyze hydrolysis reactions as mentioned before, which break chemical bonds by adding water. As a result, molecules such as starch and protein are broken down into smaller components so that they can be absorbed across the mucosal lining. Organs that release digestive secretions include the salivary glands, stomach, pancreas, gallbladder, small intestine, and large intestine. In fact, approximately 7 liters of secretion, most of which is water, are released daily into the lumen of the gastrointestinal tract. Fortunately, the body has a "recycle" system that enables much of this water to be reclaimed.

Regulation of gastrointestinal motility and secretions

Gastrointestinal motility and secretions are carefully regulated by neural and hormonal signals. These involuntary regulatory activities ensure that complex food particles are physically and chemically broken down and food mass moves along the gastrointestinal tract at the appropriate rate. The gastrointestinal tract has three regulatory control systems. The intestinal and the central nervous system provide neural control, and the intestinal endocrine system provides hormonal control.

Table 5.1 Important gastrointestinal secretions and their functions		
SECRETION	SOURCE	FUNCTION
Saliva	Mouth	Partial digesting starch with salivary amylase, lubricating food for swallowing
Mucus	Mouth, stomach, small intestine, large intestine	Protecting GI tract, lubricating food as it travels through the GI tract
Enzyme	Mouth, stomach, small intestine, pancreas	Breaking down complex foods into smaller particles for absorption
Acid	Stomach	Promoting digestion of protein, among other functions
Bile	Liver (stored in gallbladder)	Assisting fat digestion in the small intestine by suspending fat in water
Bicarbonate	Pancreas, small intestine	Neutralizing stomach acid when food mix reaches the small intestine
Hormones	Stomach, small intestine, pancreas	Stimulating production of acid, enzyme, bile, and bicarbonate, regulating peristalsis and food movement, and influencing the desire to eat

Intestinal nervous system

The gastrointestinal tract has its own local nervous system called the **enteric nervous system**. The enteric nervous system receives information from other nerves called sensory receptors, located within the gastrointestinal tract. There are two kinds of sensory receptors, **chemoreceptors** and **mechanoreceptors**, each monitoring conditions and changes related to digestive activities. Chemoreceptors detect changes in the chemical composition of the luminal contents, whereas mechanoreceptors detect stretching or distension in the walls of the gastrointestinal tract. The presence of food in the tract can stimulate both chemo- and mechanoreceptors. Information from both kinds of sensory receptor is relayed to the enteric nervous system, which responds by communicating with a variety of muscles and glands. In return, muscles and glands carry out the appropriate response to help with digestion, such as an increase in peristalsis and/or release of digestive secretions.

Central nervous system

The intestinal nervous system controls digestive functions at the local level. However, the gastrointestinal tract also communicates with the central nervous system. The central nervous system consists of the brain and spinal cord, which receive and respond to sensory input from the gastrointestinal tract. The function of both the enteric and central nervous systems keeps the digestive system and the brain in close communication. This is why sensory and emotional stimuli can affect one's digestive function. For example, sight, smell, or thought of food stimulates gastrointestinal motility and secretion. Similarly, emotional factors such as fear, sadness, anger, anxiety, and depression can cause gastrointestinal distress.

Intestinal endocrine system

The gastrointestinal tract consists of many different types of cell, some of which are hormone-producing cells referred to collectively as the **enteric endocrine system**. Hormones produced by these cells are important in providing communication in the body. Enteric hormones, which act as chemical messengers, are released into the blood in response to chemical and physical changes in the gastrointestinal tract. This information is then communicated to other organs, alerting them to the impending arrival of food. Similar to neural signals, hormones also influence the rate at which food moves through the gastrointestinal tract and the release of gastrointestinal secretions. In addition to regulating the gastrointestinal motility and secretion, some enteric hormones communicate with appetite centers in the brain, and thus influence the desire to eat. The four major enteric hormones are gastrin, secretin, cholecystokinin, and gastric inhibitory protein. The specific role of each of these hormones is discussed later in the chapter.

DIGESTION AND ABSORPTION PROCESSES

The digestive system is composed of six separate organs; each organ performs one or more specific jobs, but all of them work in a "coordinated" fashion. Because most foods we consume are mixtures of carbohydrates, lipids, and proteins, the physiology of the digestive system is designed to allow the digestion of all of those components without competition among them. The following sections of this chapter will trace a meal through all these digestive organs, from food entering the mouth to its elimination as the waste products from the large intestine.

Mouth

The mouth is the entry point for food into the digestive tract. It performs many functions in the digestion of foods. Besides chewing food to reduce it to smaller particles, the mouth also senses the taste of foods we consume. The tongue, through the use of its taste buds, identifies foods on the basis of their specific flavors. Sweet, sour, salty, and bitter constitute our primary taste sensations. In addition to these basic tastes, a compound found in the seasoning monosodium glutamate (MSG) delivers an additional taste sensation. The presence of food in the mouth stimulates the release of saliva from the salivary glands located internally at the sides of the face and immediately below and in front of the ears. Saliva contains the enzyme salivary amylase, which begins the digestion of carbohydrate. Salivary amylase can breakdown the long chains of starch into smaller segments of sugars. Saliva also lubricates the upper gastrointestinal tract and moistens the food so that it can be further tasted and easily swallowed.

Another important function performed by the mouth is chewing, which is often referred to as the first part of physical digestion. Digestive enzymes can act only on the surface of food. Therefore, chewing is important because it breaks food into small pieces, increasing the surface area in contact with digestive enzymes. Chewing also breaks apart fiber that traps nutrients in some foods. Recall from Chapter 3 that more vitamins are found in the peel or outer region of fruits and vegetables that are rich in fiber. In this context, fewer nutrients will be absorbed without an effective chewing mechanism. Adult humans have 32 teeth specialized for biting, tearing, grinding, and crushing foods. Thus missing or decayed teeth can interfere with the proper digestion of food. Tooth decay or cavities are caused by acid produced when bacteria breakdown carbohydrates.

The tongue, made primarily of muscle, assists in chewing and swallowing. As food mixes with saliva, the tongue manipulates the food mass, pushing it up against the hard, bony palate of the mouth. As we prepare to swallow, the tongue directs the soft, moist mass of food, now referred to as a **bolus**, toward the back of the mouth, an area known as the pharynx. The pharynx is the shared space between the mouth and the esophagus that connects the nasal and oral cavities. This phase of swallowing is under voluntary control, but once the bolus reaches the pharynx the involuntary phase of swallowing begins.

Esophagus

During the involuntary phase of swallowing, the soft palate rises, blocking the entrance to the nasal cavity. This helps guide the bolus into the esophagus. The esophagus is a long tube that connects the pharynx with the stomach. Near the pharynx is a flap of tissue called the **epiglottis**, which prevents the bolus of swallowed food from entering the trachea. During swallowing, food lands on the epiglottis, which folds it down to cover the opening of the trachea. Breathing also stops automatically. These responses ensure that swallowed food will only travel down the esophagus. If food travels down the trachea instead, choking may occur. The esophagus is a narrow muscular tube that passes through the diaphragm, a muscular wall separating the abdomen from the cavity where the lungs are located. At the top of the esophagus, nerve fibers release signals to tell the gastrointestinal tract that food has been consumed. This then results in an increase in gastrointestinal muscle action known as peristalsis. Continual waves of muscle contraction followed by muscle relaxation forces the food down the digestive tract from the esophagus.

To move food from the esophagus into the stomach, the food must pass through a **sphincter**, a muscle that encircles the tube of the digestive tract and acts as a valve. When muscle contracts, the valve is closed. The **lower esophageal sphincter**, located between the esophagus and the stomach, normally prevents foods from moving back out of the stomach. Heartburn occurs when some of the acidic stomach content leaks out of the stomach into the esophagus, causing a burning sensation. Vomiting is the result of a reverse peristaltic wave that causes the sphincter to relax and allows food to pass upward out of the stomach toward the mouth.

Stomach

The stomach is an expanded portion of the gastrointestinal tract that can hold up to four cups or 1 liter of food for several hours until all of the food is able to enter the small intestine. Stomach size varies individually and can be reduced surgically as a medical treatment. While in the stomach, the bolus is mixed with highly acidic stomach secretions to form a semi-liquid food mass called **chyme**. The mixing of food in the stomach is aided by an extra layer of smooth muscle in the stomach wall. As mentioned earlier, most of the gastrointestinal tract is surrounded by two layers of muscle; however, the stomach contains a third layer, enabling more powerful contractions that thoroughly churn and mix the stomach contents. Acidic secretions in the stomach help convert inactive digestive enzymes to their active form, partially digest food protein, and make dietary minerals soluble so they can be absorbed. Following a meal, the stomach contents are emptied into the small intestine over the course of 1–4 hours. The **pyloric sphincter**, located at the base of the stomach, controls the rate at which the chyme is released into the small intestine. Some digestion takes place in the stomach, but, with the exception of some water and alcohol, very little absorption takes place here.

The stomach has a capacity to accommodate large amounts of food. When empty, the stomach volume is quite small – approximately one-quarter of a cup. As food enters the stomach, its walls expand to increase its capacity to 4–8 cups or 1–2 liters. The ability to expand to this extent is due to the interior lining of the stomach, which is folded into convoluted pleats call rugae. Like an accordion, the rugae unfold and flatten, allowing the stomach to expand as it fills with food. The stretching of the stomach walls triggers mechanoreceptors to signal the brain that the stomach is becoming full. In turn, the brain causes hunger to diminish, causing a person to stop eating. The ability to recognize and respond to these internal cues is an important component of body weight regulation.

Stomach or gastric secretions are regulated by both nervous and hormonal mechanisms, as discussed earlier in this chapter. Signals from three different sites – the brain, stomach, and small intestine – stimulate or inhibit gastric secretion. Gastric secretion can be divided into three phases: cephalic, gastric, and intestinal. The **cephalic phase** occurs before food enters the stomach. During this phase, smell, sight, or taste of food causes the brain to send nerve signals that increase gastric secretion. This prepares the stomach to receive and digest food that enters the stomach. The **gastric phase** begins when food enters the stomach. The presence of food in the stomach causes gastric secretion by stretching local nerves, signaling the brain, and stimulating the secretion of the hormone **gastrin** from the upper portion of the stomach. Gastrin triggers the release of gastric juice, which is produced by gastric glands in the lining of the stomach. One of the components of gastric juice is hydrochloric acid. This strong acid stops the activity of salivary amylase and helps to begin the digestion of protein. It also serves to kill most bacteria present in food. Another component of gastric juice is pepsinogen. When pepsinogen is exposed to the acidity of stomach, it is converted into its active form, **pepsin**, which breaks protein into shorter chains of amino acids called polypeptides. The protein of the stomach wall is protected from the acid and pepsin by a thick layer of mucus. If the mucus layer is penetrated or destroyed, acid and pepsin can damage the inner-lining of the stomach, resulting in a condition called **peptic ulcers**. A peptic ulcer is the erosion in the lining of the stomach or the first part of the small intestine, known as the duodenum. One of the leading causes of stomach ulcers is acid-resistant bacteria that infect the lining of the stomach and thus destroy the mucosal layer (McManus 2000). The **intestinal phase** of gastric secretion begins by the passage of chyme into the small intestine. During this phase, stomach motility and secretion decreases to ensure that the amount of chyme entering the small intestine does not exceed the ability of the small intestine to process it. This phase also involves the action of the pyloric sphincter that regulates the rate at which chyme is released into the small intestine.

The rate of gastric emptying, or the rate at which food leaves the stomach, is influenced by several factors, including the volume, consistency, and composition of chyme. For example, large volumes of chyme increase the force and frequency of peristaltic contractions, which in turn increase the rate of gastric emptying. Thus, large meals leave the stomach at a fast rate compared to small meals. The consistency of food, i.e., liquid vs. solid, also affects the rate of gastric emptying. Because the opening of the pyloric sphincter is small, only fluids and small particles (<2 mm in diameter) can pass through. Solid foods take more time to liquefy than fluid and, therefore, remain in the stomach longer. Finally, the nutrient composition of the chyme also influences gastric emptying. A high-fat meal will stay in the stomach the longest. This is because the presence of fat in chyme causes the small intestine to release a hormone called **gastric inhibitory protein**. This hormone slows the rate of gastric emptying, enabling the small intestine to prepare for the task of fat digestion.

Small intestine

The small intestine is the primary site of chemical digestion and nutrient absorption. It is a narrow tube about 20 feet in length. It is divided into three segments. The first 12 inches are the **duodenum**; the next 8 feet are the **jejunum**; and the last 10 feet are the **ileum**. In addition to chyme, the duodenum receives secretions from the gallbladder via the common bile duct. The pancreas also releases its secretions into the small intestine. The pancreatic juice is released into the pancreatic duct, which eventually joins the common bile duct. The cells lining the small intestine also secrete enzymes that are involved in the digestion of smaller sugar (e.g., disaccharides) and polypeptides into single sugar units and amino acids, respectively.

The pancreas secretes pancreatic juice, which contains bicarbonate ions and digestive enzymes. The bicarbonate ions neutralize the acid in chyme, making the environment in the small intestine neutral rather than acidic as it is in the stomach. This neutral environment allows enzymes from the pancreas and small intestine to function. The digestive enzymes from the pancreas include pancreatic amylase, pancreatic lipase, and trypsin, a protein-digesting enzyme. These enzymes continue the job of digesting carbohydrates, lipids, and proteins started in the mouth and stomach.

The gallbladder secretes bile, but this substance is produced in the liver. Bile is a watery solution that consists primarily of cholesterol, bile acids, and a pigment that gives bile its characteristic yellowish-green color. Once bile is formed, it is transported to the gallbladder, where it is stored. Bile is necessary for fat digestion and absorption. Bile acts like detergent, dispersing large globules of fat into smaller droplets. These smaller droplets allow pancreatic lipase to more efficiently access and digest fat molecules. Without bile, it would be difficult for enzyme lipase to make direct contact with the chemical bonds. Once the lipids are absorbed, bile is reabsorbed through the ileum and returned to the liver via the hepatic portal vein. This process enables the liver to recycle many of the constituents that make up bile. Only 5% of the bile escapes into the large intestine and is lost in the feces.

As in the stomach, the lining of the small intestine contains hormone-producing endocrine cells. These cells release the enteric hormones **secretin**, **cholecystokinin** (CCK), and gastric inhibitory protein in response to conditions within the small intestine. Secretin signals the pancreas to secrete bicarbonate ions and stimulates the liver to secrete bile into the gallbladder. CCK signals the pancreas to secrete digestive enzymes and causes the gallbladder to contract and empty its contents into the duodenum via the common bile duct. As mentioned earlier, gastric inhibitory protein slows the rate of stomach motility that empties the stomach's chyme into the small intestine. Together, these enteric hormones work cooperatively to ensure that digestion and absorption in the small intestine are rapid yet effective. The roles of these and other enteric hormones in the process of digestion are summarized in Table 5.2.

The process of digestion physically and chemically liberates nutrients in food, so that nutrients are made ready for absorption. The small intestine is the primary site of absorption for virtually all nutrients, including water, vitamins, minerals, and the products of carbohydrate, lipid, and protein digestion (Table 5.3). The physical structure of the small intestine is very important to the body's ability to digest and absorb the nutrients it needs. In addition to its length, the small intestine has two other structural features that facilitate absorption. First, the intestinal walls are arranged in circular and spiral folds which increase surface area in contact with nutrients. Second, its entire inner lining is covered with finger-like projections called **villi**, and each of these villi is covered with tiny **microvilli**, often referred to as a brush border (Figure 5.4). The combined folds, villi, and microvilli in the small intestine increase its surface area

Table 5.2 Hormones that regulate digestion			
HORMONES	SOURCE	STIMULUS FOR SECRETION	MAJOR ACTION
Gastrin	Stomach	• Foods entering the stomach • Stretch of the stomach wall • Alcohol and caffeine • Smell, taste, sight	• Stimulates gastric motility • Stimulates gastric emptying • Stimulates gastric secretions
Secretin	Duodenum	• Arrival of acidic chyme into the small intestine	• Inhibits gastric motility • Inhibits gastric secretions • Stimulates release of pancreatic juice containing bicarbonate ions and enzymes
Cholecystokinin	Duodenum	• Arrival of partially digested fat and protein into the small intestine	• Stimulates gallbladder to contract and release bile • Stimulates releases of pancreatic juice
Gastric inhibitory protein	Duodenum	• Arrival of fat and glucose into the small intestine	• Inhibits gastric motility and emptying • Inhibits gastric secretions

Table 5.3 Major sites of absorption along the gastrointestinal tract	
ORGAN	PRIMARY NUTRIENTS ABSORBED
Stomach	• Alcohol (20%) • Water (minor amount)
Small Intestine	• Calcium, magnesium, iron, and other minerals • Glucose • Amino acids • Fats • Vitamins • Water (70–90% or total) • Alcohol (80% of total) • Bile acids
Large Intestine	• Sodium • Potassium • Some fatty acids • Some minerals • Some vitamins • Water (10–30% of total)

600 times beyond that of a simple tube. Each villus contains a blood vessel and a lymph vessel, which are located only one cell layer away from the nutrients in the lumen of the small intestine. Lymph vessels are known as **lacteals** and can absorb large particles such as the products of fat digestion. Nutrients must cross the mucosal layer to reach the bloodstream or lymphatic system for use by the body.

The transfer of nutrients into the mucosal cells – what is referred to as nutrient absorption – takes place by passive and active transport mechanisms: simple diffusion, facilitated diffusion, and active transport (Figure 5.5):

- Passive diffusion. When the nutrient concentration is higher in the lumen of the small intestine than in the absorptive cells, the difference in the nutrient concentration drives the nutrient into the absorptive cells by diffusion. Fats, water, fat-soluble vitamins, and some minerals are absorbed by passive diffusion.
- Facilitated diffusion. Some nutrients cannot pass freely across cell membranes even though there is a favorable concentration gradient, and they require a carrier protein to drive them into the absorptive cells. This process is called facilitated diffusion. Fructose is one example of a compound that makes use of such carrier proteins to allow for absorption.
- Active transport. In addition to the need for a carrier protein, some nutrients also require energy input to move from the lumen of the small intestine into the absorptive cells. This mechanism makes it possible for cells to take up nutrients even when they are consumed in low concentrations. Glucose and most amino acids are absorbed by this mechanism.

Figure 5.4 The small intestine contains folds, villi, and microvilli, which increase the absorptive surface area.

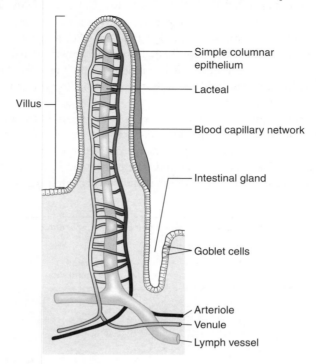

Source: Shier *et al.* 2010. Used with permission.

Large intestine

Components of chyme that are not absorbed in the small intestine pass into the large intestine, which includes the **cecum**, **colon**, and **rectum**. The cecum, the first portion of the large intestine, is a short, sac-like structure with an attached appendage consisting of lymphatic tissue called the **appendix**. On occasion, trapped materials can cause the appendix to become inflamed, which can necessitate an appendectomy – the surgical removal of the appendix. The ileocecal sphincter that separates the ileum from the cecum regulates the intermittent flow of material from the ileum to the cecum. The colon, which makes up most of the large intestine, is shaped like an inverted letter "U" and consists of the ascending colon, the transverse colon, and the descending colon (Figure 5.3). Following the descending colon is the rectum, which terminates at the anal canal, the segment of the large intestine that leads outside of the body.

When the contents of the small intestine enter the large intestine, the materials left bear little resemblance to the food originally eaten. Under normal circumstances, only a minor amount (5%) of carbohydrates, lipids, and proteins escape absorption to reach the large intestine. The large intestine differs from the small intestine in that there are no villi or digestive enzymes. The absence of villi means that little absorption takes place in the large intestine compared with the small intestine. Nutrients absorbed from the large intestine include water, some fatty acids, some vitamins, and the minerals sodium and potassium (Table 5.3). Peristalsis in the large intestine is slower than that in the small intestine. Water, nutrients, and fecal matter may spend 24 hours in the large intestine, in contrast to the 3–5 hours it takes for chyme to move through the small intestine. This slow movement favors the growth of bacteria. Unlike the small intestine, the large intestine wall has mucus-producing cells. The mucus secreted by these cells functions to hold the feces together and to protect the large intestine from the bacterial activity within it.

The large intestine is home to a large population of bacteria. Whereas the stomach and small intestine have some bacterial activity, the large intestine is the organ most heavily colonized with bacteria. In fact, over 500 species of bacteria can be found in the large intestine. The number and type of bacteria in the human colon has recently become of great interest. Research has shown that intestinal bacteria play a significant

Figure 5.5 Nutrients are absorbed from the lumen into absorptive cells by simple diffusion, facilitated diffusion and active transport.

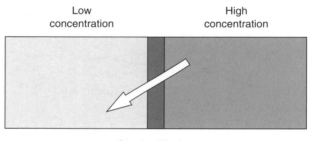

Simple diffusion

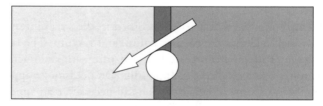

Facilitated diffusion that requires a protein carrier

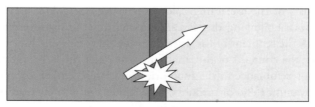

Active transport that requires energy

role in the maintenance of health, especially health of the colon. For example, some bacteria can synthesize small amounts of B vitamins and vitamin K, some of which will be absorbed. These higher levels of beneficial organisms have also been found to reduce the activity of disease-causing bacteria. Foods containing certain microorganisms, such as lactobacilli, are getting a lot of attention. The term **probiotic** is used for these microorganisms because, once consumed, they take up residence in the large intestine and lead to some health benefits, such as improving immunity and health of the intestinal tract (Madsen 2001). The probiotic microorganisms can be found in certain kinds of milk and yogurt, as well as in pill forms that are made commercially available.

Materials not absorbed are excreted as waste products in the feces. The amount of water in the feces is affected by fiber and fluid intake. Fiber retains water, so when adequate fiber and fluid are consumed, feces will have a high water content and will be easily passed. However, when material moves too quickly through the colon for sufficient water to be re-absorbed, diarrhea can occur. Diarrhea is considered beneficial when it allows the body to eliminate harmful or irritating materials quickly. However, prolonged diarrhea can result in excessive loss of fluids and electrolytes from the body, which can lead to serious complications such as dehydration. Conversely, when inadequate fiber or fluid is consumed, or too much water is removed, feces will become hard and dry, and constipation can result.

Paths of absorbed nutrients

Absorbed materials are delivered to the body cells by the cardiovascular system, which consists of the heart and blood vessels. The path by which nutrients enter the bloodstream varies with the nutrient. Amino acids from protein, simple sugars from carbohydrates, and the water-soluble products of fat, such as glycerol, are absorbed directly into the bloodstream. The products of fat digestion that are not water soluble, such as fatty acids, are taken into the lymphatic system before entering the blood.

The cardiovascular system consists of the heart and a closed vascular network through which blood is circulated. The heart is considered the engine of the cardiovascular system. It is a muscular pump with two circulatory circuits – one that delivers the blood to the lungs and one that delivers the blood to the rest of the body. The blood vessels that transport the blood toward the heart are called **veins**; those that transport blood away from the heart are called **arteries**. As arteries carry blood away from the heart, they branch many times to become smaller and smaller. The smallest arteries are called **arterioles**. Arterioles then branch to form capillaries that have thin walls, are of a narrow diameter, and are permeable to many small particles. A capillary network marks the end of the arterial blood flow to the cell and the beginning of the venous blood flow away from the cell and back to the heart. In the capillaries of the gastrointestinal tract, water-soluble nutrients diffuse across the wall of capillaries into the bloodstream that flows into the small veins, the **venules**, which converge to form larger and larger veins for return to the heart. These nutrients are then pumped out of the heart into arteries to be delivered to various parts of the body.

The intestine and the liver have a unique circulatory arrangement called the **hepatic portal circulation**. In this circulation, water-soluble nutrients such as amino acids and sugars cross the mucosal cells of the villi and enter capillaries. These capillaries merge to form venules at the base of the villi. The venules then merge to form larger veins, which eventually form the hepatic portal vein. The hepatic portal vein then transports blood directly to the liver, where absorbed nutrients are processed. This arrangement gives the liver first access to the nutrient-rich blood leaving the small intestine. Nutrients taken up by the liver can be stored or can undergo metabolic reactions. The liver also releases nutrients into the blood, which are then circulated to other parts of the body. In addition to nutrients, substances that are potentially harmful to the body, such as alcohol, are taken up and detoxified by the liver.

The liver acts as a gatekeeper between substances absorbed from the intestine and the rest of the body (Vander *et al.* 2001). Some nutrients are stored in the liver, some are changed into different forms, and others are allowed to pass through unchanged. Based on the immediate needs of the body, the liver decides whether individual nutrients will be stored, used, or delivered directly to the cells. For example, the liver modulates blood glucose by storing or releasing the absorbed glucose depending upon the level of blood glucose concentration. The liver is also responsible for the synthesis or breakdown of protein and fats. It modifies the products of protein degradation to form molecules that can be safely transported to the kidneys for excretion. The liver also helps to protect the body from toxins or to remove cholesterol from the blood and use it for making bile.

Another major structure that has a close relationship with digestion and absorption is the lymphatic system. The lymphatic system consists of a network of lymph vessels, lymph nodes, and lymph organs such as the spleen that provide protection to the body. Fluid that has accumulated in tissues drains into the lymphatic system, where it is filtered past a collection of infection-fighting cells. The cleansed fluid is then returned to the bloodstream.

Unlike water-soluble nutrients, most lipids cannot easily enter blood capillaries because they are too large and insoluble in water. Consequently, molecules such as triglycerides, fatty acids, and fat-soluble vitamins are taken up by lacteals, which are more permeable than blood capillaries. Lacteals, the smallest lymph vessels, drain these absorbed nutrients into larger lymph vessels. These larger lymph vessels, from the intestine and most other organs of the body, further drain into the thoracic duct, which empties into the bloodstream near the neck region. Thus, nutrients that are absorbed via lacteals do not pass the liver before entering the blood.

COMMON PROBLEMS WITH DIGESTION AND ABSORPTION

Each of the organs and processes of the digestive system is necessary for the proper digestion and absorption of food. However, this finely tuned organ system can develop problems. As discussed above, many factors influence gastrointestinal function. The central nervous system exerts a strong influence through diverse neural-endocrine connections with different digestive organs. Emotional state can also affect digestive function to various degrees. For example, many individuals experience intestinal cramping and a queasy stomach under stressful conditions such as a "big date" or "big game." Some individuals get an "upset stomach" at the sight of their own blood, and it is well known that emotional stress contributes to the production of gastric abnormalities. Recent evidence suggests that most gastrointestinal problems can be treated with a healthy diet and regular exercise. For example, regular exercise enhances gastric emptying, with a concomitant reduction in the incidence of liver disease, gallstones, colon cancer, and constipation.

Problems associated with any digestive organ or process can inhibit the ability to obtain adequate nutrients and thus adversely affect nutritional status. For example, dental problems can make it difficult to chew, limiting the type of food that can be consumed and reducing contact between nutrients in food and digestive enzymes. Pancreatic problems can limit the availability of enzymes needed to digest fat and proteins, and liver or gallbladder problems can interfere with fat absorption. The following section provides a brief description of some of the common digestive abnormalities. The more we know about these conditions, the more likely we are to be able to prevent or lessen them.

Lactose intolerance

Lactose intolerance is a condition that often begins after early childhood. It can lead to symptoms of abdominal pain, gas, and diarrhea after consuming lactose, especially when a large amount is eaten. Another form of the problem, secondary lactose intolerance, is a temporary condition in which the production of the enzyme lactase decreases in response to other conditions such as diarrhea. The symptoms of lactose intolerance include gas, abdominal bloating, cramps, and diarrhea. The bloating and gas are caused by bacterial fermentation of lactose in the large intestine. The diarrhea is caused by undigested lactose in the large intestine as it draws water from the circulatory system into the large intestine.

In the United States, approximately 25% of adults show signs of decreased lactose digestion in the small intestine (Lee and Krasinski 1998). Many of them are Asian-American, Africa-American, or Latino/Hispanic-American, and the occurrence increases as people age. It is considered that this digestive problem doesn't occur in some due to a genetic mutation that occurred in regions that relied on milk and dairy products as a main food source, allowing those individuals (mostly in northern Europe and the Middle East) to retain the ability to maintain high production of enzyme lactase.

Bacteria in the large intestine can breakdown lactose. Therefore, those with mild lactose intolerance symptoms can still tolerate a small amount of milk (i.e., ½–1 cup), especially when consumed with meals. Combining lactose-containing foods with other foods helps because certain properties of foods can have positive effects on the rate of digestion. For example, fat in a meal slows digestion, which then leaves more time for lactase to act. Hard cheese and yogurt are more easily tolerated than milk because much of the lactose is lost in the production process, and the bacteria cultures in yogurt can digest the lactose when they are broken apart in the small intestine. If necessary, products such as low-lactose or lactose-free milk or lactase pills can be used to assist those who are lactose intolerant.

Ulcers

A peptic ulcer can occur when the lining of the esophagus, stomach, or small intestine is eroded by the acid secreted by the stomach cells. As the stomach lining deteriorates in ulcer development, it loses its protective mucus layer, and the acid further erodes the stomach tissue. Acid can also damage the lining of the esophagus and the first part of the small intestine, the duodenum. The typical symptom of an ulcer is pain about two hours after eating. This is because the stomach acid released for digestion irritates the ulcer after most of the meal has moved from the site of the ulcer.

The further risk associated with an ulcer is that it will damage the entire stomach and intestinal wall. Following this, the gastrointestinal contents can spill into the body cavities, causing massive infection. In addition, an ulcer may erode a blood vessel, leading to substantial blood loss. For these reasons, it is important not to ignore the early warning signs of ulcer development, including a burning near the stomach that occurs immediately following a meal or awakens you at night. Other signs and symptoms of an ulcer are weight loss, nausea, vomiting, loss of appetite, and abdominal bloating.

It has long been thought that the major cause of ulcers is an excessive production of acid. As such, neutralizing and curtailing the secretions of stomach acid has been the common treatment choice. However, it has been recognized recently that although acid is still a significant player in ulcer formation, the principle causes of ulcer disease are: (1) infection of the stomach by acid-resistant bacteria such as *Helicobacter pylori*; (2) heavy use of anti-inflammatory drugs such as aspirin; and (3) other disorders that cause excessive acid production in the stomach. Stress is considered as a predisposing factor for ulcers, especially if the person is infected with *Helicobacter pylori* or has certain anxiety disorders. Cigarette smoking is also known to cause ulcers or increase ulcer complications such as bleeding.

Heartburn

Heartburn occurs when the sphincter between the esophagus and stomach relaxes involuntarily, allowing the stomach's contents to flow back into the esophagus. Unlike the stomach, the esophagus has no protective mucus lining, so acid backflow can damage it and cause pain. Other symptoms may also include nausea, gagging, cough, or hoarseness. The recurrent and therefore more serious form of the problem is called gastroesophageal reflux disease (GERD), which is characterized by the occurrence of such symptoms two or more times per week. Typically, the gastroesophageal sphincter should be relaxed only during swallowing, but in individuals with GERD, it is relaxed at other times as well.

Heartburn or GERD occurs in approximately 60% of athletes and more frequently during exercise than at rest. The mechanisms for why this condition is more prevalent among athletes or during exercise are not well understood. The many speculations include: (1) reduced gastric motility; (2) delayed gastric emptying; (3) relaxation of the lower esophageal sphincter; (4) increased intra-abdominal pressure; and (5) increased mechanical stress by the bouncing of gastrointestinal organs. Athletes involved in predominately anaerobic sports such as weight lifting experience most frequent acid reflux and heartburn, while these symptoms are found to be less frequent and milder in runners and cyclists.

Heartburn sufferers should follow the general recommendations of (1) waiting about two hours after a meal before lying down; (2) avoiding post-prandial exercise; (3) reducing meal size and fat consumption; and (4) trying to elevate the head of the bed. For an occasional heartburn, quick relief can be obtained with over-the-counter medications, such as antacids. Prescription medications are available for treating more persistent heartburn or GERD. If the proper medications are not effective at controlling the problem, surgery may be needed to strengthen the weakened esophageal sphincter.

Constipation

Constipation refers to a delay in stool movement through the colon. As fluid is increasingly absorbed during the extended time the feces stay in the large intestine, they become dry and hard. Constipation can result when people regularly inhibit their normal bowel reflexes for long periods. Another common cause is a regular consumption of a diet high in fat and low in water and fiber content. Muscle spasms of an irritated large intestine can also slow the movement of feces and contribute to constipation. In addition, calcium and iron supplements and medications such as antacids can cause constipation.

Eating foods with plenty of fiber, such as fruits and whole grain breads and cereals, along with drinking adequate fluid is the best approach to treating mild cases of constipation (Müller-Lissner et al. 2005). Fiber stimulates peristalsis by drawing water into the large intestine and forming a bulky, soft stool. Dried fruits are a good source of fiber and therefore can help stimulate the bowel. In addition, people with constipation may need to develop a habit that allows the same time each day for a bowel movement. For more severe constipation, laxatives and various other medications can be used to lessen the problem. These medications work by either stimulating peristaltic muscle contraction or by drawing more water to produce a bulky stool.

Hemorrhoids

Hemorrhoids are painful, swollen veins in the lower portion of the rectum or anus. This is often because blood vessels in this region are subject to intense pressure, especially during pregnancy and after childbirth, obesity, prolonged sitting, violent cough or sneezing, or straining during bowel movements, particularly with constipation. Such an increase in pressure causes the veins to bulge and expand, making them painful, particularly when you are sitting. Hemorrhoids may be located inside the rectum (internal hemorrhoids), or they may develop under the skin around the anus (external hemorrhoids). Internal hemorrhoids occur just inside the anus, at the beginning of the rectum. External hemorrhoids occur at the anal opening and may hang outside the anus.

Hemorrhoids are a common digestive disorder. By age 50, about half of adults have had to deal with the itching, discomfort, and bleeding that can signal the presence of hemorrhoids. Pressure from prolonged sitting or exertion is often enough to trigger the symptoms, although diet, lifestyle, and possibly hereditability play a role. Pain may be lessened by applying warm, soft compresses or sitting in a tub of warm water for 15–20 minutes. Dietary recommendations are the same as those for treating constipation, emphasizing the need to consume adequate fiber and water. Symptoms can also be alleviated by using over-the-counter creams and suppositories, although these medications should only be used for a short time because long-term use can damage the skin.

Diarrhea

Diarrhea is the condition of having three or more loose or liquid bowel movements per day. It is very common and usually not serious. It is usually accompanied by symptoms of abdominal bloating or cramps, thin or loose stools, a sense of urgency to have a bowel movement, and nausea and vomiting. Many people will have diarrhea once or twice each year. It typically lasts two to three days and can be treated with over-the-counter medicines. Diarrhea may also occur as part of irritable bowel syndrome or other chronic diseases of the large intestine. Most cases of diarrhea result from infections caused by bacteria and viruses, which can cause the intestinal cells to secrete fluid rather than absorb it. Other causes include eating foods that upset the digestive system, allergies to certain foods, certain medications, intestinal diseases, malabsorption, and alcohol abuse. Diarrhea may also follow constipation, especially for people who have irritable bowel syndrome.

Long-distance runners are more susceptible to diarrhea. Possible causes include fluid and electrolyte and altered colonic motility. However, such acute exercise-induced diarrhea is considered physiological, meaning that it does not produce dehydration or electrolyte imbalances, and tends to improve with fitness level.

Treatment of diarrhea generally requires drinking lots of fluid during the affected stage and reducing the poorly absorbed substance if that is the cause. To alleviate symptoms, those who have diarrhea may choose fruit juice without pulp, soda without caffeine, chicken broth without the fat, tea with honey, and sports drinks. Over-the-counter medicines as liquids or tablets are also available for treating mild diarrhea. Prompt treatment within 24 hours is especially important for infants and older individuals, as they are more susceptible to the effects of dehydration associated with diarrhea. If diarrhea lasts more than seven days in adults, they should be examined by a physician as it can be a symptom of more serious intestinal disease.

Irritable bowel syndrome

Irritable bowel syndrome (IBS) is a functional bowel disorder characterized by chronic abdominal pain, discomfort, bloating, and alteration of bowel habits in the absence of any detectable cause. The two IBS forms are (1) diarrhea predominant and (2) constipation predominant. In most cases, the symptoms are relieved by bowel movements. The exact cause of IBS is unknown. The most common theory is that IBS is a disorder of the interaction between the brain and the gastrointestinal tract (Andresen and

Camilleri 2006). In other words, those who suffer from IBS have altered intestinal peristalsis coupled with a decreased pain threshold for abdominal distension. IBS may begin after an infection, a stressful life event, or onset of maturity without any other medical indicators.

IBS affects 20% of the adult population and is more common in younger women than in younger men. In older adults, the ratio is closer to 50:50. Approximately 50% of patients with IBS also report psychiatric symptoms of depression and anxiety.

No cure has been found for IBS, but many options are available to treat the symptoms. For many people, careful eating reduces IBS symptoms. For example, dietary fiber may lessen IBS symptoms, particularly constipation. Whole grain breads and cereals, fruits, and vegetables are good sources of fiber. High-fiber diets keep the colon mildly distended, which may help prevent spasms. Dietary fiber also keeps water in the stool, thereby preventing hard stools that are difficult to pass. This diet intervention, however, may not help with lowering pain or decreasing diarrhea. Among other lifestyle therapies are (1) consuming meals of smaller size; (2) avoiding dairy products; (3) stress management; and (4) regular exercise. IBS can also be treated with medications used to decrease constipation, diarrhea, and intestinal muscle spasm.

Gallstones

Gallstones are pieces of solid material that develop in the gallbladder when substances in the bile, primarily cholesterol, form crystal-like particles. They may be as small as a grain of sand or as large as a golf ball. Gallstones are caused by a combination of factors, including inherited body chemistry, body weight, gallbladder motility, and diet, with excess weight being the primary factor, especially in women (Marschall and Einarsson 2007). The absence of such risk factors does not, however, preclude the formation of gallstones. Many people with gallstones have never had any symptoms. The gallstones are often discovered when having a routine x-ray, abdominal surgery, or other medical procedure. However, if a large stone blocks either the cystic duct or common bile duct, you may have a cramping pain in the middle to right upper abdomen. The pain goes away if the stone passes into the first part of the small intestine, the duodenum.

Gallstones can be treated by using medications, such as ursodeoxycholic acid, that help with dissolving the stone or by using a procedure called lithotripsy, which is a method of concentrating ultrasonic shockwaves onto the stones to break them up. However, these forms of treatment are only suitable when there are a small number of gallstones. Surgical removal of the gallbladder is the most common method for treating gallstones. Gallbladder removal has a 99% chance of eliminating the recurrence of gallstones. In most people, the lack of a gallbladder has no negative consequences.

Prevention of gallstones revolves around avoiding becoming overweight, especially for women. Avoiding rapid weight loss, substituting animal protein with plant protein, and following a high-fiber diet will help as well. Regular physical activity is also recommended, as is moderate to no caffeine and alcohol intake.

SUMMARY

- Hydrolysis reactions digest or breakdown complex molecules such as carbohydrates, lipids, and proteins into simpler forms that the body absorbs and assimilates. The reactions for hydrolysis also occur in the opposite direction and are known as condensation, a process during which individual components of the nutrients bind together to form more complex molecules.

- Enzymes are proteins that play a major role in digestion, as well as in the regulation of metabolic pathways in cells. Digestive enzymes are secreted by the mouth, stomach, small intestine, and pancreas, and function to facilitate the movement and breakdown of food molecules.

- The digestive system involves the gastrointestinal tract, consisting of a hollow tube that begins at the mouth and continues through the esophagus, stomach, small intestine, and large intestine. It also includes accessory organs, such as the liver, gallbladder, and pancreas.

- The stomach acts as a temporary storage site for food. The muscles of the stomach mix the food into a semi-liquid mass called chyme, and gastric juice containing hydrochloric acid and pepsin begins protein digestion. Little absorption occurs in the stomach except for some water and alcohol.

- The small intestine is the primary site of nutrient digestion and absorption and consists of finger-like projections called villi. In the small intestine, bicarbonate from the pancreas neutralizes stomach acid, and pancreatic and intestinal enzymes digest carbohydrate, fat, and protein. The digestion of fat in the small intestine is aided by bile from the gallbladder.

- Components of chyme that are not absorbed in the small intestine pass on to the large intestine, where some water and minerals are absorbed. The large intestine is populated by bacteria that digest some of these unabsorbed materials, such as fiber, and products from bacterial breakdown of fibers and other substances are also absorbed here.

- The water-soluble products of carbohydrate, fat, and protein digestion enter the capillaries in the intestinal villi and are transported to the liver via the hepatic portal circulation. The liver serves as a processing center, storing some of the absorbed substances in the liver, converting some of them into other forms, or allowing them to pass unchanged.

- The fat-soluble products of digestion, such as fatty acids, enter lacteals in the intestinal villi. The nutrients absorbed via the lymphatic system enter the blood circulation without first passing to the liver.

- Absorption of food across the intestinal mucosa occurs by several different processes, including simple diffusion, facilitated diffusion, and active transport. Both simple and facilitated diffusion do not require energy, but depend on a concentration gradient. Active transport requires energy, but can transport nutrients against a concentration gradient.

- Many of the common digestive disorders, such as heartburn, constipation, and irritable bowel syndrome, can be treated with diet changes. These can include increasing fiber intake and avoiding large meals high in fat. Medications are also very helpful in many cases.

CASE STUDY: UNDERSTANDING THE CONDITION OF LACTOSE INTOLERANCE ▮

Lily, a 26-year-old Asian graduate student in biomedical engineering, had been experiencing occasional discomfort after meals. The discomfort reached a new peak last Thursday evening about an hour after eating a cheeseburger and a large chocolate milkshake.

Lily spent much of that night in pain. She had abdominal cramps and diarrhea and also felt sick. Lily went to the clinic and saw a doctor the next day. The doctor asked Lily a number of questions and noted that Lily's discomfort seemed to be associated with dining out. Lily told the doctor that on most evenings she cooked for herself, usually preparing traditional Asian cuisine, and that she seldom experienced any discomfort after eating at home.

When asked if she used very much milk or cheese when preparing meals at home, Lily told the doctor that she almost never cooked with any dairy product. The doctor suspected Lily could be lactose intolerant and told Lily that she would like to have a test performed to verify her initial diagnosis. Lily could be tested that day because she had not had anything to eat or drink for two hours. At the clinic lab, Lily was given a lactose-rich fluid to drink and had her blood glucose level measured several times over the course of the next two hours. Later, her doctor informed Lily that her blood glucose level had not risen after drinking the lactose-rich fluid and therefore she was diagnosed as lactose intolerant.

Questions

- What is lactose intolerance?
- Why is this condition associated with stomach discomfort and diarrhea?
- What kinds of dietary adjustment should Lily consider in order to avoid or ease the symptoms of this condition?

▮ REVIEW QUESTIONS

1 In a biological sense, define the terms: (1) hydrolysis and (2) condensation.

2 What is peristalsis?

3 How does the structure of the small intestine aid absorption?

4 What products of digestion are transported by the lymphatic system?

5 Why is it important to maintain an acidic environment in the stomach?

6 How is the inner lining of the stomach protected from hydrochloric acid? How is the small intestine protected from acidic chyme coming from the stomach?

7 One of the ways used to treat stomach ulcers is to inhibit gastric secretion of hydrochloric acid. If such medication is used for too long, digestion of which food item will be affected? Why?

8 How is the liver related to digestion and absorption? Digestion of what type of nutrient would be affected the most if the liver were severely damaged?

9 How is the pancreas connected to the digestive tract? What enzymes does the pancreas produce that help digestion?

10 Define the terms: (1) passive diffusion; (2) facilitated diffusion; (3) active transport; and (4) osmosis.

11 What is the role of portal circulation?

12 Describe how glucose, fatty acids, and amino acids may be used once they are absorbed into the body.

SUGGESTED READING

Bi, L. and Triadafilopoulos, G. (2003) Exercise and gastrointestinal function and disease: an evidence-based review of risks and benefits. *Clinical Gastroenterology and Hepatology*, 1: 345–355.
This article evaluates the effect of the different modes and intensity levels of exercise on gastrointestinal function and disease using an evidence-based approach. It provides much-needed information, as the impact of exercise on the gastrointestinal system has seen conflicting views.

Peters, H.P., De Vries, W.R., Vanberge-Henegouwen, G.P., and Akkermans, L.M. (2001) Potential benefits and hazards of physical activity and exercise on the gastrointestinal tract. *Gut*, 48: 435–439.
Physical activity reduces the risk of colon cancer. However, acute strenuous exercise may provoke gastrointestinal symptoms such as heartburn or diarrhea. This review describes the current state of knowledge on the hazards of exercise and the potential benefits of physical activity on the gastrointestinal tract.

Williams, C. and Serratosa, L. (2006) Nutrition on match day. *Journal of Sports Science*, 24: 687–697
This article takes a practical approach to discussing how to design regular meals and dietary supplementation for a match or competition. In particular, the effect of consuming various types of carbohydrates on sports performance is discussed.

6 Energy and energy-yielding metabolic pathways

KEY TERMS

- bioenergetics
- energy
- adenosine triphosphate
- mechanical energy
- kinetic energy
- potential energy
- catabolism
- biosynthesis
- digestive efficiency
- phosphocreatine
- phosphagen system
- glycolytic system
- glycolysis
- glycogenolysis
- phosphorylation

- acetyl-CoA
- nicotinamide adenine dinucleotide
- flavin adenine dinucleotide
- oxidative phosphorylation
- uncoupling proteins
- lipolysis
- homeostasis
- steady state
- neurotransmitters
- sympathetic
- parasympathetic
- acetylcholine
- nor-epinephrine
- second messengers

ENERGY AND ITS TRANSFORMATION

Energy is required by all cells. In order for you to jump, throw, run, swim, or cycle, skeletal muscle cells must be able to extract energy from energy-containing nutrients such as carbohydrate and fat. Energy is also needed for other bodily functions such as circulation, digestion, absorption, glandular secretion, neural transmission, and bio-synthesis, to name just a few. Although the body has some energy reserves, most of its energy must be obtained through nutrition. Most cells possess chemical pathways that are capable of converting energy-containing nutrients into a biologically usable form of energy. This metabolic process is termed **bioenergetics**. During exercise, the energy requirement increases, and energy provision can become critical. In fact, the inability to transform energy contained in foodstuffs rapidly into biologically usable energy would limit sports performance. In athletes, carbohydrate depletion represents one of the most common causes of fatigue. People with a defects in energy metabolism cannot tolerate high-intensity exercise. For example, those with McArdle's diseases, which causes trouble degrading muscle glycogen for energy, will have impaired exercise capacity. The amount of food energy that is available, coupled with the ability to transform the food energy into the form that is usable by body cells is what dictates how well the body responds to physical stress. As such, it is imperative to understand what energy is and how the body acquires, converts, stores, and utilizes it.

Energy

Energy is defined as the ability to produce change and is measured by the amount of work performed during a given change. Unlike the physical properties of matter, energy cannot be defined in concrete terms of size, shape, or mass. The presence of energy is revealed only when change occurs. Energy is neither created nor destroyed. It exists in many forms that can be converted from one to another. For example, the energy in flowing water can be converted into energy for a light bulb. In the body, energy is first obtained from energy-containing nutrients in food and, in most circumstances, is then converted as potential energy stored in the body tissues. Via cellular respiration, this potential energy is converted to the high-energy compound **adenosine triphosphate** (ATP) as well as heat. The energy in ATP is used for a variety of biological work, including muscle contraction, synthesizing molecules, and transporting substances. It is one of the most important axioms of science that energy is neither created nor destroyed during any physical or chemical process – this is known as the law of the conservation of energy. In relation to the human body, this law dictates that the body does not produce, consume, or use up energy; it merely transforms energy from one state to another.

Units of energy

In the biological context, energy is measured in joules (J) or kilojoules (kJ), which are units of work, or in calories (cal) or kilocalories (kcal or Cal), which are units of heat. A kilojoule is the amount of work required to move an object of 1 kg a distance of 1 m under the force of gravity. In Europe and most parts of Asia, the joule or kilojoule is the standard measure of energy in food and the body. However, the calorie or kilocalorie is the measure most commonly used in the United States and Canada. In theory, a kilocalorie is the amount of heat required to raise the temperature of 1 kg of water by 1°C. Any measure by kilocalorie or kilojoule is 1000 times greater than those of calorie or joule, respectively. To convert calories to joules or kilocalories to kilojoules, the calories value needs to be multiplied by 4.186; i.e., 1 cal = 4.186 J or 1 kcal = 4.186 kJ.

Potential and kinetic energy

In exercise science, the form of energy that powers muscle contraction is often considered as **mechanical energy**, and activities such as walking, running, swimming, jumping, and throwing require the production of mechanical energy. This form of energy is possessed by an object due to its motion or its position or internal structure. Mechanical energy can be either **kinetic energy** (energy of motion) or **potential energy** (energy of position). For example, a book on a shelf has stored potential energy; by stretching a rubber band, you give it potential energy. Kinetic energy, on the other hand, can be illustrated by individuals performing physical activity. Thinking of a gymnast who is on the balance beam: the movements and flips she does show the kinetic energy that is being displayed while she is moving. When you are running, walking, or jumping, your body is exhibiting kinetic energy. Both forms of energy can exist at the same time, but often change from one form to another. For example, the water at the top of the waterfall has stored potential energy. Once the water leaves the top of the waterfall, the potential energy is changed into kinetic energy. Within a biological system, such a transfer of energy can be exemplified as energy stored in energy-containing nutrients being released through **catabolism**, a process in which complex substances are broken down into simpler ones. In this case, the released potential energy is transformed into kinetic energy of motion. On the other hand, **biosynthesis** may be viewed as a reverse process by which energy in one substance is transferred into other substances so that their potential energy increases.

Biologically usable forms of energy

In a living cell, ATP is the most important carrier of the energy that is necessary to perform many complex functions. This energy-containing compound stores potential energy extracted from food and can yield such energy to power various biological activities via hydrolysis, a process in which a compound is split into other compounds by reacting with water. ATP is the only form of chemical energy that is convertible into other forms of energy used by living cells. As such, ATP is often regarded as the energy currency. Fats and carbohydrate are main storage forms of energy in the body. However, energy derived from oxidation of these two fuels does not release suddenly or sufficiently fast enough to meet the energy demand of those activities that are short and explosive. It is well known that energy liberation from food is a rather complex process that is controlled by enzymes and takes place within the watery medium of the cell. But with the production of ATP, this slow energy transformation from foods is not a concern. ATP may be viewed as a temporary reservoir of energy which functions to provide instant energy to the cells whenever it is needed.

The structure of ATP consists of three main parts: (1) an adenine portion; (2) a ribose portion; and (3) three linked phosphates (Figure 6.1). The formation of ATP occurs by combining adenosine diphosphate (ADP) and inorganic phosphate (Pi) and requires a rather large amount of energy. Some of this energy is stored in the chemical bond that joins ADP and Pi. During hydrolysis, adenosine triphosphatase (ATPase) catalyzes the reaction when ATP joins with water. In the degradation of one mole of ATP, the outer-most phosphate bond splits and liberates approximately 7.3 kcal of free energy that is available for work. This then results in a production of ADP and Pi. In some cases, additional energy releases when another phosphate splits from ADP and this results in the production of adenosine monophosphate (AMP). The energy liberated during ATP breakdown transfers directly to other energy-requiring molecules. In muscle, for instance, the energy is used to energize the myosin cross-bridge, causing the muscle fiber to shorten. The splitting of an ATP molecule takes place immediately and does not need oxygen. The body can store a very limited amount of ATP. Most activities are powered by ATP mainly produced through oxidation of carbohydrate and fat. An example in which the body relies on its stored ATP is those moments of holding one's breath during a short sprint or during a heavy lift. Chemical processes in which ATP is formed from other energy fuels will be discussed in detail later in this chapter.

ENERGY CONSUMPTION

The energy needed to fuel the body comes from the food we eat as well as the energy already stored in the body. Carbohydrate, fat, and protein are the three energy-containing nutrients consumed regularly. Upon entering the body, these macronutrients undergo a series of hydrolytic reactions, including the digestion of starches and disaccharides to monosaccharides, protein to amino acids, and lipids to glycerol and fatty acids. These simpler forms of the macronutrients are then absorbed and assimilated via the hepatic portal vein, which routes blood from the capillary beds of the gastrointestinal tract into the liver. While some of these molecules are used to meet the immediate energy needs of the body, others are stored as potential energy in more complex forms, such as glycogen in muscle and liver and triglycerides in muscle and adipose tissue. The amount of energy taken in depends on the total amount of food consumed and the nutrient composition of the food.

Measurement of energy content of foods

The energy content of food can be measured by using a bomb calorimeter, which consists of a sealed steel chamber surrounded by a jacket of water (Figure 6.2). Food of known weight (e.g., 1 g) is placed in the chamber, which operates under high oxygen pressure. The reaction is started through ignition by an electrical current. As the food combusts, heat is produced and transferred through the metal wall of the chamber,

Figure 6.1 An adenosine triphosphate (ATP) molecule. The symbol "∼" represents energy stored in the phosphate bond.

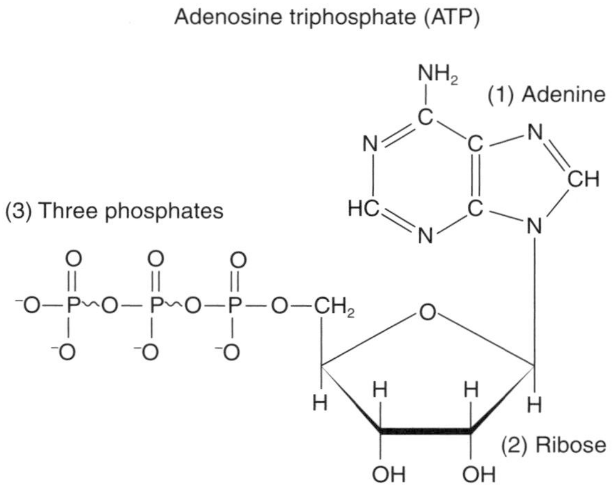

Adenosine triphosphate (ATP)

Figure 6.2 A bomb calorimeter. When dried food is combusted inside the chamber of a bomb calorimeter, the rise in temperature of the surrounding water can be used to determine the energy content of the food.

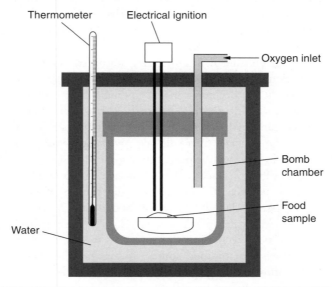

and heats the water that surrounds it. The increase in water temperature can be used to calculate the amount of energy in the food on the basis that 1 kcal is the amount of heat needed to increase the temperature of 1 kg of water by 1°C. For example, if the water volume surrounding the chamber was 5 liters and the temperature of water rises by 2°C, then the amount of energy contained in the food was $5 \times 2 = 10$ kcal (or $10 \times 4.186 = 41.86$ kJ). If the mass of the food combusted was 5 g, then the energy density of the food was $10/5 = 2$ kcal g-1.

This method determines quite accurately the total energy content in different foods. However, it is not without disadvantages. This technique is expensive to run and provides no information regarding the composition of carbohydrate, fat, and protein in the food combusted. Because the body cannot completely digest, absorb, and utilize all of the energy in a food, caloric values from this technique are often slightly higher than the amount of energy the body can actually obtain from the food. This applies particularly to proteins, because the body cannot oxidize the nitrogen component of amino acids, the building blocks of a protein. Consequently, nitrogen atoms combine with hydrogen to form urea to be excreted via the kidneys. Because energy is stored in the hydrogen bond, such a loss of hydrogen results in a reduction in energy of an amino acid that is available for use. Quantitatively, the energy the body can actually obtain from 1 g of protein consumed is, on average, about 4.6 kcal rather than 5.65 kcal, as measured by the bomb calorimeter. This represents a loss of approximately 20% of the potential energy stored in a protein molecule. As both carbohydrate and fat contain no nitrogen, the amount of fuel the body acquires from each of these two nutrients is similar to what is determined by the bomb calorimeter.

Digestive efficiency

How much energy stored in foods can become available to the body is also affected by the efficiency of the digestive process. **Digestive efficiency**, often defined as the coefficient of digestibility, represents the percentage of ingested food digested and absorbed to serve the body's metabolic needs. A coefficient of digestibility of 50 means that only half of the energy consumed was ultimately absorbed. As this digestive parameter provides information as to how much energy from the food consumed can actually arrive inside the body, it has become a major guiding factor in designing a dietary program for weight loss or maintenance. The coefficient of digestibility is relatively higher in both lipids and carbohydrates, reaching 90% and higher. However, those carbohydrate products containing dietary fiber will have lower digestibility. As such, consuming carbohydrates rich in fiber will help in reducing the amount of energy that is available to the body. According to the early data published in the *USDA Handbook* (Merrill and Watt 1973), for instance, the coefficient of digestibility of wheat bran carbohydrate is only 56%, suggesting that the body will obtain only a little over half of the energy stored in this food. Protein has a greater range of the coefficient of digestibility (i.e., 80–97%). This is due to the fact that a protein molecule can vary in terms of its constituent amino acids or its food source. In general, the coefficient of digestibility is lower in plant protein than protein from animal sources.

Table 6.1 shows different coefficients of digestibility, heats of combustion, and net energy values for nutrients in various food groups. As shown in the table, the average coefficients of digestibility for proteins, lipids, and carbohydrates are 92%, 95%, and 97%, respectively. The net energy values are identical to the product of the coefficient of digestibility and heat of combustion for lipids and carbohydrates. However, for proteins the net energy value is much lower than the coefficient of digestibility and heat of combustion (i.e., 4.05 kcal g vs. 5.20 kcal g-1). This difference is explained, as pointed out earlier, by the fact that some of the energy stored in amino acids is lost due to the production of urea, which incorporates hydrogen bonds.

Atwater General Factors

Conveniently, average net energy values can be rounded to simple whole numbers, often referred as Atwater General Factors. These factors are illustrated as follows: 1 g of carbohydrate = 4 kcal; 1 g of lipid = 9 kcal; 1 g of protein = 4 kcal. These values are named for Wibur Olin Atwater (1844–1907), an American chemist. They provide a viable and fairly accurate means of estimating the net energy consumption. They can be used to determine the caloric content of any portion of food or an entire meal from the food's composition and weight. As a result of the application of the Atwater General Factors, at present virtually all food items on the market are labeled with an overall and nutrient-specific energy content. Table 6.2 illustrates how these factors are used for calculating the caloric values of chocolate-chip ice cream.

Bodily energy stores

Energy is stored in the body primarily as fat in the form of triglycerides, though a much smaller amount is also stored as glycogen in the muscle and liver. The body

Table 6.1 Digestibility, heat of combustion, and net physiological energy values of dietary proteins, lipids, and carbohydrates

FOOD GROUP	DIGESTIBILITY (%)	HEAT OF COMBUSTION (KCAL/G)	NET ENERGY (KCAL/G)
Proteins			
Meat and fish	97	5.65	4.27
Eggs	97	5.75	4.37
Dairy products	97	5.65	4.27
Cereals	85	5.80	3.87
Legumes	78	5.70	3.47
Vegetables	83	5.00	3.11
Fruits	85	5.20	3.36
Overall average	*92*	*5.65*	*4.05*
Lipids			
Meat and eggs	95	9.50	9.03
Dairy products	95	9.25	8.79
Vegetable food	90	9.30	8.37
Overall average	*95*	*9.40*	*8.93*
Carbohydrates			
Cereals	98	3.90	3.82
Legumes	97	4.20	4.07
Vegetables	95	4.20	3.99
Fruits	90	4.00	3.60
Sugars	98	3.95	3.87
Animal food	98	3.90	3.80
Overall average	*97*	*4.15*	*4.03*

Source: Adapted from Merrill and Watt 1973.

must have a steady supply of energy, and some of it comes from glucose, the simplest form of carbohydrate. Energy is supplied by the diet we eat. Between meals, the breakdown of stored glycogen and fat help in meeting energy needs. If no food is eaten for more than several hours, the body must shift the way it uses energy to ensure that glucose continues to be available. This is accomplished by increasing the use of stored fat and by mobilizing liver glycogen. The maintenance of blood glucose is of particular importance to the survival and functioning of the central nervous system. As shown in Table 6.3, glycogen stores are limited, and for an 80-kg person the body contains approximately 500 g of glycogen, which in theory could be depleted within several hours of strenuous exercise. Where glycogen stores decrease significantly, there will be an increase in protein degradation that produces amino acids. Some amino acids are converted to glucose, while others are directly metabolized for energy. Protein is not stored as an energy fuel in the body. It serves as a structural component of muscle tissue as well as many other organ tissues. As such, a breakdown of protein for producing glucose and hence energy can result in the loss of muscle and other lean tissues.

Table 6.2 Method for calculating the caloric value of a food from its composition of macronutrients

ICE CREAM (100 G OR 3.5 OZ)	COMPOSITION (%)	WEIGHT (G)	ATWATER FACTORS	CALORIES (KCAL)
Protein	3	3	4	12
Lipid	18	18	9	162
Carbohydrate	23	23	4	92
Water	56	56	0	0

Notes:
Total calories: 266.
% kcal from lipids 162 / 299 = 61%.
Calories = Weight (g) × Atwater Factors (kcal g).

Of the three energy-containing nutrients, the fat molecule carries the largest quantity of energy per unit weight. This occurs because of the greater quantity of hydrogen in the lipid molecule. In a well-nourished individual at rest, catabolism of lipids provides more than 50% of the total energy requirement (Vander *et al.* 2001). Although most cells store small amounts of fat in their cytosol, most of the body's fat is stored in specialized cells known as adipocytes, which function to synthesize and store triglycerides during periods of food intake. As shown in Table 6.3, the potential energy stored in fat molecules for an 80-kg individual equals to 110,700 kcal. Given an energy expenditure of 100 kcal per mile, this amount of energy can fuel an individual to run over 1100 miles. This contrasts sharply to the limited 2000 kcal of stored carbohydrate, which could only fuel a 20-mile run. During prolonged energy restriction, substantial amounts of fat are used to provide energy. However, when the supply of glucose is limited, such as during starvation or in a diabetic state, fatty acids cannot be completely oxidized and chemical ketones are produced. Ketones are the by-products

Table 6.3 Availability of energy substrates in the human body

SUBSTRATES	WEIGHT (G)	ENERGY (KCAL)
Carbohydrate		
Muscle glycogen	400	1600
Liver glycogen	100	400
Plasma glucose	3	12
Total	503	2012
Lipids		
Adipose tissue	12,000	108,000
Intramuscular triglycerides	300	2700
Plasma triglycerides	4	36
Plasma fatty acids	0.4	3.6
Total	12,304	110,740

Source: Adapted from McArdle *et al.* 2009; Vander *et al.* 2001.

Note: These values were estimated based on an average 80-kg man with 15% body fat.

produced mainly in the mitochondrial matrix of liver cells when carbohydrates are so scarce that energy must be obtained from breaking down fatty acids. Ketones can be used as an energy source by many tissues. In sustained starvation, even the brain adapts to meet some of its energy needs by utilizing ketones (Powers and Howley 2001).

ENERGY TRANSFORMATION

Energy transformation is essential to our life. It occurs in both living and non-living systems. As mentioned earlier, the transformation of energy from one form to another follows the law of the conservation of energy. This law states that energy is neither created nor destroyed, but instead transforms from one state to another without being used up. For example, in photosynthesis, solar energy is harnessed by plants, which take carbon, hydrogen, oxygen, and nitrogen from their environment and manufacture carbohydrate, fat, or protein.

In the body, via cellular respiration, energy possessed by macronutrients is changed into chemical energy, which is then stored within energy substrates or converted to mechanical and heat energy. The body stores energy in a variety of chemical compounds including ATP, **phosphocreatine** (PCr), glycogen, and triglycerides. As an energy currency, ATP can be readily used to meet immediate energy needs. However, this high-energy compound is stored in a limited quantity. In fact, the body stores only 80–100 g of ATP at any one time (McArdle *et al.* 2005). This provides energy that can only sustain maximal exercise for several seconds, such as a 60-yard sprint, high and long jump, base running, and a single football play. Consequently, in most sporting events and daily physical activities, ATP is always replenished continuously through a series of chemical reactions involving energy transformation. Three distinctive energy systems have been identified to play a role in replenishing ATP: (1) the ATP–PCr system; (2) the glycolytic system; and (3) the oxidative system.

ATP–PCr system (phosphagen system)

The ATP–PCr system is also known as the **phosphagen system** because both ATP and PCr contain phosphates. This system serves as the immediate source of energy for regenerating ATP. This system has three components. First, there is ATP itself. This high-energy compound, stored in the muscles, rapidly releases energy upon the arrival of an electrical impulse. ATP is degraded to ADP by the enzyme ATPase. Because the reaction involves combination with H_2O, the splitting of ATP is often regarded as hydrolysis. This process can be illustrated as:

$$ATP \xrightarrow{\text{ATPase}} ADP + Pi + energy$$

The second player in this system is PCr. This is another high-energy compound that exists in 5–6 times greater concentration in muscle than does ATP (Brooks *et al.* 2005). Unlike ATP, energy released by the breakdown of PCr is not used directly to accomplish cellular work. Instead, PCr provides a reserve of phosphate energy used to regenerate ATP as a result of muscle contraction and to prevent ATP depletion. In this process, ADP is combined with Pi to become ATP using the bonding energy stored in PCr. This reaction is catalyzed by the enzyme creatine kinase. This process can be illustrated as:

$$PCr + ADP \xrightarrow{\text{creatine kinase}} ATP + Cr$$

The third component of this system involves ADP and the action of the enzyme adenylate kinase, or myokinase when referring to muscle. This enzyme catalyzes the production of one ATP (and one AMP) from two ADPs. This process can be illustrated as:

$$ADP + ADP \xrightarrow{\text{adenylate kinase}} ATP + AMP$$

The three components of this immediate energy system and the respective kinase enzymes are all water soluble. As such, they exist throughout the aqueous part of the cell and in close proximity to the contractile elements of the muscle outside of the mitochondria. They can be immediately available to support muscle contraction. With some ATP being re-synthesized from PCr, this system is able to fuel all-out exercise for approximately 5–10 seconds, such as a 100-m sprint. It is frequently observed that during the last few seconds of the 100-m sprint, runners often slow down. If maximal effort continues beyond ten seconds, or more moderate exercise continues for longer periods, ATP replenishment requires energy sources in addition to PCr.

Glycolytic system (glycolysis)

The **glycolytic system** uses only the energy stored in carbohydrate molecules such as glucose or glycogen for replenishing the ATP the cell needs. This system is also referred to as **glycolysis**. It contains a cascade of chemical reactions, each of which is catalyzed and regulated by a specific enzyme. As shown in Figure 6.3, glycolysis produces pyruvic acid. The production of pyruvic acid occurs regardless of whether oxygen is available. However, availability of oxygen determines the fate of pyruvic acid. When oxygen is lacking, pyruvic acid is converted to lactic acid. Glycolysis is also referred to as the Meyerhof pathway in honor of German biochemist Otto Fritz Meyerhof (1884–1951), who was awarded the Nobel Prize in Medicine in 1922 for his discovery of the pathway. Glycolysis can be summarized as follows:

$$\text{Glucose} \rightarrow 2\ ATP + 2\ \text{lactate}^- + 2\ H^+$$

Glycolysis requires 12 enzymatic reactions for the breakdown of glycogen to lactic acid (one for glycogen to become glucose, ten for glucose to become pyruvic acid, and one for pyruvic acid to become lactic acid). The complex chemical pathway involved makes this system relatively slow in generating ATP as compared to the ATP–PCr system. This energy-yielding pathway is similar to the ATP–PCr system in that both systems form ATP in the absence of oxygen and both occur in the watery medium of the cell outside the mitochondria.

Figure 6.3 The glycolytic pathway in which glucose or glycogen is degraded into pyruvic acid.

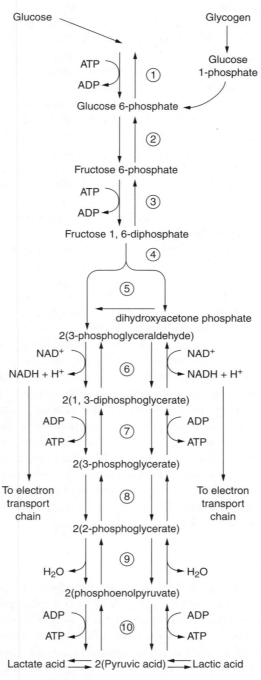

This oxygen-independent system works predominantly within skeletal muscle tissue. This is especially the case in muscles consisting primarily of fast-twitch (i.e., type IIb) muscle fibers. This type of muscle fiber contains a considerable amount of glycolytic enzyme. In muscle, glycogen is usually first broken down into glucose molecules via a process called **glycogenolysis**. These individual glucose molecules are then able to enter the glycolytic pathway. This pathway also allows entry of glucose derived from liver glycogenolysis and transported via circulation. At the onset of glycolysis, ATP is used for glucose to be converted to glucose-6-phosphate, a compound necessary for this pathway to proceed. This is then followed by another energy-requiring reaction in which fructose-6-phosphate is converted to fructose-1,6-diphosphate. During the later reactions of glycolysis, the energy released from the glucose intermediates stimulates the direct transfer of a phosphate bond to ADP. This results in production of up to four ATPs. Because two ATPs are lost in the initial steps of **phosphorylation**, which uses ATP, for each glucose molecule entering the pathway this system generates a net gain of two ATPs. The process by which energy transfers from energy substrate to ADP via a phosphate bond that does not require oxygen is called substrate-level phosphorylation.

This system is rather inefficient in terms of how much of the energy stored in a glucose molecule can result in ATP re-synthesis. In fact, the amount of ATP produced from anaerobic glycolysis is only 5% of what a glucose molecule is capable of generating. In addition, this pathway is associated with the production of lactic acid, which may be involved in the onset of fatigue. This by-product of glycolysis can release hydrogen ions that increase acidity within the muscle cell, thereby disturbing the normal internal environment necessary for maintaining muscle contraction as well as other physiological functions. This system, however, has the advantage of replenishing ATP rapidly. With this system, most cells are able to withstand very short periods of low oxygen by using anaerobic glycolysis. Consequently, this energy system plays a major role in fueling sporting events in which energy production is near maximal for 30–120 seconds, such as 200–800-m runs. There are special cases in which glycolysis supplies most, and in some case all, of the ATP that a cell needs for surviving and functioning. For example, red blood cells contain the enzymes for glycolysis but have no mitochondria; all their ATP production occurs by glycolysis. Also, as mentioned earlier, fast-twitch muscle fibers contain considerable amounts of glycolytic enzyme, but have few mitochondria. During intense exercise, these muscle fibers rely mainly on ATP derived from glycolysis.

Oxidative pathway

Most of the energy used daily comes from oxidation of carbohydrate, lipid, and, in rare cases, protein consumed in the diet. Such aerobic production of energy occurs within mitochondria, which are also known as "the powerhouses of the cell." Mitochondria are found scattered through the cytoplasm. As shown in Figure 6.4, mitochondria are oval-shaped bodies surrounded by two membranes; their internal space or matrix contains numerous enzymes that are capable of catalyzing oxidative energy transformation. As mentioned earlier, the glycolytic system captures only a very small portion of the energy stored in a glucose molecule. However, the oxidative system

Figure 6.4 Structure of a mitochondrion.

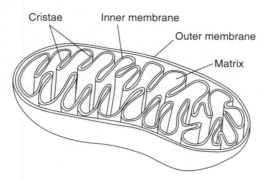

makes it possible for the remaining energy to be extracted from the glucose molecule. This is accomplished by converting pyruvate into **acetyl-CoA** rather than lactic acid, which is possible when oxygen is sufficient. Acetyl-CoA can then enter the citric acid cycle, also known as the Krebs cycle.

The oxidative pathway involves three stages (Figure 6.5).

Stage 1 is the generation of a two-carbon molecule, acetyl-CoA; note that acetyl-CoA can be formed from the breakdown of either carbohydrate, fat, or protein.

Stage 2 is the oxidation of acetyl-CoA in the Krebs cycle. In this process acetyl-CoA combines with oxaloacetate to form citrate. What follows is a series of reactions to regenerate oxaloacetate and two molecules of CO_2, which then allows the pathway to begin again. The primary function of the Krebs cycle is to remove hydrogens and associated energy from various intermediates involved in the cycle using **nicotinamide adenine dinucleotide** (NAD) and **flavin adenine dinucleotide** (FAD) as hydrogen carriers. As a result, NADH and FADH are formed. The importance of hydrogen removal is that hydrogen atoms, by virtue of the electrons that they possess, contain the potential energy stored in the food molecules. Both NADH and FADH then proceed through a series of oxidative reactions collectively called the electronic transport chain, which is stage 3 of the oxidative pathway.

In stage 3, energy stored in these molecules is used to combine ADP and Pi to form ATP. Oxygen does not participate in the reactions in the Krebs cycle, but is the final hydrogen acceptor at the end of the electron transport chain, which produces water. Because ATP is formed via the use of oxygen within mitochondria, this energy-yielding process is also termed **oxidative phosphorylation**. Using glucose as an example, this system can be summarized as:

$$C_6H_{12}O_6 + O_2 \rightarrow 32\ ATP + 6CO_2 + 6H_2O$$

The mechanism of oxidation of NADH and FADH is coupled to the phosphorylation of ADP. This can be further explained by the chemiosmotic hypothesis postulated by Peter Mitchell in 1961. He proposed that electron transport and ATP synthesis are coupled by a proton gradient across the inner mitochondrial membrane (Stryer 1988). In his model, the energy released as electrons is transferred along the respiratory chain and leads to the pumping of protons, H^+, from the matrix to the other side of the inner mitochondrial membrane. As a result, there is a higher concentration of H^+ within the intermembrane space compared to that in the matrix. This then generates an electrical potential which serves as a source of energy to be captured. Mitchell proposed that it is this proton-motive force that drives the synthesis of ATP. Mitchell's hypothesis that oxidation and phosphorylation are coupled by a proton gradient has been validated by a wealth of evidence. In 1978 he was awarded the Nobel

Figure 6.5 The three stages of the oxidative pathway of ATP production.

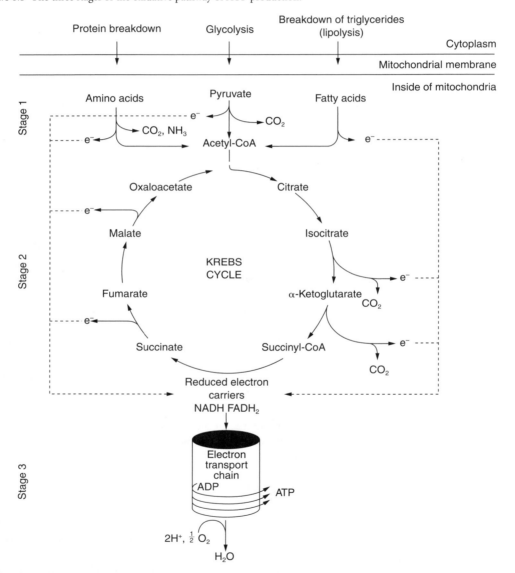

Source: Adapted from Mathews, VanHolde, and Ahern 2000.

Prize in Chemistry due to his extraordinary contribution to our understanding of the fundamental mechanisms of bioenergetics.

According to the chemiosmotic theory as discussed above, cellular energy production takes place across the inner mitochondrial membrane. In this process, ADP is phosphorylated to adenosine triphosphate ATP using energy associated with a gradient of protons that is generated during electron transport. If protons leak back, the gradient is abolished and heat is produced instead of useful energy. This disruption of the connection between food breakdown and energy production is known as "uncoupling." It was long thought that energy metabolism was fully coupled to ATP production, which can then be stored or used in support of various cellular functions. However, with discovery of **uncoupling proteins** (UCP), it is now known that this notion is incorrect. The proton gradient can be diminished by the action of UCP. In fact, in living cells a significant proportion of mitochondrial respiration is normally not coupled to the phosphorylation of ADP, and energy that fails to be coupled to ATP synthesis is dissipated as heat.

Unlike the glycolytic pathway, which only applies to carbohydrates, the aerobic pathway allows oxidation of not only carbohydrates, but also lipids and proteins. Lipids that normally participate in energy metabolism are triglycerides. Via **lipolysis**, triglycerides are broken down into fatty acids and glycerol. Fatty acids can then undergo a series of reactions to form acetyl-CoA. Although in the liver glycerol can be converted into an intermediate of glycolysis, which later becomes pyruvate and then acetyl-CoA, this does not occur to a great extent in skeletal muscle. Therefore, glycerol is not an important muscle fuel source during exercise (Gollnick 1985; Holloszy and Coyle 1984). Fat oxidation is a rather slow process due to the complexity of its metabolism. Nevertheless, it has the ability to yield a large amount of energy. For example, oxidation of the fatty acid palmitate, which contains 18 carbons, can liberate 129 ATP, nearly four times more than the amount of ATP produced from oxidation of a glucose molecule.

Protein is not considered a major energy source, since it contributes less than 15% of the energy produced during exercise (Dolny and Lemon 1988; Gollnick 1985; Lemon and Mullin 1980). However, it can be crucial in maintaining energy continuity and glucose homeostasis under special circumstances such as starvation or prolonged strenuous exercise where bodily carbohydrate decreases significantly. Protein can enter bioenergetic pathways in many different places, but first needs to be cleaved into amino acids. What happens next depends on which amino acids are involved. For example, some amino acids can be converted to glucose or pyruvate, some to acetyl-CoA, and still others to Krebs cycle intermediates. Before an amino acid can be used, the nitrogen residue must be removed. This is accomplished by switching the nitrogen to some other compound, a process known as transamination, or by removing nitrogen through oxidation, a process known as deamination. The energetic roles of carbohydrate, fat, and protein during exercise are discussed in Chapter 7.

The oxidative system confers energy using oxygen, which differs from the ATP–PCr and anaerobic glycolysis systems. Due to its potential of extracting energy from all three macronutrients, this system produces the majority of energy throughout the day. The operation of both the Krebs cycle and electron transport chain take place in mitochondria. As such, the ability to generate energy aerobically depends in part on the size and content of mitochondria. Other factors such as myoglobin content and capillary density can also modulate the effectiveness of this system. This energy system is used primarily in sports emphasizing endurance, such as distance running ranging from 5 km to marathons and beyond.

Energy transformation in sports and physical activity

ATP–PCr, glycolysis, and aerobic pathways are the three energy systems possessed by each individual. There are two inherent limits to the energetic processes: the maximal rate (power) and the amount of ATP that can be produced (capacity) (Sahlin *et al.* 1998). The power and capacity vary drastically among the three energy systems, with both the ATP–PCr and glycolysis systems having more power but a lower capacity than the aerobic system. Brooks *et al.* (2005) have attempted to classify athletic activities into one of the three groups: power, speed, and endurance. Such a classification has the advantage of allowing us to identify a predominant energy system for many different athletic activities. This leads to properly designed training aimed at enhancing the performance of such energy systems. According to this classification, the intramuscular, high-energy phosphate compounds ATP and PCr supply most of the energy for power events such as short-distance sprinting and weight lifting. For rapid, forceful exercises that last about 1 min or so, muscle depends mainly on glycolytic energy sources. Intense exercise of longer duration (i.e., >2 min), such as middle-distance running and swimming requires a greater demand for aerobic energy transfer. Table 6.4 illustrates energy sources of muscular work for various types of athletic activities.

Table 6.4 Energy source of muscular work for different types of sporting events

	POWER	SPEED	ENDURANCE
Event	Shot put Discus Weight lifting High jump 40 yard dash Vertical jump 100 m sprint	200–800 m run 100–200 m swim	1500 m run 10 km run 400–800 m swim Cross-country Road cycling Marathon
Duration of event	0–10 s	10 s–2 min	>2 min
Major sources of energy	ATP PCr	ATP PCr Muscle glycogen	Muscle glycogen Liver glycogen lipids
Energy system involved	ATP–PCr	Glycolysis	Aerobic pathway
Rate of process	Very rapid	Rapid	Slower
Oxygen required	No	No	Yes

Source: Adapted from Brooks *et al.* 2005.

It should be noted that the activities listed in Table 6.4 are primarily track and swimming events in which exercise lasts continuously for a given time period. The fact that these individual events differ only in duration has enabled us to estimate energy expenditure and fuel utilization using laboratory instruments. It is difficult to draw a general conclusion on energy metabolism in team sports such as soccer, field hockey, and lacrosse. This is because energy and fuel requirements for performing these stop-and-go sports can vary depending upon field position and duration of each burst of exercise. Using soccer as an example, it is likely that those who play midfield positions run longer and therefore derive proportionally more of the total energy from aerobic sources. Conversely, those who play forward positions often sprint and thus use the majority of the total energy that is coming from ATP–PCr system.

The three energy systems can also be classified according to whether the operation of the system requires a proper supply of oxygen. In this context, both the ATP–PCr and glycolysis systems are regarded as anaerobic in that they operate outside of mitochondria and energy transfer does not require oxygen. The oxidative system utilizes oxygen as the electron acceptor so that energy transfer can proceed. Most sporting events or physical activities are categorized as to whether they are anaerobic or aerobic. This classification has made it easier to convey to the public whether the activity is tolerable. Generally, activities that place demands primarily on the aerobic system are less intense but require more endurance, such as walking, jogging, cycling, and swimming. Conversely, activities that require anaerobic sources of energy are generally intense, fast-moving, and more explosive, such as sprinting and jumping. They can also be resistance exercises in which muscle tension increases significantly once contracted.

How the three energy systems respond during exercise of changing intensity is a complex issue. This is because as exercise intensity increases, a transition from one energy system to another will occur. It must be kept in mind that for most activities, energy needed is not provided by simply turning on a single energetic pathway, but rather a mixture of several energy systems operating in a sequential fashion but with considerable overlap. Such a mixed use of energy systems may be particularly manifested during (1) rest-to-exercise transition; and (2) incremental exercise in which intensity rises progressively. In the transition from rest to light or moderate exercise, oxygen consumption increases progressively to reach a **steady state** within 1–4 min. The fact that oxygen consumption does not increase instantly to the desired level suggests that energy systems other than the oxidative pathway contribute to the overall production of ATP at the beginning of exercise. There is evidence to suggest that at the onset of exercise, the ATP–PCr system is the first bioenergetic pathway being activated, followed by glycolysis, and finally aerobic energy production. However, after a steady state is reached, the body's ATP requirement can be met primarily via aerobic metabolism.

CONTROL OF ENERGY TRANSFORMATION

It must be kept in mind that the increased rate of energy metabolism does not always occur and it happens only if there is an increase in energy demand. In this context, questions remain as to how such demand-driven energy metabolism comes about and how the body modulates the rate of energy metabolism. The following sections are designed to deal with issues related to the intrinsic regulation of energy metabolism using carbohydrate, fat, and protein. Particular emphasis is placed on the examination of how energy metabolism of each of the three macronutrients is influenced by exercise-induced alterations in the energy substrate and subsequent hormonal secretion. Muscular exercise can be considered a dramatic test of the body's homeostatic control systems. This is because exercise has the potential to disturb many homeostatic variables. For example, heavy exercise results in large increases in muscle oxygen (O_2) requirement, and large amounts of carbon dioxide (CO_2) being produced. These changes must be corrected by increases in breathing and blood flow to increase O_2 delivery to the exercising muscle and remove metabolically produced CO_2, which will otherwise increase the body's acidity. In addition, as heavy exercise begins, there is an immediate increase in the use of ATP. As a result, ATP storage decreases. The body's energy systems must respond rapidly to replenish ATP from substrates such as PCr and carbohydrate so that a continuous energy supply exists and thus energy homeostasis can be maintained.

Homeostasis and steady state

In 1857 a French physiologist, Claude Bernard, became the first person to recognize the central importance of maintaining a stable internal environment. This concept was further elaborated and supported in 1932 by the American physiologist Walter Cannon, who emphasized that such stability could be achieved only through the operation of a carefully coordinated physiological process. The activities of cells, tissues, and organs must be regulated and integrated with each other in such a way that any change in the internal environment initiates a reaction to minimize the change. As such, Cannon described the term **homeostasis** as the maintenance of a constant or unchanging internal environment. It must be noted that changes in the composition of the internal environment do occur, but the magnitude of these changes are small and are kept within narrow limits via multiple coordinated homeostatic processes.

A similar term, steady state, is often used by exercise scientists to denote a steady physiological environment. Although the terms "steady state" and "homeostasis" are often used interchangeably and both result from compensatory regulatory responses, homeostasis generally refers to a relatively constant environment during unstressful conditions such as rest, whereas a steady state does not necessarily mean that the internal environment is completely normal, but simply that it is unchanging (Vander *et al.* 2001). In other words, a steady state only reflects a stability of the internal environment that is achieved by balancing the demands placed on the body and the body's responses to those demands. An example that helps to distinguish the two terms is the case of oxygen consumption during exercise. As shown in Figure 6.6, upon the commencement of moderate-intensity exercise, oxygen uptake reaches a plateau level within a few minutes. This plateau of oxygen uptake represents a steady-state metabolic rate specific to the exercise. However, this constant oxygen uptake occurs at a rate that is greater than the resting level of metabolism, and thus does not reflect a true homeostatic condition.

The fact that the internal environment, such as body temperature, blood pressure, plasma glucose, or acidity, is always maintained as relatively constant in most circumstances suggests that the body operates many control systems that work to

Figure 6.6 The time course of oxygen uptake (VO$_2$) in the transition from rest to submaximal exercise.

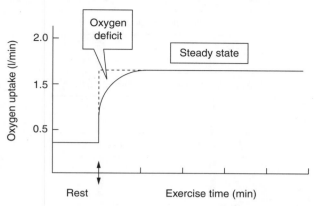

maintain homeostasis on a regular basis. Indeed, every one of the fundamental processes performed by any single cell must be carefully regulated. What determines how much glucose enters a cell? Once inside the cell, what determines how much of this glucose is used for energy and how much is stored as glycogen? To answer these questions, it is important for us to understand not only the metabolic processes, but also the mechanisms that control them.

Control system and its operation

The body has hundreds of different control systems that regulate some physiological variables at or near a constant value. A control system within the organism can be defined as a series of interconnected components that maintain physiological and chemical parameters of the body at a near constant value. The general components of the system are (1) receptors, (2) the afferent pathway, (3) the integrating center, (4) the efferent pathway, and (5) effectors. Figure 6.7 represents the schematic of such a control system. A receptor is capable of detecting the unwanted change or disturbance in the environment and sends the message to the integrating center, which assesses the strength of the stimulus and amount of the response needed to correct the disturbance. The pathway traveled by the signal between the receptor and the integrating center is known as the afferent pathway. The integrating center then sends an appropriate output message to an effector, which is responsible for correcting the disturbance, and causes the stimulus to be removed. The pathway along which this output message travels is known as the efferent pathway.

Most control systems of the body operate via negative feedback. Negative feedback is defined as the working process in which a change in the variable being regulated brings about responses that tend to push the variable in the direction opposite to the original change. An example of negative feedback can be seen in the respiratory control of CO$_2$ concentration in the extracellular fluid. In this case, an increase in extracellular CO$_2$ above the normal level triggers a chemical receptor, which sends information to the respiratory control center in the brain stem to increase breathing. Effectors in this

Figure 6.7 Schematic illustration of a biological control system.

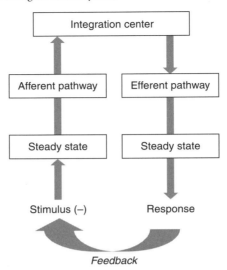

example are respiratory muscles, and an increase in their contraction will reduce extra-cellular CO_2 concentration back to normal, thereby re-establishing homeostasis.

There is another type of feedback, known as positive feedback, in which an initial disturbance in a system sets off a series of events that increases the disturbance event further. Apparently, the positive feedback does not favor the maintenance of the internal environment.

Traditionally, the concept of control systems was restricted to situations in which the first four of the components are all parts of the nervous system. However, this concept is no longer so narrowly focused and recognizes that the principles are essentially the same when blood-borne messengers such as hormones, rather than nerve fiber, serve as the afferent or – much more commonly – the efferent pathway, when an endocrine gland serves as the integrating center. For example, in the case of thermo-regulation when the body temperature drops, the integrating centers in the brain not only send signals by way of nerve fibers to the muscle in order to trigger contraction, but also cause the release of hormones that travel by the blood to many target cells, producing an increase in thermogenesis. Although hormones play an integral role in maintaining homeostasis, a control system that involves hormones could lack a receptor and an afferent pathway. For example, the release of parathyroid hormone is triggered by a fall in plasma calcium concentration. This hormone then functions to increase the release of calcium from bone into the blood. Likewise, the release of insulin is caused by a rise in plasma glucose concentration. This hormone then functions to increase cellular glucose uptake from the blood. In both examples, the objective of the control system involved is to maintain a normal plasma concentration, either of calcium or glucose. However, neither control process involves a receptor or an afferent pathway. This is because glandular cells themselves are sensitive to the change in chemical concentration of the blood supplied to them (Vander *et al.* 2001).

Neural and hormonal control systems

In light of the previous discussion, it is clear that both the nervous and endocrine systems are involved in the control and regulation of various functions in order to maintain homeostasis. Both are structured to be able to sense information, organize an appropriate response, and deliver the message to the proper organ or tissue. The two systems often work together to maintain homeostasis. However, they differ in that in order to deliver the output message the endocrine system relies on hormonal release, whereas the nervous system uses neurotransmitters.

With respect to nervous control, the autonomic nervous system uses **neurotransmitters**, which act as the efferent branch of the nervous system and are most directly related to the regulation of the internal environment (Brooks 2005). The autonomic nerves innervate glands, blood vessels, cardiac muscle, and smooth muscle found in the respiratory and gastrointestinal systems. As such, the system operates below the conscious level. The autonomic nervous system can be further divided into **sympathetic** and **parasympathetic** divisions. The parasympathetic division controls resting functions and has effects such as slowing the heart rate and stimulating digestion. It is comprised of neurons that release **acetylcholine** (ACh). On the other hand, the sympathetic division controls fight-or-flight responses. Unlike the parasympathetic division, this division is comprised of two types of neuron. The first neuron releases ACh, but the second, which directly innervates the cell, releases **nor-epinephrine**. These neurotransmitters bind to the receptors in the cell membranes of target tissues, altering the membrane permeability to certain ions. For example, in the heart ACh promotes entry of Cl^- to reduce the occurrence of action potential, whereas nor-epinephrine stimulates entry of Na^+ and Ca^{++} to facilitate the production of action potential. Consequently, ACh slows heart rate, whereas nor-epinephrine speeds heart rate.

The endocrine glands release hormones directly into the blood, which carries the hormone to a tissue to exert an effect. The hormone exerts its effect by binding to a specific protein receptor. In doing so, the hormone can circulate to all tissues, but only affect the tissues that have the correct receptor. As mentioned earlier, hormonal secretion from endocrine glands is regulated by feedback mechanisms – i.e., a hormone is released in response to a change in the internal environment. However, the secretion of the hormone will diminish and eventually stop if a particular end result of the hormonal action is achieved. Of many endocrine glands, both the pancreas and adrenal glands are perhaps most relevant to exercise metabolism.

The pancreatic hormones are proteins secreted by the islets of Langerhans, clusters of endocrine cells in the pancreas. Islets of Langerhans contain several distinct types of cells, of which both α and β cells have been highly investigated. The β-cells secrete insulin, which stimulates glucose and amino acid uptake by many cells, of which muscle and adipose tissue are quantitatively the most important. This will then be followed by increased synthesis of glycogen and protein in muscle and triglycerides in adipose tissue (Table 6.5). High levels of circulating insulin also inhibit hepatic glucose output and thus promote glycogen as well as triglyceride synthesis in the liver. The α-cells of the pancreas secrete glucagon. While insulin promotes removal of glucose from the blood if it is too high, glucagon functions to raise the blood glucose level if it is too low. Unlike insulin, glucagon exerts its effect primarily on the liver. It enhances both glycogenolysis and gluconeogenesis, the two processes that generate free glucose (Table 6.5). An increase in gluconeogenesis is achieved via glucagon's role of stimulating hepatic amino acid uptake. Both insulin and glucagon function together to help maintain a relatively stable blood glucose concentration.

The adrenal gland contains two sections: the adrenal medulla and the adrenal cortex. The adrenal medulla releases both epinephrine and nor-epinephrine, which are collectively known as catecholamines. These two hormones are not only involved in activating energy metabolism in order to meet the demand of exercise, but also in maintaining blood glucose concentration (Table 6.5). They are also important in regulating cardiovascular and respiratory responses in an effort to facilitate energy homeostasis. The adrenal medulla is innervated by the sympathetic nervous system. As such, sympathetic activity stimulates the secretion of these hormones from the adrenal medulla. The adrenal cortex, the outer part of the adrenal gland, produces cortisol, aldosterone, and sex hormones, of which only cortisol is directly related to energy metabolism. Cortisol contributes to the maintenance of plasma glucose by stimulating lipolysis in adipose tissue and gluconeogenesis in the liver (Table 6.5). Unlike catecholamines, whose release is controlled by sympathetic nerves, cortisol secretion is subject to the action of the stimulating hormones secreted by the hypothalamus and is regulated by a negative feedback mechanism.

Table 6.5 Selected hormones and their catabolic role in maintaining energy homeostasis

ENDOCRINE GLAND	HORMONE	CATABOLIC ACTION	CONTROLLING MECHANISM	STIMULI
Anterior pituitary gland	Growth hormone	• Mobilization of FFA • Gluconeogenesis	• Hypothalamic GH-releasing hormone	• Exercise stress • Low-plasma glucose
Pancreatic β-cells	Insulin	• Uptake of glucose, amino acids and FFA into tissue	• Plasma glucose concentration • Autonomic nervous system	• Elevated plasma glucose • Decreased epinephrine and nor-epinephrine
Pancreatic α-cells	Glucagon	• Mobilization of FFA and glucose • Gluconeogenesis	• Plasma glucose concentration • Autonomic nervous system	• Low-plasma glucose • Elevated epinephrine and nor-epinephrine
Adrenal cortex	Cortisol	• Mobilization of FFA gluconeogenesis	• Hypothalamic adrenal cortex stimulating hormone	• Exercise stress • Low-plasma glucose
Adrenal medulla	• Epinephrine • nor-Epinephrine	• Glycogenolysis • Mobilization of FFA	• Autonomic nervous system	• Exercise stress • Low plasma glucose

In order to produce their action, the catecholamines interact with two receptors, referred to as α- and β-receptors, located on the cell membrane surface. Nor-epinephrine affects mainly the α-receptors, whereas epinephrine affects both α- and β-receptors. The β-receptors can be further subdivided into $\beta1$ and $\beta2$ receptors. In general, the $\beta1$ receptors influence cardiac function, while $\beta2$ is related to tissue metabolism. The actions of these receptors are listed in Table 6.6.

Both α- and β-receptors are also considered to be adrenergetic in that they can be activated by epinephrine and nor-epinephrine. These receptors, once bound to either hormone, will cause changes in cellular activity by increasing or decreasing the cyclic AMP or Ca^{2+}, which are often referred to as **second messengers**. Second messengers can be viewed as intracellular molecules or ions that are regulated by extracellular signaling agents such as neurotransmitters and hormones (first messengers). The second messenger then activates another set of enzymes called protein kinases, which trigger various cellular events in response to the original stimulus. Unlike catecholamines, which are comprised of peptides, cortisol is a lipid hormone and can diffuse easily through the cell membrane and become bound to a protein receptor in the cytoplasma of the cell. The hormone–receptor complex enters the nucleus and binds to a specific protein linked to DNA. This then leads to the synthesis of proteins necessary to alter the metabolism. This process does not involve the production of second messengers. It takes longer for the action of cortisol to be turned on, but its effect will last longer as compared to catecholamines.

SUMMARY

- Energy is defined as the ability to perform work. It is neither created nor destroyed, but instead transforms from one state to another without being used up. The two major interchangeable forms of energy as related to human movement are kinetic and potential energy.
- ATP serves as the body's energy currency, although its quantity is very limited. The free energy yielded from splitting of the phosphate bond of ATP powers all forms

Table 6.6 Interaction of epinephrine and nor-epinephrine with adrenergic receptors

RECEPTOR TYPE	INTRACELLULAR MEDIATOR	EFFECT
α	Cyclic AMP and Ca^{++}	• Vasoconstriction • Gastrointestinal relaxation
$\beta1$	Cyclic AMP	• Increased heart rate • Increased cardiac contraction • Increased lipolysis • Increased glycogenolysis
$\beta2$	Cyclic AMP	• Vasodilation • Bronchodilation

Source: Adapted from Tepperman and Tepperman 1987.

of biological work. In most activities, ATP is generated instantly from degradation of carbohydrate and fat.

- Carbohydrate, fat, and protein represent the three energy-containing nutrients consumed daily. As compared to fat, carbohydrate stored as glycogen is rather limited. However, it is a more preferable source of energy. Protein contains energy, but contributes little to energy metabolism. Carbohydrate and protein each provide about 4 kcal of energy per gram, compared with about 9 kcal per gram for fat.

- The potential energy stored in nutrients is captured through three energy-yielding systems: (1) the ATP–PCr system; (2) the glycolytic system; and (3) the oxidative system. The operation of these systems is essential to the continual supply of ATP in support of various biological functions.

- The three energy systems differ considerably in terms of rate and capacity of ATP production; their contribution will vary depending upon intensity and duration of an activity. However, such differences among the three energy systems provide the ability for the body to derive energy under various circumstances, whether generating explosive power, enduring a long-distance event, or simply performing a household activity.

- The oxidative system involves breakdown of fuels with the use of oxygen. Compared with the ATP–PCr and glycolytic system, operation of the oxidative system is slowest in generating ATP. However, it is the most capable of extracting energy stored in energy-containing nutrients. The oxidative system also represents a "common" pathway shared by carbohydrate, fat, and protein for use as an energy source.

- For most activities energy needed is not provided by simply turning on a single energetic pathway, but rather a mixture of several energy systems operating concurrently. However, the percentage contribution of each system differs depending upon intensity and duration of the activity.

- The term "homeostasis" is defined as the maintenance of a constant internal environment. It differs from the term "steady state" in that the latter represents a constant internal environment achieved under stressful conditions such as exercise.

- The maintenance of a constant internal environment is achieved by many biological control systems that operate mainly in a negative feedback manner and are capable of detecting, processing, and making appropriate adjustments to correct the changes.

- Both the nervous and endocrine systems often work together as part of a control system. They are structured to be able to sense information, organize an appropriate response, and deliver the message via neurotransmitters or hormones to the proper organ or tissue in order to exert their actions.

- Hormones involved in energy metabolism exert their effect by first combining with protein receptors and then activating the enzymes necessary to catalyze the intended chemical reactions. Specifically, for those peptide hormones such as catecholamines, binding with receptors takes place on the cell membrane, which triggers the production of second messenger within the cell that helps complete the action of the hormone. This process differs from lipid-like hormones such as cortisol, which always diffuse across cell membranes and bind to a receptor within the cell before exerting their actions.

CASE STUDY: DO ENERGY DRINKS REALLY PROVIDE A SOURCE OF ENERGY?

Rhonda just landed the job of her dreams as a writer for *Runner's World* magazine. Since high school, where she had excelled in cross-country, Rhonda had been a consistent runner. Her first assignment is to write a report on the efficacy of the energy drink XS Citrus Blast®. It was required that to write this report she must be very accurate in her analysis.

Rhonda knew that XS Citrus Blast® was used by athletes to provide "fuel" as they practice and compete. She also saw other people using it more casually as a way to become "energized." However, she is confused about the labeling of the drink. For example, XS Citrus Blast® boasts that it has no calories but still provides energy. That made no sense, based on what Rhonda knew about biological energy! Rhonda decided to find out more about the drink before she wrote her report. She found out the following information:

- Ingredients: carbonated water, taurine, glutamine, citric acid, adaptogen blend, natural flavors, acesulfame potassium, caffeine, sodium benzoate, potassium sorbate, sucralose, niacin, pantothenic acid, pyridoxine HCl, yellow 5, cyanocobalamin.
- Nutrition facts: serving size: 8.4 fl. oz; servings per container: 1; calories: 8; fat: 0 g; sodium: 24 mg; potassium: 25 mg; total carbs: 0 g; sugars: 0 g; protein: 2 g; vitamin B-3: 100%; vitamin B-6: 300%; vitamin B-5: 100%; vitamin B-12: 4900%.

Questions
- What is a biological definition of energy?
- When we say that something gives us "energy," what does that mean?
- Why is the XS Citrus Blast® that contains only 8 kcal considered an "energy booster"?
- What ingredients and nutrients provide energy? How do they do that?

REVIEW QUESTIONS

1 Define the term "energy." What is the law of the conservation of energy?
2 List five bodily functions that depend on the use of energy.
3 Define the term "coefficiency of digestibility." Why does protein have the lowest coefficiency of digestibility?
4 Define kinetic and potential energy and provide an example that illustrates the transformation between these two forms of energy.
5 How much glycogen does an 80-kg person, on average, possess? How is glycogen distributed between the muscle and the liver?
6 What is the total energy stored in food containing 50 g of carbohydrate, 15 g of fat, and 8 g of protein?
7 Compare the three energy systems in terms of complexity, cellular location, end products, oxygen requirements, rate and capacity of ATP production, and sporting events supported by the system.
8 Define the terms: (1) glycerol; (2) pyruvate; (3) acetyl-CoA; and (4) NADH.
9 State the chemiosmotic hypothesis.

10 Define the term "homeostasis." How does it differ from the term "steady state"?

11 What are the components of a biological control system? Provide an example that illustrates the operation of a control system.

12 Describe how each of the following hormones affects carbohydrate, fat, and protein utilization during exercise: (1) growth hormone; (2) insulin; (3) glucagon; (4) cortisol; (5) epinephrine; and (6) nor-epinephrine.

SUGGESTED READING

Burke, L.M. (2001) Energy needs of athletes. *Canadian Journal of Applied Physiology*, 26: S202–S219.

This article provides practical advice about how athletes should use their energy budget to choose foods that fulfill macronutrient and micronutrient needs for optimal health and performance.

Fitts, R.H. (1996) Muscle fatigue: the cellular aspects. *American Journal of Sports Medicine*, 24 (6): S9–S13.

This article addresses exercise-induced cellular changes that may lead to muscle fatigue and discusses how diet and fluid replacement may help counteract such changes and thus prevent or delay fatigue.

Gastin, P.B. (2001) Energy system interaction and relative contribution during maximal exercise. *Sports Medicine*, 31: 725–741.

This article provides a more contemporary overview of how the three distinctive energy systems operate during exercise. Some of the misconceptions with regard to how these energy systems are affected by exercise intensity and duration are also discussed.

7 Nutrients metabolism during exercise

KEY TERMS

- hypoglycemia
- hepatic glucose output
- gluconeogenesis
- lactate threshold
- lipase
- β-oxidation
- carnitine
- carnitine palmitoyl transferase (CPT)
- branched-chain amino acids
- nitrogen balance

- endogenous
- glycogen phosphorylase
- phosphofructokinase
- isocitrate dehydrogenase
- pyruvate dehydrogenase
- hyperglycemia
- malonyl-CoA
- somatomedins,
- insulin-like growth factors

CARBOHYDRATE METABOLISM

Exercise poses a serious challenge to the bioenergetic pathways in the exercising muscle. For example, during heavy exercise, the body's total energy output may increase 15–20 times above that of the resting condition. Most of this increase in energy production is used to provide ATP for contracting skeletal muscle, which may increase their energy utilization 200 times over utilization at rest. Therefore, it is apparent that skeletal muscles have a great capacity to produce and use large quantities of ATP during exercise. Such a large increase in ATP production is made possible by our ability to extract the energy stored in carbohydrates, lipids, and proteins we consume daily. In this context, a strong tie exists between nutrition and sports performance. Compared to lipids and proteins, carbohydrates remain the preferred fuel during high-intensity exercise because it can rapidly supply ATP both aerobically and anaerobically. Therefore, emphasizing a sufficient consumption of carbohydrate daily should be an integral part of the training regimen for most athletes. Lipids represent another potential source of energy. However, their catabolism normally results in a lower energy turnover that cannot quite match the energy demand imposed by most sporting events. Given that excess lipids can have negative impacts upon one's health, it is equally important to understand the unique characteristics associated with lipid metabolism, so that an effective lifestyle strategy can be developed to facilitate fat loss.

Carbohydrate: an ideal energy source during exercise

Two main macronutrients provide energy for replenishing ATP during exercise: (1) muscle and liver glycogen; and (2) triglycerides within adipose tissue and exercising muscle. To a much lesser degree, protein or amino acids within skeletal muscle can donate carbon skeletons, thereby furnishing energy. During prolonged exercise, carbohydrates such as muscle glycogen and blood glucose derived from liver glycogenolysis are the primary energy substrates. Glycogen is a readily mobilized storage form of glucose. It is a very large branched polymer of glucose residues, as mentioned in Chapter 2 (see also Figure 2.4). Glycogen undergoes a process of glycogenolysis that yields free glucose molecules. This glucose can then enter the glycolytic pathway in which energy is transformed. The importance of their availability during exercise is demonstrated by the observation that fatigue is often associated with muscle glycogen depletion and/or **hypoglycemia**. With respect to energy provision, carbohydrates are superior to fat in that (1) they can be used for energy with and without oxygen; (2) they provide energy more rapidly; (3) they must be present in order to use fat; (4) they are the sole source of energy for the central nervous system; and (5) they can generate 6% more energy per unit of oxygen consumed.

Carbohydrate utilization at onset of exercise

As mentioned in Chapter 6, phosphocreatine (PCr) is the primary energy substrate available for replenishing ATP during very intense muscular exercise of short duration (i.e., ≤10 s). This idea was initially supported both by theoretical calculations of the energy required for production of muscle force and the rapid decline in PCr found during very intense exercise. Consequently, it has long been assumed that the provision of energy via a particular metabolic pathway is linked sequentially – that is, during intense exercise, PCr stores are almost depleted in the initial 10 s and further contractile activity is then sustained by the metabolism of muscle glycogen. However, it is now understood that carbohydrate, particularly glycogen and glucose, will take a fair share of energy provision at the onset of exercise. Boobis *et al.* (1982) found that with all-out bicycle ergometer exercise aimed to accomplish as much work as possible in 6 s, a 35% decrease in PCr occurred along with a 15% reduction in glycogen. When such exercise was performed for 30 s, a 65% decrement in PCr was found, concomitant with a 25% reduction in glycogen. Similar results have been reported for short-duration maximal treadmill running and cycling (Cheetham *et al.* 1986; McCartney *et al.* 1986). Collectively, these studies indicate that PCr breakdown and glycogenolysis occur concomitantly from the onset of exercise.

Influence of exercise intensity and duration

During very intense exercise when oxygen consumption fails to meet energy demands, stored muscle and liver glycogen becomes the primary energy source because energy transfer from carbohydrate can occur without oxygen. With the reintroduction of the needle biopsy technique in the early 1960s, considerable effort has been devoted to the study of glycogen utilization during exercise, as well as the re-establishment of glycogen stores after exercise. A landmark study by Gollnick *et al.* (1974) revealed that muscle glycogen breakdown is most rapid during the early stages of exercise, with its rate of utilization being exponentially related to exercise intensity. They have also found that slow-twitch muscle fibers were the first to lose glycogen. This was then followed by increased glycogen utilization in fast-twitch muscle fibers. As exercise continues, the rate of muscle glycogen utilization declines and this is then accompanied by an increased contribution of blood-borne glucose degraded from liver glycogen as a metabolic fuel. It is estimated that a two-hour vigorous workout depletes glycogen in the liver and exercised muscle.

During moderate-intensity exercise, utilization of PCr as an energy source is relatively mild, even at the onset of exercise. Glycogen stored in active muscle supplies almost all of the energy in the transition from rest to exercise. As soon as a steady state is attained, energy is provided through a mixed use of carbohydrates and lipids. Typically, liver and muscle glycogen supply 40–50% of the energy requirement, whereas the remainder is furnished via oxidation of lipids. Such an energy mixture may vary depending upon the intensity of exercise, although it can also be influenced by the training status of an individual and dietary intake of carbohydrate. For example, a trained individual is able to use proportionally more fat as energy at a submaximal workload, and those who consume a diet low in carbohydrate may force their body to use relatively more fat instead of carbohydrate. As exercise continues, muscle glycogen stores diminish progressively. Consequently, as blood glucose from liver glycogen becomes the major supplier of carbohydrate energy, relative contribution of fat to the total energy provision also increases. With glycogen depletion, the maximal steady-state exercise intensity that can be sustained decreases accordingly. This is mainly caused by a slower rate of energy production via fat metabolism. As bodily glycogen decreases significantly, blood glucose also falls, because the liver's glucose output fails to match the rate of glucose uptake by exercising muscles. Hypoglycemia is said to occur when blood glucose concentration is lower than $2.5\,\text{mmol}\,l^{-1}$ or $45\,\text{mg}\,dL^{-1}$. This condition can occur during strenuous exercise that lasts for close to or more than two hours. Table 7.1 illustrates the percentage of energy derived from the four major sources of fuel during prolonged moderate-intensity exercise (i.e., 65–75% $VO_2\text{max}$) in trained individuals.

Table 7.1 Percentage of energy derived from the four major sources of fuel during moderate-intensity exercise at 65–75% VO_2max

ENERGY SOURCES	% OF ENERGY EXPENDITURE				
	ONSET OF EXERCISE	1ST HOUR	2ND HOUR	3RD HOUR	4TH HOUR
Muscle glycogen	45	35	22	12	0
Blood glucose	5	13	23	30	40
Plasma-free fatty acids	25	32	39	46	52
Muscle triglycerides	25	20	16	12	8

Note: Data are the estimated percentages of energy expenditure based upon studies that used endurance athletes.

Liver sources of carbohydrate

During exercise, when utilization of carbohydrate accelerates, an increased release of glucose from the liver is functionally important to maintain blood glucose homeostasis and to possibly attenuate muscle glycogen depletion. It remains debated as to whether the increased availability of blood glucose helps in sparing the use of muscle glycogen. However, it appears that once hypoglycemia is induced, muscle glycogen utilization is likely to accelerate. The increased contribution of liver glycogen is a universal phenomenon that was revealed nearly 50 years ago. Early studies have demonstrated that release of glucose from liver glycogen was accelerated to 3–6 times that of resting values during muscular work. Over the following 2–3 decades there has been repeated evidence suggesting that **hepatic glucose output** can increase by two- to three-fold during moderate exercise and up to seven- to ten-fold during vigorous exercise.

The intensity and duration of an exercise are the important factors that determine the source and quantity of glucose released by the liver. During moderate exercise (<60% VO_2max), the blood glucose level remains relatively constant despite an increase in glucose utilization by exercising muscle. At this intensity, a drop in blood glucose will not occur unless exercise is prolonged for several hours. The level of blood glucose reflects a balance between hepatic glucose output and muscular glucose utilization. Hence, at moderate intensity, an exercise-induced rise in hepatic glucose output is able to match the increased glucose utilization. In contrast, if exercise becomes more intense (>60% VO_2max), blood glucose concentrations increase, especially during the early phase of exercise. Such an increase is more pronounced at higher exercise intensities (Hargreaves and Proietto 1994). This mismatch may be due to the fact that hepatic glucose output exceeds glucose uptake by working muscles. It has been considered that the production of glucose from the liver is not totally regulated by feedback mechanisms, which are fundamentally important in maintaining homeostasis. In the case of over-production of hepatic glucose, the finding has been attributed to increased efferent signals of the central nervous system, which regulate hepatic glucose metabolism (Kjær *et al.* 1987).

Gluconeogenesis: generating glucose in the liver

An increase in liver glucose output can be brought about by an enhancement of glycogenolysis and **gluconeogenesis**. While glycogenolysis is a relatively simple process that involves glycogen breakdown into glucose, gluconeogenesis entails a rather complex pathway that involves the use of non-glucose molecules such as amino acids or lactate for the production of glucose in the liver. This "new" glucose can then be released into the blood and transported back to skeletal muscles to be used as an energy source. This internal production of glucose may be viewed as the secondary resort by which the body obtains glucose. Figures 7.1 and 7.2 demonstrate the two different processes of gluconeogenesis in which glucose is generated from its precursors, lactate and alanine, respectively. In general, glycogenolysis appears to respond more quickly and contribute more of the total hepatic glucose output during exercise. During the first 30 min of exercise of either moderate or heavy intensity, most of the glucose released by the liver is derived from hepatic glycogenolysis. Gluconeogenesis, however, seems to occur more as the exercise continues; it plays a more important role during the later phase of prolonged exercise. In a series of experiments using dogs, Wasserman *et al.* (1988) found that the relative contribution of gluconeogenesis to the total hepatic glucose output was only 15% during the first 60 min of exercise. However, it reached 20–25% when exercise continued for 90 min.

Exercise intensity can influence the type of gluconeogenic precursors used to produce glucose in the liver. Glycerol, lactate, and amino acid are the three major

Figure 7.1 An example of gluconeogenesis during which the muscle-derived alanine is converted to glucose and this newly formed glucose then circulates back to muscle.

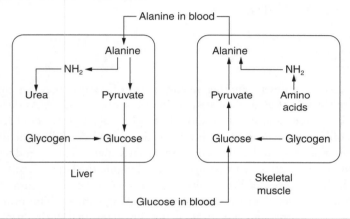

Figure 7.2 An example of gluconeogenesis during which the muscle-derived lactate is converted to glucose and this newly formed glucose then circulates back to muscle.

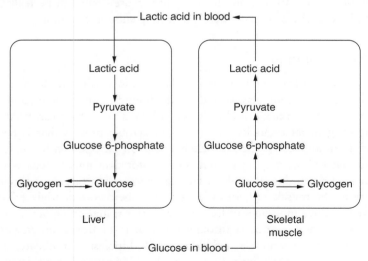

precursors that can be converted to glucose via gluconeogenesis. It is generally believed that when exercise is performed at an intensity level below **lactate threshold**, an intensity above which the production of lactate will increase sharply, glycerol is the primary molecule used for gluconeogenesis. As exercise intensity approaches and exceeds the lactate threshold, more lactate becomes available for producing glucose in the liver. Such different uses of gluconeogenic precursors at different exercise intensities makes sense in that glycerol is a product of lipolysis and fat utilization increases during exercise of low to moderate intensity. However, as intensity increases, more glycogen is degraded, thereby producing more lactic acid. An increased contribution of amino acids to gluconeogenesis would be seen particularly during vigorous exercise that lasts for a prolonged period of time. In this case, both muscle and liver glycogen stores decrease significantly and there is a need to produce more glucose in order to prevent the occurrence of hypoglycemia.

LIPID METABOLISM

Triglycerides represent another major source of energy; they are stored primarily in adipose tissue, although they are also found in muscle tissue. As discussed in Chapter 2, a triglyceride molecule is comprised of a glycerol and three fatty acids; it can vary in terms of how many carbons each fatty acid contains. A triglyceride is split via lipolysis to form a glycerol and three independent fatty acids. These products can then enter metabolic pathways for energy production. Despite the large quantity of lipid available as fuel, the processes of lipid utilization are slow to be activated and proceed at rates significantly lower than the processes controlling carbohydrate utilization. However, lipids are an important element of energy substrates used during prolonged exercise or during extreme circumstances such as fasting or starvation, when carbohydrate stores decline significantly. Even small increases in the ability to use lipids as fuel during exercise can help slow muscle glycogen and blood glucose utilization, delaying the onset of fatigue. An increase in the ability to use lipids can be realized by improved oxidative capacity of skeletal muscle following endurance training.

Energy sources from lipids

Three lipid sources supply energy: (1) fatty acids released by the breakdown of triglycerides; (2) circulating plasma triglycerides bound to lipoproteins; and (3) triglycerides within the active muscle itself. Unlike carbohydrate, which can yield energy without using oxygen, fat catabolism is purely an aerobic process that is best developed in heart, liver, and slow-twitch muscle fibers. Most fat is stored in the form of triglycerides in fat cells or adipocytes, but some is stored in muscle cells as well. The major factor that determines the role of fat as an energy substrate during exercise is its availability to the muscle cell. In order for fat to be oxidized, triglycerides must first be cleaved to three molecules of free fatty acid (FFA) and one molecule of glycerol. This process – lipolysis – occurs through the activity of **lipase**, an enzyme found in the liver, adipose tissue, muscle, and blood vessels. Lipolysis is modulated by the hormones epinephrine and nor-epinephrine. As such, this process is considered to be intensity dependent because the release of these hormones increases as exercise intensity increases. It must be noted that lipolysis and oxidation are the two separate processes of fat utilization. The latter process is facilitated during low- to moderate-intensity exercise in which production of lactic acid is low.

Preparatory stages for fat utilization

Following lipolysis, two additional processes must also happen before FFA can be combusted: (1) mitochondria transfer; and (2) β-**oxidation**. The oxidation of fatty acids occurs within the mitochondria. However, long-chain fatty acids are normally unable to cross the inner mitochondrial membrane due to their molecular size. This would then require a membrane transport system consisting of protein carriers. The carrier molecule for this system is **carnitine**, which is synthesized in humans from the amino acids lysine and methionine and is found in high concentration in muscle. Under the assistance of carnitine and an enzyme called **carnitine palmitoyl transferase** (CPT), fatty acids can be brought from cytoplasm into mitochondria. Upon entry into the mitochondria, fatty acids undergo another process called β-oxidation. β-oxidation is a sequence of reactions that reduce a long-chain fatty acid into multiple two-carbon units in the form of acetyl-CoA (Figure 7.3). This process may be viewed as being analogous to glycolysis, the first stage of the oxidative pathway for glucose in which a glucose molecule is converted into two molecules of acetyl-CoA. Once formed, acetyl-CoA then becomes a fuel source for the Krebs cycle and leads to the production of ATP within the electron transport chain.

Influence of intensity and duration on fat utilization

Fat oxidation is influenced by exercise intensity and duration. Romijn *et al.* (1993) found that during exercise at 25% VO_2max, 90% of the total energy is furnished via oxidation of plasma FFA and muscle triglycerides. The relative contribution of fat to total oxidative metabolism decreases as exercise intensity increases. However, such a decrease in relative contribution of fat is relatively minor compared with an increase in oxygen consumption. Therefore, despite a decrease in relative contribution, there is actually an increase in the amount of fat being oxidized until the intensity reaches a value close to one's lactate threshold or ~60% VO_2max. A number of recent studies have found that the intensity at which the highest fat oxidation is observed ranges from 55% to 65% VO_2max (Achten *et al.* 2002; Achten and Jeukendrup 2004). For example, by testing with multiple levels of intensity, Achten *et al.* (2002) found that exercise at 60–65% VO_2max would help in eliciting the maximal rate of fat oxidation. We recently also observed that the intensity of maximal fat oxidation corresponds well with the lactate threshold, suggesting that in order to obtain the maximal fat oxidation, a comparatively more intense exercise ought to be chosen (Kang *et al.* 2007). When exercise is performed at an intensity above the lactate threshold, fat oxidation decreases significantly. This is because increased carbohydrate utilization and/or accumulation of lactic acid can serve to inhibit fat utilization.

As shown in Table 7.1, as a steady-state exercise of light to moderate intensity continues, the contribution of fat to the total oxidative metabolism increases progressively.

Figure 7.3 An illustration of β-oxidation.

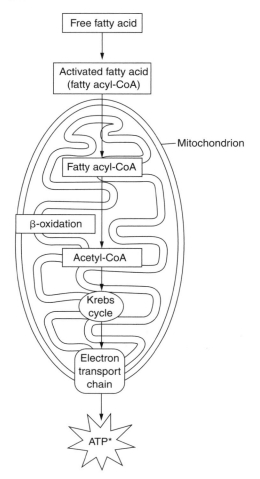

Source: Powers and Howley 2007. Used with permission.

Using a prolonged exercise protocol in which exercise lasted for several hours, earlier studies have demonstrated a progressive decrease in respiratory exchange ratio, signifying a steady increase in fat combustion (Edwards *et al.* 1934) (see Chapter 10 for further details on the concept and application of the respiratory exchange ratio). It has been estimated that the relative contribution of fat can account for as much as 80% of the total energy expenditure during prolonged exercise. The progressive increase in fat utilization over time is due to a concomitant decrease in glycogen stores as a result of prolonged exercise. This reduction in carbohydrate energy substrates will trigger a release of glucoregulatory hormones such as glucagon, cortisol, and growth hormone. These hormones function to stimulate the breakdown of lipids in response to reduced carbohydrate stores. Please refer to the later sections of this chapter for more information with regard to hormonal regulation of fuel utilization.

Interaction between carbohydrate and fat utilization

Utilization of carbohydrate and fat is not two separate processes. Instead, the two forms are coordinated, and utilization of one substrate is affected by the availability of the other. This can be proven by the fact that fat utilization increases in accordance with a decrease in glycogen stores during prolonged exercise. It has been suggested that carbohydrate availability during exercise modulates the level of lipolysis and fat oxidation. Recent studies found that ingesting high-glycemic carbohydrates prior to exercise significantly blunts release of fatty acids from adipose tissue and thus oxidation of long-chain fatty acids by skeletal muscle (Coyle *et al.* 1997; De Glisezinski *et al.* 1998). As carbohydrate substrates decline, such a suppressive effect of carbohydrate on fat utilization is withdrawn. Similarly, it has long been found that elevated FFA is able to suppress glucose utilization. This can be explained by the well-recognized theory of the glucose–fatty acid cycle that was conceived by British biochemist Philip Randle (it is also known at the Randle cycle). This cycle explains how an increased availability of fatty acids slows down the utilization of glucose. Randle's hypothesis has been used to explain how the mitochondrial adaptations resulting from endurance training help promote lipid oxidation and thus spare glycogen utilization in skeletal muscle during exercise. Further details regarding this cycle and its function can be found later in this chapter.

PROTEIN AND AMINO ACID METABOLISM

Skeletal muscle constitutes approximately 40% of the body weight and is the second largest store of potential energy in the body after fat. However, protein and amino acids serving as energy substrates are a rather uncommon topic. This is because amino acids contribute only a minor portion (5–15%) of the total energy consumed during exercise. Unlike carbohydrate and fat, which can be stored as energy substrates, there are virtually no inert amino acids that are designated for such a purpose. However, it must be recognized that during fasting and starvation, catabolism of proteins to amino acids and conversion of amino acids into energy are very important processes in maintaining the levels of blood glucose essential for brain and kidney function. It has been reported that gluconeogenesis that uses amino acids increases every morning in response to the fall in glycogen stores. In the last decade or so, researches have realized that even a minor increase in protein consumption is important in conditions of high energy demands over a prolonged period of time. There is growing evidence, especially with more recent research on **branched-chain amino acids** such as leucine, valine, and isoleucine, to suggest that protein serves as energy fuel to a much greater extent than previously believed.

Protein degradation and synthesis

Protein metabolism includes its degradation and synthesis. Protein degradation often occurs to a lesser extent than that seen in carbohydrate and fat. However, it can be increased significantly when exercise is performed at high intensity for a prolonged period of time. There are two classes of protein in skeletal muscle: contractile and non-contractile. While contractile-related proteins are those that are responsible for muscle contraction, non-contractile-related proteins are those that are essential for other cellular functions. In humans, contractile and non-contractile proteins comprise 66% and 34% of the total muscle protein, respectively. Remember, a protein molecule is comprised of chains of amino acids. As such, the amino acids tyrosine and phenylalanine have been used as indicators of non-contractile protein degradation. In an early study in which the experimental protocol entailed 40 min of exercise performed at different intensities, Felig and Wahren (1971) demonstrated a greater release of tyrosine and phenylalanine, as well as alanine, during exercise compared with rest, with an enhanced efflux of metabolites being greater at higher exercise intensity. Later, Babij *et al.* (1983) also observed a direct linear relationship between exercise intensity and oxidation of leucine, one of the three branched-chain amino acids. In terms of metabolism of contractile proteins, the measurement of 3-methyhistidine (3-MH) in the urine has been the most widely used approach in reflecting the degradation of contractile proteins, although this parameter can also be determined via blood. Through a thorough review of the literature, Dohm *et al.* (1987) came to the conclusion that the production of this catabolic index of contractile protein decreases during exercise. However, there are studies reporting an increase in the efflux of 3-MH during recovery. Taken together, these findings suggest that the integrity of contractile protein remains unaffected during exercise when muscle contraction is in demand; however, this is not the case during recovery. The mechanism responsible for the divergent response in 3-MH between exercise and recovery is unclear.

Assessment of protein degradation and synthesis will provide an idea as to whether those who perform exercise will need extra protein in order to prevent a loss in lean body mass. Such an assessment can be accomplished by determining **nitrogen balance**. Protein contains nitrogen, and the body cannot oxidize the nitrogen component. Consequently, nitrogen atoms combine with hydrogen to form urea to be excreted via the kidneys. Nitrogen balance involves assessing the relationship between the dietary intake of protein and protein that is degraded and excreted. Nitrogen balance is said to occur when protein intake equals the amount excreted. A positive nitrogen balance suggests that protein intake exceeds protein output and the excessive protein may have been used to repair damaged tissue and/or synthesize new tissue. The positive nitrogen balance is expected in children, pregnant women, and body builders (Table 7.2).

Table 7.2 Expected nitrogen balance status among various individuals

EXAMPLES	NITROGEN INTAKE (G)	NITROGEN OUTPUT (G)	NITROGEN BALANCE (G)
Individuals on weight loss diet or with poor nutrition	6.4	8.0	−1.6
Healthy individuals with normal diet	11.2	11.2	0
Pregnant women, children, body builders	12.8	10.4	+2.4

A negative nitrogen balance indicates that protein loss is greater than its intake. This type of nitrogen imbalance is often manifested in individuals who are on a weight-loss diet or those suffering from poor nutrition or eating disorders (Table 7.2). A negative nitrogen balance may also occur in athletes who are overly trained, because the protein that is lost may have been degraded and used for energy due to exercise. Protein synthesis decreases during exercise and this finding has been universally demonstrated. This decreased protein synthesis, together with increased protein degradation, clearly suggests that those who train heavily would experience an augmented protein loss and thus require a higher protein intake on a regular basis. Lemon *et al.* (1992) administered two levels of dietary protein in a group of novice body builders who underwent a month of resistance training. They found that the majority of those who were on the lower protein intake ($0.99 \, \mathrm{g\,kg^{-1}}$ per day) experienced a negative nitrogen balance, whereas all of those who were on the high protein intake ($2.62 \, \mathrm{g\,kg^{-1}}$ per day) achieved a positive nitrogen balance. It was their further calculation that nitrogen balance occurs at $1.43 \, \mathrm{g\,kg^{-1}}$ per day. The RDA for protein is $0.8 \, \mathrm{g\,kg^{-1}}$ per day for a healthy adult. However, in light of augmented protein catabolism associated with heavy exercise, it is suggested that those undertaking endurance training should consume protein at $1.2–1.4 \, \mathrm{g\,kg^{-1}}$ per day and those who resistance train may benefit from consuming $\sim 1.6 \, \mathrm{g\,kg^{-1}}$ per day (Fielding and Parkington 2002).

Energy metabolism of amino acids

There are three principal sources of amino acids for energy metabolism: (1) dietary protein; (2) the free amino acid pool; and (3) **endogenous** tissue protein. Dietary protein is a relatively minor source of amino acids because it is not a common practice to consume a large protein meal prior to exercise. The free amino acid pool that exists in muscle and blood is also very small compared with amino acids derived from degradation of tissue protein. It has been estimated that the intra-muscular amino acid pool constitutes less than 1% of the metabolically active amino acids. Consequently, the most important source of amino acids comes from endogenous protein breakdown (Dohm 1986).

The catabolism of amino acids requires the removal of the amino group (the nitrogen-containing portion) by transamination or oxidative deamination. Transamination is a common route for exchange of nitrogen in most tissues, including muscle, and involves the transfer of an amine from an amino acid to another molecule. A typical example is transfer of the amine from glutamate to pyruvate to produce alanine, which can then be utilized to produce glucose in the liver via a process called the alanine cycle, as shown in Figure 7.2. The process of deamination occurs in the liver and is responsible for converting the nitrogen residue into the waste product urea, which can be excreted from the kidneys.

The remaining carbon skeleton can then be converted to various intermediates of the Krebs cycle, which is common to both carbohydrate and fat metabolism. As shown in Figure 7.4, amino acids can give rise to pyruvate, acetyl-CoA, and Krebs cycle intermediates, such as oxaloacetate, fumarate, succinyl–CoA, α-ketoglurate, all of which can be oxidized via the Krebs cycle. Another way by which amino acids contribute to energy metabolism is to be converted to glucose via gluconeogenesis; this glucose is then used for generating energy or preventing hypoglycemia. This process has been discussed above in the context of deamination and transamination. As shown in Figure 7.2, alanine is first produced from pyruvate via transamination in active skeletal muscle and then travels to the liver via circulation. Upon entry to the liver, alanine becomes pyruvate via deamination. Gluconeogenesis then converts the remaining carbon skeleton of alanine into glucose, which then enters the blood for use by active muscle. This gluconeogenic process helps in maintaining blood glucose homeostasis during fasting and starvation. It also assists in prolonged exercise as additional energy fuel. It is estimated that the alanine–glucose cycle can generate up to 15% of the total energy requirement during prolonged exercise (Paul 1989).

Metabolic role of branched-chain amino acids

Leucine, isoleucine, and valine are the three branched-chain amino acids (BCAA) that have drawn a great deal of attention in terms of their role in bioenergetics. They are essential amino acids that cannot be synthesized in the body. Thus, they must be replenished via the diet. BCAAs are unique in that they are catabolized mainly in the skeletal muscle. Like other amino acids, the first step in the metabolism of BCAA is the removal of the amino group so that the remaining carbon skeleton can be further utilized. This transamination results in the production of glutamate, which can then

Figure 7.4 Major metabolic pathways for various amino acids following removal of a nitrogen group by transamination or deamination.

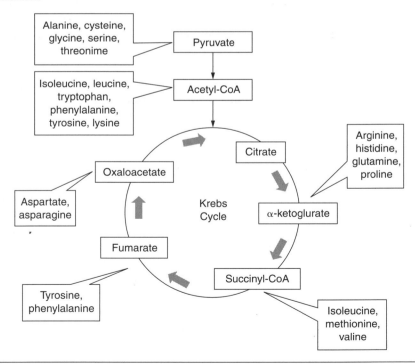

donate a nitrogen-containing portion to pyruvate to form alanine. Supplementation of BCAA has been claimed to enhance exercise performance in a variety of ways. They can: (1) serve as additional energy fuel; (2) prevent or attenuate the excessive loss of protein; and (3) help in improving function of neurotransmitters, thereby reducing the feeling of fatigue.

Recently, Eric Newsholme, a biochemist at Oxford University, proposed the central fatigue hypothesis. He postulated that high levels of serum free-tryptophan (fTRP) in conjunction with low levels of BCAA, or a high fTPR : BCAA ratio, may be a major factor that causes fatigue during prolonged endurance exercise. This contention was developed based upon the fact that fTRP is used for the production of serotonin, which is believed to play a key role in the onset of fatigue. BCAA, on the other hand, can compete against fTRP for the carrier-mediated entry in the central nervous system, thus mitigating the production of serotonin. The central fatigue hypothesis also predicts that ingestion of BCAA will raise the plasma BCAA concentration and thus reduce transport of fTRP into the brain. The subsequent reduced formation of serotonin may alleviate sensations of fatigue, thereby improving endurance performance. If this central fatigue hypothesis is true, then the opposite must also be correct; that is, consumption of tryptophan before exercise should reduce the time to exhaustion, thereby hampering performance. Nevertheless, a study by Stensrud *et al.* (1992) demonstrated no differences in exhaustive running performance between those who were on tryptophan and those who were on placebo. It appears that further research is still needed in order to substantiate the claims associated with BCAA.

REGULATION OF SUBSTRATE METABOLISM DURING EXERCISE

In response to an increase in exercise intensity, utilization of muscle glycogen increases. As a bout of exercise lasts for a prolonged period of time, there will be a gradual shift toward an increased use of liver glycogen, a change that is of necessity in terms of maintaining energy supply and blood glucose homeostasis. As carbohydrate energy stores diminish, both degradation and oxidation of triglycerides increase to ensure energy homeostasis. Such a pattern of substrate utilization has been well outlined in the previous sections. How do these changes come about in terms of an increase in utilization and shift between different pools of energy substrates? Chapter 6 covered various metabolic pathways through which carbohydrate and fat are degraded for yielding energy necessary to match increased energy demand imposed by exercise. It was also noted in Chapter 6 that the speeds at which these pathways operate are regulated directly by a series of enzymes that catalyze reactions, and the extent to which these enzymes are activated is in turn controlled by release of certain catabolic hormones. The following sections are devoted to illustrating more specifically the factors thought to modulate fuel utilization during exercise.

Regulation of muscle glycogenolysis

Muscle glycogen is the primary carbohydrate fuel for most types of exercise; the greater the exercise intensity, the faster glycogen is degraded. Interestingly, the increase in plasma epinephrine, a hormone that facilitates glycogenolysis, has been found to correlate positively with exercise intensity. It is generally considered that epinephrine plays the most important role in mediating glycogen degradation. Such a catabolic effect is believed to be initiated by second messengers, which activate protein kinases needed for glycogenolysis. Plasma epinephrine is responsible for the formation of cyclic AMP when bound with β-adrenergic receptors (Hargreaves 2006).

Muscle glycogenolysis is also regulated by the activity of **glycogen phosphorylase**, the key enzyme for glycogenolysis. Glycogen phosphorylase catalyzes the first step of glycogen breakdown and is responsible for supplying individual glucose molecules to the glycolytic pathway for producing ATP. In the resting state, this enzyme exists primarily in the inactive b form, the activity of which can be stimulated by an increase in ADP and AMP or inhibited by an increase in ATP. In response to muscle contraction or stimulation by epinephrine, the phosphorylase b inactive form is converted to the phosphorylase a active form. However, under the influence of insulin, activated phosphorylase can become deactivated. This will then reduce the availability of glucose.

Substrate availability will also affect the rate of glycogen degradation. Early studies have shown that increases in pre-exercise muscle glycogen result in enhanced muscle glycogen utilization during exercise. It was generally believed that glycogen can bind to glycogen phosphorylase and, in doing so, increase its activity. There is also evidence that alterations in the availability of blood-borne substrates such as glucose may also influence muscle glycogenolysis. Coyle *et al.* (1991) found that an increase in blood glucose as a result of intravenous infusion of glucose resulted in a decrease in muscle glycogen utilization. However, when blood glucose was brought down, no alteration in glycogen utilization was observed.

Regulation of glycolysis and the Krebs cycle

Glycolysis and the aerobic pathway containing the Krebs cycle are the two possible pathways by which glucose is further metabolized with or without oxygen. Each process consists of a series of sequential chemical reactions, and each reaction is catalyzed by a specific enzyme. The rate of glycolysis is controlled by the activity of **phosphofructokinase** (PFK). PFK catalyzes the third step of glycolysis. When exercise begins, increases in ADP and Pi levels activate PFK, thereby accelerating glycolysis. PFK is also activated by an increase in cellular levels of hydrogen ions and ammonia. The Krebs cycle, like glycolysis, is also subject to enzymatic regulation. Among numerous enzymes involved, **isocitrate dehydrogenase** (IDH) is thought to be the rate-limiting enzyme in the aerobic pathway. This enzyme catalyzes a reaction during the early phase of the Krebs cycle. Similar to PFK, the enzyme is stimulated by ADP and Pi and inhibited by ATP. IDH is also sensitive to the change in cellular levels of calcium. McCormack and Denton (1994) have found that an increase in Ca^{++} levels in mitochondria stimulates IDH. This finding is congruent with the concept that an increase in Ca^{++} in muscle is essential to initiate muscle contraction that requires energy. It is intriguing to note that the two aforementioned rate-limiting enzymes (PFK and IDH) are found early in the metabolic pathway. This is an important phenomenon because unwanted products of the pathway could accumulate if these enzymes were located near the end of the pathway.

Effect of fatty acid availability on carbohydrate utilization

In light of the preceding discussion, it can be concluded that carbohydrate utilization is regulated by the energy needs of the exercising muscle. Such a linkage between carbohydrate degradation and substrate availability may also be explained by the fact that an increase in fatty acids reduces carbohydrate utilization. By infusing triacylglycerol (Intralipid), Costill *et al.* (1977) found that muscle glycogen breakdown was reduced with elevated plasma fatty acids following 60 min of moderate-intensity exercise. A similar finding was also observed in a more recent study in which subjects were fed with fat in conjunction with heparin, which helps to facilitate lipolysis (Vukovich *et al.* 1993). This interaction between carbohydrate and fat may be explained by the classical glucose–fatty acids cycle discovered by Philip Randle. As shown in Figure 7.5, with an increase in plasma fatty acid concentration, there is an increase in fatty acid entry into the cell and subsequent increase in β-oxidation in which fatty acids are broken down to acetyl-CoA. An increased concentration of acetyl-CoA will then inhibit the **pyruvate dehydrogenase** (PDH) that breaks down pyruvate to acetyl-CoA. In addition, the increased production of acetyl-CoA also increases the

Figure 7.5 A schematic illustration of the glucose–fatty acid cycle (or Randle cycle).

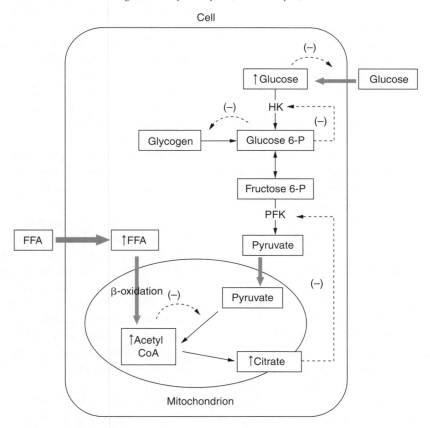

PFK = phosphofructokinase. HK = hexokinase.

concentration of citrate, an intermediate of the Krebs cycle. This increased citrate level will then inhibit PFK, a rate-limiting enzyme of glycolysis, as well as hexokinase, which regulates cellular glucose uptake. Together, these reduced enzymatic activities decrease carbohydrate utilization. An increase in fat utilization was also found to decrease cellular levels of AMP and Pi and, therefore, mitigate carbohydrate break-down (Dyck *et al.* 1993, 1996). As discussed earlier, increases in cellular AMP and Pi reflect a low-energy state and as a result will stimulate the activity of PFK in an attempt to maintain energy homeostasis.

Regulation of liver glycogenolysis

Glucose released from the liver plays an important role in maintaining blood glucose during exercise. As utilization of blood glucose increases in exercising muscle, there is a concurrent increase in glucose output from the liver so that symptoms associated with hypoglycemia are prevented. In principle, the regulation of hepatic glucose output during exercise is believed to be accomplished through a negative feedback mechanism. Jenkins *et al.* (1985, 1986) found that, via glucose infusion, an increase in blood glucose concentration inhibits hepatic glucose production during moderate-intensity exercise. Given that a change in blood glucose is linked to the production of insulin and glucagons, it appears that this feedback mechanism may be hormone-related. Indeed, both insulin and glucagon can influence the activity of the liver as the target organ. By manipulating insulin and glucagon levels with the infusion technique, studies have demonstrated favorable responses of hepatic glucose output (Wasserman *et al.* 1984, 1989). In other words, when plasma insulin was made to decrease or when plasma glucagon was made to increase, there was a resultant increase in hepatic glucose production. The greatest effect on hepatic glucose uptake was observed when there was a simultaneous decrease in glucagons and increase in insulin (Marker *et al.* 1991).

Hepatic glucose output can outpace the rate at which blood glucose is taken up by the tissue, thereby causing **hyperglycemia**. Indeed, an increase in plasma glucose was observed during intense exercise. As this mismatch between hepatic glucose production and peripheral glucose utilization is more pronounced when exercise intensity increases, Kjær (1995) suggested that the regulation of glucose production is in part mediated by activity in motor centers in the central nervous system in a feed-forward fashion; that is, the greater the exercise intensity, the greater the central command, and thus the greater hepatic glucose output. Glucose production from the liver may also be subject to control by autonomic/adrenergetic activities. In a study that used leg and combined arm-and-leg exercise, Kjær *et al.* (1991) observed a positive correlation between plasma catecholamines and hepatic glucose output. In addition, in an animal study in which the adrenal medulla was removed, Ritchter *et al.* (1981) and Sonne *et al.* (1985) found a reduced liver glycogenolysis and hepatic glucose output. These findings suggest that epinephrine and nor-epinephrine also play a role in the exercise-induced increase in glucose output from the liver.

Exercise-induced hyperglycemia is a transit phenomenon. Upon completion of exercise, blood glucose will be restored to a normal level quickly, although this may not be the case for diabetics, who have diminished insulin secretion.

Regulation of lipolysis

In order to be metabolized, stored fat must first undergo lipolysis in which triglycerides are degraded to fatty acids and glycerol. The resulting fatty acids are then converted into acetyl-CoA and enter the Krebs cycle for further metabolism. Adipose tissue lipolysis is controlled by the hormone-sensitive enzyme, lipase, which breaks the bonding of triglycerides so that fatty acids and glycerol are formed. This enzyme is regulated by hormones of insulin, glucagon, catecholamines, cortisol, and growth hormone, which is released from the pituitary gland. Of these hormones, insulin is the only one that inhibits lipolysis, whereas all the others function as stimulators. Of those stimulating hormones, catecholamines are the most potent stimulators of lipolysis during exercise. Catecholamines are bound with β-adrenergic receptors, causing production of a second messenger cyclic AMP. This later product then triggers a series of chemical events through which the activity of hormone-sensitive lipase becomes activated. Insulin can reverse the effects of lipolytic hormones. The mechanism underlying the anti-lipolytic action of insulin remains uncertain. It is thought that insulin suppresses lipolysis by either decreasing cyclic AMP concentration or inhibiting enzymes needed to activate hormone-sensitive lipase.

Just as fatty acid availability affects carbohydrate utilization, plasma glucose concentration can also influence adipose tissue lipolysis. Wolfe *et al.* (1987) found that in healthy subjects, glucose infusion suppressed the rate of appearance of glycerol, a common measure of lipolysis. From this study, however, it is difficult to discern whether decreased lipolysis is due to an increase in plasma glucose or a resultant increase in plasma insulin. In a later study in which insulin concentration was kept unchanged, Carlson *et al.* (1991) showed that appearance of glycerol decreases by 32% due to hyperglycemia, in which plasma glucose was raised up to ~10 mM. This finding suggests that plasma glucose regulates fat mobilization independently of changes in plasma hormones.

Regulation of lipid oxidation

Regulation of fat utilization requires fatty acids to be transported into the mitochondria, where fat can be oxidized. In this context, the mass of mitochondria and oxygen delivery are very important factors in determining rates of fat utilization. Tissues such as the heart and liver are highly adapted for fat utilization, whereas brain cells and red blood cells rely almost exclusively on glycolysis for energy. In skeletal muscle, the ability to use fat as an energy source varies depending on the muscle fiber type. Fast-twitch muscle fiber is limited in its use of fat due to its low volume of mitochondria, as well as a less than optimal blood supply. In contrast, slow-twitch muscle fiber is highly capable of oxidizing fat as it is rich with mitochondria and capillaries.

The process by which fatty acids are transported across the mitochondrial membrane is mainly controlled by the activity of carnitine palmitoyl transferase (CPT). As discussed earlier, CPT is part of the transport system needed for long-chain fatty acids to enter the mitochondrion, where oxidation takes place. In this context, it makes sense that CPT plays an important role in the control of fat oxidation. CPT is regulated by **malonyl-CoA**, an intermediate in fatty acid synthesis. An increase in malonyl-CoA content inhibits the activity of CPT, thereby reducing fat utilization. During high-intensity exercise, the high rate of glycogenolysis increases the amount of acetyl-CoA in the muscle cell, and some of this acetyl-CoA is converted to malonyl-CoA. This increased malonyl-CoA then suppresses CPT and thus reduces the transport of fatty acids into the mitochondria. Conversely, as carbohydrate energy sources deplete, the inhibitive effect from malonyl-CoA upon CPT attenuates due to reduced glycogenolysis. Consequently, the activity of CPT is augmented and fat utilization is enhanced. Besides malonyl-CoA, Starritt *et al.* (2000) suggested that a reduction in pH (or an increase in acidity) associated with high-intensity exercise would also serve as an inhibitor to the activity of CPT.

Carbohydrate and fat together represent most, if not all, of the energy expenditure in most situations. However, compared with carbohydrate in terms of regulation, fat utilization is not as tightly controlled. For example, as mentioned earlier, the rate of carbohydrate utilization is closely related to the changes in markers such as ADP and AMP that reflect the energy needs of the exercising muscle. This is not the case with respect to fat utilization. The rate of fat oxidation appears to be influenced mainly by the oxidative capacity of the tissue, the availability of fatty acids, and the rate of carbohydrate utilization.

Regulation of protein synthesis and degradation

Protein synthesis and degradation are mainly regulated by hormonal secretion, which can be influenced by many circumstances such as exercise, stress, and diet. Glucagon, cortisol, and catecholamines are found to be mainly associated with protein degradation, whereas insulin, growth hormones, and testosterone are primarily linked to protein synthesis. Thus far, cortisol has been considered the most potent stimulator of protein catabolism or degradation (Graham and Maclean 1992; Rooyackers and Nair 1997). It degrades tissue protein to yield amino acids for glucose synthesis in the liver via gluconeogenesis. As this action helps to generate new glucose units, its role during exercise has been thoroughly investigated. Cortisol secretion does not increase much during early phases of exercise even when exercise is performed at a strenuous level (e.g., >75% VO_2max). However, as a strenuous exercise persists, the blood cortisol level begins to increase. An increase in the blood cortisol level has been reported to occur at 30 min or longer into exercise. This rise in cortisol appears to occur just in time, because muscle glycogen is likely to reduce significantly during strenuous exercise that lasts for more than 30 min. When muscle glycogen is low, an increase in cortisol secretion will assist in maintaining a continuous energy supply throughout the entire period of exercise.

Insulin, growth hormone, and testosterone appear to be the most influential hormones mediating protein synthesis. Insulin is known for its role in regulating plasma glucose homeostasis in response to hyperglycemia. However, in addition to its affect on stimulating blood glucose uptake by peripheral tissues, insulin also promotes the entry of circulating amino acids into certain cells such as skeletal muscle fibers where protein synthesis takes place. As such, insulin is also regarded as an anabolic hormone in terms of protein synthesis. Insulin release can be blunted during exercise when intensity and duration surpass certain thresholds. This is an appropriate response in that a decrease in insulin favors the mobilization of glucose from the liver and fatty acids from adipose tissue, both of which are necessary to maintain the plasma glucose concentration. If exercise were associated with an increase in insulin, blood glucose would be taken up into tissues at a faster rate, leading to immediate hypoglycemia.

The effect of growth hormone on protein synthesis is carried out by increasing the membrane transport of amino acids into cells, synthesis of RNA and ribosomes, activity of ribosomes, and all other events essential to protein synthesis. More recent literature also reveals a linkage between growth hormone administration and a diminished amino acid oxidation (Rooyackers and Nair 1997). Growth hormone can act indirectly via enhanced hepatic release of **somatomedins**, which are carried by the blood to target tissues where they induce growth-promoting effects, particularly in cartilage and bone. Because the somatomedins are structurally and functionally similar to insulin, they are referred to as **insulin-like growth factors**. Particularly intriguing is that the blood level of growth hormone increases during vigorous exercise and remains elevated for a period of time after exercise. This elevation in growth hormone during exercise has been found to function similarly to cortisol, serving as a lipolytic hormone to maintain blood glucose homeostasis. Growth hormone stimulates fat breakdown and indirectly suppresses carbohydrate utilization. A low plasma glucose concentration can serve to stimulate the release of growth hormone by the anterior pituitary gland.

Testosterone and testosterone analogs such as anabolic steroids are well known for their anabolic effect on protein metabolism. However, the mechanism of how this hormone regulates protein metabolism still remains elusive. The fact that testosterone enhances protein synthesis is well documented. It remains debated as to whether the increased protein synthesis is brought about by an increase in cellular amino acid uptake. It has been argued that testosterone increases protein synthesis by mainly enhancing the utilization of amino acids from the intracellular pool. Thus far, little is known about the effect of estrogen and progesterone on protein metabolism, although it is evident that menopause in women is associated with accelerated muscle loss.

SUMMARY

- Carbohydrate and fat are the two primary sources of energy. Compared to fat, carbohydrate provides energy more quickly, can be used regardless of whether there is oxygen, and serves as the sole source of energy for the central nervous system. It must also be available in order for the body to use fat. As such, carbohydrate is the main source of fuel for most sporting events.
- Muscle glycogen serves as an initial source of energy upon the start of a strenuous exercise. As exercise continues, degradation of liver glycogen will increase its contribution by providing additional glucose for use by muscle and to prevent hypoglycemia. The liver is also capable of manufacturing new glucose in an effort to maintain glucose homeostasis.
- Preparing carbohydrate and fat molecules for final entry into a metabolic pathway is an important step of energy metabolism. In order to be oxidized, both glucose and fatty acid need to be converted into acetyl-CoA. This is accomplished through glycolysis for glucose and β-oxidation for fatty acid. Both glycolysis and β-oxidation can be viewed as being similar in that these pathways function to ultimately produce such "common" molecules of acetyl-CoA.

- Oxidizing fat depends upon the level of exercise intensity. Contrary to common belief, the maximal fat oxidation rate occurs at moderate rather than low intensity. This is because fat oxidation is also the function of absolute caloric expenditure. Intensity near one's lactate threshold or around 60–65% VO_2max will elicit maximal fat oxidation.

- Protein does not normally participate in energy metabolism and therefore there is no such storage form of protein used for energy as there is for carbohydrate or fat. However, under the condition where there is a significant decrease in bodily carbohydrate, protein can be used as a fuel. Protein contributes to energy provision by first being degraded into amino acids; amino acids will then be converted into glucose or various intermediates of the Krebs cycle in order to fulfill their energetic role.

- Leucine, isoleucine, and valine are the three branched-chain amino acids, and their potential roles include: (1) serving as additional energy fuel; (2) preventing or attenuating the excessive loss of protein; and (3) helping in improving the function of neurotransmitters, thereby reducing the feeling of fatigue.

- A key enzyme that regulates muscle glycogen degradation is glycogen phosphorylase. This enzyme is further influenced by the level of ATP, epinephrine concentration, and glycogen stores. Increases in ADP and AMP levels, epinephrine release, and muscle glycogen concentration have been found to stimulate glycogen phosphorylase and therefore also glycogenolysis.

- Phosphofructokinase (PFK) and isocitrate dehydrogenase (IDH) are the rate-limiting enzymes that control glycolysis and the Krebs cycle, respectively. Rate-limiting enzymes are ones found earlier in the metabolic pathway and are sensitive to the level of energy substrate available.

- An increase in fatty acid availability can reduce the dependence upon carbohydrate energy sources. The inhibitive effect of fat upon carbohydrate utilization can be explained by the Randle cycle proposed more than 50 years ago. In brief, the theory suggests that an increase in acetyl-CoA derived from fat degradation will suppress a number of enzymes necessary for converting glucose into energy.

- The primary function of glucose output from the liver is to maintain normal blood glucose concentration, although some of this can also be taken by muscle tissue for use as energy. The hepatic glucose output is well regulated by insulin and glucagon, so there will not be a mismatch between glucose output and utilization under normal circumstances. However, during high-intensity exercise hepatic glucose output can outpace utilization, thereby causing hyperglycemia.

- Utilization of fat is not as tightly controlled as that of carbohydrate. While carbohydrate utilization is closely regulated by changes in AMP and ADP, which reflect the energy needs of the cell, the rate of fat oxidation appears to be influenced mainly by the oxidative capacity of the tissue, the availability of fatty acids, and the rate of carbohydrate utilization.

- Glucagon, cortisol, and catecholamines are found to be mainly associated with protein degradation. Insulin, growth hormone, and testosterone are the most influential hormones mediating the anabolic process, although growth hormone can function catabolically during exercise. Both cortisol and growth hormone serve as a stimulus to lipolysis and gluconeogenesis during prolonged exercise when carbohydrate stores decrease significantly. Such catabolic action is considered desirable in that it helps prevent hypoglycemia and muscle glycogen depletion.

CASE STUDY: FUEL UTILIZATION DURING EXERCISE ■

Steve is a 49-year-old man who is in the process of training for his first marathon to celebrate his fiftieth birthday. He had some previous recreational running experience and had done several 10-km races and two half-marathons. At 5' 7" (170 cm) and 184 lb (84 kg), he realizes that losing weight and body fat will help him accomplish his goal of running a marathon. He participated in an indirect calorimetry study at a nearby university. In this study, he underwent a series of metabolic tests at rest and while running on a treadmill to learn more about his energy expenditure and fuel utilization at different running paces. The following are his results from the study.

TABLE B

RUNNING PACE (MILE/HR)	HEART RATE (BPM)	RESPIRATORY EXCHANGE RATIO	PERCENTAGE ENERGY FROM FAT	PERCENTAGE ENERGY FROM CHO	TOTAL ENERGY OUTPUT (KCAL/MIN)	FAT USE (KCAL/MIN)	CHO (KCAL/MIN)
Rest	70	0.77	77.2	22.8	1.5	1.2	0.3
6.0	130	0.87	42.5	57.5	11.8	5.0	6.8
6.5	139	0.89	35.8	64.2	13.5	4.8	8.7
7.0	145	0.91	29.2	70.8	14.4	4.2	10.2
7.5	155	0.93	22.6	77.4	15.3	3.5	11.8
8.0	166	0.95	16.0	84.0	16.4	2.6	13.8

Questions

- How did HR and RER change as running pace increased? Do these changes make sense? Why?
- The data shows that as Steve moved from rest to running at 6 miles/hr, the percentage of energy from fat decreased, while kilocalories per min from fat use increased. Why is there such a divergent response?
- If Steve was able to maintain a running pace of 7.5 mph throughout the entire marathon, according to this data how much carbohydrate would he have used both in terms of kilocalories per minute and grams per minute?

■ REVIEW QUESTIONS

1. Why is carbohydrate often referred to as the most preferable source of energy?
2. How is the use of energy fuels influenced by exercise intensity and duration?
3. In what circumstance will protein be used as an energy fuel?
4. Define the term "gluconeogenesis." How is this process related to the Cori and glucose–alanine cycles?
5. What is β-oxidation? How does this process differ from lipolysis?
6. Describe how protein enters energy metabolism. Why are branched-chain amino acids considered ergogenic and used widely in sports?

7 Briefly describe the theory of the glucose–fatty acid (or Randle) cycle.

8 Epinephrine, insulin, glucagon, cortisol, and growth hormone are the major hormones involved in energy metabolism. What specific role does each hormone play with regard to energy metabolism during exercise?

9 Blood glucose concentration is generally maintained throughout exercise. How does this come about? When can hypoglycemia and hyperglycemia occur?

10 How would you prescribe exercise intensity and duration that will help maximize energy expenditure and fat utilization?

11 What would be the major energy fuel used under each of the following conditions:
   • after meal?
   • between meals?
   • prolonged starvation?
   • exercise at low intensity?
   • exercise at high intensity?
   • intense exercise for prolonged duration?

12 What would be the major energy system used during each of the following events:
   • 100-m sprint run?
   • 100-m swimming?
   • 800-m run?
   • 10-km run?
   • marathon?
   • wrestling?
   • soccer game?

SUGGESTED READING

Achten, J. and Jeukendrup, A.E. (2004) Optimizing fat oxidation through exercise and diet. *Nutrition*, 20: 716–727.

Interventions aimed at increasing fat metabolism could potentially reduce the symptoms of metabolic diseases such as obesity and type-2 diabetes and may have tremendous clinical relevance. This article is written to help readers understand various factors, including those associated with exercise and diets, that increase or decrease fat oxidation.

Hargreaves, M.H. and Snow, R. (2001) Amino acids and endurance exercise. *International Journal of Sport Nutrition and Exercise Metabolism*, 11: 133–145.

Protein degradation during exercise is an area that is under-addressed and subject to much debate. This article provides a comprehensive review of how amino acids are degraded in order to generate energy from a source other than carbohydrate and fat. Ergogenic properties of some amino acids are also discussed.

Holloszy, J.O., Kohrt, W.M. and Hansen, P.A. (1998) The regulation of carbohydrate and fat metabolism during and after exercise. *Frontiers in Bioscience*, 3: D1011–1027.

This classic review complements the textbook in that it provides more detailed and evidence-based information that can help understand how carbohydrate and fat are metabolized and how these metabolic processes are regulated during and after exercise.

8 Guidelines for designing a healthy and competitive diet

▮ KEY TERMS

- dietary reference intakes
- estimated average requirements
- recommended dietary allowances
- adequate intake levels
- tolerable upper intake levels
- estimated energy requirement
- acceptable macronutrient distribution
- Dietary Guidelines for Americans
- adequacy
- balance

- nutrient density
- moderation
- variety
- discretionary calories
- daily value
- nutrient content claims
- health claims
- female athlete triad
- glycogen supercompensation

▮ NUTRITION FOR HEALTH AND PERFORMANCE

Food is simply a vehicle for nutrients, which provide the building blocks and energy for all of the body's structures and functions. An optimal diet supplies required nutrients in adequate amounts for tissue maintenance, repair, and growth without excess energy intake. Poor nutrition can lead to suboptimal physiological function and poor health. Consuming too little of a nutrient can cause a nutritional deficiency, which can be serious and sometimes fatal. Conversely, consuming too much of some nutrients can also be unhealthy. For example, under-consumption of dietary fiber in conjunction with over-consumption of simple sugar has been linked to many chronic conditions such as obesity, diabetes, and heart disease. Under- and over-nutrition make up the two opposite ends of what is called the nutritional status continuum; both are examples of malnutrition.

Health correlates of dietary carbohydrate

There is evidence that a high sugar intake can adversely affect blood lipid levels, thereby increasing the risk of heart disease. However, diets high in whole grains and fibers may reduce blood cholesterol levels and thus protect against heart diseases and strokes (Kushi *et al.* 1999; Trumbo *et al.* 2002). Studies indicate that soluble fibers from foods such as legumes, oats, pectin, and flax seed are particularly effective in lowering blood cholesterol level. The cholesterol-lowering effect of increased dietary fiber has been attributed to the ability of soluble fibers to bind cholesterol and bile acids, which are made from cholesterol, in the digestive tract. When bound to fiber, cholesterol and bile acids are excreted in feces rather than being absorbed and used. The liver uses cholesterol from the blood to produce new bile acids. It has been found that the bacterial by-products of fiber fermentation in the colon also inhibit cholesterol synthesis in the liver.

High-fiber foods play a key role in reducing the risk of type-2 diabetes (Fung *et al.* 2002). As discussed in Chapter 2, a diet high in refined starches and added sugars causes greater glycemic responses and therefore increases the amount of insulin needed to maintain normal blood glucose levels. Ample evidence exists that long-term consumption of high fibers and low sugars decreases the risk of developing type-2 diabetes (Liu *et al.* 2000; Meyer *et al.* 2000). It is believed that when viscous fibers trap nutrients and delay their digestion, glucose absorption is slowed, and this helps prevent the glucose surge and rebound. So although a diet high in simple carbohydrates and refined starches does not cause diabetes, it does increase the demand for insulin required to maintain normal blood glucose levels and may increase the risk of developing diabetes later on.

To many weight-loss enthusiasts, carbohydrates are viewed as the "fattening" nutrient. Indeed, studies comparing weight loss associated with low- vs. high-carbohydrate diets show that a greater weight loss is achieved on the lower carbohydrate diet at end of the sixth month (Foster *et al.* 2003; Brehm *et al.* 2003). However, it should be emphasized that the weight loss associated with low-carbohydrate diets is caused by a reduced caloric intake, rather than by alterations in macronutrient composition of the diet. In other words, limited food choice and increased satiety associated with high-protein, high-fat foods may have caused people to eat less and therefore lose weight. There is no evidence that carbohydrate restriction causes the body to burn fat more efficiently. Although low-carbohydrate diets appear effective in the short term, little, if anything, is known about their long-term effects (Astrup *et al.* 2004). Carbohydrate is no more fattening than any other energy source, and gram for gram it contains less than half of the calories provided by fat. In fact, diets that are low in fat and protein and high in carbohydrate have long been considered effective in terms of weight loss and weight maintenance. However, one should strive to maintain an adequate amount of the total energy intake and to minimize the consumption of simple sugar while maximizing the consumption of unrefined and complex carbohydrates that are rich in fiber. Foods rich in fiber tend to be low in fat and added sugars and can therefore promote weight loss by delivering less energy per bite. In addition, as fiber absorbs water from digestive juices, it swells, creating feeling of fullness and delaying hunger.

Carbohydrate and athletic performance

Adequate bodily carbohydrate reserves are required for optimal athletic performance. As the most efficient fuel for exercising muscles, carbohydrates are the primary source of energy during high-intensity activities. Extensive research confirms the major role carbohydrate plays in endurance (aerobic) exercise as well as strength and power events. A particular challenge, however, is that unlike protein and fat, the body has limited carbohydrate reserves. Dietary carbohydrates are stored in the body as glycogen, primarily in the muscle and liver. During activity, the body relies on this stored glycogen to be released and used by the muscles and brain for energy. The body's limited glycogen stores can be depleted in a single bout of exercise of sufficient intensity and duration. Thus, daily carbohydrate intake is necessary to maintain these glycogen stores. If muscle and liver glycogen stores become depleted during exercise, the muscles will be left without fuel and fatigue will set in – a condition known as "hitting the wall."

There are other important reasons why athletes or physically active individuals should emphasize adequate carbohydrate consumption. Carbohydrates are the major fuel source for the brain and nervous system. If blood glucose and glycogen levels are low, athletes may feel irritable and tired, and may lack concentration, which could interfere with even simple performance-related tasks. Carbohydrates also aid in fat metabolism. The body requires the presence of carbohydrate in order to utilize fat for energy. Carbohydrates provide a "protein-sparing effect," helping athletes to maintain the muscle mass they worked so hard to develop. As previously mentioned, the brain requires a constant and significant amount of carbohydrate. When glycogen stores become depleted and dietary carbohydrates are not consumed, the body will turn to protein (from muscle tissue) to "make" carbohydrate in a process known as gluconeogenesis. By consuming a diet with adequate carbohydrates and calories, the body will be less likely to have to make carbohydrate at the expense of muscle tissue.

For many athletes during intense training, multiple stressors can impair immunity. These stressors include lack of sleep, mental stress, poor nutrition, weight loss, and inflammation from exercise. Of these stressors, inadequate carbohydrate can contribute to decreased immunity and increased possibility of getting sick. Lancaster *et al.* (2005) found that taking 30–60 g of carbohydrate per hour during 2.5 hours of high-intensity cycling prevented the decline of interferon-y, an important virus-fighting substance. Other researchers found improved levels of various antibodies when carbohydrate was taken during exercise. An adequate intake of carbohydrate can also reduce the release of the hormone cortisol and free up some key amino acids to help with immune function.

Like many other people, athletes can fall prey to the latest diet and/or nutrition fad in their efforts to gain a competitive advantage. Today, that fad is the low-carbohydrate diet. Unfortunately, a low-carbohydrate diet is just the opposite of what the athlete's body needs for optimal performance, because carbohydrates are the primary fuel for the exercising muscles and therefore absolutely essential for supporting an athlete's training and performance. Thus, the well-balanced performance diet is one that provides sufficient energy, mostly in the form of carbohydrates, with the balance of energy as proteins and fats.

Dietary fat: health implications

Lipids contribute to the texture, flavor, and aroma of our food. It is the high fat content of ice cream that gives it its smooth texture and rich taste. Olive oil imparts a unique taste to salad, and sesame oil gives Chinese food its distinctive aroma. In some cases, "fat-free" also means tasteless. The fats in foods contribute to their appeal, but also increase their caloric content.

Adequate amounts of essential fatty acids are required in the diet to maintain normal body function. However, diets high in fat, particularly some types of fats, are associated with an increased risk for many chronic diseases. The development of cardiovascular disease has been linked to diets high in cholesterol, saturated fat, and trans fat (Krauss *et al.* 1996; Shikany and White 2000). The risk of certain types of cancer, including breast, colon, and prostate, has been associated with a high fat intake. Obesity is also associated with diets high in fat because these diets are usually high in energy and promote storage of body fat. Excess body fat is associated with an increased risk of diabetes, cardiovascular disease, and high blood pressure.

Many dietary lipids have been implicated in influencing risk for cardiovascular disease, including total lipids, type of lipids, specific fatty acids, and cholesterol. Populations that consume a diet high in saturated fatty acids, trans fatty acids, and cholesterol have a high incidence of heart disease. Populations that consume a diet high in unsaturated fatty acids, especially omega-3, have a low incidence of heart disease. For example, in Mediterranean countries, where the diet is high in monounsaturated fat such as olive oil as well as grains, fruits, and vegetables, the mortality rate from heart disease is only half of that in the United States, according to American Heart Association statistics.

When saturated fat in the diet is replaced by any type of polyunsaturated fat, there is a beneficial decrease in LDL cholesterol, which is often regarded as "bad" cholesterol that promotes plaque build-up in coronary arteries. One such example of polyunsaturated fat is omega-3 fatty acids. Regular consumption of omega-3 fatty acids has been found to reduce LDL cholesterol levels, while possibly increasing HDL cholesterol level (Katan *et al.* 1995; Stone 1997; Connor *et al.* 1997). It has been considered that replacing some of the fat in the diet with omega-3 fatty acids reduces the incidence of health disease. This is because omega-3 fatty acids may reduce heart disease risk by preventing the growth of atherosclerotic plaque and by affecting blood clotting, blood pressure, and immune function. The beneficial effects are greater when the omega-3 fatty acids are consumed from seafood, such as salmon and tuna, rather than supplements. It is recommended by the American Heart Association that one should consume two 3-oz servings of fish per week in order to protect against cardiovascular diseases.

Obesity is a medical condition in which excess body fat has accumulated to the extent that it may have an adverse effect on health, leading to reduced life expectancy and/or increased health problems. Although the etiology of obesity is complex, nutrient intake is a major contributor. As fatty acids provide more than twice as many calories per gram as carbohydrate and protein, fat intake is likely an important piece of the obesity puzzle. Regardless of cause, obesity is a major public health concern worldwide and is associated with increased risk for many diseases such as cardiovascular disease, type-2 diabetes, and some forms of cancer. It has been recommended that to reduce the risk for obesity we limit our fat consumption. In response to the obesity epidemic and consumer demand for reducing the prevalence of obesity, many food manufacturers produce low-fat and fat-free products, as well as foods that contain fat substitutes. These alternative products replicate the taste, texture, and cooking properties of fat, but contribute less energy.

Cancer is the second leading cause of death in the United States, and it is estimated that 30–40% of those cancers are directly linked to dietary choices. As with cardiovascular disease, there is a body of epidemiological evidence correlating diet and lifestyle with the incidence of cancer. For example, in populations where the diet is high in fat and low in fiber, the incidence of breast cancer is high. In populations where the typical fat intake is low, the incidence is lower and the survival rate of those with the disease is better. Epidemiology has also correlated the incidence of colon cancer with high-fat, low-fiber diets. The correlation is stronger for diets high in animal fat, especially those from red meats. The mechanism by which a high intake of dietary fat increases the incidence of various cancers is less well understood than the relationship between dietary fat and cardiovascular disease. However, dietary fat has been suggested to be both a tumor promoter and tumor initiator.

As is the case with heart disease, the type of fat is as important as the total amount of fat in determining cancer risk. For example, the incidence of breast cancer in Mediterranean women who rely on olive oil – which is high in monounsaturated fat – as a source of dietary fat is low despite a total fat intake similar to that seen in the United States. Epidemiological studies also support a protective effect from an increased intake of omega-3 fatty acids from fish, such as in the native Eskimos of Alaska and Greenland. On the other hand, a high intake of trans fatty acids found in foods such as stick margarines may increase the risk of breast cancer.

USING DIETARY REFERENCE STANDARDS FOR CONSTRUCTING A DIET

It is not enough to simply know how much energy and nutrients a person consumes; a complete assessment of one's nutritional status must go one step further and determine whether these amounts are likely to be adequate. For this purpose, the Institute of Medicine of the National Academy of Science has developed a set of nutritional standards to be used for assessing the adequacy of a person's diet. These standards are collectively known as the **dietary reference intakes** (DRIs). These standards were first published in 1943, when malnutrition in the United States was generally due to under-nutrition, and nutrition deficiencies were common. DRIs were established by highly qualified scientists and represent our best knowledge of recommended intakes for all the essential nutrients. Because nutrient requirements differ by sex, age, and life status, such as pregnancy and lactation, DRI values are stratified by each of these variables. Both the United States and Canada recognize the DRIs as their official set of dietary reference standards.

Dietary reference standards

The DRIs represent a set of four types of nutrient intake reference standards used to assess and plan dietary intake: (1) **estimated average requirements** (EARs); (2) **recommended dietary allowances** (RDAs); (3) **adequate intake levels** (AIs); and (4) **tolerable upper intake levels** (UIs). The DRIs also include calculations for **estimated energy requirement** (EERs), which can be used to assess whether one's energy intake is sufficient, and **acceptable macronutrient distribution** (AMDRs), which provide a recommended distribution of the macronutrients in terms of energy consumption. Note that the DRIs are only estimates of average nutrient requirements in a healthy population, so your level may be less or greater than the average. DRIs are provided in Appendix C. The following is a brief description of each of these specific types of nutrient intake reference standards.

Estimated average requirement

The EAR for a particular nutrient is the average daily amount that will maintain a specific biochemical or physiological function in half the healthy people of a given age and gender group. In other words, if a woman consumes the EAR value for a particular nutrient, she is consuming the amount that meets the requirement of about 50% of all women in the same age group. A look at enough individuals reveals that their requirements fall into a symmetrical and normal distribution, with most near the midpoint or the mean – as shown in Figure 8.1 – and only a few at the extremes.

The EARs are useful in research settings to evaluate whether a group of people is likely to consume adequate amounts of a nutrient. However, it may not be appropriate to use the EARs as recommended dietary intake goals for a specific individual. This is because even though the EARs are differentiated by age and gender, the exact requirements of people of the same age and gender are likely to vary.

Recommended dietary allowance

The RDA represents the average daily amount of a nutrient considered adequate to meet the known nutrient needs of nearly all healthy people of a given age and gender group. If people were to follow the EARs and consumed exactly the average requirement of a given nutrient, half of the population would develop deficiencies of the nutrient. Recommendations should be set high enough above the EAR to meet the needs of most healthy people. Small amounts above the daily requirement do no harm, whereas amounts below the requirement lead to health problems. Therefore, to ensure that the nutrient RDAs meet the needs of as many people as possible, the RDAs are set near the top end of the range of the population's estimated requirements.

As shown in Figure 8.1, the RDA is set near the right end of the curve. Such a point can be calculated mathematically so that it covers ~98% of a population. Almost everybody, including those whose needs are higher than the average, would be included if they met this dietary goal. In this context, RDAs have been used as nutrient intake goals for all individuals. They include a built-in safety margin to help ensure adequate nutrient intake in the population. Unless specifically noted, RDAs do not distinguish between whether the nutrient is found in foods, added to foods, or consumed in supplement form.

Adequate intake level

AI is the average daily amount of a nutrient that appears sufficient to maintain a specific criterion. It is a value used as a guide for nutrient intake when scientific evidence is insufficient to establish an EAR and thus to accurately set an RDA. In other words, the establishment of an AI instead of an RDA for a nutrient means that more research is needed. Similar to the RDAs, AIs are meant to be used as nutrient intake goals for

Figure 8.1 Comparison of estimated average requirements (EAR) and recommended dietary allowances (RDA).

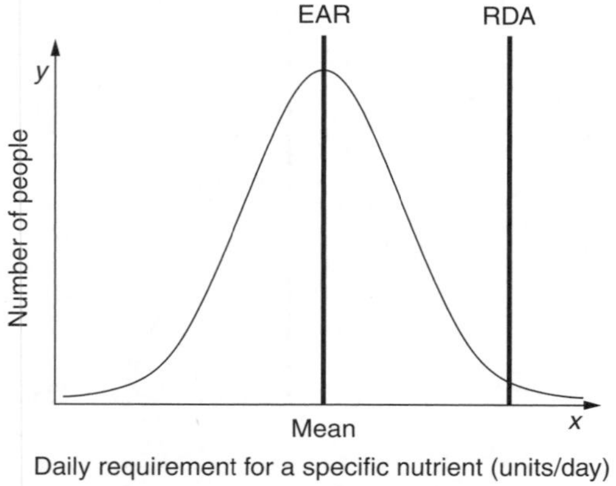

individuals. However, the differences between an AI and an RDA are noteworthy. An RDA for a given nutrient is supported by enough scientific evidence to expect that the needs of almost all healthy people will be met. An AI, on the other hand, is determined based primarily on scientific *judgments* because sufficient scientific evidence is lacking. The percentage of people covered by an AI is unknown; an AI is expected to exceed average requirements, but it may cover more or fewer people than an RDA would if an RDA could be determined. An example of a nutrient with an AI instead of an RDA is calcium. More research is required to be able to establish RDAs for this mineral. You can see which nutrients have AI rather than RDA values by examining the DRI tables in Appendix C.

Tolerable upper intake levels

The RDA and AI values have been established to prevent deficiencies and decrease the risk of chronic diseases. However, avoiding the other end of the nutritional status continuum – nutrient over-consumption or toxicity – is also important. As such, ULs have been established as the maximum daily amount of a nutrient that appears safe for most healthy people, and beyond which there is an increased risk of adverse health effects. ULs are not to be used as target intake levels or goals. Instead, they provide limits for those who take supplements or consume large amounts of fortified foods, because some nutrients are harmful at very high intakes. Note that scientific data are insufficient to provide UL values for all nutrients. The lack of ULs for a particular nutrient indicates the need for caution in consuming high intakes of that nutrient; it does not mean that high intakes pose no risk.

Estimated energy requirements

EERs represent the average energy intakes needed to maintain weight in a healthy person of a particular age, sex, weight, height, and physical activity level. EERs are similar to EARs in that they are set at the average of the population's estimated

Figure 8.2 Estimated energy requirement (EER). Note the similarity between EER and EAR shown in Figure 8.1.

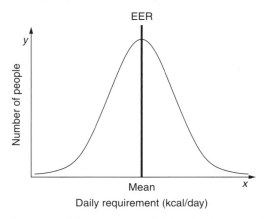

requirements (Figure 8.2). In contrast to the RDA and AI values for nutrients, the recommendation for energy is not generous. Balance is the key to the energy recommendation. Enough energy is needed to sustain a healthy and active life, but too much energy can result in weight gain and obesity. Because any amount in excess of need results in weight gain, there is neither RDAs nor ULs for energy.

The EER equations for adults of healthy weight are provided as follows; others can be found in Appendix D.

Adult man:

$$EER = 662 - [9.53 \times \textit{age y}] + PA \times [15.91 \times \textit{wt (kg)} + 539.6 \times \textit{Ht (m)}]$$

Adult woman:

$$EER = 354 - [6.91 \times \textit{age (y)}] + PA \times [9.36 \times \textit{wt (kg)} + 726 \times \textit{Ht (m)}]$$

In this equation, PA refers to physical activity level, which is categorized as sedentary, low active, active, or very active. Table 8.1 shows examples of these activity categories and their corresponding values.

The following is a sample calculation for determining an EER:

John Doe is a 35-year-old man. He weighs 154 lb, is 5'9" tall, and has a low activity level, which is equal to 1.12, according to Table 8.1. His EER is calculated as follows:

Age = 35 years
Physical activity (PA) = 1.11
Weight (wt) = 154 lb = 70 kg (154 ÷ 2.2)
Height (Ht) = 5'9" = 69 inches = 1.75 m (69 × 0.0254)

Table 8.1 Physical activity categories and values			
ACTIVITY LEVEL OF CATEGORY	PHYSICAL ACTIVITY VALUE		MAJOR ACTION
	MEN	WOMEN	
Sedentary	1.00	1.00	No physical activity aside from that needed for independent living
Low activeness	1.11	1.12	1.5–3 miles/day at 2–4 miles/hour in addition to the light activity associated with typical day-to-day life
Active	1.25	1.27	3–10 miles/day at 2–4 miles/hour in addition to the light activity associated with typical day-to-day life
High activeness	1.48	1.45	10+ miles/day at 2–4 miles/hour in addition to the light activity associate with typical day-to-day life

$$EER = 662 - [9.53 \times age\ (y)] + PA \times [15.91 \times wt\ (kg) + 539.6 \times Ht\ (m)]$$
$$= 662 - [9.53 \times 35] + 1.11 \times [15.91 \times 70 + 539.6 \times 1.75]$$
$$= 662 - 333.55 + 1.11 \times [1113.7 + 944.3]$$
$$= 662 - 333.55 + 2284.38$$
$$= 2613\ kcal/day$$

Acceptable macronutrient distribution ranges

People don't eat energy directly; they derive energy from energy-containing nutrients: carbohydrates, lipids, and proteins. Each of these nutrients contributes to the total energy intake, and those contributions vary in relation to each other. The AMDRs reflect the ranges of intakes for each class of energy source associated with reduced risk of chronic disease while providing adequate intakes of essential nutrients. The AMDRs, which are expressed as percentages of total energy intake, are listed below and can be found in Appendix C:

- carbohydrate: 45–65% of total energy
- protein: 10–35% of total energy
- lipids: 20–35% of total energy

To meet daily energy and nutrient needs while minimizing risks for developing chronic diseases such as heart disease and type-2 diabetes, an average adult should obtain 45–65% of total calories from carbohydrates. This relatively wide range provides for flexibility, in recognition that both the high-carbohydrate/low-fat diet of Asian peoples and the relatively high-fat diet of people from the Mediterranean region contribute to good health. Acceptable lipid intake ranges between 20% and 35% of caloric intake. This range is consistent with the 30% limit set by the American Heart Association, American Cancer Society, and National Institutes of Health. It is believed that very low fat intake combined with high intake of carbohydrate tends to lower

HDL-cholesterol and raise triglyceride levels. On the other hand, high intake of dietary fat coupled with increased total caloric intake contributes to obesity and its related medical complications. Moreover, high-fat diets are usually associated with an increase in saturated fatty acid intake and LDL-cholesterol, which further increases coronary heart disease risk. Recommended protein intake ranges between 10% and 30% of total calories; this range is broad enough to cover the protein needs of all individuals, regardless of their age, gender, and training status.

The DRIs have many uses. They provide a set of standards that can be used to plan diets, to assess the adequacy of diets, and to make judgments about deficient or excessive intakes for individuals and populations. For example, they can be used as a standard for meals prepared for schools, hospitals, and government feeding programs for the elderly. They can be used to determine standards for food labeling and to develop practical tools for diet planning. They can also be used to evaluate the nutritional adequacy of the food consumed by an individual or population that may be of health concern. Each of the DRI categories serves a unique purpose. For example, the EARs are most appropriately used to develop and evaluate nutrition programs for groups such as school-age children. The RDAs or AIs (if an RDA is not available) can be used to set goals for individuals. The ULs help to guard against the overconsumption of nutrients and to keep nutrient intakes below the amounts that increase the risk of toxicity.

Planning a nutritious diet using MyPyramid

The first publication with regard to official dietary guidelines can be traced back more than a century, to when the US Department of Agriculture (USDA), which works both to optimize the nation's agricultural productivity and to promote a nutritious diet, published its first set of nutritional recommendations for Americans. Since this publication, there has been a succession of versions, all designed to translate nutrient intake recommendations into guidelines for dietary planning. In 1980, the USDA and the US Department of Health and Human Services (DHHS) jointly issued a new form of dietary recommendations called **Dietary Guidelines for Americans**, which provided specific advice about how good dietary habits can promote health and reduce risk of major chronic diseases. These guidelines are revised about every five years; the latest version was published in 2005 and accompanied by an updated version of the federal food guide system, called MyPyramid, aimed to help people put the recommendations of the dietary guidelines into practice.

Dietary Guidelines for Americans

The 2005 Dietary Guidelines for Americans differ from previous versions in that their development relied heavily on mathematical equations that predicted the "best diet" for reducing risk of the major chronic diseases: overweight and obesity, hypertension, type-2 diabetes, heart disease, cancer, and osteoporosis. The recommendations placed strong emphasis on monitoring one's caloric intake and increasing physical activity because more people are becoming overweight each year. They were also devised to help boost several nutrients that are often low in the American diet, including vitamin

C, vitamin D, vitamin E, vitamin A, calcium, magnesium, potassium, iron, and dietary fiber. In addition, the current edition of the dietary guidelines includes advice related to alcohol intake and food-borne illness.

The 2005 Dietary Guidelines for Americans identify 41 key recommendations, of which 23 are for the general public and 18 are for special populations. They are grouped into nine general topics:

- adequate nutrient intake within calorie needs
- weight management
- physical activity
- specific food groups to encourage
- fats
- carbohydrates
- sodium and potassium
- alcoholic beverages
- food safety.

The 2005 Dietary Guidelines for Americans are available at www.healthierus.gov/dietaryguidelines. In general, the dietary guidelines recommend that we:

- consume a variety of nutrient-dense foods and beverages within and among the basic food groups identified in the new version of the food guide pyramid, while choosing foods that limit the intake of saturated and trans fats, cholesterol, added sugars, salt, and alcohol (if used). Foods to emphasize are vegetables, fruits, legumes, whole grains, and fat-free or low-fat milk or equivalent milk products.
- maintain body weight in a healthy range by balancing calorie intake from foods and beverages with calories expended. For the latter, engage in at least 30 minutes of moderate-intensity physical activity, above usual activity, at work or home, on most days of the week.
- practice food handling when preparing food. This includes cleaning hands, food contact surfaces, and fruits and vegetables before preparation; and cooking foods to an adequate temperature to kill microorganisms.

To follow the 2005 Dietary Guidelines for Americans, one must keep in mind the five diet-planning principles: (1) adequacy; (2) balance; (3) nutrient density; (4) moderation; and (5) variety.

Adequacy means that the diet provides sufficient energy and enough of all the nutrients to meet the needs of healthy people. Each day people lose some iron, so it must be replaced by eating foods that contain iron. Otherwise, people may develop the symptoms of iron-deficient anemia, such as feeling weak, tired, and having frequent headaches.

Balance involves consuming enough, but not too much, of each type of food. For example, meats, fish, and poultry are rich in iron, but poor in calcium. Conversely, milk and milk products are rich in calcium but poor in iron. Therefore, one should use a balanced approach: consume some meats and some milk products in order to obtain both essential minerals; and save some space for other foods, such as grains, vegetables, and fruits, since a diet consisting of meat and milk alone would not be adequate.

Nutrient density is defined as the amount of nutrients that are in a food relative to its energy content. Foods with high nutrient densities (nutrient-dense foods) provide high amounts of essential nutrients relative to the amount of calories. For example, consider foods that contain calcium: you can get about 300 mg of calcium from either 1.5 oz of cheddar cheese or a cup of fat-free milk, but the cheese delivers about twice as much food energy as the milk. The fat-free milk is therefore twice as calcium-dense as the cheddar cheese, and it offers the same amount of calcium at half the calories. Both foods are excellent choices in terms of adequacy, but to achieve adequacy while controlling calories, the fat-free milk is a better choice. Foods that are notably low in nutrient density, such as potato chips, candies, and colas, are sometimes referred to as empty calorie foods in that they deliver only energy with little or no essential nutrients.

Moderation, or not consuming too much of a particular food, is also important, especially when it comes to controlling caloric intake and maintaining a healthy body weight. For example, foods rich in fat and sugar provide enjoyment and energy, but relatively few nutrients. They promote weight gain when eaten in excess. A person practicing moderation would eat such foods only on occasion and would regularly select foods low in fat and sugar. This practice also improves nutrient density.

Variety emphasizes the importance of consuming different foods within each food group, because it helps ensure that a person is getting adequate amounts of all the essential nutrients. People should select foods from each of the food groups daily and vary their choices within each food group for several reasons. First, different foods within the same group contain different arrays of nutrients. Second, no food is guaranteed to be entirely free of substances that, in excess, could be harmful. Third, variety is the spice of life, as most of us would agree. Eating nutritious meals need never be boring!

MyPyramid: a menu-planning tool

To help people put the recommendations of the Dietary Guidelines into practice, the USDA has established its newest food guidance system, called MyPyramid, which replaces the USDA Food Guide Pyramid published in 1992 (www.mypyramid.gov). The current version of the pyramid is titled "Steps to a Healthy You," which is reflected in the image of a person climbing the pyramid (Figure 8.3). It provides a more "individualized" approach to improving diet and lifestyle than previous guides. The MyPyramid symbol represents the recommended proportion of foods from each food group to create a healthy diet. Physical activity is a new element in the pyramid. It sends a clear message that consumers should choose the right amount and kinds of food to balance their daily physical activity.

Several key elements of the MyPyramid symbol are worth noting. As mentioned earlier, physical activity is emphasized. Balancing energy intake with energy expenditure is a major component of the MyPyramid. Six of the food groups – grains, vegetables, fruits, oils, dairy products, and meats and beans – are represented in different colors, each with different widths on the MyPyramid symbol. The MyPyramid recommends that we choose foods in approximate proportion to the base widths of the bands. A moderate intake of solid fats and added sugars is represented by the

Figure 8.3 MyPyramid: steps to a healthier you.

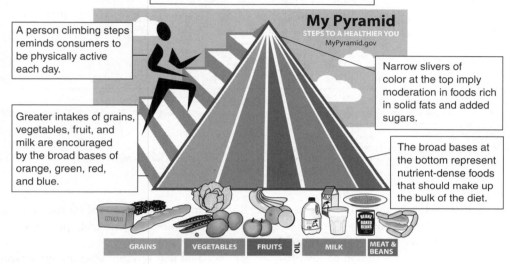

Colors of the pyramid illustrate variety: each color represents one of the five food groups, plus one for oils. Different band widths suggest the proportional contribution of each food group to a healthy diet.

A person climbing steps reminds consumers to be physically active each day.

Greater intakes of grains, vegetables, fruit, and milk are encouraged by the broad bases of orange, green, red, and blue.

Narrow slivers of color at the top imply moderation in foods rich in solid fats and added sugars.

The broad bases at the bottom represent nutrient-dense foods that should make up the bulk of the diet.

My Pyramid
STEPS TO A HEALTHIER YOU
MyPyramid.gov

GRAINS   VEGETABLES   FRUITS   OIL   MILK   MEAT & BEANS

Source: USDA.
Note: color version available at www.MyPyramid.gov

narrowing of each food group's stripe from bottom to top. The narrowing pattern of the color bands from bottom to top also indicates that the more active you are, the more of these foods you can fit into your diet.

To put MyPyramid into action, first you need to estimate your calorie needs using the method for determining EER. Once you have determined the calorie allowance appropriate for you, you can use Table 8.2 to discover how that calorie allowance corresponds to the recommended numbers of servings from each food group. Overall, MyPyramid translates the latest nutrition advice into 12 separate pyramids based on calorie needs (1000–3200 kcal).

MyPyramid provides serving sizes of foods for the various food groups in household units, as shown below. Close attention should be paid to the stated serving size for each choice when following MyPyramid. This will help control the portion size and total caloric intake. See Figure 8.4 for a convenient guide to estimating common serving size measurements.

- Grains – 1 slice of bread; 1 cup of ready-to-eat breakfast cereal; ½ cup of cooked rice, pasta, or cereal.
- Vegetables – one cup of raw or cooked vegetables or vegetable juice; two cups of raw leafy greens.
- Fruits – one cup of fruit; 100% fruit juice; ½ cup of dried fruit.

Table 8.2 MyPyramid recommendations for daily food consumption based on calorie needs											
CALORIE LEVEL	1000	1200	1400	1600	1800	2000	2200	2400	2600	2800	3000	3200
Fruits	1 cup	1 cup	1.5 cups	1.5 cups	1.5 cups	2 cups	2 cups	2 cups	2 cups	2.5 cups	2.5 cups	2.5 cups
Vegetables[1]	1 cup	1.5 cups	1.5 cups	2 cups	2.5 cups	2.5 cups	3 cups	3 cups	3.5 cups	3.5 cups	4 cups	4 cups
Grains[2]	3 oz	4 oz	5 oz	5 oz	6 oz	6 oz	7 oz	8 oz	9 oz	10 oz	10 oz	10 oz
Meats and beans	2 oz	3 oz	4 oz	5 oz	5 oz	5.5 oz	6 oz	6.5 oz	6.5 oz	7 oz	7 oz	7 oz
Milk[3]	2 cups	2 cups	2 cups	3 cups	3 cups	3 cups	3 cups	3 cups	3 cups	3 cups	3 cups	3 cups
Oils[4]	3 tsp	4 tsp	4 tsp	5 tsp	5 tsp	6 tsp	6 tsp	7 tsp	8 tsp	8 tsp	10 tsp	11 tsp

Notes:
1 Vegetables are divided into five subgroups: dark green, orange, legumes, starchy, and other. A variety of vegetables should be eaten, especially green and orange vegetables.
2 At least half of the grain servings should be whole-grain varieties.
3 Most of the milk servings should be fat free or low fat.
4 Limit solid fats such as butter, stick margarine, shortening, and meat fat, as well as foods that contain these.

- Milks – one cup of milk or yogurt; 1.5 oz of natural cheese; 2 oz of processed cheese.
- Meat and beans – 1 oz of meat, poultry, or fish; one egg; one tablespoon of peanut butter; ¼ cup of cooked dry beans; 0.5 oz of nuts or seeds.
- Oils – a teaspoon of any plant- or fish-based oil that is liquid at room temperature.

If interested, you can consider using the website www.mypyramid.gov. This website provides interactive technology that allows consumers to obtain dietary recommendations that are specific to their age, gender, height, weight, and level of physical activity. It includes several important features: MyPyramid Menu Planner, Inside MyPyramid, MyPyramid Tracker, and MyFoodapedia. MyPyramid Menu Planner provides a quick estimate of what and how much food the individual should eat from the different food groups based upon each individual profile. Inside MyPyramid provides in-depth information for every food group, including recommended daily amounts in commonly used measures such as cups and ounces, with examples and everyday tips. The section also includes recommendations for choosing healthy oils, **discretionary calories**, and physical activity. Discretionary calories refer to calories allowed from food choices rich in added sugars and solid fat. MyPyramid Tracker allows users to assess their diet quality and physical activity status by comparing one day's worth of foods eaten to the guidance provided by MyPyramid. Messages with regard to nutrition and physical activity are provided based on the need to maintain current weight or to lose weight. MyFoodapedia gives users quick access to search for calories and the MyPyramid food group for a particular food. This section also allows comparison between any two foods.

Use of food labels in assisting dietary planning

Food labels are another tool that can be used in diet planning. They are designed to help consumers make healthy food choices by providing information about the

Figure 8.4 A convenient guide to estimating common serving size measurements.

Grains	≈ 2 ounces (*Bagel, muffin*)
Vegetables	≈ 1 cup (*Green beans*)
Fruits	≈ $\frac{1}{2}$ cup (*Medium/small apple*)
Oils	≈ 2 tablespoons (*salad dressing, margarine*)
Milk	≈ 1 ounce (*cheese*)
Meat and beans	≈ 3 ounces (*Meat, chicken, fish*)

nutrient composition of foods and about how a food fits into the overall diet. Today, nearly all foods sold in stores must be in a package that has a label containing the following information: (1) the product name; (2) the name and address of the manufacturer; (3) the amount of product in the package; (4) the ingredients listed in descending order by weight; and (5) a Nutrition Facts panel. Of special interest to many people is the Nutrition Facts panel. Understanding how to read a food label is important in making healthy food choices.

Nutrition Facts

The nutrition information section of the label is entitled "Nutrition Facts" (Figure 8.5). In this section, the serving size is listed in common household and metric measures, and is based on a standard list of serving sizes designed to be representative of the serving sizes people choose. In other words, serving sizes on the Nutrition Facts panel must be consistent among similar foods. The use of standard serving sizes allows comparisons to be made easily between products. For example, comparing the energy content of different types of crackers is simplified because all packages list energy values for a standard serving size of about 30 g and tell you the number of crackers per serving.

The serving size on the label is followed by the number of servings per container. The label must then list the total kilocalories (or Calories on food labels), kilocalories from fat, total fat, saturated fat, cholesterol, sodium, total carbohydrate,

Figure 8.5 The sample Nutrition Facts panel.

Nutrition Facts

Serving Size 1 cookie (28g)
Serving Per Container 15

Amount Per Serving

Calories 120	Calories from Fat 45

	% Daily Value*
Total Fat 5g	8%
Saturated Fat 3g	15%
Cholesterol 25mg	8%
Sodium 100mg	4%
Total Carbohydrate 18g	6%
Dietary Fiber less than 1 gram	3%
Sugars 11g	
Protein 1g	

Vitamin A 4%	•	Vitamin C 0%
Calcium 2%	•	Iron 4%

*Percent Daily Values are based on a 2,000 calorie diet. Your daily values may be higher or lower depending on your calorie needs

	Calories:	2,000	2,500
Total Fat	Less than	65g	80g
Saturated Fat	Less than	20g	25g
Cholesterol	Less than	300mg	300mg
Sodium	Less than	2,400mg	2,400mg
Total Carbohydrate		300g	375g
Dietary Fiber		25g	30g

Calories per gram:
Fat 9 • Carbohydrate 4 • Protein 4

dietary fiber, sugars, and protein. The amounts of these nutrients are given per serving, and most are listed as a percentage of a standard called the **daily value**. The percentage of the daily value (%DV) is usually given for each nutrient per serving and is based on a 2000-kcal diet. For example, if a label states that a food provides 10% of the daily value for dietary fiber, then the food provides 10% of the recommended daily intake for dietary fiber in a 2000-kcal diet. Daily values may not be as applicable to those who require considerably more or less than 2000 kcal per day. Daily values are mostly set at or close to the highest RDA value or related nutrient standard seen in the various age and gender categories for a specific nutrient (Appendix E).

Many manufactures list the daily values set for dietary components such as fat, cholesterol, and carbohydrate on the Nutrition Facts panel. This can be useful as a reference point. As noted, they are based on a 2000-kcal diet; if the label is large enough, amounts based on a 2500-kcal diet are listed as well for total fat, saturated fat, cholesterol, sodium, fiber, total carbohydrate, and dietary fiber. Daily values help consumers determine how a food fits into their overall diet.

Exceptions to food labeling

Foods such as raw fruits and vegetables, fish, meats, and poultry currently are not required to have a Nutrition Facts label. However, many grocers and some meat packers have voluntarily chosen to provide their customers with information about these products. Protein deficiency is not a public health concern in the United States. Therefore, disclosure of the %DV for protein is not mandatory on foods for people over four years of age. If the %DV for protein is given on a label, the Food and Drug Administration (FDA) requires that the product be analyzed for protein quality. This procedure is expensive and time-consuming, so many companies opt not to list a %DV for protein. However, labels on food for infants and children under four years of age must include the %DV for protein.

Nutrient content claims

Food Labels can also contain additional nutrition-related information. For example, **nutrient content claims** describe the level of a nutrient in a food. These include phrases such as "sugar free," "low sodium," and "good source of …." The use of these terms is regulated by the FDA, and some of the approved definitions are provided as follows:

- light or lite – if 50% or more of the calories are from fat, fat must be reduced by at least 50% as compared to a regular product. If less than 50% of calories are from fat, fat must be reduced at least 50% or calories reduced at least one-third compared to a regular product;
- reduced calories – at least 25% fewer calories per serving compared to a regular product;
- calorie free – less than 5 kcal per serving;
- fat free – less than 0.5 g of fat per serving;
- low fat – 3 g or less of fat per serving;
- saturated fat free – less than 0.5 g of saturated fat per serving;
- low in saturated fat – 1 g of saturated fat per serving and containing 15% or less of calories from saturated fat;
- cholesterol free – less than 2 mg of cholesterol per serving. Note that cholesterol claims are only allowed when food contains 2 g or less saturated fat per serving;
- low in cholesterol – 20 mg of cholesterol or less per serving;
- sodium free – less than 5 mg of sodium per serving;
- low in sodium – 140 mg of sodium per serving;
- sugar free – less than 0.5 g of sugars per serving;
- high, rich in, or excellent source of … – contains 20% or more of the daily value to describe protein, vitamins, minerals, dietary fiber, or potassium per serving;
- fresh – a raw food that has not been frozen, heat processed, or otherwise preserved;
- fresh frozen – food that was quickly frozen while still fresh.

Health claims

Food labels are also permitted to include a number of **health claims** if they are relevant to the product. Health claims refer to a relationship between a nutrient or a food and the risk of a disease or health-related condition. They can be used on conventional foods or dietary supplements, and can help consumers choose products that will meet their dietary needs or health goals. For example, low-fat milk a good source of calcium, might include on the package label a statement indicating that a diet high in calcium will reduce the risk of osteoporosis. Health claims are permitted only after the scientific evidence is reviewed and found to be valid, and must be approved by the FDA. More complete information regarding nutrient claims and health claims can be found on the FDA website: www.fda.gov/Food/LabelingNutrition/default.htm. The claims allowed at this time may show a link between:

- a diet with enough calcium and a reduced risk of osteoporosis;
- a diet low in total fat and a reduced risk of some cancers;
- a diet low in saturated fat and cholesterol and a reduced risk of cardiovascular or heart disease;
- a diet rich in fiber and a reduced risk of some cancers;
- a diet low in sodium and high in potassium and a reduced risk of hypertension and stroke;
- a diet rich in fruits and vegetables and a reduced risk of some cancers;
- a diet adequate in the synthetic form of vitamin folate or folic acid and a reduced risk of neural tube defects;
- use of sugarless gum and a reduced risk of tooth decay;
- a diet rich in fruits, vegetables, and grain products that contain fiber and a reduced risk of cardiovascular disease;
- a diet rich in whole grain foods and other plant foods, as well as low in total fat, saturated fat, and cholesterol, and reduced risk of cardiovascular disease and certain cancers;
- a diet low in saturated fat and cholesterol that also includes 25 g of soy protein and a reduced risk of cardiovascular disease;
- fatty acids from oils present in fish and a reduced risk of cardiovascular disease.

DIETARY CONSIDERATIONS FOR PHYSICALLY ACTIVE INDIVIDUALS AND ATHLETES

Adequate nutrition is essential to fitness and performance. Diet must provide sufficient energy from adequate sources to fuel activity, enough protein to maintain muscle mass, and water to transport nutrients and cool the body. In general, while there should be an increase in the total energy intake in order to meet the energy demand imposed by physical activity and training, research in sports nutrition indicates that those who exercise or train regularly to keep fit and competitive do not require additional nutrients beyond those obtained by consuming a nutritionally well-balanced diet. For example, the Dietary Guidelines suggest that one should maintain at least 50% of the total energy intake being derived from carbohydrate. This recommendation should apply to every physically active individual, as well as the majority of athletes. Remember, as the total caloric intake increases, the absolute quantity of carbohydrate consumed also increases in the same proportion. However, modifications to the Dietary Guidelines may help enhance performance for certain athletic endeavors, especially those that challenge the body's limit.

Calorie needs

The amount of energy needed for an activity depends on not only the characteristics of the exerciser, such as body size and body composition, but also the duration, intensity, and frequency of the activity (Table 8.3). A small person may need a daily intake of only 1800 kcal to sustain normal daily activities without losing body weight, while a large, muscular man may need 4000 kcal. For a casual exerciser, the energy needed for activity may increase energy expenditure by a few hundred kilocalories per day. However, for an endurance athlete, such as a marathon runner, the energy needed for training may increase expenditure by 2000–3000 kcal per day. Therefore, some athletes may need as much as 6000 kcal or more daily to maintain body weight while training. In general, the more intense the activity, the more energy it requires. For example, riding a bicycle involves less work than running the same distance and therefore requires less energy. Similarly, the more time spent exercising, the more energy it requires. Riding a bicycle for 60 min requires six times the energy needed to ride for 10 min. If an athlete experiences daily fatigue, the first consideration should be whether he or she is consuming enough food. Up to six meals per day may be needed, including one before each workout.

Body weight and composition can affect athletic performance. Athletes involved in activities in which a small, light body offers an advantage, such as ballet, gymnastics, and certain running events, may restrict energy intake to maintain a low body weight. While a slightly leaner physique may be beneficial, dieting to maintain an unrealistically low body weight can be harmful to health and performance. An athlete who needs to lose weight should do so in advance of the competitive season to prevent the restricted diet from affecting performance. In addition, to preserve lean body mass and enhance fat loss, weight loss should occur at a rate of 1–2 lb per week. This can be accomplished by lowering food intake by 200–500 kcal per day while maintaining a regular exercise program. On the other hand, in sports such as American football and rugby, in which being large and heavy is advantageous, an increase in body weight may be desirable. If an athlete needs to gain weight, increasing food intake by 500–1000 kcal per day should be undertaken (Position of the American Dietetic Association, Dietitians of Canada, and the American College of Sports Medicine 2000). In addition, strength training should accompany weight gain to promote an increase in lean body mass (Kraemer *et al.* 1999).

Carbohydrate, fat, and protein needs

The source of dietary energy is often as important as the amount of energy. In general, the diets of physically active individuals and most athletes should contain the same proportion of carbohydrate, fat, and protein as is recommended for the general public: about 45–65% of total energy as carbohydrate, 20–35% of energy as fat, and 10–35% of energy as protein (Position of the American Dietetic Association, Dietitians of Canada, and the American College of Sports Medicine 2000).

ACTIVITY/SPORT	50KG	60KG	70KG	80KG	90KG
Aerobic dance	270	310	350	380	420
American football	240	270	305	340	370
Aquarobics	235	290	310	360	400
Archery	190	220	250	270	300
Badminton	270	310	350	385	420
Baseball/softball	220	250	285	315	350
Basketball (half-court)	240	270	305	340	370
Basketball (competition)	480	545	610	670	740
Body building	375	427	480	530	585
Bowling	215	230	275	305	335
Boxing (sparring)	190	220	250	270	300
Calisthenics	190	220	250	270	300
Canoeing/kayaking (4 mph)	240	270	350	385	420
Circuit training	263	300	335	375	410
Climbing (mountain)	480	545	610	680	745
Cricket (fielding)	240	270	350	385	420
Cross-country skiing	560	635	715	790	870
Cycling (moderate speed)	165	190	214	240	260
Dance (social)	223	255	285	315	350
Fencing	240	270	305	340	370
Golf (walking with bag)	200	230	255	280	310
Gymnastics	255	280	315	350	380
Hockey	430	490	550	610	670
Horse-riding	190	220	250	270	300
Ice hockey	280	270	355	395	435
Jogging (9 kmh)	520	590	660	735	806
Martial arts	250	280	315	350	385
Orienteering	520	590	660	735	806
Rope jumping (continuous)	560	635	715	790	870
Rowing (recreational)	190	220	250	270	300
Rugby	430	490	550	610	670
Running (16 kmh)	719	820	920	1016	1116
Skiing (downhill)	480	545	610	680	745
Soccer	430	490	550	610	670
Squash	480	545	610	680	745
Swimming (fast)	426	512	639	767	853
Swimming (slow)	349	419	524	629	698
Tennis	335	380	430	475	520
Table tennis	236	283	354	424	472
Volleyball	280	270	355	395	435
Walking (brisk)	240	280	315	350	385
Weight training	375	427	480	530	585

Table 8.3 Energy expenditure in kilocalories per hour based on body mass

Source: Adapted Kent 2003; McArdle *et al.* 2009.

Carbohydrate needs

Carbohydrate represents the most important source of energy. Carbohydrate is needed to maintain blood glucose levels during exercise and to replace glycogen stores after exercise. In general, the diets of physically active individuals should contain the same proportion of carbohydrate, fat, and protein as recommended to the general public, as mentioned earlier. However, athletes should be encouraged to aim at the high end of the percentage range for carbohydrate (~60%). For a 2000-kcal diet, this translates into 300 g of carbohydrate (1 g of carbohydrate = 4 kcal) and 4.3 g per kg of body weight (if using a body weight of 70 kg). It is recommended that although the amount of carbohydrate needed depends on the total energy expenditure, type of sport, gender, and environment, it should be about 6–10 g per kilogram of body weight per day (Position of the American Dietetic Association, Dietitians of Canada, and the American College of Sports Medicine 2000). People engaged in aerobic training and endurance activities of about 60-min duration per day may need 6–7 g per kilogram of body weight. When exercise duration approaches several hours per day, carbohydrate intake may increase to 10 g per kilogram of body weight. In other words, triathletes and marathon runners should consider eating about 500–600 g of carbohydrate daily, or even more if necessary, in order to prevent chronic fatigue and to load the muscles and liver with glycogen.

For athletes and physically active individuals, most of the carbohydrate in their diet should be complex carbohydrates from whole grains and starchy vegetables, with some naturally occurring simple sugars from fruits and milk. These foods provide the necessary vitamins, minerals, phytochemicals, fibers, and energy. Their focus is to include high-carbohydrate foods while moderating concentrated fat sources. Sports nutritionists emphasize the difference between a high-carbohydrate meal and a high-carbohydrate/high-fat meal. Before endurance events, such as marathons or triathlons, some athletes seek to increase their carbohydrate reserves by eating foods such as potato chips, French fries, and pastries. Although such foods provide carbohydrate, they also contain a lot of fat. Better carbohydrate choices include pasta, rice, potatoes, bread, fruit and fruit juices, and many breakfast cereals. Consuming a moderate rather than high amount of fiber during the final day of training is a good precaution to reduce the chances of bloating and intestinal gas during the athletic event.

Fat needs

A diet containing up to 35% of calories from fat is generally recommended for athletes. Dietary fat supplies fat-soluble vitamins and essential fatty acids, as well as an important source of energy. Body stores of fat provide enough energy to support the needs of even the longest endurance events. No performance benefits have been associated with diets containing less than 15% fat. On the other hand, excess dietary fat is unnecessary and excess energy consumed as fat, carbohydrate, and protein can cause an increase in body fat and thus weight gain. In addition, consumption of saturated fat and trans fat should be limited.

Protein needs

Protein is essential to maintaining muscle mass and strength. The RDA for protein for non-athletes is 0.8 g per kilogram of body weight. Therefore, a person weighing 70 kg requires 56 g (2 oz) of protein daily. Assuming that even during exercise, relatively little protein loss occurs through energy metabolism, this protein recommendation remains adequate for most active individuals.

Although a diet that contains the RDA for protein (0.8 g per kilogram of body weight) provides adequate protein for most active individuals, competitive athletes participating in endurance and strength/power sports may require more protein. In endurance events such as marathons, protein is needed for energy and to maintain blood glucose, so these athletes may benefit from consuming 1.2–1.4 g of protein per kilogram of body weight per day. Strength and power athletes who require amino acids to synthesize muscle proteins may benefit from 1.4–1.6 g per kilogram of body weight per day. In fact, the protein intake for most athletes often exceeds the protein RDA, and their diet usually contains 2–3 times the recommended protein. For example, if an 85-kg man consumes 3000 kcal, 18% of which is from protein, his protein intake is 135 g (1 g of protein = 4 kcal) or 1.6 g of protein per kilogram of body weight.

Any athlete not specifically on a low-calorie regimen can easily meet the protein recommendations by eating a variety of foods. To illustrate, a 57-kg (125-lb) woman performing endurance activity can consume 68 g of protein (57 × 1.2) during a single day by including 3 oz of chicken (e.g., a chicken breast), 3 oz of beef (e.g., a small, lean hamburger), and two glasses of milk in her diet. Similarly, a 77-kg (180-lb) man who wants to gain muscle mass through strength training needs to consume only 6 oz of chicken (e.g., a large chicken breast), 0.5 cups of cooked beans, a 6 oz can of tuna, and three glasses of milk to achieve an intake of 125 g of protein (77 × 1.6) in a day. For both athletes, their calculations do not even include protein present in the grains and vegetables they will also eat. It is clear that by meeting calorie needs, many athletes consume much more protein than is required. There are hundreds of protein supplements that are commercially available. Although certain types of exercise do increase protein needs, the protein provided by these expensive supplements will not necessarily meet an athlete's needs any better than the protein found in a balanced diet.

Vitamin and mineral needs

Although vitamins and minerals are essential nutrients, the amount of vitamins we need to prevent deficiency is small. In general, humans require a total of about 28 g (1 oz) of vitamins for every 70 kg (150 lb) of food consumed. Given such a small requirement, with proper nutrition from a variety of food sources, the physically active person or competitive athlete need not consume vitamin and mineral supplements. However, about 40% of adults in the United States take vitamin and/or mineral supplements on a regular basis, some at unsafe levels. They are spending $15 billion annually on supplements. The health-related value of this practice is debated.

Adequate vitamin and mineral intake is essential to optimal performance. In addition, the need for some micronutrients may be increased by exercise. Generally speaking, vitamin and mineral needs are the same or slightly higher for athletes compared with those of sedentary adults. However, athletes or physically active individuals usually have high caloric intakes, so they tend to consume plenty of vitamins and minerals. The only exceptions are: (1) they consume low-calorie diets, such as seen with female athletes participating in events in which maintaining a low body weight is crucial; (2) they are vegetarians who eliminate one or more food groups from their diet; and (3) they consume a large amount of processed foods and simple sugars with low nutrient density. In these adverse situations, a multivitamin and mineral supplement at the recommended dosage can upgrade the nutrient density of the daily diet.

Vitamin needs

Of many vitamins, B vitamins are among those that are often chosen as supplements because of the important roles they can play during exercise. Most B vitamins function as coenzymes and they are involved in energy production. They are also required for red blood cell synthesis, protein synthesis, and tissue repair and maintenance. Supplementing vitamin C and E is another common interest among athletes. Both serve as antioxidants, and their deficiencies have been related to impaired synthesis of collagen, production of neurotransmitters, and anemia, all of which have high implications for exercise performance. However, no exercise benefit exists for vitamins with intakes above recommended values. Supplementing for four days with a highly absorbed derivative of thiamin, a component of the pyruvate dehydrogenase that catalyzes movement of pyruvate into the Krebs cycle, offered no advantage over a placebo on measures of oxygen uptake, lactate accumulation, and cycling performance during exhaustive exercise (Webster *et al.* 1997). In addition, studies using high-potency multivitamin mineral supplementation for well-nourished, healthy individuals have failed to demonstrate any beneficial effect on aerobic fitness, muscular strength, and neuromuscular function following prolonged running or athletic performance (Gauche *et al.* 2006). The lack of efficacy of vitamin supplementation can be attributed to the fact that vitamin status in those who are physically active or in highly trained athletes does not differ from that of untrained individuals, despite large differences in daily physical activity level.

Given the important roles vitamins play, any physically active individual or athlete should be cognizant of an adequate intake of these nutrients in order to maximize micronutrient density in their diet. The use of large doses of vitamins requires more study and is not currently recommended as an accepted part of dietary guidance for athletes. Experts suggest consuming a diet containing foods rich in B vitamins and antioxidants such as fruits, vegetables, whole grain breads and cereals, and vegetable oils. There is evidence that antioxidant function in the body enhances as exercise training progress. This would suggest that a physically active lifestyle coupled with a sound nutrition plan will be an ultimate solution to success in fitness and performance. Multivitamin and mineral supplementation may be considered for athletes who restrict energy intake or use severe weight-loss practices, eliminate one or more of the food groups, or consume high-carbohydrate and low-micronutrient diets.

Mineral needs

The use of mineral supplements is not recommended unless prescribed by a physician or registered dietitian, because of potential adverse consequences. A well-balanced diet with an adequate intake of total energy will provide more than enough of both major and minor minerals for all individuals. However, some minerals may be worth mentioning due to their greater loss in high-intensity training among athletes. For example, loss of water and accompanying mineral salts – primarily sodium, chloride, and potassium – in sweat poses an important challenge during prolonged exercise, especially in hot weather. Excessive water and electrolyte loss impairs heat tolerance and exercise performance, and can cause dysfunction in the form of heat cramps, heat exhaustion, and heat stroke. The number of annual heat-related deaths during summer football practice tragically illustrates the importance of fluid and electrolyte replacement. During a practice or game, an athlete may lose up to 5 kg (~10 lb) of water by sweating. This corresponds to a loss of about 8 g of salt, because each kilogram (or liter) of sweat contains about 1.5 g of salt. Therefore, replacement of water and salt lost through sweat become a crucial and immediate need. One can achieve proper supplementation by drinking a 0.1–0.2% salt solution (i.e., adding one-third of a teaspoon of table salt per liter of water). Further discussion on water and electrolyte loss and their replacement strategies is presented in Chapter 14.

Iron is involved in red blood cell production, oxygen transport, and energy production, so a deficiency of this mineral can detract from optimal athletic performance. For most individuals, exercise does not increase iron needs. However, in athletes, especially female athletes, a reduction in the amount of stored iron is common. Poor iron status may be caused by inadequate iron intake, increased iron needs, increased iron losses, or a redistribution of iron due to exercise training. Dietary iron intake may be limited in athletes who are attempting to keep their body weight low, or in those who consume a vegetarian diet and therefore do not eat meat – an excellent source of readily absorbed iron. Iron needs may be increased in athletes because exercise stimulates the production of red blood cells, so more iron is needed for hemoglobin synthesis. Iron is also needed for the synthesis of muscle myoglobin and iron-containing proteins used for ATP production in mitochondria. An increase in iron losses with prolonged training, possibly because of increased urinary and sweat losses, also contribute to increased iron needs in athletes.

Calcium is another important mineral that deserves extra attention, particularly among women athletes who try to lose weight by restricting their intake of dairy products rich in calcium. Calcium is needed to maintain blood calcium levels and promote and maintain bone density, which in turn reduces the risk of osteoporosis. In general, exercise, especially weight-bearing exercise, increases bone density. However, in female athletes with extremely low body weight and body fat, their calcium status can be at risk. These athletes are found to also have a high risk of developing eating disorders and amenorrhea. The combination of disordered eating, amenorrhea, and osteoporosis is referred to as **female athlete triad**. Female athletes who strive to reduce their body weight to achieve an ideal body image and to meet the performance goals set by coaches, trainers, or parents are at increased risk of developing this syndrome of interrelated disorders. The extreme energy restriction that occurs in eating disorders can create a physiological condition similar to starvation, and can contribute to menstrual abnormalities. High-intensity exercise can also affect the menstrual cycle by increasing energy demands or by causing a decrease in female reductive hormones, particularly estrogen (Otis *et al.* 1997). When combined, energy restriction and excessive exercise can contribute to amenorrhea, the delayed onset of menstruation or the absence of three or more consecutive menstrual cycles. Loss of regular menstrual cycles in female athletes stems from a reduction of estrogen. A low level of estrogen has other negative consequences on the body: it reduces calcium absorption and, when combined with poor calcium intake, leads to premature bone loss and increased stress fractures.

Female athletes experiencing symptoms of the female athlete triad, such as irregular menstrual periods and/or stress fractures, should consult a physician to determine the cause. Decreasing the amount of training or increasing energy intake and body weight often restore regular menstrual cycles and stabilize bone mass. A physician may prescribe a multivitamin and mineral supplements, as well as calcium supplements as needed to maintain an intake of at least 1200 mg per day. If irregular menstrual cycles persist, severe bone loss and osteoporosis can result. What is more alarming is that such bone loss cannot be completely reversed by either increasing dietary calcium or by performing more weight-bearing exercise. Therefore, early prevention is crucial, and it is encouraged that teachers, coaches, health professionals, and parents educate female athletes about the triad and its health consequences.

Special planning for before, during, and after competition

For most of us, a trip to the gym requires no special dietary planning beyond that needed to consume a balanced diet as described earlier. However, for athletes competing in athletic events, foods eaten in preparation for competition can mean the difference between victory and defeat. For this reason, specialized dietary advice has been developed with regard to what athletes should consume before, during, and after the competition. For example, it has been considered that spaghetti, muffins, bagels, and pancakes with fresh fruits are good food choices for a pre-game meal. Liquid meal-replacement formulas such as Carnation instant breakfast can also be used. Foods especially rich in fiber should be eaten the previous day to empty the colon before an event, but they should not be eaten the night before or in the morning before the event. Experts have developed a number of sports nutrition recommendations, which will be discussed in the following sections.

Pre-competition meal

The pre-competition meal should provide adequate carbohydrate energy and ensure optimal hydration. Therefore, fasting before competition or intense training makes no sense physiologically because it rapidly depletes liver and muscle glycogen. Muscle glycogen is depleted by exercise, but liver glycogen is used to supply blood glucose and can be depleted even during rest if no food is ingested. As a general rule, the meal should be high in carbohydrate and low in fat and protein. High-fiber foods should be avoided to prevent feeling bloated during competition. As the increased stress and tension that usually accompany competition decrease blood flow to the digestive tract, depressing intestinal absorption, the meal should be consumed 3–4 hours before the event. The amount of calories may vary depending upon gender and size of the athlete, but it should be within a tolerable range (300–500 kcal). As for food choices, athletes should select those that they prefer and/or believe will give them a winning edge. Some athletes find that, in addition to a pre-competition meal, a small high-carbohydrate snack or beverage consumed shortly before an event may enhance performance (Coyle 1995). Because foods affect people differently, athletes should test the effects of their choices during training, not during competition. It must be emphasized that the pre-competition meal is only valid if the athlete maintains a nutritionally sound diet throughout training. Pre-competition feedings cannot correct any existing nutritional deficiencies or inadequate nutrient intake in the weeks before competition. Table 8.4 shows sample pre-game meals that reflect portion size and composition requirements.

Glycogen supercompensation

While carbohydrate intake in the hours before competition will mainly optimize liver glycogen, carbohydrate consumption in the days leading up to competition allows muscle glycogen stores to be fully replenished. In 1967 Scandinavian scientists discovered that muscle glycogen could be supercompensated by changes in diet and exercise (Bergström and Hultman 1967). In a series of studies, these scientists developed a so-called **glycogen supercompensation** or carbohydrate-loading protocol, which has been found to be able to increase muscle glycogen stores 20–40% above the level that would be achieved via a typical diet. This diet and exercise regimen involves depleting glycogen stores by exercising strenuously and then replenishing glycogen by consuming a high-carbohydrate diet for a few days before competition, during which time only light exercise is performed. For example, consider the glycogen supercompensation schedule of a 25-year-old man preparing for a marathon. His typical calorie needs are about 3500 kcal per day. Six days before the competition, he completes a final, hard, 60-min workout. On that day, carbohydrates contribute to 50% of his total caloric intake. As he goes through the rest of the week, the duration of his workout decreases to 40 min, and then to about 20 min by the end of the week. Meanwhile, he increases the amount of carbohydrate in his diet to reach at least 70% of the total caloric intake as the week continues. The total caloric intake should decrease as exercise time decreases. On the final day before competition, he rests while maintaining the high carbohydrate intake.

Table 8.4 Sample pre- and post-exercise meals

PRE-EXERCISE MEALS

	FOOD	SERVING	CALORIES AND NUTRIENTS
OPTION 1	Cheerios	¾ cup	450 kcal
	Reduced-fat milk	1 cup	90 g (80%) carbohydrate
	Blueberry muffin	1	
	Orange juice	4 oz	
OPTION 2	Fruit yogurt	1 cup	482 kcal
			85 g (70%) carbohydrate
	Bagel	Half	
	Apple juice	4 oz	
	Peanut butter	1 tbsp	

POST-EXERCISE MEALS

	FOOD	SERVING	CALORIES AND NUTRIENTS
OPTION 1	Bagel	1	562 kcal
	Peanut butter	2 tbsp	75 g carbohydrate 22 g protein
	Fat-free milk	8 oz	
	Banana	1 medium	
OPTION 2	Instant breakfast	1 packet	438 kcal
	Fat-free milk	8 oz	70 g carbohydrate 17 g protein
	Banana	1 medium	
	Peanut butter	1 tbsp	

The regimen that was proposed is generally referred to as the "classic supercompensation protocol." It has been used successfully by several top endurance athletes. In fact, many marathon runners still use this method to optimize their performance. Although the supercompensation protocol has been very effective in increasing muscle glycogen concentration, it also has several potential disadvantages, of which athletes should be aware. The main problem may be the incidence of gastrointestinal problems associated with this regimen. Diarrhea has often been reported on the days when the high-protein, high-fat diet is consumed. During the first three days, athletes may also experience hypoglycemia, and they may not recover very well from the exhausting exercise bout when no carbohydrate is ingested. Some athletes also feel muscle stiffness while following the regimen. This is because an increase in muscle glycogen will require additional water to be incorporated into the muscle. In addition, as the protocol would require an athlete to reduce their training volume, most athletes won't feel comfortable and may develop mood disturbances which can have a negative effect on their mental preparation for an event.

Because of the numerous disadvantages of the classic supercompensation protocol, studies have focused on a more moderate protocol that would achieve similar results. Sherman *et al.* (1981) studied three types of muscle glycogen supercompensation regimens in runners. The subjects slowly reduce their training over a six-day period from 90 min running at 75% VO_2max to complete rest on the last day. During each taper, they ingested one of the following three diets: (1) a mixed diet with 50% carbohydrate; (2) a low-carbohydrate diet (25%) for the first three days, followed by three days of a high-carbohydrate diet (70%) – the classic supercompensation protocol; and (3) a mixed diet for the first three days (50% carbohydrate), followed by three days of a high-carbohydrate diet (70%) – the moderate supercompensation protocol. The study revealed that both the classic and moderate protocols produced the same increase in muscle glycogen content, suggesting that a moderate-carbohydrate to high-carbohydrate diet is just as effective as the classic supercompensation diet. This modified supercompensation regimen is shown in Table 8.5.

A carbohydrate supercompensation regimen increased time to exhaustion on average by about 20% and reduced the time to complete a set task by 2–3% (Hawley *et al.* 1997). However, it seems that ergogenic benefits of this protocol can only be demonstrated in events that last more than 90 min. Such carbohydrate loading appears to have no effect on sprint performance, while high-intensity exercise up to about 30 min compares with normal diets (i.e., 50% carbohydrate). This finding is not unexpected, because for this exercise duration, glycogen depletion is not a major

Table 8.5 A modified regimen to supercompensate muscle glycogen stores							
DAYS	1	2	3	4	5	6	7
Exercise duration (min)	90	40	40	20	20	Rest	Competition
Exercise intensity (% VO$_2$max)	75	75	75	75	75	Rest	
Diet (% of calories from carbohydrate)	50% (4 g kg)	50% (4 g kg)	50% (4 g kg)	70% (10 g kg)	70% (10 g kg)	70% (10 g kg)	

performance-limiting factor. Carbohydrate supercompensation has also been reported to improve performance in team sports involving high-intensity intermittent bouts of exercise such as soccer and hockey (Balsom *et al.* 1999).

Carbohydrate feeding during exercise

For sporting events that last longer than 60 min, carbohydrate feeding during exercise can also improve athletic performance. This is because prolonged exercise depletes muscle glycogen stores and low levels of muscle glycogen and blood glucose lead to fatigue, both physical and mental. When endogenous carbohydrate stores, such as muscle glycogen, run low, athletes often complain of "hitting the wall," the point at which maintaining a competitive pace seems impossible. However, consuming about 60 g of liquid or solid carbohydrates every hour has been shown to reverse this obstacle. It must be mentioned that such a benefit of carbohydrate feeding has been generally demonstrated during exercise in which intensity exceeds 70% VO_2max. As mentioned in Chapter 6, sustained exercise at or below 60% VO_2max places much less demand on carbohydrate breakdown. This level of exercise does not tax glycogen reserves to a degree that would limit endurance. Nevertheless, glucose feedings provide supplementary carbohydrate during intense exercise when demand for glycogen increases significantly. The mechanisms by which carbohydrate feeding during exercise may improve endurance performance include:

- maintaining blood glucose and high levels of carbohydrate oxidation;
- sparing liver and possibly muscle glycogen;
- promoting glycogen synthesis, especially during low-intensity periods of intermittent exercise;
- enhancing the function of the central nervous system.

Studies have also addressed questions of which carbohydrates are most effective, what is the most effective schedule, and what is the optimal amount of carbohydrate to consume. The timing of carbohydrate feedings seems to have little effect on the use of ingested carbohydrate, which is typically measured by determining exogenous carbohydrate oxidation rates. Studies in which a large dose (100 g) of carbohydrate in solution was given produced similar exogenous carbohydrate oxidation rates to studies in which 100 g of carbohydrate were ingested over regular intervals. Knowing the amount of carbohydrate that needs to be ingested to attain optimal performance while producing no side effect is important. In theory, the optimal amount should be the amount that results in the maximal exogenous carbohydrate oxidation rate. Through an analysis using a large number of studies, Jeukendrup and Jentjens (2000) found that carbohydrate feeding at rates of about 1.1–1.2 g per minute produce the maximal exogenous carbohydrate oxidation rate. This finding suggests that athletes who adopt an ingestion rate of 70 g per hour can expect an optimal carbohydrate delivery. This amount of carbohydrate can be found in the following sources: 1 liter of sports drink (e.g., Gatorade, PowerAde, Isostar); 600 ml of cola drink; 1.5 Power Bars or Gatorade energy bars; or three medium bananas. As far as the type of carbohydrate is concerned, results appear to be mixed. It appears that glucose is oxidized at much higher rates than fructose and galactose, because the latter would have to be converted into glucose in the liver before they can be metabolized. However, the oxidation rates of maltose, sucrose, and glucose polymers (maltodextrins) are comparable to those of glucose. In addition, starches with a relatively large amount of amylopectin are rapidly digested and absorbed and their oxidation occurs at a similar rate as glucose, whereas those with high amylase content have a relatively slow rate of hydrolysis. Therefore, in order to obtain a quick energy contribution from carbohydrate feedings, one may choose to use glucose, maltose, sucrose, maltodextrins or amylopectin as they can be digested, absorbed, and oxidized more rapidly.

Carbohydrate intake during recovery

Replenishing glycogen stores following exhaustive exercise is another important consideration. When exercise ends, the body must shift from the catabolic state of breaking down glycogen to the anabolic state of restoring glycogen, so that athletes can be ready for the next competition or training session. Often, the time to recover between successive athletic competitions or training sessions is very short. In such cases, rapid glycogen repletion becomes even more important. The timing of carbohydrate intake can have an important effect on the rate of muscle glycogen synthesis during recovery. In one study, Ivy *et al.* (1988a) observed that when carbohydrate intake is delayed until two hours after exercise, the amount of muscle glycogen restored was only half of what was achieved when carbohydrate intake took place immediately following exercise. Immediately after exercise is when glycogen synthesis is greatest, because the muscles are more insulin-sensitive at this point. It is also recommended that within this two-hour post-exercise period, athletes should consume carbohydrate in an amount of 1.0–1.5 g per kilogram of body weight every hour (Ivy *et al.* 1988b).

Since timing is important, carbohydrate foods listed with a moderate to high glycemic index should be chosen to ensure that more glucose molecules will be available in the blood right after exercise. Athletes can consume sugar candies, sugared soft drinks, fruit or fruit juice, or a sport-type carbohydrate supplement such as a sports-type bar or gel right after training. Later, they can choose enriched bread, mashed potatoes, and short-grain white rice or spaghetti noodles. See Table 2.4 for the glycemic index and glycemic load for various foods. Because certain amino acids have a potent effect on the secretion of insulin, adding an appropriate amount of protein (e.g., a ratio of 3 g of carbohydrate to 1 g of protein) during recovery can be especially helpful to maximize glycogen repletion. For a 70-kg athlete, this corresponds to about 75 g of carbohydrate and 25 g of protein. Table 8.4 shows sample post-exercise meals of this composition.

SUMMARY

- A person's nutritional state can be categorized as: desirable nutrition, in which the body has adequate stores for times of increasing needs; under-nutrition, which may be present with or without clinical symptoms; and over-nutrition, which can lead to toxicities and various chronic diseases.

- A strong link exists between nutrition and chronic diseases. There is evidence that a high sugar intake can adversely affect blood lipid levels, thereby increasing the risk of heart disease. However, diets high in whole grains and fibers may reduce blood cholesterol levels and thus protect against heart diseases and strokes.

- Diets high in cholesterol, saturated fat, and trans fat have been linked to the development of many chronic diseases, including cardiovascular diseases and some types of cancer, and overweight and obesity. Obesity is also associated with diets high in fat; excess body fat in turn contributes to an increased risk of diabetes, cardiovascular disease, and high blood pressure.

- When saturated fat in the diet is replaced by any type of polyunsaturated fat, there is a beneficial decrease in LDL-cholesterol, which is often regarded as "bad" cholesterol that promotes plaque build-up in coronary arteries. One such example of polyunsaturated fat is omega-3 fatty acids. Regular consumption of omega-3 fatty acids has been found to reduce LDL-cholesterol levels, while possibly increasing HDL-cholesterol levels.

- Nutrition recommendations made to the public for health promotion and disease prevention are based on available scientific knowledge. Dietary standards such as the dietary reference intakes (DRIs) provide recommendations for intakes of nutrients and other food components that can be used to plan and assess the diets of individuals and populations.

- The DRIs represent a set of four types of nutrient intake reference standards used to assess and plan dietary intake: (1) estimated average requirements (EARs); (2) recommended dietary allowances (RDAs); (3) adequate intake levels (AIs); and (4) tolerable upper intake levels (UIs).

- The DRIs also include calculations for estimated energy requirement (EERs), which can be used to assess whether one's energy intake is sufficient, and acceptable macronutrient distribution (AMDRs), which provide a recommended distribution of the macronutrients in terms of energy consumption.

- To follow the 2005 Dietary Guidelines for Americans, one must keep in mind the five diet-planning principles: (1) adequacy; (2) balance; (3) nutrient density; (4) moderation; and (5) variety. Adequacy means that diet provides sufficient energy and enough of all the nutrients to meet the needs of healthy people; balance involves consuming enough, but not too much, of each type of food; nutrient density concerns the amount of nutrients that are in a food relative to its energy content; moderation means not consuming too much of a particular food; and variety emphasizes the importance of consuming different foods within each food group.

- MyPyramid is designed to translate nutrient recommendations into a food plan that exhibits variety, balance, and moderation. This meal-planning tool incorporates one's physical activity pattern and emphasizes the balance of energy intake and energy expenditure. It entails six food groups: grains, vegetables, fruits, oils, dairy products, and meats and beans, which are represented in different colors, each with different widths.

- Anyone who exercises regularly should consume a diet that meets calorie needs and is moderate to high in carbohydrates and fluid and adequate in other nutrients such as iron and calcium. In general, the diet of an active individual should contain about 45–65% of total energy as carbohydrate; 20–35% of energy as fat; and about 10–35% energy as protein.

- Plenty of carbohydrates should be in the pre-event meal, especially for endurance athletes. High glycemic index carbohydrates should be consumed by an athlete within two hours of a workout in order to begin restoration of muscle glycogen stores. Mixing with some protein in the post-exercise meal may make glycogen restoration more effective.
- Competitive endurance athletes may utilize glycogen supercompensation regimens to maximize glycogen stores before an event.
- Athletes should consume enough fluid and ultimately restore pre-exercise weight. Sports drinks help replace fluid, electrolytes, and carbohydrate lost during workouts. Their use is especially appropriate and important when continuous activity lasts beyond 60 min.

CASE STUDY: PLANNING A TRAINING DIET PROPERLY

Tom is training for a triathlon event that takes place in three weeks' time. He has read a lot about sports nutrition, and especially about the importance of eating a high-carbohydrate diet while in training. He has been struggling to keep his weight in a range at which he feels he can perform his best. Consequently, he is trying to eat as little as possible. Over the past two weeks, Tom has been feeling weak and his workouts in the afternoon have not met his expectations. His run times are slower and he shows signs of fatigue after just 20 minutes into his training program.

Tom's dietary records reveal that his breakfast yesterday was a large bagel, a small amount of cream cheese, and a cup of apple juice. For lunch he had a small salad with low-fat dressing, a large plate of pasta with marinara sauce and broccoli, and a diet soda. For dinner, he had a small broiled chicken breast, a cup of rice, some carrots and a glass of ice tea. Later, he snacked on fat-free pretzels with a glass of water.

Questions

- Tom is on the high-carbohydrate diet, but why did he experience weakness and fatigue?
- Provide some changes that should be made in Tom's diet, including some specific foods that should be included.
- Are there any other recommendations that might help Tom maintain his training quality and prevent fatigue?

REVIEW QUESTIONS

1 Discuss why improper nutrition can contribute to the development of many chronic diseases.
2 Describe the four types of standards that make up the dietary reference intakes.
3 How would you explain the concept of nutrient density?
4 Explain the various symbols associated with MyPyramid.

5 What dietary changes would you need to make to meet the MyPyramid guidelines on a regular basis?

6 What are the recommended daily allowances for carbohydrate for an average adult male and an average adult female? How much fiber should they be consuming?

7 What are the recommended daily allowances for protein for (1) an average individual; (2) an adolescent male or female; and (3) an athlete involved in heavy resistance training or prolonged endurance training?

8 What information must be provided on a food label? How are percentage daily values determined? How would you interpret these values?

9 Estimate your energy requirement using the EER formula. What are the major factors that can affect this estimate?

10 What is glycogen supercompensation or carbohydrate loading?

11 What are the dietary recommendations for replenishing muscle glycogen following exhaustive exercise?

12 What are the mechanisms responsible for carbohydrate feedings during exercise improving performance?

SUGGESTED READING

Brooks, G.A., Butte, N.F., Rand, W.M., Flatt, J.P., and Caballero, B. (2004) Chronicle of the Institute of Medicine physical activity recommendation: how a physical activity recommendation came to be among dietary recommendations. *American Journal of Clinical Nutrition*, 79: S921–S930.

This article reviews the scientific literature regarding macronutrients and energy and develops estimates of daily intake that are compatible with good nutrition throughout the life span and that may decrease the risk of chronic disease. The article emphasizes the concept of energy balance and suggests that physical activity recommendations must consider one's energy intake status.

Burke, L.M., Cox, G.R., Culmmings, N.K., and Desbrow, B. (2001) Guidelines for daily carbohydrate intake: do athletes achieve them? *Sports Medicine*, 31: 267–299.

This article discusses the dietary guidelines designed for athletes to achieve high carbohydrate intakes. It provides recommendations for routine carbohydrate intake, but also analyzes the current status of athletes in meeting their energy needs and dietary goals.

Jeukendrup, A.E. (2004) Carbohydrate intake during exercise and performance. *Nutrition*, 20: 669–677.

This article discusses various strategies of consuming or supplementing carbohydrate aimed at enhancing athletic performance. Among the major intriguing issues discussed are the type and dose of carbohydrate, the rate of ingestion, and the timing of supplementation.

Seal, C.J. (2006) Whole grains and CVD risk. *Proceedings of the Nutrition Society*, 65: 24–34.

This article provides a thorough review of the literature on the protective effect of whole grains against cardiovascular diseases. Of particular interest is that this article also discusses the underlying mechanisms of such protective effects.

9 Ergogenic aids and supplements

KEY TERMS

- ergogenic
- sports supplements
- doping
- bicarbonate
- boron
- caffeine
- L-carnitine
- chromium
- coenzyme Q10
- creatine

- dehydroepiandrosterone
- androstenedione
- ephedrine
- glutamine
- glycerol
- β-hydroxy-β-methylbutyrate
- inosine
- ergolytic
- phosphate loading

ERGOGENIC AIDS: AN AREA OF COMPLEXITY AND CONTROVERSY

By nature, athletics demands a competitive attitude. The athletes may desire to out-perform the opponent, or the athlete may compete with oneself while striving to maximize personal potential. This drive to succeed has fueled a sustained growth in the market of sports supplements. Many men and women at all levels of prowess use pharmacologic and chemical agents, believing that a specific substance positively influences strength, power, or endurance. Such a quest for reaching maximum physical performance or desirable aesthetics can be traced back to ancient times. For example, ancient athletes of Greece reportedly used hallucinogenic mushrooms and ground dog testicles for ergogenic purposes, while athletes of the Victorian era routinely used caffeine, alcohol, nitroglycerine, heroin, cocaine, and rat poison (strychnine) to gain a competitive edge. Today's athletes are perhaps more likely than their predecessors to experiment with purported ergogenic aids even though most of them may not have been substantiated. Two key factors important to athletic success are genetic endowment and state of training. At high levels of competition, athletes generally have similar athleticism and have been exposed to similar training methods. Therefore, they are fairly evenly matched. Given the emphasis on winning, many athletes are always searching for a "magic" ingredient that provides them with that winning edge. When such ingredients are harmless, they are only a waste of money, but when they impair performance or harm health, they can waste athletic potential and cost lives. This chapter will review some of the commonly used nutritional substances and ergogenic products that have been claimed to affect basal metabolism, food consumption, energy transformation, fat utilization, and/or sports performance.

What is an ergogenic aid?

The word **ergogenic** is derived from the Greek words "ergo" (meaning work) and "gen" (meaning production of), and is defined as increasing work or potential to do work. Ergogenic aids consist of substances or procedures that improve physical work capacity, physiological function, or athletic performance. An ergogenic aid does not need to be nutritional or pharmacological; it can also be mechanical, psychological, or physiological.

Mechanical aids are designed to increase energy efficiency, thereby improving mechanical advantage. One example of using such an ergogenic aid is runners wearing lightweight racing shoes so that less energy is needed to move the legs.

Psychological aids are designed to enhance psychological process or mental strength during sports competition. One example of using such an ergogenic aid is the mental conditioning through hypnosis that some athletes use to minimize distractions, thereby enhancing their performance.

Physiological aids, which will be further discussed along with nutritional or pharmacological aids in this chapter, are designed to augment natural physiological processes to increase physical power. Blood transfusion and bicarbonate loading are the two common examples of such aids for enhancing performance.

This chapter will be mostly devoted to the discussion of various nutritional and pharmacological ergogenic aids. These ergogenic aids, collectively regarded as **sports supplements**, are concerned with the use of nutrients or chemical compounds or drugs thought to be effective in enhancing physiological or psychological functions. For example, anabolic steroids, drugs that mimic the actions of the male sex hormone, testosterone, may increase muscle size and strength; however, to avoid the potential side effects of taking this drug, strength-training athletes may consider using a nutritional approach by taking protein supplements to achieve the same goal.

Why are ergogenic aids popular?

Weight loss and muscle gain are important concerns for many athletes, as well as for individuals not involved in athletic training. Because achieving these goals is very difficult with conventional methods such as increasing energy expenditure through physical activity, using supplements that may potentiate or replace the effect of training becomes an attractive option.

Many athletes believe that certain foods may possess "magic" qualities. As sports supplements are not regulated by government agencies, such as the FDA, the media becomes consumers' leading source of nutrition information, but many news reports of nutrition research often provide inadequate depth for consumers to make wise decisions. For example, isolated nutrition facts may be distorted, or results of a single study or studies from non-peer-reviewed journals are used to market a specific product.

Many of these products are endorsed by professional athletes, giving the product an aura of respectability. Specific supplements also may be recommended by coaches and fellow athletes. However, research studies reveal that many coaches have poor backgrounds in nutrition, suggesting that misconceptions adopted by coaches can be perpetuated in their athletes. It has been estimated that more than 50% of all athletes have used some form of nutritional or pharmacological supplements, and some athletes use several supplements at the same time and in very high doses (Burke and Reed 1993). Use of dietary supplements has also been found to be prevalent among high-school and collegiate athletes, military personnel, and fitness club members.

Non-regulation of sports supplements

In contrast to prescription drugs, which are carefully regulated, nutrition supplements and ergogenic aids receive very little government oversight, and manufacturers and retailers have enormous freedom in making claims to promote their products. The FDA strictly regulates the clinical testing, advertising, and promotion of foods and drugs, so that those products that fail clinical trials or are marketed by unproven claims will not be allowed to sell. Drugs are extensively tested for safety before they can be sold, but nutritional supplements are not. The FDA regulates nutritional supplements under a different set of regulations than those covering "conventional" foods and drug products. In accordance with the Dietary Supplement Health and Education Act of 1994, the FDA requires that the dietary supplement manufacturer is responsible for ensuring that a dietary supplement is safe before it is marketed, and will take action against any unsafe dietary supplement product after it reaches the market. Generally, manufacturers do not need to register their products with the FDA nor get FDA approval before producing or selling dietary supplements. The dietary supplements being referred to by the FDA are vitamins, minerals, herbs and botanicals, amino acids, dietary substances intended to supplement the diet by increasing the total dietary intake (e.g., enzyme or tissue), or any concentrate, metabolite, constituent or extract.

Legality of ergogenic aids

The use of pharmacological agents to enhance performance in sport has been prohibited by the governing bodies of most organized sports. The use of drugs in sports is known as **doping**, and the Medical Commission of the International Olympic Committee (IOC) has provided an extensive list of drugs and doping techniques that have been prohibited (Table 9.1; also see www.usada.org). The specific substances that are banned by both the IOC and the National Collegiate Athletic Association (NCAA) are provided in Appendix F. At the present time, all essential nutrients are not classified as drugs and are considered to be legal for use in conjunction with athletic competition. Most other food substances and constituents sold as dietary supplements are also legal. However, some dietary supplements are prohibited, such as **dehydroepiandrosterone** (DHEA) and **androstenedione** because they are classified as anabolic steroids. Others may have prohibited substances included. For example, many weight-loss products contain ephedrine or amphetamine, a stimulant that is considered as an illegal drug by many sports organizations. It is hopeful that, with pending legislation,

219

Table 9.1 International Olympic Committee Medical Commission doping categories	
Doping classes	• Stimulants • Narcotics • Anabolic agents • Diuretics • Peptide and glycoprotein hormones and analogs
Doping method	• Blood doping • Pharmacological, chemical, and physical manipulation
Classes of drugs subject to certain restrictions	• Alcohol • Marijuana • Local anesthetics • Corticosteroids • β-blockers • Specified $\beta2$-agonists

all ingredients will be listed in correct amounts on dietary supplement labels. In the meantime, athletes should consult with appropriate authorities before using any sports nutrition supplements marketed as performance enhancers, although the use of sports supplements is completely at the athlete's own risk, even if the supplements are "approved" or "verified." Readers may wish to visit www.supplementwatch.com, which provides the most-current scientific reviews for many supplements.

CRITICAL EVALUATION OF ERGOGENIC AIDS

Manufacturers expend considerable money and effort to show a beneficial effect of an ergogenic aid. Often, however, a placebo effect, not the aid per se, improves the performance because of psychological factors. In other words, the individual performs at a higher level because of the suggestive power of believing that a substance or procedure should work. Athletes and others must critically examine claims made by the dietary supplements industry, including the scientific evidence that supports the claims. The following sections discuss six areas for questioning the validity of research claims concerning the efficacy ergogenic aids.

Rationale and justification

Does the study have a sound rationale and clear hypothesis that a specific treatment or supplement should produce an effect? A well-designed study has a clear hypothesis and a strong theoretical basis for the expected outcomes. For example, a theoretical basis exists to believe that ingesting carbohydrate solution will provide an extra energy source to improve endurance performance. However, no rationale exists to hypothesize that carbohydrate loading should enhance short-term power performance such as in the 100-m and 200-m dash. In addition, some studies are designed with a "shotgun" approach that lacks a clear hypothesis. This type of study often winds up measuring many different variables, some of which bear no theoretical link to the supplement being examined. The more variables that are examined, the greater chance that some of them will change.

Subjects

Was the study conducted using cells, muscles, animals, or humans? Often, results are extrapolated from findings in the cell cultures. These *in vitro* experiments can help our understanding of molecular interactions at the cellular level. However, *in vivo* situations may be very different. Muscle cells in the body may behave differently to isolated muscle cell preparations. Even if living animals are used, the physiology and metabolism of animals can be very different to those of humans. Compared with humans, rats have a relatively small store of intramuscular triglycerides. In addition, high-fat diets in rats have been shown to improve exercise performance, but no evidence indicates that high-fat diets improve performance in humans. Even within humans, caution is needed when generalizing findings across populations of different ages, genders, training levels, and nutrition and health status. Alcohol seems to impact women more profoundly than men due to body size difference. Coenzyme Q_{10} supplementation improves VO_2max and exercise capacity in cardiac patients, but has no such benefits in healthy individuals. In addition, supplemental iron enhances aerobic capacity in a group with iron-deficient anemia. However, one cannot generalize that iron supplementation will benefit all individuals.

Research design

Were the experimental trials randomized? Was the study double-blind placebo controlled? How were extraneous variables controlled? These are important questions in terms of research design and must be considered by investigators prior to the start of data collection. If subjects "self-select" into a treatment group, a question should arise as to whether the result is produced by the treatment per se or by a change due to a subject's motivation. For example, desire to enter a weight-loss program may elicit behaviors that produce weight loss independently of the treatment. Randomization reduces the confounding effects of variables that were not controlled or could not be controlled. However, great difficulty exists in assigning truly random samples of subjects into a treatment and a control group. When a small number of subjects (e.g., $n = 10$) are used, a so-called counter-balanced design is preferred. A counter-balance procedure is to assign subjects into either a treatment or a control condition, but the decision as to which five of the ten subjects will take part in which condition is random. In this procedure, each group receives treatment in a different order. In other words, half of the subjects will take the supplement first; the other half will take the placebo first. Failure to randomize treatments in a study may confound the outcome and hence make any conclusion untrustworthy.

The ideal experiment to evaluate the ergogenic effect of a supplement requires that treatment and control subjects remain unaware or "blinded" to the substance administered. To achieve this goal, while all subjects should receive a similar quantity and/or form of the proposed aid, the control group receives an inert compound or placebo. The placebo treatment evaluates the possibility of subjects performing well simply because they receive a substance they believe should benefit them. If subjects have prior knowledge or expectations with respect to a treatment or supplement, their performance could be affected. To further prevent bias from influencing the experimental outcomes, those who administer the treatment and measure the outcomes must also be unaware of which subject receives the treatment or placebo. In such a double-blind experiment, both investigator and subjects remained unaware of the treatment condition. It must be noted that with some nutrition interventions, matching placebos, especially those that produce the same taste, are difficult to find. Therefore, despite the use of double-blind, placebo-control procedures, some studies can still bear the limitation associated with the fact that subjects may be aware of what they receive.

In an ideal study, all variables and conditions should be made as identical as possible, so the only difference between the experimental trials is the treatment – whether supplement or placebo – each group receives. In doing so, all observed changes can be ascribed with greater confidence to the treatment.

Testing and measurement

Reproducible, objective, and valid measurement tools must be used to evaluate research outcomes. For example, it would not be ideal to use a step test to determine one's aerobic capacity or to use infrared interactance to estimate one's body composition because these tools have a relatively large margin of error, especially if the change to be detected is rather small. If a treatment or supplement is said to have no effect, perhaps the particular method used in the study was not sensitive enough to pick up the small differences. A small change in performance (e.g., <3%) that is undetectable in a laboratory setting may determine success or failure in a sporting event.

Conclusions

The conclusions of a research study must logically follow the research outcomes and be supported by statistical analysis. Sometimes, investigators who study ergogenic aids extrapolate conclusions beyond what their data suggests. The implication and generalization of research findings must remain within the subjects studied, the context of measurements made, and the magnitude of the response. For example, increases in testosterone levels as a result of a dietary supplement reflect just that; they do not necessarily indicate an increase in muscle size and contractile function.

Correct interpretation of statistical analysis can be another obstacle to many consumers. Investigators must ensure that the appropriate inferential statistical analysis is used to quantify the potential that chance caused the research outcome. The finding of statistical significance of a particular treatment only means a high probability exists that the result did not occur by chance. One must also evaluate the magnitude of an effect for its impact on actual performance. For example, a reduction in the time taken to run 100 m by 0.5 s may not reach statistical significance, yet it could mean a difference between first and last place in a race.

Dissemination

One way to ensure that a research study is of high quality is to use a peer-review system. Most scientific journals require that reports of studies be reviewed by two or three experts in the field who did not take part in the research that is being evaluated. Before an article can be published in the journal, these scientists must agree that the experiments were well designed and conducted and that the results were analyzed and interpreted correctly. Peer review provides a measure of quality control over scholarship and interpretation of research findings. Publications in popular magazines or online journals do not undergo the same rigor of evaluation as peer review. High-quality nutrition articles can be found in peer-reviewed journals, such as the *American Journal of Clinical Nutrition, European Journal of Clinical Nurition, Journal of Nutrition, Journal of the American Dietetic Association, New England Journal of Medicine*, and *International Journal of Sports Nutrition and Metabolism.*

SPORTS FOODS

Products manufactured by companies such as MET-Rx, EAS, Power Bar, and Gatorade are some of the biggest sellers of sports foods in this market. Sport foods come in the form of bars, shakes, drinks, and gels, and they tend to be complex and food-like, and often contain one or more kinds of nutrient. The most salient feature of these products is that they contain energy sources, such as carbohydrate. Some of these products are consumed before, during, or after exercise, and others are meant to serve as partial or full meal replacements.

Sport bars

Sport bars represent one of the fastest-growing areas of the sport food industry. These products provide energy as well as other essential nutrients. Their composition can vary tremendously based on the intended consumer and purpose of the sports food. For example, some sports bars are marketed as a quick-release, high-energy source, so that they can be used before or after intense training or sports competition. Others are designed to be meal replacements that contain a high amount of protein and/or fiber. The energy and nutrient formulation for some of the more popular sport bars are presented in Table 9.2.

Carbohydrate is usually the energy foundation of sport bars that are to be consumed before and after exercise. In these products, corn syrup or high fructose corn syrup (HFCS) is the common carbohydrate ingredient, and both are based on partially digested cornstarch. Other carbohydrate ingredients include fruit juice concentrates and dried fruit, oat bran, brown rice, and rice crisps. Many sport bars also contain fiber. Having fiber allows a slower and more even absorption of carbohydrate, thereby producing a lower insulin response. However, sport bars rich in fiber should not be used just before or during exercise because it can cause intestinal discomfort.

The protein component of sport bars is largely based on proteins isolated from milk and/or egg whites because of their higher biological values (see Chapter 2).

Table 9.2 Nutrient composition of selected top-selling sports bars

SPORTS BAR	ENERGY (KCAL)	CARBOHYDRATE (G)	FIBER (G)	PROTEIN (G)	FAT (G)
Balance	200	22	<1	14	6
Promax	270	39	1	20	4.5
Ironman	230	20	0	16	7
Power Bar Protein Plus	300	38	1	23	6
Myoplex Deluxe	340	37	1	30	9
Power Bar Performance	230	45	3	10	2
Clif Bar	250	45	5	10	5
Zone perfect	210	21	1	15	7
Snickers/Marathon	220	32	2	10	7

Amino acids, such as branched-chain amino acids, are often added to create a more desirable composition. Manufacturers of sport bars often trademark their protein/amino acid source as a proprietary blend. Fat contributes energy, flavor, and sensory aspects of sport bars. However, it is not the focus among the energy nutrient ingredients.

The energy–nutrient ratio varies among sport bars depending on their purpose. Some sport bars have a carbohydrate–protein ratio of approximately $4:1$ or $3:1$, which is ideal for use during recovery. Some sports bars derive more than 60% of the energy from carbohydrate, which makes them a perfect choice for a pre-game meal. In some other sport bars, protein accounts for more than 50% of the energy, a value higher than carbohydrate. These bars are not designed to be an energy source, but rather used for enhancing protein synthesis and possibly muscle size.

Vitamins and minerals are typically added to sport bars, especially those directly involved in energy metabolism, such as B vitamins, magnesium, zinc, and iron. Adding nutrients such as vitamins C and E, as well as copper, iron, lipoic acid, and glutathione often reflects an attempt to optimize antioxidant status. In addition, some sport bars contain other ergogenic substances such as creatine, carnitine, and β-hydroxy-β-methylbutyrate (HMB).

Sports drinks

Sports drinks are popular among a broad range of athletes. Research has demonstrated that carbohydrate-containing sport drinks can enhance performance during endurance and intermittent high-intensity exercises and may also benefit competitive weight lifters. Sports drinks may be broken into two categories: (1) fluid and electrolyte replacement drinks in which the carbohydrate content is relative low; and (2) drinks that contain a higher carbohydrate formulation. The former is more appropriate for use during exercise, whereas the latter is better for consumption after training or in preparation for an upcoming event. This second category is also referred to as recovery or loading beverages. The composition of the two sport drink categories is listed in Table 9.3.

NUTRITION FACTS (240 ML OR 8 OZ)	FLUID AND ELECTROLYTE REPLACEMENT DRINK (GATORADE)	HIGH-CARBOHYDRATE ENERGY DRINK
Total energy	50 kcal	210 kcal
Total carbohydrate	14 g	52 g
Sugar	14 g	28 g
Other carbohydrate	0 g	24 g
Protein	0 g	0 g
Fat	0 g	0 g
Sodium	110 mg	135 mg
Potassium	30 mg	70 mg

Table 9.3 Comparison of energy and carbohydrate content of Gatorade and an energy drink

In sport drinks, carbohydrate is typically provided as glucose, sucrose, fructose, corn syrup, maltodextrins, and glucose polymers. These carbohydrates usually make up about 4–8% of a fluid/electrolyte replacement drink and >10% of a recovery/loading beverage. One of the most important considerations with regard to carbohydrate percentage is how it influences the rate of gastric emptying. As carbohydrates exceed 8% of the solution, the gastric emptying begins to slow down. Therefore, the fluid/electrolyte replacement drinks are often used during endurance events such as distance running and cycling, as well as intermittent sports such as soccer, field hockey, lacrosse, tennis, and hockey. This is because during exercise gastrointestinal motility and absorption decreases. As shown in Table 9.3, the amount of carbohydrates in the fluid/electrolyte replacement drinks is only one-quarter of what exists in the recovery/loading type drinks.

The purpose of fluid/electrolyte replacement drinks is to not only replace carbohydrate and water, but also provide sodium and chloride, which are the main electrolytes found in sweat and frequently subject to heavy losses during prolonged exercise. This type of sports drink also contains potassium, though in a smaller quantity. Potassium, along with sodium and chloride, plays an important role in neuromuscular function. Other ingredients that can be found in both types of sports drink include phosphorus, chromium, calcium, magnesium, iron, caffeine, and certain vitamins.

Whether or not carbohydrate consumption in amounts typically provided in sports drinks (4–8%) improves performance in events lasting one hour or less is controversial. Current research supports the benefit of this practice especially in athletes who exercise in the morning after an overnight fast, when liver glycogen is low. Providing exogenous carbohydrate under these conditions helps maintain blood glucose levels and improve performance. Accordingly, performance advantages in short-duration activities may not be apparent when exercise is done in the non-fasting state. For longer events, consuming 0.7 g of carbohydrate per kilogram of body weight per hour (approximately 30–60 g per hour) has been shown to extend endurance performance (Coggan and Coyle 1991; Currell and Jeukendrup 2008). Ingesting carbohydrate during exercise is even more important in situations in which athletes have not carbohydrate-loaded, consumed pre-exercise meals, or have restricted energy intake for weight loss. To be more effective, ingestion of carbohydrate should be done at 15–20-min intervals throughout exercise (McConell *et al.* 1996). If the same total amount of carbohydrate and fluid is ingested, the form of carbohydrate does not seem to matter – some athletes may prefer to use a sport drink, whereas others may prefer to eat a solid or gel and consume water.

Several factors may influence the rate of absorption of sports drink ingredients, including the temperature and concentration or osmolarity of a sport drink. With regard to temperature, cooler solutions (e.g., 5–15°C) may empty from the stomach more quickly than warmer or hot solutions. In addition, cooler drinks are more enjoyable and therefore may promote greater consumption. This is why sport drinks are often kept in coolers and poured into cups for athletes to consume during many sporting events. However, one should keep in mind that melting ice in a cooler dilutes the drink. Osmolarity is a measure of solute concentration of a solution and its tendency to draw water, and the greater the osmolarity of a solution, the greater the ability of this solution to attract water. As the particle concentration within the stomach and small intestine exceeds that in the blood or extracellular fluid, water is drawn in by osmotic force. This can in turn reduce the intestinal absorption of fluid, as well as carbohydrates. Sport drinks designed for use during exercise often contain lower concentrations of carbohydrate (4–8%) aimed to facilitate intestinal fluid absorption.

SPORTS SUPPLEMENTS

The term "sport supplements" is used here to be inclusive of both nutritional and pharmacological ergogenic aids. Indeed, of possibly more than 500 supplements on the market, some of them are common nutrients or their derivatives, whereas others can be considered chemical agents purported to enhance sport performance. Despite such a large quantity of ergogenic aids, the fundamental mechanism of each may be explained by one or more of the following: (1) acts as a central or peripheral nervous system stimulant; (2) increases storage or availability of a limiting substrate; (3) acts as a supplemental fuel source; (4) reduces performance-inhibiting metabolic by-products; (5) facilitates recovery; and (6) enhances tissue synthesis. The following includes a more in-depth discussion on some of the commonly used ergogenic aids. Readers can consult Table 9.4 for a quick reference in terms of their description, action, and major claims. This section will not include supplements that were discussed previously, such as branched-chain amino acids (Chapter 7), high glycemic index carbohydrates (Chapter 8), and carbohydrate loading (Chapter 8).

Arginine, ornithine, and lysine

Lysine is an essential amino acid and arginine is considered conditional or semi-essential because it may become essential during periods of growth. Ornithine is a non-essential amino acid not found in proteins but important for efficient nitrogen removal in the urea cycle. This amino acid is supposed to also enhance the efficiency of intestinal absorption. These amino acids can be purchased as individual supplements or in a combination sometime marketed as "natural growth hormone." This is

Table 9.4 Description of selected sports supplements and their ergogenic claims

ERGOGENIC AIDS	DESCRIPTION	ACTIONS/CLAIMS
Arginine, ornithine, and lysine	Lysine: an essential AA; arginine: semi-essential AA; ornithine: non-essential AA not found in proteins but important for nitrogen removal	Increases growth hormone release, thus muscle development
Biocarbonate	A chemical buffer found primarily in extracellular fluid	Maintains acid–base balance by buffering hydrogen ions produced from working muscle, thereby improving anaerobic performance and performance of intense endurance events
Boron	A trace mineral involved in bone mineral metabolism, steroid hormone metabolism, and membrane functions	Increases testosterone production that helps with tissue building and anabolic actions
Caffeine	Naturally occurring substance found in coffee, tea, and chocolate	Improves cognitive function, increases fat use and spares muscle glycogen, which benefits both anaerobic and aerobic performance
Chromium	A trace mineral found in foods such as brewer's yeast, cheese, broccoli, wheat germ, nuts, liver, and egg yolk	Potentiates insulin action and thus helps with anabolic tissue building
Carnitine	A substance found in relatively high quantities in meat	Functions as a carrier protein that transports long-chain fatty acids into mitochondria, thereby improving endurance performance
Coenzyme Q10	Also referred to as ubiquinone and an integral component of the mitochondrion's electron transport system	Plays an important role in oxidative phosphorylation, thereby improving aerobic capacity and endurance performance
Creatine	A nitrogen-containing molecule that is produced from the liver, kidneys, and pancreas	Augments PCr levels and buffers hydrogen ions, thereby improving performance of ultra-short-/short-term events
DHEA and androstenedione	Precursors to testosterone and produced mainly in the adrenal glands	Improves lean body mass, thereby improving strength. DHEA may also enhance immune function and protect against cardiovascular diseases
Ephedrine	Also referred to as Ma-Huang and naturally occurred in some botanicals	Functions as a stimulant like catecholamine to improve body composition and both aerobic and anaerobic performance
Glutamine	A naturally occurring non-essential AA and the most abundant AA in human muscle and plasma	Enhances protein synthesis and protects against infections or illness associated with exhaustive exercise
Glycerol	A component of the triglyceride molecule and an important constituent of the cells' phospholipid plasma membrane	Induces hyperhydration, decreases heat stress, and improves performance
HMB	A metabolite of the essential AA leucine; also found in red meats, catfish, asparagus, cauliflower, and grapefruit	Reduces muscle damage and suppresses protein degradation associated with intense physical effort
Inosine	A nucleoside comparable to adenine, which is one of the structural components of ATP	Increases ATP stores, favoring strength and power athletes, and is involved in the production of 2,3-DPG, which facilitates the delivery of oxygen from oxyhemoglobin compound
Phosphate	A major mineral found mainly in bones, and also a component of ATP and 2,3-DPG	Increases ATP synthesis and oxygen extraction in muscle cells, thereby improving performance of both anaerobic and aerobic events

because supplementation with these three amino acids has been proposed to augment the growth hormone level in circulation, leading to greater muscle development (Chromiak and Antonio 2002).

Such a claim is based on early studies involving individuals who had suffered significant burns. For example, it was found that when large doses of ornithine were given to burn patients intravenously, their blood growth hormone levels were increased (Donati *et al.* 1999). It was also found that these patients established a positive nitrogen balance more quickly in the days that followed the treatment (Donati *et al.* 1999; De Bandt *et al.* 1998). An increased release of growth hormone was also demonstrated in healthy but untrained subjects who underwent oral supplementation of combined arginine and lysine (Isidori *et al.* 1981; Suminski *et al.* 1997). However, studies that involved weight lifters and body builders failed to show an increase in blood growth hormone concentrations when these amino acids were given orally (Fogelholm *et al.* 1993; Lambert *et al.* 1993). These athletes regularly experience a natural increase in growth hormone due to their training bouts. This may have limited the ability of these athletes to benefit further from amino acid supplementation. An important consideration regarding the efficacy of amino acid supplementation is that even in the studies that observed an increase in growth hormone, the research protocols did not extend to assessing changes in lean body mass, strength, or anaerobic performance. Studies involving burn patients provided the three amino acids intravenously and at high dosages. Therefore, these amino acids, once entered into circulation, can exert their effect directly. On the other hand, oral ingestion of these amino acids requires that they enter the liver first before being further used. It is likely that the liver can metabolize most of these amino acids, with the remainder being insufficient to produce an action.

Biocarbonate loading

Dramatic alterations in acid–base balance of the intracellular and extracellular fluids occur when maximal exercise is performed between 30 s and several minutes, such as 400-m, 800-m, and 1500-m running, track cycling, and speed skating events. This is because muscle fibers rely predominately on anaerobic energy transfer. As a result, significant quantities of lactate accumulate, with a concurrent fall in intracellular pH. An increased accumulation of H^+ in muscle cells can reduce the calcium sensitivity of the contractile proteins, thereby impairing muscle function (Chin and Allen 1998; Street *et al.* 2005).

The body has several systems to adjust and regulate acid–base balance. Chemical buffers provide a very effective and rapid way of normalizing the H^+ concentration. Other systems include exhalation of CO_2 via pulmonary ventilation and excretion of H^+ via the kidneys. The primary chemical buffers in the muscle are phosphates and tissue proteins. The most important buffers in the blood are proteins, hemoglobin, and bicarbonate. During intense exercise, as intracellular buffers are insufficient to buffer all the hydrogen ions formed, efflux of H^+ into the circulation increases. In this context, maintaining high levels of extracellular **bicarbonate** can facilitate the release of H^+ from the cells and thus delay the onset of intracellular acidosis. The mechanism by which bicarbonate supposedly exerts its action is through the buffering of H^+ in the blood, not in the muscle as is often claimed. The buffering of H^+ in the blood, however, increases the efflux of H^+ from the muscle. The following illustrates the process by which bicarbonate (HCO_3^{1-}) acts against excessive acid production:

$$HCO_{3-}^1 + H^+ \leftrightarrow H_2CO_3 \leftrightarrow H_2O + CO_2$$

Research in this area has produced conflicting results, but they appear to stem from diverse doses of bicarbonate or different types of exercise used to evaluate the ergogenic effects of bicarbonate loading. It appears that a minimal dose of bicarbonate ingestion is needed to improve performance (Horswill 1995). A dose of 200 mg or higher per kilogram of body weight, ingested 1–2 hours before exercise, seems to improve performance in most studies that used exercises that lasted longer than 1 min, whereas doses less than 100 mg per kilogram of body weight do not affect performance. A 300 mg kg dose seems to be optimum from a performance perspective, but doses higher than 300 mg kg can be accompanied by gastrointestinal problems, including bloating, abdominal discomfort, and diarrhea.

No ergogenic effect emerges for typical resistance exercises or exercises that last less than 1 min (e.g., squat, bench press, and jump). This can be attributed to the fact that these ultra-short-term activities generally have lower absolute anaerobic metabolic load compared with continuous, maximal whole-body activities. Bicarbonate loading with all-out effort of less than 1 min improves performance only with repetitive exercise bouts, which in accumulation can produce high intracellular H^+ concentrations (Bishop *et al.* 2004). Bicarbonate loading does not benefit low-intensity, aerobic exercise because pH and lactate remain near resting levels. However, some research indicates there are benefits in aerobic exercise of high intensity (McNaughton *et al.* 1999; Potteiger *et al.* 1996). For example, using a 30-km time trial, Potteiger *et al.* (1996) found that the race times of trained male cyclists were better after consuming a buffering solution (500 mg per kilogram of body weight) before exercise than in placebo trials, and this increase in performance was accompanied by an elevated level of blood pH throughout the exercise.

Boron

Boron is an essential trace mineral involved in bone mineral metabolism, steroid hormone metabolism, and membrane functions. It is found in non-citrus fruits, leafy vegetables, nuts, and legumes. Boron has been studied in relation to osteoporosis. One of these studies found that 48 days of boron supplementation increased estrogen and testosterone levels of post-menopausal women (Nielsen *et al.* 1987). It also decreased the excretion of calcium, phosphorous, and magnesium in urine. Therefore, it appears that boron supplementation helps improve bone mineral density. The finding that boron supplementation increased testosterone level has been singled out disproportionally and extrapolated to the claim that it may improve muscle growth and strength. However, what is often overlooked is that the participants of the study were post-menopausal women who were on a boron-deficient diet for four months. In addition, one must be aware that boron supplementation raised serum estrogen levels as well. In studies involving athletic populations, boron supplementation has not been proven to increase testosterone levels. For example, male body builders who took 2.5 mg of boron daily for seven weeks had no changes in measures of testosterone, lean body mass, or strength as compared to the placebo condition (Ferrando and Green 1993). On the basis of these studies, boron supplementation does not appear to confer any ergogenic benefit.

Caffeine

Caffeine occurs naturally in a variety of beverages and foods, including coffee, tea, and chocolate (Table 9.5). It is consumed by most adults in the world. It has been estimated that the daily intake of caffeine for an average adult is approximately 3 mg kg^{-1}, 80% of which is consumed in the form of coffee (Barone and Roberts 1996). Caffeine is recognized as a food and a drug in both the scientific and regulatory domains. It is, however, not a typical nutrient and is not essential for health. There are several over-the-counter medications containing 30–100 mg of caffeine. These include cold medicines, diuretics, weight-loss products, and preparations that help people stay awake. Caffeine can be rapidly absorbed in the digestive tract and distributed to all tissues. It can also easily cross the blood–brain barrier to reach the tissue of the central nervous system.

Unlike ephedrine, the existing literature appears to be more definitive in support of ergogenic effects of caffeine on cognitive and physical performance. On the other hand, caffeine alone is not as effective in producing weight loss. Caffeine's cognitive and behavioral effects have been documented in a number of well-controlled studies using males, females, young and elderly volunteers. Effects on particular aspects of cognitive function, as well as effects on mood state, are generally consistent with common perception of caffeine as a compound that increases mental energy and performance. For example, using a double-blind, placebo-control protocol, Fine *et al.* (1994) reported that a single dose of 200 mg of caffeine improves visual vigilance in rested volunteers. Similar effects have been documented with doses equivalent to a

Table 9.5 Caffeine content of some common foods, beverages, and medicines

COMMON PRODUCT	SERVING SIZE	CAFFEINE (MG)
Baking chocolate	28 g (1 oz)	45
Chocolate candy	57 g (2 oz)	45
Chocolate milk	237 ml (8 oz)	48
Mello Yello	355 ml (12 oz)	51
Mountain Dew	355 ml (12 oz)	54
Cola beverages	355 ml (12 oz)	32–65
Instant coffee	177 ml (6 oz)	54–75
Brewing coffee	177 ml (6 oz)	150–200
Ice tea	355 ml (12 oz)	150
Hot tea	177 ml (6 oz)	65–105
Aspirin products	Standard dose	30–125
Vivarin tablets	1 tablet	200

single serving of a cola beverage (~40 mg) up to multiple cups of coffee (Lieberman *et al.* 1987). However, with a very high dose of caffeine (~500 mg), cognitive performance was reported to decrease (Kaplan *et al.* 1997). This finding indicates that a dosage that is achievable via diet would be sufficient in order to procure cognitive benefits of caffeine. Caffeine has also been shown to enhance cognitive performance such as reasoning and memory when an individual's mental ability has decreased due to sleep deprivation (Penetar *et al.* 1993).

With respect to sports performance, the most consistent observation is that caffeine can increase the time to exhaustion during submaximal exercise bouts lasting approximately 30–60 min, though results are more inconsistent when activities of shorter duration are examined. Caffeine as an ergogenic substance has been evaluated for several decades. Despite an overwhelming agreement on caffeine's ability to improve endurance performance, the precise mechanism as to how this compound exerts its ergogenic effect remains elusive. For years it has been postulated that caffeine causes glycogen sparing and therefore prolongs endurance performance. However, in terms of the exercise duration for which caffeine was found to be effective, it is unlikely that muscle glycogen would be depleted and thus serve as a limiting factor. In a recent study by Graham *et al.* (2000), who employed a muscle biopsy technique, it was found that the ingestion of caffeine at 5 mg kg^{-1} did not alter muscle glycogen utilization. Most studies have used a rather small sample size (approximately eight subjects), which may have reduced the statistical power needed to detect the difference. Nevertheless, by pooling together the studies conducted in the same laboratory but at different times that involved a total of 37 subjects, Graham *et al.* (2001) failed to observe a significant difference in muscle glycogen utilization between the caffeine and placebo conditions.

Another metabolic claim associated with caffeine is that it increases fat utilization. This notion was derived from early studies in which caffeine resulted in a larger decrease in muscle triglycerides concomitant with less utilization of muscle glycogen (Essig *et al.* 1980). It has been speculated that caffeine increases fat oxidation by augmenting lipolysis, a process stimulated by the release of epinephrine and nor-epinephrine. However, according to more recent studies by Raguso *et al.* (1996) and Graham *et al.* (2000), variables reflecting intracellular utilization of fatty acids such as fat oxidation and uptake of fatty acids by muscle remained unaffected by caffeine. These findings provide little support for the theory that caffeine increases fat oxidation.

Carnitine (L-carnitine)

L-Carnitine, a substance found in relatively high quantities in meat, has received a lot of attention over the previous 20 years. As a supplement, it has been very popular among athletes, especially after the rumors circulated that it helped the Italian national soccer team to become world champions in 1982. L-carnitine can be synthesized from lysine and methionine, and synthesis occurs in the liver and kidneys. Therefore, even when dietary sources are insufficient, the body can produce enough from lysine and methionine to maintain normal storage. As discussed in Chapter 7, carnitine functions as a carrier protein that helps transport long-chain fatty acids into the mitochondrial matrix so they can be oxidized. In this context, carnitine is often regarded as a "fat burner." It is assumed that oral ingestion of carnitine increases the muscle carnitine concentration. This will then cause an increase in fat oxidation and a gradual loss of the body fat stores. However, several studies showed that oral ingestion of carnitine up to 14 days produces no change in muscle carnitine concentration.

L-carnitine supplementation has also been thought to be ergogenic in improving endurance performance. This belief is based on the similar assumption mentioned earlier that oral ingestion of carnitine increases the total carnitine concentration in the muscle and that the increase in muscle carnitine increases the oxidation rate of plasma fatty acids and intramuscular triglycerides. This increased fat oxidation will in turn reduce muscle glycogen breakdown and postpone fatigue. However, the results of nearly all of the experimental trials have not revealed a positive influence on either fatty acid utilization or glycogen sparing, nor does carnitine supplementation delay fatigue (Heinonen 1996; Wagenmakers 1999). In addition, direct measurements of muscle following 14 days of carnitine supplementation failed to show increases in muscle carnitine concentration (Barnette *et al.* 1994; Vukovich *et al.* 1994). It has been found that carnitine helps in preventing lactic acid from increasing via its effect on the maintenance of the acetyl-CoA:CoA ratio (Bremer 1983). Nevertheless, Trappe *et al.* (1994) found that lactate accumulation, acid–base balance, or performance in five 100-yard swims with 2-min rest intervals did not differ between competitive swimmers who consumed 2 g of L-carnitine twice per day for seven days and swimmers who consumed the placebo.

L-carnitine also acts as vasodilator in peripheral tissues, thus possibly enhancing regional blood flow and oxygen delivery. In one study, each subject took either L-carnitine supplements (3 g per day for three weeks) or an inert placebo to evaluate the effectiveness of L-carnitine supplementation on delayed onset of muscle soreness (DOMS) (Giamberardino *et al.* 1996). They then performed eccentric muscle actions to induce muscle soreness. Compared with the placebo condition, subjects who took the treatment experienced less post-exercise muscle pain and tissue damage as indicated by lower plasma levels of creatine kinase. These findings suggest that the vasodilation property of L-carnitine might improve oxygen supply to injured tissue and promote clearance of muscle damage by-products, thus reducing DOMS.

Chromium

Chromium is a trace mineral that is present in foods such as brewer's yeast, cheese, broccoli, wheat germ, nuts, liver, apple with skin, asparagus, mushroom, and egg yolk. As discussed in Chapter 4, chromium potentiates insulin action and insulin stimulates the glucose and amino acid uptake by muscle cells. The stimulated amino acid uptake is thought to increase protein synthesis and thus muscle mass. For this reason, chromium is considered ergogenic in that it helps athletes improve their strength and power. Chromium supplements are marketed mainly as chromium picolinate. Picolinate is derived from the amino acid tryptophan and binds with chromium in order to enhance intestinal absorption of chromium. Chromium exists mainly as a positively charged ion (Cr^{+3}). Thus, combining with picolinate would decrease the potential interaction between chromium and other negatively charged substances in food such as phytates. Touted as a "muscle builder," chromium represents one of the best-selling mineral supplements in the United States, second only to calcium.

The ergogenic impact of chromium supplementation was demonstrated in earlier studies in which college students and football players were given 200 μg of chromium picolinate or a placebo each day for 40 days while they were on a resistance-training program (Evans 1989). It was found that those with chromium supplementation gained significantly more lean body mass as compared to the control group, although lean body mass was estimated from skinfold thickness. However, later studies (Clancy *et al.* 1994; Hallmark *et al.* 1996; Hasten *et al.* 1992; Lukaski *et al.* 1996) that used more sophisticated techniques to determine body composition have not been able to confirm the results of Evans (1989). For example, in a study by Lukaski *et al.* (1996), chromium supplements in nearly the same amount as used by Evans (1989) were provided to a group of young men who were also in a resistance-training program, but no change in body composition and strength were observed between the treatment and placebo groups. In another study by Clancy *et al.* (1994), football players received 200 μg of chromium picolinate for each day for nine weeks during a strength-training program. Here, again, supplementation failed to independently affect percentage body fat, lean body mass, and muscular strength. Clearly, the ergogenic benefits of chromium supplementation remain questionable.

Caution must be exercised in the use of chromium supplements. No studies have evaluated the safety of long-term supplementation with chromium picolinate or ergogenic efficacy of supplementation in individuals with less-than-optimal chromium status. Concerning the bioavailability of trace minerals in the diet, excessive dietary chromium can inhibit zinc and iron absorption. In an extreme case, this could result in iron-deficient anemia, blunt the ability to train intensely, and negatively affect the performance of exercises that depend on a high level of oxygen supply. Based upon laboratory studies of cultured cells, an accumulation of chromium picolinate in cells was shown to cause chromosome damage (Stearns *et al.* 1995), although this finding has yet to be demonstrated in humans.

Coenzyme Q_{10} (ubiquinone)

Coenzyme Q_{10} is often referred to as CoQ_{10} or ubiquinone. It is found primarily in meats, peanuts, and soybean oil. It functions as an integral component of the mitochondrion's electron transport system and, therefore, plays an important role in oxidative phosphorylation. This lipid-soluble natural component of all cells exists in high concentrations within myocardial tissues. CoQ_{10} has been used clinically to treat cardiovascular diseases and promote recovery from cardiac surgery on account of its antioxidant properties, which promote scavenging of free radicals (Vasankari *et al.* 1997). Due to its positive effect on oxygen uptake and exercise performance in cardiac patients, CoQ_{10} has been considered as an ergogenic aid for endurance performance. The rationale behind this consideration is that supplementation could increase the flux of electrons through the respiratory chain and thus augment aerobic re-synthesis of ATP.

Based on the current literature, it appears that supplementation of CoQ_{10} increases serum CoQ_{10} levels, but it does not improve aerobic capacity and endurance performance (Braun *et al.* 1991; Snider *et al.* 1992; Zuliani *et al.* 1989). For example, a study that provided male cyclists with 100 mg of ubiquinone everyday for eight weeks failed to demonstrate improvements in cycling performance, VO_2max, or lipid peroxidation (Braun *et al.* 1991). Likewise, when triathletes were given 100 mg of ubiquinone in a daily supplement that also contained vitamin E, inosine, and cytochrome c for four weeks, no differences in endurance performance and blood glucose, lactate, and free fatty acid concentrations at exhaustion were observed between treatment and placebo conditions (Snider *et al.* 1992). In a study in which the tissue CoQ_{10} level was measured, Svensson *et al.* (1999) reported that ingestion of 120 mg of CoQ_{10} daily for 20 days resulted in marked increases in plasma CoQ_{10} concentration. However, the muscle CoQ_{10} levels as well as other cellular activities such as lipid peroxidation and mitochondrial function remained unaltered. This finding may explain why most studies failed to demonstrate ergogenic values of CoQ_{10} supplementation despite the fact that supplementation increased plasma CoQ_{10} concentration.

Creatine

Creatine is a nitrogen-containing molecule that is produced by the liver, kidneys, and pancreas. Because creatine can be synthesized internally, it is not considered an essential nutrient. In normal, healthy individuals, creatine is synthesized at about 1–2 g per day. At approximately the same rate, creatine is broken down into creatinine and excreted in the urine. Under normal circumstances, the degradation of creatine is matched by its synthesis. In strength and power athletes, however, the rate of creatine breakdown is expected to be much higher. Therefore, adequate dietary creatine becomes important for obtaining required amounts of this compound. Because creatine is found in meat products such as meat, poultry, and fish, vegetarians are at a distinct disadvantage in obtaining ready sources of exogenous creatine.

Creatine is synthesized from arginine, glycine, and methionine. Once synthesized, creatine is transported via the blood from the liver and kidney to the muscle. Muscle takes up creatine against the concentration gradient by an active transport process. A 70-kg (154-lb) man may have a creatine pool of roughly 120 g or a little more than 0.25 lb, 95% of which is in muscle tissues. Skeletal muscle represents the largest reservoir of creatine among three types of muscle tissues (skeletal, smooth, and cardiac muscle tissue). About 40% of the total exists as free creatine; the remainder combines readily with phosphate to form phosphocreatine (PCr). Type II or fast-twitch muscle fibers store about 30% more creatine than type I fibers.

As discussed in Chapter 6, the body has limited storage of ATP. During maximal exercise the ATP stores can only provide energy for several seconds. With ATP falling by 30%, the muscle fatigues (Hultman *et al.* 1991). PCr serves as the "energy reservoir" to provide rapid phosphate-bond energy to re-synthesize ATP and to maintain ATP concentration close to resting levels. In doing so, the maximal work of muscle can be sustained for a longer period of time. This process occurs during the first few seconds of high-intensity exercise, and thus allows time for other more sustained glycogen breakdown and glycolysis to speed up to the required rate. Another important function of PCr is that it helps with buffering capacity for hydrogen ions. Hydrogen ions are used during ATP regeneration. Therefore, as PCr continues to be used to phosphorylate ADP to ATP, less free hydrogen ions will be available. The increased availability of PCr also lessens reliance on energy transfer from anaerobic glycolysis that results in lactate formation.

Creatine became a popular supplement after the 1992 Olympics in Barcelona. Gold medal winners Linford Christie in the men's 100-m dash and Sally Gunnell in the women's 400-m hurdles supposedly used creatine supplements. By the Olympics of Atlanta 1996, approximately 80% of all athletes used creatine. Since then, creatine has become one of the most popular ergogenic aids used by athletes worldwide. Creatine is usually marketed in the form of creatine monohydrate. Most studies use a creatine-loading regimen of 20–25 g per day in four portions of 5–6 g each, given at different times of the day for 5–7 days. This is because a similar loading regimen has been shown to increase the muscle creatine concentration by 20% (Hultman *et al.* 1996). Hultman *et al.* (1996) also found that a subsequent dose of 2 g per day was enough to maintain the high total creatine concentration for 35 days, whereas stopping creatine supplementation after six days caused a slow, gradual decline of the creatine concentration in muscle. Another approach suggested by this same research group is to take creatine at a constant dose of 3 g per day, but in order to achieve the similar total muscle creatine concentration one would need to maintain daily consumption for about one month. It was also found that creatine taken in conjunction with carbohydrate would produce greater creatine retention compared with creatine alone (Green *et al.* 1996). It is considered that by adding carbohydrate the supplement would trigger a greater release of insulin, which would then increase tissue uptake of creatine.

Research regarding the effect of creatine supplementation on performance in humans began in the early 1990s. Since then it has quickly proliferated. Of the studies published, a majority of them have been in favor of this ergogenic aid. In general, oral supplements of creatine monohydrate at 20–25 g per day increases muscle creatine and performance of both men and women in intense exercise, particularly repeated intense bouts of muscular effort. Even daily doses as low as 6 g for five days elicits improvement in repeated power performance. The exercises used in the studies that observed the ergogenic effect of creatine supplements are mainly sprint or repeated sprint events that involve running, cycling, and swimming. Supplementation has not been shown to improve endurance performance, but if a competition involves sprinting such as cycling breakaway or final running sprints, creatine may be of benefit. It has been postulated that creatine could facilitate aerobic energy production and enhance endurance performance because PCr provides a shuttle system for the transfer of high-energy phosphate groups from mitochondria, where ATP is produced, to contractile myofibrils, where energy is used for causing muscle contraction. However, this theory remains to be tested. Endurance athletes must also consider the potential weight gain that may accompany creatine supplementation.

Several studies have reported an increase in total body mass and lean body mass and a corresponding decrease body fat percentage. It is likely that creatine causes water retention in skeletal muscle cells because of an increase of the intracellular osmolarity of the muscle cells. Many believe that the creatine-induced increase in water retention leads to increased protein synthesis, although in the short term (5–6 days), the increase in protein synthesis may not be of great magnitude. Such favorable change in body composition has also been attributed to high-quality training due to creatine supplementation. The increase in body mass may be beneficial. However, in sports that involve weight-bearing activities, such as running and gymnastics, the weight gain caused by creatine supplementation could have a negative impact on performance. Figure 9.1 illustrates the proposed mechanisms for how elevating intramuscular creatine and phosphocreatine enhances intense, short-term performance and improves body composition.

DHEA and androstenedione

Dehydroepiandrosterone (DHEA) and its sulfated metabolite (DHEAS) are produced in the body by the adrenal glands. Both DHEA and DHEAS circulate to peripheral tissues, where they are converted into androgens (including testosterone) and/or estrogen (Figure 9.2). DHEA is derived from cholesterol and can be converted to an antrostenedione and then to testosterone. The ability of tissue to covert DHEA to androstenedione, to testosterone, and/or estrogen relies on the presence of steroidogenic and/or a metabolizing enzyme system. Both androstenedione and testosterone can be converted to estrogen via aromatase. Many tissues produce these converting enzymes, including gonads (testes and ovaries), liver, kidneys, and adipose tissue. This allows the conversion of both hormones to be regulated at the tissue level. Skeletal muscle lacks the ability to convert androstenedione to testosterone.

Body levels of DHEA are high in young adulthood, with peak production occurring between 18 and 25, and gradually decrease with age. In contrast to the glucocorticoid and mineralocorticoid (i.e., aldoesterone) adrenal steroids, whose plasma levels remain relatively high with aging, a long and slow decline in DHEA begins after 30. This has fueled speculation that plasma DHEA might play a role in biological aging and its related dysfunctions or diseases. As such, it is believed that supplementing with DHEA blunts the negative effects of aging by raising the plasma level of DHEA to

Figure 9.1 Possible mechanisms of how creatine supplementation works in improving performance and body composition.

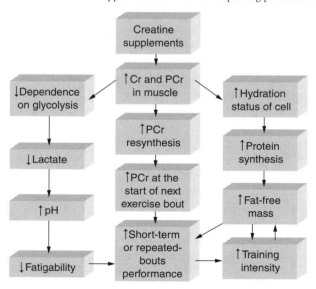

more "youthful" concentrations (Percheron *et al.* 2003). Popular claims for using DHEA include that it: (1) blunts aging; (2) facilitates weight loss; (3) boosts immune function; (4) protects against heart disease; and (5) increases muscle mass.

Treatment with DHEA has been shown to have beneficial effects in preventing diseases, such as cancers, atherosclerosis, viral infections, obesity, diabetes, and enhancing immune function and life span, but results come from studies that used rodents. Scientists have argued that the findings from research on rats and mice who produce little, if any, DHEA, do not necessarily apply to healthy humans. Results from human studies are mixed. Evidence that supports the use of DHEA comes from the cross-sectional comparisons relating levels of DHEA to risk of death from heart disease. However, it was found that a high DHEA level conferred cardio-protective effects only in men, not in women (Jakubowicz *et al.* 1995). In the context of ergogenic aids, the DHEA-induced improvement in body composition, immune function, and muscular strength have been reported in studies involving middle-aged and elderly adults (Villareal and Holloszy 2006). However, research in young men who were given DHEA at comparable doses failed to demonstrate any positive effect on serum testosterone, lean body mass, and muscular strength (Percheron *et al.* 2003; Wallace *et al.* 1999). Therefore, despite its popularity among exercise enthusiasts, no data exist concerning ergogenic effects of DHEA supplements on young adults. Use of DHEA supplements for the elderly seems to be more promising, but its safety requires further investigation.

Androstenedione and related compounds such as androstenediol and norandrostenediol function similarly to DHEA. Androstenedione is an intermediate or precursor hormone between DHEA and testosterone (Figure 9.2). Normally produced by the adrenal glands and gonads, it converts to testosterone by 17-β-hydroxysteroid

Figure 9.2 Metabolic pathways for producing DHEA and androstenedione.

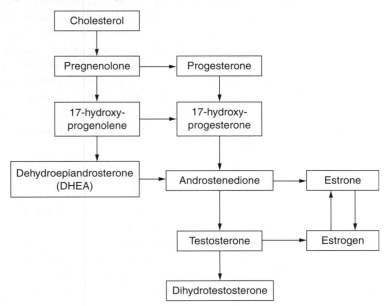

dehydrogenase, found in diverse tissues. Androstenedione also serves as an estrogen precursor. Originally developed by East Germany in the 1970s to enhance performance of their elite athletes, androstenedione was first made commercially available in the United States in 1996. Androstenedione received considerable notoriety during the 1998 baseball season when Mark McGwire, who established a home run record at that time, acknowledged using the supplement. Subsequently, androstenedione-related products flooded the marketplace for resistance-training individuals, even though no reputable research was available supporting the beneficial effect.

It remains controversial whether androstenedione supplementation can increase the level of testosterone in the blood. Leder *et al.* (2000) observed that although a week-long supplementation of androstenedione at 100 mg per day did not alter serum testosterone levels, at 300 mg it elevated testosterone levels by 24%. In this study, however, no performance was measured. On the other hand, when supplementing subjects with androstenedione at 300 mg per day for eight weeks, King *et al.* (1999) failed to observe any significant effects on serum testosterone, body fat, lean body mass, muscle fiber diameters, or muscle strength, although serum androstenedione levels increased 100% in the treatment group. Of particular interest is that they noted a significant increase in serum estradiol and estrogen concentrations, suggesting an increased aromatization of ingested androstenedione to estrogen instead of testosterone. Based on these findings, it appears that androstenedione supplementation does not predictably elevate serum testosterone levels. Even though some researchers found significant increases in testosterone levels in the blood, these increases were not accompanied by favorable changes in protein synthesis, muscle mass, or strength.

Concerns exist about the effect of long-term DHEA and androstenedione supplementation on body function and overall health. Converting these anabolic hormone precursors into potent androgens like testosterone promotes facial hair growth in females and alters normal menstrual function. As with exogenous anabolic steroids, DHEA or androstenedione through supplementation might stimulate the growth of the prostate gland, which can lead to a tumor. They may also accelerate the growth of cancer cells if cancer is present. In addition, a lowering of HDL-cholesterol has been reported in a few studies (Broeder *et al.* 2000; Brown *et al.* 2000). A low level of HDL-cholesterol is considered a risk factor for heart disease. The IOC and NCAA have placed both DHEA and androstenedione on their banned substance list at zero tolerance level. Supplementation of these substances may influence the ratio of testosterone to epitestosterone (T/E), which is used to screen for steroid doping by organizations such as the IOC and NCAA.

Ephedrine

Ephedrine is a sympathomimetic agent that is structurally related to catecholamine. It is found in several species of the plant ephedra and has been used for thousands of years as a herbal medicine. Ephedra, which is also called Ma-Huang, is the dry stem of a plant that is indigenous to China, Pakistan, and northwestern India. While ephedrine occurs naturally in some botanicals, it can also be synthetically derived. Ephedrine is widely available in over-the-counter remedies for nasal congestion and hay fever. It has also been used as a central stimulant to treat depression and sleep disorders.

Ephedrine is a potent chemical stimulant with a variety of peripheral and central effects. It acts by enhancing the release of nor-epinephrine from sympathetic neurons and is also a potent β-adrenergic agonist. In this regard, ephedrine has been known for its effect of stimulating bronchodilation, increasing heart rate and cardiac contraction, and augmenting energy expenditure and fat oxidation. Ephedrine will also function to suppress appetite and food intake through adrenergic pathways in the hypothalamus.

Ephedrine is recognized as an anti-obesity drug mainly due to its stimulating effect upon the sympathetic nervous system. Ephedrine's potential for weight loss was first reported in 1972. It was noted that asthmatic patients being treated with a compound containing ephedrine, caffeine, and phenobarbitol experienced unintentional weight loss. This unexpected discovery has led to a series of investigations aimed to authenticate whether ephedrine is indeed effective in facilitating weight loss and whether it warrants safety concern despite its efficacy. Among the early studies in which ephedrine was the only substance tested, results of energy expenditure and weight loss are controversial. Astrup *et al.* (1985) noticed a sizable increase in resting oxygen consumption and a significant reduction in body weight in obese women who were treated with ephedrine at 60 mg per day. However, Pasquali *et al.* (1987a) found that administration of ephedrine at a dose of 75 mg or 150 mg per day produced essentially no effect compared to the placebo condition in obese individuals. In this study, side effects such as agitation, insomnia, headaches, palpitations, giddiness, tremors and constipation were reported in the group treated with 150 mg per day. This same research group, however, observed that when ephedrine was given at 150 mg per day in conjunction with a more stringent hypocaloric diet (~1000 kcal per day), obese women did experience significant weight loss (Pasquali *et al.* 1987b). It appears that the efficacy is questionable, especially when ephedrine is given alone.

Ephedrine has also been promoted as a performance-enhancing or ergogenic aid. However, most studies have not demonstrated any kind of improvement in athletic performance following ingestion of ephedrine alone at a dose generally considered safe (<120 mg). DeMeersman *et al.* (1987) found no significant effects of ephedrine administered in a dose of 40 mg on sustained aerobic exercise. Swain *et al.* (1997) also failed to observe increases in VO_2max and endurance time to exhaustion following consumption of pseudoephedrine at doses of 1–2 mg kg^{-1}. Ephedrine also seems to be ineffective with respect to its impact on anaerobic performance. Chu *et al.* (2002) reported that ingestion of pseudoephedrine of 120 mg two hours before testing did not improve muscular strength as measured by intermittent isometric contraction and anaerobic performance as measured by the Wingate cycling test (a 30-s maximal cycle against a pre-determined resistance). It may be argued that a threshold dosage level exists for the ergogenic effects of ephedrine to manifest, as the peak weight-lifting performance was improved when taking a high dose (180 mg) of pseudoephedrine (Gill *et al.* 2000). Nevertheless, in a more recent study in which subjects were fed with 240 mg of pseudoephedrine, Chester *et al.* (2003) failed to demonstrate an ergogenic effect, although the testing protocol of this study involved aerobic rather anaerobic performance.

More recently, a great deal of research has been undertaken to investigate the ergogenic effect of ephedrine in combination with caffeine. There is a general consensus that the combined use of ephedrine and caffeine is of greater ergogenic benefit than each compound used alone. For instance, Bell *et al.* (2001) demonstrated a significant increase in power output during a 30-s Wingate test following the ingestion of ephedrine (1 mg kg^{-1}) alone and in combination with caffeine (5 mg kg^{-1}) as compared with caffeine alone and placebo. Using the same treatment paradigm, this same research group also found a significant reduction in completion time for a 10-km race following ingestion of ephedrine (0.8 mg kg^{-1}) alone and in combination with caffeine (4 mg kg^{-1}) as compared with caffeine alone and placebo (Bell *et al.* 2002). The mechanism responsible for the performance advantages of this combined approach is unclear. It may be related to the hypothesis that caffeine serves to prolong the ephedrine-induced adrenergic effect. It should be noted that the amount of ephedrine used by Bell *et al.* is about twice as potent as the pseudoephedrine used in many early studies. This may also have contributed to positive findings shown by most of Bell's publications.

The most common side effects for ephedrine are agitation, insomnia, headaches, palpitations, dizziness, tremors, and constipation, many of which are quite similar to those associated with sympathomimetic drugs such as Phentermine and Fenfluramine. Ephedrine can also trigger cardiovascular events such as tachycardia, cardiac arrhythmias, angina, and vasoconstriction with hypertension, although these cardiovascular side effects have been found to be rather infrequent and tend to diminish on repeated dosing. Because of this, ephedrine and other sympathomimetics are banned by the IOC and other organizations such as the NCAA and NFL.

Glutamine

Glutamine is a naturally occurring non-essential amino acid and is the most abundant amino acid in human muscle and plasma. It is important as a constituent of protein and as a means of nitrogen transport between tissues. It is also important in acid–base regulation and as a precursor of the antioxidant glutathione. Its alleged ergogenic effects with glutamine supplementation can be classified as anabolic and protective. Glutamine supplementation is considered to promote a positive protein balance by enhancing protein synthesis and counteracts decline in protein synthesis. This claim, however, remains to be substantiated. In a study that used female rats, infusing glutamine was found to inhibit the down-regulation of myosin synthesis and atrophy. Nevertheless, several human studies have not been able to confirm this anabolic effect. For example, Zachwieja *et al.* (2000) infused amino acid mixtures with and without glutamine directly into the blood of male and female subjects for several hours while estimating the rate of muscle protein synthesis. Although protein synthesis was enhanced by the amino acid infusion, there was no additional enhancement with the mixture that included glutamine. In another study in which a group of young adults were given glutamine supplements at 0.9 g per kilogram of lean body mass daily while also undergoing resistance training for six weeks, glutamine supplementation had no significant effect on muscle performance, body composition, or muscle protein degradation indices (Candow *et al.* 2001). Data also indicate that glutamine supplementation promotes muscle glycogen synthesis during recovery, perhaps by serving as a gluconeogenic substrate in the liver (Varnier *et al.* 1995). However, practical application of these findings for promoting glycogen replenishment following exercise requires further research.

The protective aspect of glutamine concerns its use as an energy fuel for nucleotide synthesis by disease-fighting cells, particularly the lymphocytes and macrophages that defend against infection. Injury, burns, surgery, and endurance exercise lower glutamine levels in plasma and skeletal muscle due to increased demand by the liver, kidneys, gut, and immune system. It has been suggested that a lowered plasma glutamine concentration contributes, at least in part, to the immune suppression that occurs with extreme physical stress (Smith and Norris 2000). For example, prolonged exercise at 50% and 70% VO_2max causes a 10–30% fall in plasma glutamine concentration that may last for several hours during recovery. This fall in plasma glutamine coincides with the time period during which athletes are more susceptible to infections following prolonged exercise. Will glutamine supplementation reverse this stress-related perturbation in immune function? Although supplementing glutamine has been presumed to be able to prevent the impairment of immune function after exhaustive exercise, insufficient data exists to substantiate this claim. In fact, most recent studies have failed to demonstrate any effect of glutamine supplementation on preserving immune function following exhaustive exercise. The supplementation appears to prevent the exercise-induced fall in plasma glutamine levels, but did not prevent the fall in lymphocyte proliferation and lymphocyte-activated killer cell activity (Castell *et al.* 1996; Gleeson and Bishop 2000).

Glycerol

Glycerol is a component of the triglyceride molecule and an important constituent of the cell membrane. This three-carbon molecule can be a substrate for gluconeogenesis and, as such, could provide fuel during exercise. We ingest a fairly large amount of glycerol as part of triglycerides or dietary fats on a daily basis. Glycerol is also released into the bloodstream from lipolysis, which breaks down triglycerides into glycerol and three free fatty acids. Thus, during exercise, when lipolysis is stimulated, plasma glycerol concentrations will increase. Glycerol has been touted as a possible ergogenic aid for two reasons. First, it is a substrate for gluconeogenesis and thus may become a necessary resource for glucose production during prolonged exercise. Second, supplemented glycerol distributes evenly throughout the body fluid and may provide an osmotic influence that could help an athlete hyperhydrate prior to competition in a warmer environment.

Studies that have investigated the efficacy of glycerol as a fuel have not been positive. In well-controlled research, glycerol feedings did not prevent either hypoglycemia or muscle glycogen depletion. In addition, the contribution of glycerol to the overall energy expenditure was found to be relatively small. Glycerol cannot be oxidized directly in very large amounts in the muscle. Therefore, glycerol must be converted into a glucose molecule via gluconeogenesis before being used as a fuel. Unfortunately, the rate at which the human liver converts glycerol to glucose is not rapid enough for glycerol to be an effective energy source during prolonged strenuous exercise.

Glycerol as a hyperhydrating agent has received a great deal of attention. When consumed with 1–2 liters of water, glycerol facilitates water absorption from the intestine and causes extracellular fluid retention, mainly in the plasma fluid compartment. This action may occur because glycerol moves through various tissues at a relatively slow rate; this then creates an osmotic effect that draws fluid into extracellular space, including plasma. The hyperhydration effect of glycerol supplementation reduces overall heat stress during exercise as reflected by an increased sweating rate. This lowers heart rate and body temperature during exercise and enhances endurance performance under heat stress (Lyons *et al.* 1990). Several studies have found no effect of glycerol on thermoregulation (Inder *et al.* 1998; Latzka *et al.* 1997, 1998). However, in these studies the volume of water used (500 ml) may have been too small. It is recommended that the ingestion of 1 g per kilogram of body weight of glycerol with 1–2 liters of water would be effective in protecting against heat stress. Glycerol supplementation may have a positive effect on hyperhydration. However, users should be aware that glycerol has significant side effects including nausea, dizziness, headaches, bloating, and cramping (Wagner 1999).

HMB (β-hydroxy-β-methylbutyrate)

HMB is a metabolite of the essential amino acid leucine. Depending on the amount of HMB contained in foods, the body synthesizes 0.2–0.4 g per day, of which a very small percentage (~5–10%) derives from dietary leucine catabolism. Foods rich in HMB are red meats, asparagus, cauliflower, catfish, and grapefruit. HMB is available to consumers as an independent supplement or as an ingredient of combination supplements or sports food. Because of its nitrogen-retaining effect, many resistance-training athletes supplement directly with HMB to prevent or reduce muscle damage and to suppress proteolysis or protein degradation associated with intense physical effort. The assumption that HMB supplementation would have a positive impact on muscle metabolism may be based on *in vitro* animal studies in which researchers noted a marked decrease in protein breakdown and a slight increase in protein synthesis in muscle tissue of rats and chicks exposed to HMB.

The number of studies to test the efficacy of HMB as an ergogenic aid is rapidly growing. However, findings remain divided at present. In those studies that supported the use of HMB as an ergogenic aid, a decrease in muscle damage and protein synthesis was observed, along with an increase in strength performance following consumption of HMB supplements. For example, in one such supportive study, Nissen *et al.* (1996) examined the ergogenic effect of HMB using two separate randomized trials. In trial 1, the authors provided 1.5 g or 3 g of HMB daily to untrained males for three weeks, during which time subjects also participated in a resistance-training program three days per week for three weeks. In trial 2, subjects consumed either 0 g or 3 g of HMB per day and weight lifted for 2–3 hours, six days per week for seven weeks. It was found from trial 1 that HMB supplementation depressed the exercise-induced rise in muscle proteolysis as reflected by urine 3-methylhistidine (3-MH) and plasma creatine kinase levels during the first two weeks of training. 3-MH is a marker of contractile protein breakdown, and creatine kinase is an indicator of muscle damage. In addition, this group lifted more total weight than the placebo group during each training week, with the greatest effect in the group receiving the largest dose of HMB supplement. With regard to trial 2, it was found that subjects who received the HMB supplement had higher fat-free mass than the unsupplemented subjects at week 2 and weeks 4–6 of training.

Not all research shows beneficial effects of HMB supplementation with resistance training. For example, a study involving collegiate football players taking 3 g of HMB daily, whose strength-training program was monitored by their strength and conditioning coach, failed to demonstrate a positive effect on strength and body composition (Ransone *et al.* 2003). One of the hypotheses is that HMB supplementation would reduce muscle damage and suppress protein degradation associated with intense training. Nevertheless, Kreider *et al.* (1999) reported that 28 days of HMB supplementation at a dose of 3 g or 6 g per day during resistance training did not reduce catabolism in experienced resistance-training males. Paddon-Jones *et al.* (2001) also found that HMB supplementation at a similar dose did not reduce the symptoms associated with muscle soreness induced by eccentric contraction. There appears to be some evidence to support the use of HMB as an ergogenic aid. However, more studies are needed, especially to examine its efficacy among trained individuals.

Inosine

Inosine is a nucleoside, a purine base comparable to adenine, which is one of the structural components of ATP. It is found naturally in brewer's yeast and organ meats. Inosine is not considered an essential nutrient. The body synthesizes inosine from precursor amino acids and glucose. Inosine, in the form of nucleotide inosine monohydrate (IMP), is used to make adenine monophosphate (AMP), which in turn can be phosphorylated to the high-energy phosphate compound ATP. Strength and power athletes supplement with inosine in the belief that it increases ATP stores, thereby improving training quality and competitive performance. Inosine is also thought to improve endurance performance. It has also been theorized that inosine participates in the formation of 2,3-diphosphoglycerate (2,3-DPG), a substance in red blood cells that facilitates the release of oxygen from hemoglobin to the tissue. Other claims regarding inosine include its role in: (1) stimulating insulin release to enhance glucose delivery; (2) augmenting cardiac contractility; and (3) acting as a vasodilating agent. It is mainly because of these theoretical considerations that inosine has been extolled as an ergogenic supplement to improve both anaerobic and aerobic performance.

Objective data do not support the ergogenic role of inosine supplementation in improving either aerobic or anaerobic performance. In fact, it has been suggested that this supplement may have **ergolytic** effects under some conditions. In one carefully conducted study, trained men and women were administered 6 g per day of inosine or placebo for two days, but no change was observed in three-mile treadmill run time, VO_2max, or perceived exertion (Williams *et al.* 1990). After a 30-min break, subjects performed another run in which speed was kept constant but the treadmill grade increased gradually. It was found that time to exhaustion in this run was actually longer during the placebo trial, suggesting a possible negative effect of inosine supplementation. In another study, male competitive cyclists received either a placebo or 5 g per day of oral inosine supplement for five days (Starling *et al.* 1996). They then performed a Wingate cycle test, a 30-min self-paced bicycle endurance test, and a constant load, supramaximal cycling sprint to fatigue. No significant differences occurred in any of the criterion variables in terms of performance or blood 2,3-DPG concentrations between placebo and treatment conditions. Similarly, this study also showed that cyclists fatigue nearly 10% faster on the supramaximal sprint test when they consumed inosine than without it, again indicating that inosine can be detrimental to performance.

Phosphate

Although it is uncommon, some athletes have tried to enhance performance by ingesting gram amounts of phosphate shortly before strenuous training or competition. This practice is called **phosphate loading**. Phosphate and phosphorus are often used interchangeably. Technically, phosphorus is an element (P), whereas phosphate is a molecular anion (PO_4^{3-}), part of phosphoric acid (H_3PO_4). Phosphate is a component of high-energy compounds, such as ATP and PCr, as well as 2,3-DPG, a molecule that facilitates oxygen release from hemoglobin for use by body tissues. Therefore, it has been postulated that phosphate supplementation may increase ATP synthesis and improve oxygen extraction in muscle cells because of elevations of 2,3-DPG in erythrocytes.

Despite these appealing rationales, the ergogenic benefits of phosphate loading have not been consistently demonstrated. Some studies show improvement in VO_2max, anaerobic thresholds, and endurance performance, and/or decreased lactate concentration at submaximal workloads (Cade *et al.* 1984; Kreider *et al.* 1990, 1992; Stewart *et al.* 1990), while others failed to do so (Bredle *et al.* 1988; Duffy and Conlee 1986; Galloway *et al.* 1996; Mannix *et al.* 1990). The inconsistencies in these findings may be related to the differences in the experimental protocol – e.g., dosage and duration of supplementation, type of subjects, exercise mode and intensity, and pre-test diets. Also, most studies used very small numbers of subjects, which can make relatively small changes in exercise performance difficult to detect.

At present, little reliable scientific evidence exists to recommend exogenous phosphate as an ergogenic aid. On the negative side, chronic phosphate loading can alter the calcium–phosphate ratio, thereby affecting the rigidity of bones. In addition, excess plasma phosphate can stimulate secretion of parathyroid hormone. Excessive production of this hormone increases release of calcium from bones, causing loss of bone mass.

SUMMARY

- "Ergogenic" is defined as increasing work or potential to do work. Ergogenic aids consist of substances or procedures that improve physical work capacity, physiological function, or athletic performance. An ergogenic aid does not need to be nutritional or pharmacological; it can also be mechanical, psychological, or physiological.

- In contrast to prescription drugs, which are carefully regulated, nutrition supplements and ergogenic aids receive very little government oversight, and manufacturers and retailers have enormous freedom in making claims to promote the product.
- All ergogenic aids need to be critically evaluated. Often, a "placebo effect," not the aid per se, improves the performance because of psychological factors. Athletes and others must carefully examine claims made by the dietary supplement industry, including the scientific evidence that supports the claims. A solid understanding of research and development and experimental procedure is important in judging the validity of an ergogenic aid.
- Sports bars provide energy as well as other essential nutrients. Their composition can vary based on the intended consumer and purpose of the sports food. Some sports bars are marketed as a quick-release, high-energy source, so that they can be used before or after intense training or sports competition. Others are designed to be meal replacements that contain a high amount of protein and/or fiber.
- Carbohydrate is usually the energy foundation of many sport bars that are to be consumed before and after exercise. Many sports bars also contain fiber and protein, as well as vitamins and minerals that are directly involved in energy metabolism, such as B vitamins, magnesium, zinc, and iron.
- Sports drinks can be separated into two categories: (1) fluid and electrolyte replacement drinks in which the carbohydrate content is relative low; and (2) drinks that contain a higher carbohydrate formulation. The former is more appropriate for use during exercise, whereas the latter is more suitable for consumption after training or in preparation for an upcoming event.
- In sport drinks, carbohydrate is typically provided as glucose, sucrose, fructose, corn syrup, maltodextrins, and glucose polymers. They also contain plenty of electrolytes, including sodium, potassium, and chloride. The carbohydrates usually make up about 4–8% of a fluid/electrolyte replacement drink and >10% of a recovery/loading beverage. Such a diluted concentration will help facilitate gastric emptying, thereby enhancing fluid/electrolyte replacement.
- Despite a large quantity of ergogenic aids, the working mechanism of each may be explained by one or more of the following: (1) acts as a central or peripheral nervous system stimulant; (2) increases storage or availability of a limiting substrate; (3) acts as a supplemental fuel source; (4) reduces performance-inhibiting metabolic by-products; (5) facilitates recovery; and (6) enhances tissue synthesis.
- Arginine, ornithine, lysine, bicarbonate, boron, caffeine, carnitine, chromium, coenzyme-Q10, creatine, DHEA, androstenedione, ephedrine, glutamine, glycerol, HMB, inosine, and phosphate are the popular ergogenic choices. However, most of them still remain controversial in terms of their efficacy and/or potential risks. The existence of such inconsistencies may be in part related to the differences in the experimental protocol – e.g., dosage and duration of supplementation, number and type of subjects, exercise mode and intensity, testing and measurement, and control of diet and physical activity.

CASE STUDY: MAKING A SOUND DECISION ON AN ERGOGENIC AID ▍

Andy is a running back on the college football team. He also competes for the wrestling team. Andy is always very conscious about his diets. He eats a balanced diet and takes multivitamins and mineral supplement each day. He would like to improve his strength and power as well as his body composition, and decides to experiment with some ergogenic aids. Based on the articles and advertisements he has read in sports magazines, he selects creatine for improving his strength and power and HMB (β-hydroxy-β-methylbutyrate) to improve his muscle mass and body composition. But before he begins taking these, he wants to explore their risks and benefits as well as their working mechanisms.

Questions

- What is the exact working mechanism by which creatine improves performance? What about HMB?
- Do these supplements live up to their promoters' promises?
- Where should Andy look if he needs more authentic information on these products?
- Would you recommend that Andy takes these supplements? Why?

▍ REVIEW QUESTIONS

1 What is a double-blind study? What are the advantages associated with this research design?

2 Why is it necessary to use a placebo in research studies on ergogenic aids?

3 What is a research hypothesis? How does a research hypothesis differ from a research objective?

4 Provide one example of an ergogenic aid that fits each of the following claims: (1) acts as a central or peripheral nervous system stimulant; (2) increases storage or availability of a limiting substrate; (3) acts as a supplemental fuel source; (4) reduces performance-inhibiting metabolic by-products; (5) facilitates recovery; and (6) enhances tissue synthesis.

5 List all ergogenic effects of creatine supplementation. What are potential risks associated with this ergogenic aid?

6 Explain how caffeine works in improving performance.

7 What are the claims associated with DHEA? Does research prove its efficacy?

8 Explain why glycerol is used for enhancing hydration?

9 Discuss β-hydroxy-β-methylbutyrate (HMB) in terms of its origin and food sources. Why is this compound considered ergogenic?

10 How is phosphate related to enhanced performance?

SUGGESTED READING

American College of Sports Medicine (1987) American College of Sports Medicine position stand on the use of anabolic-androgenic steroids in sports. *Medicine & Science in Sports & Exercise*, 19: 534–539.

This article allows readers to learn the position of the American College of Sports Medicine on the use of anabolic-androgenic steroids in sports.

American College of Sports Medicine (1987) American College of Sports Medicine position stand on blood doping as an ergogenic aid. *Medicine & Science in Sports & Exercise*, 19: 540–543.

This article allows readers to learn the position of the American College of Sports Medicine on the use of blood doping as an ergogenic aid for athletic competition.

Armstrong, L.E. (2002) Caffeine, body fluid–electrolyte balance, and exercise performance. *International Journal of Sport Nutrition and Exercise Metabolism*, 12: 189–206.

In this review, the author critiques several controlled investigations regarding the effects of caffeine on dehydration and athletic performance. The article also analyzes the potential consequences of consuming caffeinated beverages on fluid–electrolyte balance and exercise capacity in both athletes and recreational enthusiasts.

Graham, T.E. (2001) Caffeine, coffee and ephedrine: impact on exercise performance and metabolism. *Canadian Journal of Applied Physiology*, 26: S103–S119.

This article addresses areas where there is controversy regarding caffeine as an ergogenic aid and also identifies topics that have not been adequately addressed, such as using caffeine in conjunction with ephedrine.

10 Nutrient metabolism in special cases

KEY TERMS

- indirect calorimetry
- respiratory exchange ratio
- *vastus lateralis*
- follicular
- estrogen
- progesterone
- corpus luteum
- placenta
- testosterone
- lipoprotein lipase
- luteal
- exogenous
- adenylate cyclase
- gestational diabetes
- placental lactogen
- insulin resistance
- prolactin
- thermogenesis
- glucose tolerance
- hyperlipidemia
- infancy
- childhood
- adolescence
- puberty
- oxygen deficit
- co-morbidities
- glucose transporters
- euglycemic
- hyperinsulinemic glucose clamp
- insulin sensitivity
- insulin responsiveness
- visceral
- subcutaneous
- portal circulation
- metabolic inflexibility
- post-absorptive state

GENDER AND AGE DIFFERENCES IN SUBSTRATE METABOLISM

In the not-so-distant past, our society was influenced by the notion that boys were meant to be active and athletic, whereas girls were weaker and thus less well suited to physical activity. In fact, women were prohibited from running any race longer than 800 m until the 1960s (Wilmore and Costill 2004). This notion is no longer common, and girls and women are given equal access to most athletic activities. Due to increased involvement in physical activity and training, there has been a tremendous decrease in the gender gap in terms of athletic performance. In events other than those requiring muscular strength and power, performance differences between genders are no more than 15%. The current body of knowledge regarding metabolic responses to exercise and training is based largely on the responses of young adult males. This is because much of the previous research in the area of exercise metabolism has primarily been conducted using male subjects. Due to the ever-increasing involvement of women in sports and leisure and occupational physical activities, there has been a steady increase in research that aims to compare exercise-induced metabolic responses and adaptations between genders. The proliferation of research in this regard will serve to improve our understanding of gender differences in exercise metabolism. It will also help establish guidelines to be used for designing a gender-specific training regimen and exercise prescription.

Substrate utilization of males and females

Although there is some disagreement, perhaps the most repeatedly stated evidence of metabolic difference between genders is that, compared to men, women are able to derive proportionally more of the total energy expended from fat oxidation during aerobic exercise. This conclusion was drawn primarily from studies using **indirect calorimetry**, a method that calculates the heat living organisms produce from their consumption of oxygen. In these studies, a lower **respiratory exchange ratio** (RER) during submaximal endurance exercise was found in females as compared to males (Friedlander *et al.* 1998; Tarnopolsky *et al.* 1990, 1995, 1997). As discussed further in Chapter 11, RER is a qualitative indicator of which fuel (carbohydrate or fat) is being metabolized to supply the body with energy, and the lower the RER, the greater the percentage of energy derived from fat. In these studies, an effort was made to match male and female subjects to their VO_2max or training status. This approach was used to preclude the potential confounding effect of fitness on the gender-related difference in exercise metabolism. The matching is a time-consuming process because investigators have to recruit sedentary males and females, expose them to the same training program, and then test them at the same relative and absolute exercise intensity following training. In these studies, oxygen uptake was also normalized relative to lean body mass in order to minimize the gender differences in energy metabolism that is attributable to percentage fat.

The evidence from indirect calorimetry that indicates that women oxidize less carbohydrate and more fat during exercise is consistent with recent investigations that have involved more sophisticated laboratory techniques – i.e., muscle biopsy and isotopic tracer methods. For example, Tarnopolsky *et al.* (1990) found that ***vastus lateralis*** glycogen concentration was less depleted in women following exercise. In this study, authors had six males and six equally trained females run for more than 90 min at 65% VO_2max following three days on a controlled diet. Muscle glycogen utilization was calculated from pre- and post-exercise needle biopsies of the *vastus lateralis*. Using a different analytical approach, Carter *et al.* (2001b) also observed a lower utilization of muscle glycogen in women than men during endurance exercise both before and after endurance training. In these studies, lipid oxidation as determined by indirect calorimetry was found uniformly higher in women than men during exercise at the same relative intensity.

The evidence that supports the gender differences in substrate metabolism has not always been demonstrated. For example, Tarnopolsky *et al.* (1997), who previously observed lower glycogen utilization in women during running, failed to confirm such a finding when exercise was conducted on a stationary cycle ergometer. Thus, it remains to be determined whether such a discrepancy can be attributed to the differences in muscle recruitment between sexes during running as compared to cycling. Romijn *et al.* (2000) found that RER, as well as rates of fat and carbohydrate oxidation, were similar between trained men and women during exercise at either low or high intensity. These findings disagree with those reported by Friedlander *et al.* (1998), who used the same experimental approach. It must be noted that Romijn *et al.* did not control for the effect of menstrual phases, whereas most studies tested exercising women in the mid-**follicular** phase. There are indications that the menstrual cycle can affect the metabolic response to exercise; however, such an impact of menstrual phases on exercise metabolism appears to disappear when exercise is performed at higher intensities.

Underlying mechanisms: role of sex hormones

There has been very little investigation into the potential mechanism of why women place greater priority than men on lipid oxidation versus carbohydrate oxidation during exercise. This is in part because the concept of gender differences in exercise metabolism has not been universally accepted. According to the currently available research, gender differences in exercise metabolism seem to be mediated primarily by sex hormones such as **estrogen** and **progesterone**, which present in small quantities in men as well. Progesterone, released from the **corpus luteum**, **placenta**, and adrenal glands, is considered a precursor to the male and female sex hormones, testosterone and estrogen, respectively. Estrogen is a collective term for a group of 18-carbon steroid hormones. The most biologically active estrogen is 17-β-estradiol (E_2); there are also other, less potent, estrogens, such as estrone (E_1) and estriol (E_3). Estrogens are secreted mainly by the ovaries and, to a lesser extent, the adrenal glands. Estrogens are also synthesized from androgens such as **testosterone** in blood or other organs such as adipose and muscle tissues.

Animal studies of estrogen and progesterone

A number of animal studies have been undertaken to examine the impact of estrogen on the utilization of glycogen. In these studies, the experimental approach is to alter the hormonal environment by injecting estrogen and then to evaluate the metabolic consequences. For example, Kendrick *et al.* (1987) administered E_2 to rats in doses sufficient to achieve blood levels of estrogen in the physiological range, and showed decreased utilization of glycogen stored in skeletal muscle as well as in the heart and liver.

The role of progesterone in exercise metabolism is less clear. It has been reported that this hormone increases liver glycogen content and suppresses hepatic gluconeo-genesis, and that these effects can be enhanced by concurrent administration of E_2. In this context, it appears that the two female hormones may work additively or synergis-tically in reducing carbohydrate utilization during exercise.

High levels of E_2 have also been found to increase the availability of free fatty acids (FFA) during exercise in rats. For example, Ellis *et al.* (1994) observed that during exer-cise E_2 increased lipolysis in adipose tissue and enhanced distribution of FFA to the muscle. This increased availability of FFA can be further attributed to the alterations in activity of **lipoprotein lipase** (LPL), which regulates fat metabolism. In this same study, Ellis *et al.* demonstrated a decreased activity of adipocyte LPL, which promotes fat syn-thesis, and an increased activity of muscle LPL, which promotes fat utilization. Of par-ticular interest is that estrogen and progesterone would play an opposing role in regulating fat metabolism, which is not the case in terms of their actions on carbohydrate metabolism. Hansen *et al.* (1980) found that the rate of fatty acid synthesis was lower in E_2-treated rats compared with progesterone-treated rats. More recently, Campbell and Febbraio (2001) also observed increased activity in several key enzymes involved in fatty acid oxidation as a result of estrogen supplementation, while such an effect was reversed with concurrent administration of progesterone. The roles estrogen and progesterone play in regulating fat and carbohydrate metabolism are illustrated in Table 10.1.

Observations with human subjects

The effect of sex hormones on energy metabolism has also been examined using humans. Hackney *et al.* (1990) performed muscle biopsies on the *vastus lateralis* of ten healthy women in both the follicular and **luteal** phase of the menstrual cycle. They found that under resting conditions, muscle glycogen content was higher in the luteal than follicular phase. This same research group later also reported a lower carbohy-drate oxidation during the luteal phase during exercise at 35% and 60% VO_2max (Hackney *et al.* 1994). As the luteal phase is when production of both estrogen and progesterone is higher, this finding is consistent with the conclusion of animal studies: estrogen attenuates the utilization of carbohydrate. The inhibitive role of estrogen on

Table 10.1 The actions of estrogen and progesterone on carbohydrate and fat metabolism

ACTIONS	ESTROGEN	PROGESTERONE
Carbohydrate metabolism		
Muscle glycogenolysis	Inhibiting	Inhibiting
Liver glycogenolysis	Inhibiting	Inhibiting
Glucose transport into muscle	Inhibiting	Inhibiting
Fat metabolism		
Adipose tissue lipolysis	Stimulating	Inhibiting
Fatty acids transport into mitochondria	Stimulating	Inhibiting

Source: Adapted from Deon and Braun, 2002.

carbohydrate utilization was also evidenced in studies in which subjects were supplemented with **exogenous** estrogen. By providing 17-β-estradiol (E_2) to a group of amenorrheic females, Ruby *et al.* (1997) observed altered carbohydrate metabolism. In this study, an isotopic tracer method was used so investigators were able to determine muscle glucose utilization and hepatic glucose production. It was found that the release of glucose from the liver was reduced as a result of increased E_2 levels during exercise, while glucose utilization by muscle remained similar between E_2 and placebo groups. Such a reduction in hepatic glucose output due to E_2 supplementation was also observed by Carter *et al.* (2001a), who administered E_2 to a group of men, although this research group found a decrease in muscle glucose utilization. In both studies, no differences in whole-body substrate oxidation was found between the experimental and placebo group. It appears that despite the indication from animal studies that estrogen may mitigate muscle glycogen utilization, such a role of estrogen in humans is less conclusive.

Both Ruby *et al.* (1997) and Carter *et al.* (2001a) have reported a decrease in blood concentration of epinephrine during exercise in subjects with estrogen treatment. This observation has led to a hypothesis that observed gender differences in substrate metabolism may be mediated by changes in circulating levels of, or tissue sensitivity to, epinephrine. Epinephrine plays a key role in regulating carbohydrate metabolism during exercise. In muscle, epinephrine activates **adenylate cyclase**, the second messenger, and thus stimulates glycogenolysis and glycolysis. In the liver, epinephrine also stimulates glycogenolysis and thus glucose output. The level of epinephrine during submaximal exercise is generally lower in women than men. This may explain why women are better able to use fat instead of carbohydrate as energy fuel.

PREGNANCY

Pregnancy places unique demands on women's metabolism. It affects the metabolic cost and physiological strain imposed by exercise. An investigation in the 1970s studied 13 women from six months of pregnancy to six weeks after gestation (Knuttgen and Emerson 1974). It was found that during walking heart rate and VO_2 increased progressively despite an unchanged exercise intensity. However, these two parameters remained constant throughout steady-state cycle exercise. These findings suggest that due to an increase in body mass, including fetal tissue, there would be an increase in energy cost during weight-bearing activities like walking, jogging, and running. In addition to this added energy cost during exercise, it has also been demonstrated that pregnant women will have an increased resting metabolism, especially during the later stages of pregnancy. Table 10.2 provides a comparison in caloric cost of common household activities between pregnant and non-pregnant women. It has been estimated that throughout the entire pregnancy there would be an additional demand of 75,000 kcal required to build new tissues and to meet the higher energy cost of daily activities. This figure represents an extra expenditure of 250 kcal per day during a ten-month pregnancy period. Despite the increased energy cost of weight-bearing activities, it has been assumed that such an increase in energy cost is offset by a decrease in the amount of time spent in weight-bearing activities, as well as by the

Table 10.2 Comparisons of energy cost of household activities in pregnant and non-pregnant women

ACTIVITY	ENERGY COST (KCAL MIN^{-1})	
	PREGNANT	NON-PREGNANT
Lying quietly	1.11	0.95
Sitting	1.32	1.02
Sitting, combing hair	1.36	1.22
Sitting, knitting	1.55	1.47
Standing	1.41	1.12
Standing, washing dishes	1.63	1.33
Standing, cooking	1.66	1.41
Sweeping with broom	2.90	2.50
Bed making	2.98	2.66

Sources: Brooks *et al.* 2005. Used with permission.

relaxed and economical fashion in which pregnant women move. Consequently, the net increase in energy expenditure associated with pregnancy may only reflect an increase in resting metabolism as a result of growing of both maternal and fetal tissues.

Substrate metabolism during pregnancy

Although pregnancy consists of a series of small, continuous physiological adjustments, the alterations in substrate metabolism appear to occur mostly during the later phase of pregnancy. From a metabolic standpoint, pregnancy may be divided into phases, such as first and second halves. The first half of pregnancy is primarily a time of preparation for the demands of rapid fetal growth that occurs later in pregnancy. During this period, there is a continuous increase in the production of estrogen and progesterone. The presence of these hormones can help not only mobilize fat for energy, but also stabilize plasma glucose at relatively high levels in order to meet the needs of the fetus. There is evidence that perhaps as a means of protecting the fetus from hypoglycemia, pregnancy reduces the ability of the mother to metabolize carbohydrate (Clapp *et al.* 1987). This metabolic alteration could limit pregnant women from performing anaerobic or strenuous aerobic exercises in which carbohydrate is a primary fuel.

For pregnant women, measurement of insulin sensitivity is often used to detect a possibility of **gestational diabetes**. This is because estrogen and **placental lactogen** have been considered diabetogenic hormones due to their inhibitive effects on insulin-mediated glucose uptake by various tissues. A number of studies have reported that during the early phase of pregnancy, the sensitivity of peripheral tissues to insulin was either normal or increased slightly (Buch *et al.* 1986; Gatalano *et al.* 1991). However, longitudinal studies of glucose tolerance have shown that as gestation continues, there would be a progressive increase in insulin response to a given dose of glucose (Sivan *et al.* 1997). This greater than normal insulin response is consistent with the phenomenon of **insulin resistance** and suggests that pregnant women can potentially diminish their ability to handle glucose with insulin. This deficiency is especially the case in obese pregnant women who have a high risk of developing diabetes even without pregnancy (Sivan *et al.* 1997). The reduced insulin sensitivity is thought to be secondary to the gestation-induced changes in hormones, including estrogen, progesterone, cortisol, and **prolactin**, although the precise mechanism remains unclear. From a fetal standpoint, a certain degree of insulin resistance is considered desirable in that it can serve to shunt ingested nutrients to the fetus.

Exercise during pregnancy

The metabolic reserve available for performing exercise during pregnancy is diminished owing to increased resting metabolism and a blunted sympathetic response to physical activity. However, during the early stages of pregnancy, light to moderate activity can be pursued safely, given that blood glucose is carefully monitored to prevent hypoglycemia. Regular exercise during pregnancy counteracts the effects of deconditioning. It attenuates pregnancy-related fatigue and helps maintain muscular strength, which may speed delivery. It can also prevent excessive weight gain, insulin resistance, and type-2 diabetes. Based upon the literature, it appears that ordinary pregnant women are able to tolerate light- to moderate-intensity exercise sessions of up to 30 min in duration four times per week, although exercise tolerance can be affected by environmental condition, as well as the fitness level of the mother.

Caution should be taken in selecting appropriate exercise modality. As pregnancy advances, the capacity for exercise, especially those activities that occur against gravity, decreases. Therefore, during the later stages of pregnancy, it is helpful to introduce weight-supported activities such as cycling, swimming, and water aerobics. Exercise can be dangerous if excessive. Important contraindications to vigorous exercise include hypoglycemia, intrauterine growth retardation, premature labor and/or ruptured membrane, placental injury or dysfunction, an incompetent cervix, pregnancy-induced hypertension, and blood poisoning (American College of Obstetricians and Gynecologists 1994).

THE ELDERLY

The elderly (those who reach and pass the age of 65), make up the fastest-growing segment of today's society. For example, approximately 35 million (nearly 12%) Americans in the United States today exceed 65 years of age. It is predicted that this figure will climb to 70 million (22%) by 2030. Such trends seen in the United States also apply in many Western nations and may soon emerge also in developing countries. Aging refers to the normal yet irreversible biological changes that occur throughout an individual's life span. It involves a diminished capacity to regulate the internal environment in order to meet external challenges. As shown in Table 10.3, such a reduced ability can be further attributed to a series of attenuated or impaired metabolic functions, which ultimately reduce one's ability to generate energy needed.

Aging is influenced by genetics. This may be attested by observations that the life span of twins is remarkably similar. Identical twins usually die within 2–4 years of each other, whereas non-identical twins within 7–9 years. Aging is also affected by

Table 10.3 Aging-related metabolic changes and their physiological consequences

METABOLIC CHANGE	PHYSIOLOGICAL CONSEQUENCES
Myosin–ATPase↓	Reduced muscle contractility
Lactate dehydrogenase↓	Reduced glycolysis
Succinic dehydrogenase↓	Reduced oxidative capacity
Malic dehydrogenase↓	Reduced oxidative capacity
Cytochrome oxidase↓	Reduced oxidative capacity
Mitochondria size and number↓	Reduced oxidative capacity
Type II muscle fibers↓	Reduced muscle strength and power
Capillary density↓	Reduced oxygen delivery
Glucose tolerance↓	Increased risk of diabetes and heart diseases
Blood insulin↑	Increased risk of diabetes and heart disease
Insulin sensitivity↓	Increased risk of diabetes and heart disease
Sympathetic stimulation↓	Reduced maximal heart rate and lipolysis
Muscle mass↓	Reduced basal metabolism and fat oxidation

lifestyle factors. It is considered to be the process associated with an accumulation of wear-and-tear that leads to gradual loss of the ability to respond to stress. Although it may not halt the aging process, regular physical activity and a healthy diet can help improve quality of life and prolong life expectancy.

Reduced aerobic capacity and energy expenditure

Reductions in physical capacity can be characterized by a decrease in aerobic power or VO_2max, which was observed more than half a century ago. With cross-sectional comparisons, Robinson (1938) demonstrated that in men VO_2max declines an average of 0.44 ml kg^{-1} min^{-1} per year up to age 75. This is translated to be about 1% per year or 10% per decade. For women between the ages of 25 and 65 years, Åstrand (1960) showed a decline of 0.38 ml kg^{-1} min^{-1}, or 0.9% per year. Since these early observations, there have been numerous cross-sectional and, to a lesser extent, longitudinal studies attempting to further characterize such age-related decline in VO_2max and its metabolic consequences. The rate of decline in VO_2max found in these studies in general agrees with what was reported initially by Robinson in 1938. A reduction in VO_2max due to aging has been further ascribed to a decrease in maximal heart rate, maximal cardiac output, and maximal ability of working muscle to utilize oxygen for energy transfer.

Goran and Poehlman (1992) observed a significant correlation between VO_2max and total daily energy expenditure. The authors of this study also found a modest relationship between the total daily energy expenditure and the level of physical activity. These findings suggest that those with greater aerobic fitness tend to be more physically active and therefore have greater daily energy expenditure. However, such a linkage between fitness level and energy expenditure provides no information on cause and effect. In other words, it is unclear whether the increased total energy expenditure associated with a physically active lifestyle leads to a higher VO_2max, or alternatively, the individuals with a higher VO_2max engage in physical activities more frequently and intensely because of the higher work capacity.

A reduction in total energy expenditure has been well documented in the elderly (Margaret-Mary and Morley 2003; Elia *et al.* 2000). Interestingly, many normal-weight healthy older men and women decrease their energy intake well below their energy expenditure and thus lose weight. In addition to the lack of physical activity, the age-related decrease in energy expenditure has also been linked to reductions in basal metabolic rate and diet-induced **thermogenesis**. More detailed discussion on these two energy components can be found in Chapter 13.

Changes in enzymes of bioenergetic pathways

As mentioned earlier, mitochondria serve to allow biologically usable energy to be generated via oxidative pathways. Therefore, the frequently examined markers of mitochondrial function in skeletal muscle have been activities of selected oxidative enzymes and rate of ATP production. Numerous studies have all reported age-related decline in enzymes, such as citrate synthase, succinate-dehydrogenase, and cytochrome c oxidase. Furthermore, with data provided by Holloszy *et al.* (1991), it appears that such reduction in activity of oxidative enzymes occurs primarily in red (slow-twitch muscle), predominantly oxidative muscle than in white (fast-twitch muscle) glycolytic muscle. As a result of these enzymatic changes, mitochondrial oxidative capacity is impaired. With the use of muscle samples, Papa (1996) observed a decreased ability of mitochondria to consume oxygen for generating energy. This *in vitro* observation was later confirmed by an *in vivo* study in which Conley *et al.* (2000) used nuclear magnetic resonance techniques and found that the average rate of ATP formation in the quadricep muscles of older subjects between the ages of 65 and 80 was approximately half that of young subjects between the ages of 25 and 48.

Alterations in carbohydrate and fat metabolism

Another metabolic hallmark of the aging process is the impairment of carbohydrate and fat metabolism. Substantial evidence has been provided showing that increasing age is associated with decreased **glucose tolerance**. The glucose tolerance test measures the body's ability to metabolize glucose. It is performed after an overnight fast. During the test, a patient drinks a solution containing a known amount of glucose. Blood is obtained before the patient drinks the glucose solution, and drawn again every 30–60 min after the glucose is consumed, lasting 2–3 hours. Blood glucose levels above normal limits at the times measured can be used to diagnose type-2 diabetes or gestational diabetes. It has been estimated that the two-hour plasma glucose level during an oral glucose tolerance test rises on average $5.3\,mg\,dl^{-1}$ per decade and the fasting plasma glucose rises on average $1\,mg\,dl^{-1}$ per decade (Davidson 1979). Insulin, a hormone produced by the pancreas to move glucose from the bloodstream into cells, has also been found to be higher in many older individuals. This observation suggests that as one ages he/she may lose the ability to respond to insulin effectively and therefore require extra insulin to maintain normal blood glucose levels.

Impairment of fat metabolism is another age-related metabolic disorder. Aging has been associated with reduced fat oxidation at rest (Nagy *et al.* 1996) and following a meal (Roberts *et al.* 1996). Sial *et al.* (1998) also demonstrated age-related reduction in fat oxidation during aerobic exercise. It is thought that these reductions in fat utilization play an important role in mediating the age-related increase in adiposity, especially in the abdominal compartment. Sial *et al.* (1998) also reported a greater carbohydrate oxidation, a finding that was thought to result from the impaired fat utilization. All of these age-related metabolic changes will gradually deprive the elderly of the ability to use their energy fuels efficiently during exercise as compared to younger individuals.

Fat oxidation is mainly a function of two processes, the release of fatty acids from adipose tissue and the capacity of respiring tissue to oxidize fatty acids. Previous studies using aging rats and humans have demonstrated a diminished sympathetic stimulation of lipolysis (Lönnqvist *et al.* 1990). However, when examined in relation to the needs of the metabolically active tissue, the release of free fatty acids was found to be greater in older compared to younger individuals (Toth *et al.* 1996). The age-related reduction in fat utilization is therefore considered to be primarily due to the loss of the size and/or oxidative capacity of metabolically active tissues such as skeletal muscle. Those fatty acids that are released but not metabolized could have adverse metabolic effects such as **hyperlipidemia** and insulin resistance. In this regard, aerobic training becomes particularly important because it has the ability to maintain or increase the size and function of skeletal muscle, thereby improving the health status of elderly individuals.

CHILDREN AND ADOLESCENTS

The period of life from birth to the start of adulthood may be divided into three phases: **infancy**, **childhood**, and **adolescence**. Infancy is defined as the first year of life. Childhood spans from the first birthday to the beginning of adolescence. The period of adolescence is more difficult to determine, but is often considered to begin at the onset of **puberty** and terminate as growth and development are completed. Research on metabolism with regard to this early stage of life spectrum is rather limited. This is mainly due to ethical considerations and methodological constraints in studying children and adolescents. For example, there are very few investigators who would puncture a child's artery or take a needle biopsy of a child's muscle. In addition, there is still an on-going effort to find instruments and protocols that are age- and/or size-appropriate. Consequently, our understanding of children's metabolic responses to exercise has been based upon a limited number of investigations. Many conclusions regarding exercise metabolism in children and adolescents are derived primarily from measurements of cardiorespiratory parameters, such as oxygen uptake and respiratory exchange ratio.

Children and adolescents should not be regarded as miniature adults. The age-related functional deficiency in children and adolescents is not always attributable to the fact that they are smaller in size. It is generally true that children are less capable of performing a given task compared to adults, yet their physiological function increases as they grow older and bigger. However, only some gains in physiological function are proportional to changes in size. For example, muscle strength increases in direct proportion to its cross-sectional area. Many changes in function have been found to be either partially related to or completely independent of changes in size. For example, anaerobic capacity depends on activity of some key anaerobic enzymes in addition to muscle size. Additionally, it has been found that some physiological parameters such as blood concentration of oxygen and glucose remain unchanged despite a gain in body size. It is important to understand the patterns of the function–size relationship in growing individuals. This will help in making proper interpretations of age-related physiological differences and in determining whether there is a need to normalize a physiological parameter for body size prior to age-related comparisons.

Aerobic capacity

As mentioned earlier, VO_2max reflects the highest metabolic rate made available by aerobic energy transfer; this parameter can be expressed in both $l\,min^{-1}$ and $ml\,kg^{-1}\,min^{-1}$. As shown in Table 10.4, which depicts the chronological changes in VO_2max in $l\,min^{-1}$ reported by previous studies involving boys ($n=2180$) and girls ($n=1730$), VO_2max increases continuously until the age of 16–18 in boys, but increases minimally beyond the age of 14–15 in girls. Such gender difference in VO_2max can be ascribed in part to the differences in muscle mass between boys and girls (Davies *et al.* 1972). As a result of ratio scaling in which VO_2max in $l\,min^{-1}$ is divided by body mass, however, the average VO_2max in $ml\,kg^{-1}\,min^{-1}$ is still somewhat higher in boys than girls, especially during later periods of adolescence (Table 10.4). This finding suggests that the increase in VO_2max can also be explained by other factors, such as those involved in oxygen transport and utilization, that are gender specific. Over years, VO_2max in $ml\,kg^{-1}\,min^{-1}$ remains essentially unchanged in boys and slightly declines among girls. Thus, when comparing aerobic capacity using VO_2max already adjusted for body mass, one should expect no differences between children and adults. The decline in this relative VO_2max seen in girls may be due to a progressive increase in body fat in girls during adolescence.

Oxygen deficit and respiratory exchange ratio

Among other gas-exchange parameters that have received a great deal of attention for children are VO_2 kinetics and respiratory exchange ratio (RER). VO_2 kinetics assesses the integrated responses of oxygen requirements and supply at the onset of and during exercise of varying intensity. VO_2 kinetics at the onset of exercise can be characterized

Table 10.4 Average maximal aerobic power in children and adolescents

VO_2max		AGE (YEARS)					
		6	8	10	12	14	16
VO_2max in $l\,min^{-1}$	Boys	1.0	1.3	1.6	2.1	2.7	3.5
	Girl	0.9	1.2	1.4	1.6	1.8	2.0
VO_2max in $ml\,kg^{-1}\,min^{-1}$	Boys	47	50	51	51	51	51
	Girl	46	46	45	43	42	40

Source: Adapted from Bar-Or and Rowland 2004.

by a phenomenon of **oxygen deficit**, which is defined as a lag in oxygen supply in relation to oxygen demand. A number of studies have attempted to examine VO_2 kinetics at the onset of aerobic exercise in children (Armon *et al.* 1991; Hebertseit *et al.* 1998; Sady 1981). In general, these studies agreed upon the observation that children demonstrated a faster increase in oxygen uptake at the onset of exercise than adults. It appears that children have the ability to adapt their oxidative metabolism faster to meet the energy demand imposed by exercise. Aside from having a more prompt activation of the aerobic system, children are also found to be able to derive proportionally more of their total energy from fat oxidation. A number of studies have reported a lower RER in children than adults (Martinez and Haymes 1992; Rowland *et al.* 1987). This contention, however, needs to be further evaluated as Rowland and Rimany (1995) and Macek *et al.* (1976) failed to observe this age-related difference.

Metabolic efficiency

When performing aerobic exercise, children are found to be less efficient than adults. This is manifested by a greater mass-specific oxygen uptake or VO_2 expressed in $ml\,kg^{-1}\,min^{-1}$ in adults than is observed in children. This metabolic feature is particularly the case during weight-bearing activities such as walking and running (Fawkner and Armstrong 2003). Sallis *et al.* (1991) have attempted to quantify the excessive metabolic cost of walking and running by compiling data from various studies. As shown in Figure 10.1, on average, a five-year-old child would expend about 35–40% more oxygen than an adult who performed the same task. This excess, however, decreases with age. It is suggested that the low economy of locomotion in children is caused by multiple factors, including high resting metabolic rate, high stride frequency, mechanically wasteful locomotion style, and excessive co-contraction of antagonist muscles (Bar-Or and Rowland 2004).

Carbohydrate storage and utilization

Much less is known in regard to carbohydrate utilization during exercise in children compared to adults. This lack of information is attributable to ethical concerns over the taking of muscle biopsies from children. Based upon very limited data of muscle

Figure 10.1 Excess oxygen cost of walking and running per kilogram of body mass in children of various ages compared with young adults.

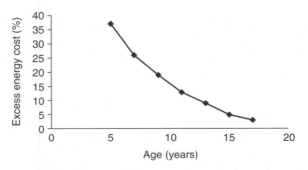

biopsies performed on children (Eriksson *et al.* 1973), it appears that there is a lower glycogen content at rest and reduced rate of glycogenolysis during exercise in children as compared to adults. Glycogen utilization is regulated by the activity of such enzymes as phosphofructokinase, and this enzyme has been found to be less active in the muscle cells of boys compared to adults (Eriksson *et al.* 1971, 1973; Bell *et al.* 1980). These findings suggest that it would be a disadvantage for children to compete in prolonged strenuous events that are glycogen dependent. Endurance capacity may not be limited by the volume of mitochondria in children, as Bell *et al.* (1980) reported a similar mitochondrion–muscle fiber ratio in pre-pubertal and adult muscle tissues.

Anaerobic capacity

Despite the fact that children often perform activities in an intermittent fashion, their anaerobic capacity is lower than that of adults. This lower anaerobic capacity may be manifested particularly in short-term events that last 1–2 min, such as 400–800-m running or 100–200-m swimming. This is because children are less able to store glycogen, as well as to extract energy from glycogen via glycolysis, as mentioned earlier. Such a lower glycolytic ability is consistent with the evidence that the peak lactate concentration is lower in children than adults (Eriksson *et al.* 1971). The reduced anaerobic capacity in children may also be explained by decreased sympathetic activity, which functions to stimulate glycogenolysis and glycolysis. Pullinen *et al.* (1998) showed that adolescent males aged 15 ± 1 have lower levels of blood catecholamines during resistance exercise compared to adult males aged 25 ± 6 years. Interestingly, unlike glycogen, both storage and utilization of ATP and creatine phosphate were found to be comparatively similar between children and adults (Zanconato *et al.* 1993; Eriksson *et al.* 1971). In these adult–child comparisons, the level of energy substrates was expressed relative to muscle mass in order to account for differences in body size.

OBESITY

Obesity is defined as an excess accumulation of body fat (i.e., body mass index $\geq 30 \, \text{kg} \, \text{m}^2$). It refers to the over-fat condition that is associated with a number of **comorbidities**, including: glucose intolerance, insulin resistance, dyslipidemia, non-insulin dependent (type-2) diabetes, hypertension, and increased risk of coronary heart disease and cancer. Obesity can be simply attributed to energy imbalance in which energy intake chronically exceeds energy expenditure. Disruption in energy balance often begins in childhood, and those who are overweight in childhood will have a significantly greater chance of becoming obese adults. Childhood obesity has been in part ascribed to parental obesity – if parental obesity exists, the child's risk of obesity in adulthood is 2–3 times that of normal-weight children without obese parents. The ages range 25–45 represents another dangerous period in which there is a progressive weight gain over time (Crawford *et al.* 2000). There are reports indicating that despite a progressive decrease in food consumption, a 35-year-old male will gain an average of 0.5 kg (1 lb) of fat each year until the sixth decade of life. It remains unclear as to whether this "creeping" obesity results from alterations in lifestyle or reflects a normal biological pattern.

Energy consumption

Do obese people eat more than lean individuals? This question remains equivocal at the present, in part because the techniques used to determine food intake have so far not been sufficiently accurate in quantifying how much energy obese individuals consume compared to their lean counterparts. Most surveys of nutritional histories suggest that energy intake is significantly lower in the overweight than those of normal weight (Baecke *et al.* 1983). However, one major concern is the validity of studies in which data are based on self-reports. By repeating dietary surveys on obese individuals over a three-month period, Bray *et al.* (1978) demonstrated an apparent under-reporting of food intake in the initial interview. With direct observations, on the other hand, it was found that obese people tend to choose larger meals than did lean people (Stunkard and Kaplan 1977). Nevertheless, this over-eating behavior appears to primarily occur in a public place. Stunkard and Waxman (1981) found that obese adolescent boys ate more than their siblings at school, but not at home. In general, the energy intake declines with age, with the peak values occurred in the second decade of life in both sexes (Bray 1983). In this context, obesity must also be attributed to a greater reduction in energy expenditure.

Energy expenditure

As will be covered in Chapter 13, the total energy expenditure can be partitioned into energy expended via: (1) the resting metabolic rate; (2) the thermal effect of food; and (3) physical activities. Of these three energy components, both the resting metabolic rate and energy cost due to physical activity appear to receive the most attention in terms of studying the etiology of obesity. This may be because thermogenesis associated with food consumption constitutes a very small portion of the daily energy expenditure. Resting metabolism is the energy required by the body in a resting state. It can be influenced by age, gender, drugs, climate, body weight, and body composition, and accounts for a majority of daily energy expenditure. There is a gradual decline in resting metabolism as one ages. In addition, those with greater lean body mass will have greater resting metabolism. However, as related to obesity, several studies have failed to prove that this energy component is responsible for obesity. For example, Seidell *et al.* (1992) and Weinsier *et al.* (1995) reported that a gain in body weight occurred independently of changes in resting metabolism over ten and four years, respectively. In fact, obese individuals were found to have an expanded lean body mass and thus greater resting metabolic rate (Bray 1983).

Energy output associated with physical activity can vary tremendously. As such, this energy component has been the center of the majority of research dealing with obesity and its prevention and treatment. Despite some controversies, there has been a popular belief that a reduced level of physical activity leads to the development of obesity. Indeed, Andersen *et al.* (1998) and Gortmaker *et al.* (1996) have demonstrated a positive relationship between time spent watching television and incidence of obesity in children. It should be noted that those who are obese tend to expend relatively more energy for any given movement (Bray 1983). Consequently, despite reduced physical activity, the resulting energy expenditure may not necessarily be less in obese individuals as compared with lean individuals. This raises a question as to whether or to what extent a decrease in physical activity actually contributes to occurrence of obesity. It has been recently suggested that the impact of energy expenditure on the cause of obesity can vary from individual to individual and can also have different effects within individuals at different stages of development. Perhaps future studies aimed at the etiology of obesity should be devoted to examining the impact of energy balance over time using relatively homogeneous groups in terms of age, gender, fitness, and severity of obesity.

DIABETES MELLITUS

Diabetes mellitus is defined as abnormally high levels of blood glucose due to the inability to manufacture or respond to insulin. Worldwide, 100–120 million people have this chronic condition. In the United States its prevalence currently stands at about 16 million people, nearly half of whom do not realize they have the disease.

Insulin-dependent and non-insulin-dependent diabetes mellitus

The major types of diabetes are insulin-dependent diabetes mellitus (IDDM) and non-insulin-dependent diabetes mellitus (NIDDM). IDDM, also called type-1 diabetes, usually emerges before age 30, and tends to come on suddenly. NIDDM, also referred to as type-2 diabetes, is far more common than IDDM. It usually starts after age 30, and the majority of those who have the disease are obese. Recently, there has been a steady increase in cases in which those that are younger (<30 years of age) are also diagnosed with NIDDM, particularly if they are overweight. The onset of NIDDM tends to be more gradual than that of IDDM, and blood glucose levels remain more stable.

IDDM is an autoimmune disease in which the body produces antibodies that attack and damage the pancreatic β-cells. At first, the ability of the β-cells to secrete insulin is merely impaired, but usually within a year or so these cells stop producing or produce little insulin. People with IDDM must inject insulin daily as the function of their body tissues in responding to insulin remains normal. Although heredity plays some role in IDDM, there is no known family history of diabetes in most cases.

NIDDM, on the other hand, begins with the impairment of the body's tissues in responding to insulin. Therefore, in order to get the cells the glucose they need, the β-cells must increase their production of insulin. Diabetes results when the β-cells are unable to secrete enough extra insulin to overcome the tissue's resistance to insulin. Most people with NIDDM can be treated with oral drugs aimed to improve insulin sensitivity, or simply with lifestyle intervention that promotes weight loss. About 30–40% of NIDDM patients need insulin to achieve adequate control of their blood glucose. Heredity plays an important role in NIDDM and those with NIDDM are highly likely to have at least one relative with diabetes.

Cellular defects in glucose metabolism

Diabetes is regarded as a metabolic disorder in that it impairs the way the body utilizes glucose due to a deficiency of insulin. Every cell needs a regular supply of glucose. The cells absorb glucose from the blood and use some of it immediately for various metabolic functions. The rest of the glucose is converted to glycogen in the liver and muscles and stored there for future use. However, the body's ability to store glycogen is limited, and glucose that is not used immediately or stored as glycogen will be converted to triglycerides stored in adipose tissue. Insulin is the key regulator of glucose in the body. As blood glucose levels rise – such as after a meal – the pancreas produces insulin, which is then transported via circulation to target organs such as muscles and the liver. Insulin then attaches to sites on the surface of cells called receptors. Binding of insulin to these receptors causes carrier proteins or **glucose transporters** (GLUTs) to move from inside the cell to the cell's surface. GLUTs travel back and forth across the cell membrane, picking up glucose from the blood and dropping it off inside the cell. In diabetes, insufficient insulin production or tissues' insensitivity to insulin results in elevated blood glucose, which, if it remains uncontrolled, can cause many chronic complications including cardiovascular diseases, kidney damage, neuropathy, and diabetic foot.

GLUTs are a family of membrane proteins found in most mammalian cells that are responsible for transporting glucose across cell membranes. There have been multiple isoforms of GLUT proteins (e.g., GLUT 1, GLUT 2, GLUT 3, GLUT 4) identified; they are distributed differently throughout different body tissues. Much of the research in this area has been related to GLUT 4, in part because it is found primarily in skeletal muscle, which is considered a major depot for storing carbohydrate. In addition, unlike other GLUTs, the function of GLUT 4 can be affected by insulin. The working mechanism of GLUTs remains hypothetical at this point. It is thought that the binding of insulin to its receptors on the cell membrane triggers a series of events involving second messengers, which leads to translocation of GLUT 4 from inside the cell to the cell surface. These GLUT 4 proteins then bind with glucose molecules, and such binding provokes a conformational change associated with transport, and thus releases glucose to the other side of the membrane (Hebert and Carruthers 1992; Cloherty *et al.* 1995).

CONDITIONS OF INSULIN RESISTANCE

Many individuals with obesity and diabetes mellitus share the same common condition: **insulin resistance**. Insulin is an anabolic hormone that promotes synthesis of glycogen and triglycerides. Insulin resistance is defined as decreased ability of insulin to stimulate cellular glucose uptake and storage and to suppress hepatic glucose production. This condition is also associated with a reduced ability of insulin to suppress fat mobilization and thus increased levels of circulating fatty acids. Insulin resistance can occur in individuals without NIDDM, but most of those that suffer from it are obese. It has been suggested that insulin resistance can be influenced by many factors, including obesity, physical activity, and dietary composition.

Testing for insulin resistance

Insulin resistance can be assessed by using an oral glucose tolerance test. In this test, 75–100 g of glucose in water is given orally to a fasting subject. Blood levels of glucose and insulin are then measured at intervals over a period of 2–3 hours. During this measurement period, the blood level of glucose rises initially and then falls due to the action of insulin (Figure 10.2). The response curve for blood level of insulin generally follows a similar but lagged time course. Insulin resistance is therefore judged from the insulin response compared to the glucose response. Those who demonstrate a high insulin response in the face of a normal or high glucose response will be considered insensitive to insulin or insulin resistant.

Although simplistic, this technique has its weakness, in that it is difficult to interpret because blood glucose concentration depends not only on insulin sensitivity of the liver and peripheral tissues, but also on many other factors such as glucose absorption, insulin secretion, and insulin clearance. In this test, concentrations of glucose and insulin are mutually interrelated, and changes on one variable can simultaneously result in changes to the other and vice versa. Thus, a causal effect of insulin on glucose metabolism cannot be determined, at least in a sequential manner.

The shortcomings associated with the oral glucose tolerance test have promoted the development of a more precise but invasive technique called **euglycemic, hyperinsulinemic glucose clamp** (DeFronzo *et al.* 1979). With this technique, blood glucose concentration is kept constant by glucose infusion that is regulated according to repeated, rapid blood glucose measurements. The blood insulin concentration is

Figure 10.2 Sample plasma glucose and insulin responses during a three-hour oral glucose tolerance test before and after aerobic training.

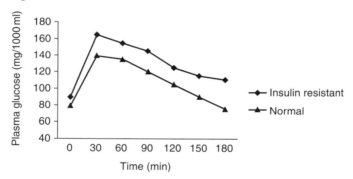

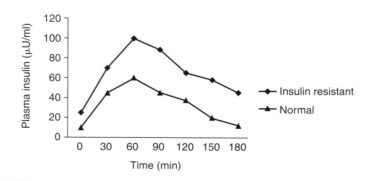

initially raised and then maintained at a constant via a prime-continuous infusion of insulin. Under these steady-state conditions of euglycemia and hyperinsulinemia, the glucose infusion rate equals to glucose disposal or uptake by the cell, which can then be used to determine the severity of insulin resistance. This technique, if conducted in conjunction with indirect calorimetry or an isotopic tracing technique, can also allow partitioning of the amount of glucose taken up by tissues into those being oxidized and those being stored. Such information is valuable in exploring cellular mechanisms that account for insulin resistance.

The glucose clamp technique has been widely used in most clinical studies designed to investigate glucose metabolism at a given insulin concentration. With the use of multiple levels of insulin concentration, it becomes possible to generate a dose–response relationship between insulin concentration and its effect on glucose disposal, which allows more complete examination of insulin action (Figure 10.3). Two terms have been derived as a result of this analytical approach: **insulin sensitivity** and **insulin responsiveness**. Increased insulin sensitivity is defined as a reduction in the insulin concentration that produces half of the maximal response, whereas increased insulin responsiveness is defined as an increase in the maximal response to insulin.

Insulin resistance and body fat distribution

Insulin resistance is associated with not only the overall accumulation of fat in the body, but also with how body fat is distributed. There is considerable evidence suggesting that excess accumulation of fat in the upper body, or truncal region, is a strong predictor of insulin resistance. For example, Banerji (1995) observed that variance in **visceral** adiposity accounted for much of the inter-individual variation in insulin resistance among individuals with type-2 diabetes. In addition, a weight-loss intervention study conducted by Goodpaster *et al.* (1999) revealed that among non-diabetic obese

Figure 10.3 Hypothetical dose–response relation between hormone concentration and its biological effect. Insulin responsiveness is the maximal response of glucose disposal. Insulin sensitivity is the hormone concentration eliciting half of the responsiveness.

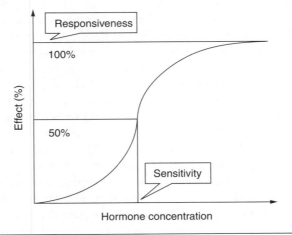

subjects, the decrease in visceral adiposity was the body composition change that best predicted the improvement in insulin sensitivity after weight loss. Such a distinctive role of different patterns of fat distribution is also supported by several *in vitro* studies that have examined metabolic heterogeneity of adipose tissue (Richelsen *et al.* 1991; Jansson *et al.* 1990). The general experimental approach of these studies was to isolate adipose tissue from abdominal and lower body **subcutaneous** regions so that lipolytic activity of adipose tissue could be compared between the two regions. Collectively, these studies revealed that adipose tissue from the abdominal region is metabolically more active and has a greater tendency to be broken down into FFAs. As FFAs formed due to lipolysis in this central region are directly released into the **portal circulation**, it is considered that in those with central obesity, their liver may have been exposed to high concentrations of FFAs, which can ultimately decrease hepatic insulin sensitivity. According to Randle *et al.* (1963), an excess of fatty acids in the systemic circulation derived from the abdominal region can also inhibit skeletal muscle glucose metabolism, and this has been considered a cause of insulin resistance manifested in the peripheral regions such as skeletal muscle.

Subcutaneous adipose tissue in the legs has been generally regarded as a relatively weak marker of insulin resistance. However, there is a growing amount of research that has attempted to examine the impact of peripheral adiposity on insulin resistance. Goodpaster *et al.* (2000) used computed tomography imaging to measure the quantity and distribution of adipose tissue in the thigh. Via a novel approach of subdividing adipose tissue into that present above the fascia lata (termed subcutaneous adipose tissue) and that present below the fascia lata (termed subfascial adipose tissue), these authors observed that variance in the amount of adipose tissue beneath muscle fascia correlated with insulin resistance, whereas no correlation was found between insulin sensitivity and the subcutaneous adiposity of the legs. These findings suggest that the amount of fat contained beneath the fascia, as well as within the muscle tissue in the lower extremities, is a key determinant of insulin resistance.

Effect of insulin resistance on glucose and fat utilization

Numerous studies have been attempted to examine mechanisms underlying the impairment in insulin-mediated glucose utilization seen in obese and NIDDM individuals. These studies have generally used the glucose clamp technique in conjunction with muscle sampling, indirect calorimetry, and/or the isotope tracer method so that metabolic fates of glucose taken by skeletal muscle can be further divided into glucose oxidation, glucose storage, and non-oxidized glycolysis. This experimental approach allows the examination of mechanisms responsible for insulin resistance at the cellular level. An early study by DeFronzo *et al.* (1985) revealed an approximately 45% reduction in insulin-stimulated leg glucose uptake in non-obese diabetic subjects. Using obese diabetic subjects, Kelley *et al.* (1992) observed a 60% decrease in insulin-stimulated leg glucose uptake. Of this deficit of glucose uptake, 66% was due to decreased leg glucose storage, whereas 33% was due to decreased leg glucose oxidation. This finding, together with the fact that these patients had lower than normal activity of glycogen synthase (Kelley *et al.* 1992), suggests that the reduced insulin-mediated glucose uptake can be attributed mainly to decreased leg glucose storage. It is now widely believed that glycogen synthesis is the metabolic pathway in skeletal muscle most severely affected by insulin resistance and is primarily responsible for decreased rates of glucose utilization. From a practical standpoint, individuals with obesity and NIDDM are at a disadvantage with regard to physical activity due to insulin resistance and its associated reduction in glycogen storage. These individuals may avoid performing sustained strenuous exercise that depends heavily on muscle glycogen as source of energy.

Reduced muscle glycogen content will lead to an increase in fat utilization in order to maintain an adequate energy supply. This homeostatic adaptation has been well documented in healthy individuals. However, recent studies have suggested that the skeletal muscle in those with insulin resistance were unable to make such a switch easily from carbohydrate to fat utilization (Kelley 2005). In other words, as a result of insulin resistance, fat utilization is also lower in addition to impaired glucose metabolism. This conclusion was derived from observations that during fasting conditions respiratory quotient (RQ) across the tissue bed of the leg was comparatively higher in people who were obese and insulin resistant than in metabolically healthy individuals. As will be discussed in Chapter 11, a high RQ represents a greater reliance on carbohydrate oxidation. Recently, Ukropcova *et al.* (2005) found that fat oxidation of skeletal muscle increased in subjects with improved insulin sensitivity, leanness, and aerobic fitness. An inability to increase the reliance upon fat oxidation in the face of reduced muscle glycogen has been termed as **metabolic inflexibility**. As shown in Figure 10.4, healthy individuals predominately use fat as a source of energy during fasting, but are able to shift efficiently to glucose oxidation upon insulin stimulation. However, those with obesity and NIDDM demonstrate a constrained adjustment to the transition

Figure 10.4 Schematics of metabolic inflexibility associated with insulin resistance.

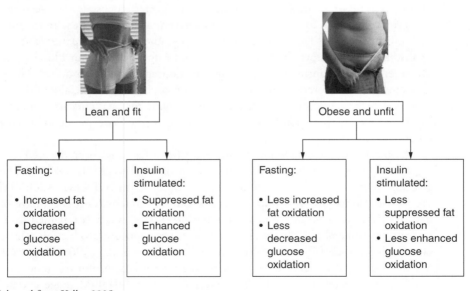

Source: Adapted from Kelley 2005.

between fasting and insulin-stimulation conditions. Their metabolic responses are blunted in terms of fat utilization during fasting and carbohydrate utilization upon insulin stimulation. The phenomenon of metabolic inflexibility emphasizes the importance of physical activity as a part of intervention for treating obesity and NIDDM. As discussed in Chapter 7, regular aerobic exercise training will increase the ability of skeletal muscle to oxidize fat and thus delay the consumption of glycogen.

METABOLISM DURING EXERCISE IN OBESITY AND DIABETES

Exercise has long been regarded as a beneficial treatment for obesity and diabetes. However, it was not until the last few decades that the interaction between exercise and these metabolic disorders was studied extensively. Research in this area has served to provide insight to the unique characteristics of exercise responses related to these metabolic disorders. This information will ultimately help in developing an effective exercise program aimed at treating or preventing obesity and diabetes.

Blood glucose

In those early investigations that used patients with IDDM, it was generally found that a bout of aerobic exercise will cause a fall in blood glucose to a normal level if these patients have been treated regularly with insulin and had mild hyperglycemia. This observation suggests that regular exercise training may be an effective aid to glucose regulation. Plasma glucose concentration reflects a balance between glucose uptake by peripheral tissues, mostly muscles, and glucose production by the liver. Wahren *et al.* (1975) found that in insulin-withdrawn diabetic subjects, muscle glucose uptake was greater, while hepatic glucose production was similar as compared to healthy controls during exercise. According to the authors, this greater glucose uptake seen in IDDM was primarily driven by the action of hyperglycemia. Consequently, more glucose can rush into tissue due to a greater concentration gradient. In these patients, it is the lack of insulin that prevents their tissues from drawing glucose from blood successfully. However, muscle contraction can allow this desired response to occur. The glucose-lowering effect of exercise, however, was not always the case, especially in patients with more severe presentations of IDDM. It was found that patients with more marked hyperglycemia may respond to exercise with a further rise in blood glucose levels (Wahren *et al.* 1978).

Within the last few decades, considerable efforts have also been devoted to examining blood glucose responses during exercise in patients with NIDDM. The favorable glycemic response as seen in IDDM was also observed in NIDDM. Additionally, the mechanism underlying this response appears to be similar. Researchers have observed that in NIDDM, a fall in blood glucose during exercise was accompanied by a greater increased peripheral glucose uptake, whereas hepatic glucose production remained the same between diabetics and healthy controls (Colberg *et al.* 1996; Kang *et al.* 1999; Martin *et al.* 1995). NIDDM is characterized by marked insulin resistance in skeletal muscle. The greater increase in glucose uptake seen in NIDDM suggests that insulin resistance does not substantially impede the cellular uptake of blood glucose during exercise. In fact, it seems possible that muscle contraction abetted by the hyperglycemia and hyperinsulinemia is able to provide an additive or synergistic effect upon glucose uptake. This contention is underscored by the findings of DeFronzo *et al.* (1981), who reported that exercise in combination with experimentally induced hyperinsulinemia produced glucose uptake that was greater than that following either treatment alone in non-diabetic individuals.

Muscle glycogen

Muscle glycogen represents a major depot of energy for use during exercise, and its utilization increases as exercise intensity increases. Early studies using IDDM subjects have shown that the rates of glycogen utilization during exercise are no different in diabetics as compared with healthy controls (Saltin *et al.* 1979; Maehlum *et al.* 1977). Furthermore, the glycogen depletion pattern in the different fiber types during exercise is also similar in diabetics and healthy controls (Saltin *et al.* 1979). However, there is some indirect evidence to suggest that patients with IDDM may use less muscle glycogen. This contention was derived from the observation that diabetics have reduced rates of total carbohydrate oxidation concomitant with increased rates of muscle glucose uptake and was based upon the assumption that all glucose molecules taken up by muscle are oxidized. Re-synthesis of muscle glycogen during post-exercise recovery is an insulin-dependent process. Maehlum *et al.* (1977) found that in the absence of insulin injection, muscle glycogen repletion during recovery following exercise is minimal in diabetic patients, while with insulin, the rate of repletion is the same as in healthy subjects.

The reduced ability of skeletal muscle to use glycogen during exercise has been more uniformly reported in patients with NIDDM and obesity (Colberg *et al.* 1996; Kang *et al.* 1999; Goodpaster *et al.* 2002). By having three groups of healthy, obese, and NIDDM subjects exercise at a mild intensity, Colberg *et al.* (1996) found that while utilization of glycogen was lower in both the obese and NIDDM groups, it was only half as much in NIDDM as compared with that in healthy controls. This finding was confirmed by Kang *et al.* (1999) and Goodpaster *et al.* (2002), who used patients with NIDDM and obesity, respectively. To date, it remains uncertain as to what may have caused this reduced glycogen utilization to occur. In these studies, muscle glycogen was determined indirectly by subtracting the rate of glucose uptake from the rate of total carbohydrate oxidation and a decrease in muscle glycogen utilization was accompanied by an increase in plasma glucose uptake. In this context, it is possible that this reduced utilization of muscle glycogen may be secondary to a compensatory response resulting from greater glucose utilization. Both NIDDM and obesity have been associated with lower muscle glycogen content. Therefore, it is also likely that the reduced utilization of muscle glycogen during exercise can be brought about by lower muscle glycogen content prior to exercise.

Fatty acids and triglycerides

Circulating fatty acids and intramuscular triglycerides are the two major sources of fat energy utilized during exercise. Studies in exercising men have estimated that intramuscular triglycerides generally contribute more to the total fat oxidation than circulating fatty acids. However, circulating fatty acids will become more important oxidative fuels during prolonged exercise. The relationship between intramuscular triglycerides and circulating fatty acids appears to resemble that of muscle glycogen and plasma glucose – that is, while the intramuscular substrate stores are relatively more important at the start of the work, fuels supplied via blood are the predominant substrates during prolonged work. In patients with IDDM in which insulin is completely absent, Wahren *et al.* (1984) found a greater increase in fat oxidation during exercise as compared to healthy controls. These authors also observed a more exaggerated exercise-induced rise in plasma nor-epinephrine. Nor-epinephrine is a lipolytic hormone that helps mobilize fatty acids from adipose tissue. In this regard, it is thought that the increased fat oxidation seen in patients with IDDM may be due to increased lipolysis occurring in adipose tissue. The utilization of intramuscular sources of triglycerides was also found to be higher in the diabetic state. This finding was reported by studies using both IDDM patients and depancreatized dogs (Standl *et al.* 1980; Issekutz and Paul 1968).

More recently, greater fat oxidation was also reported during exercise in patients with NIDDM and obesity (Goodpaster *et al.* 2002; Horowitz and Klein 2000; Blaak *et al.* 2000). However, this increased fat oxidation can only be explained by an increased oxidation of intramuscular triglycerides, because oxidation of blood-borne fatty acids was found to be either lower or the same in NIDDM or obese patients as compared to healthy controls. The greater utilization of intramuscular sources of fat has been attributed to an increased accumulation of intramuscular triglycerides frequently found in these patients during the **post-absorptive state,** and is considered a cause of insulin resistance. The post-absorptive state is the period during which the gastrointestinal tract, such as the stomach and small intestine, is empty of nutrients and body stores must supply required energy. It appears that patients with NIDDM or obesity are generally not limited in their ability to oxidize fatty acids during exercise, although they tend to accumulate excessive intramuscular triglycerides. Given that intramuscular triglycerides have been related to insulin resistance, exercise is considered ideal and necessary for these patients in that it can help stimulate a greater fat utilization, and in the case of NIDDM and obesity may help alleviate or prevent insulin resistance. Table 10.5 provides a summary of altered carbohydrate and fat utilization in patients with IDDM and NIDDM, as discussed in the preceding paragraphs.

Table 10.5 Substrate utilization during aerobic exercise in patients with IDDM and NIDDM as compared to healthy controls

SUBSTRATE	IDDM	NIDDM
Carbohydrate		
Muscle glycogen	Same	Decreased
Plasma glucose	Increased	Increased
Fat		
Muscle triglycerides	Increased	Increased
Plasma fatty acids	Increased	Decreased/same

ROLE OF EXERCISE IN IMPROVING INSULIN SENSITIVITY

Exercise can produce many favorable responses with respect to carbohydrate and fat metabolism. For example, glucose uptake by peripheral tissues such as muscle increases during exercise. Such an increase in glucose uptake helps correct hyperglycemia. Exercise also allows a greater utilization of fatty acids and thus is considered a viable option for reversing insulin resistance. Because of these beneficial patterns of metabolism, which clearly suggest a potential benefit of exercise in treating or preventing metabolic diseases, there has been another growing stream of research that has attempted to directly examine the role exercise has in reversing diabetes- and obesity-related symptoms such as insulin resistance. Unlike studies discussed in the preceding sections, the scientific goal in this regard was generally pursued by examining the insulin sensitivity or insulin-mediated glucose disposal sometime after the cessation of a single bout of exercise or a period of regular exercise training. Much of the pertinent literature is related to NIDDM and obesity because insulin resistance is the hallmark of these metabolic disorders. This may be in part because previous studies have generally failed to demonstrate significant improvements in glycemic control assessed by glycosylated hemoglobin (HbA1c) and fasting glucose concentration in IDDM after physical training. For patients with IDDM, attempts should be focused on devising regimens that will allow regular and safe exercise to avoid hypoglycemia and hyperglycemia and not necessarily promoting exercise as a means of improved glycemic control.

Acute improvement in insulin-mediated glucose disposal

The impact of prior aerobic exercise on subsequent insulin sensitivity or insulin-mediated glucose disposal has been typically examined using the euglycemic, hyperinsulinemic glucose clamp. This procedure is often performed during the recovery period in order to examine the acute effect of prior exercise, although in some studies it occurs on the next day or two. With the selection of multiple levels of insulin to be infused, a dose–response curve can be generated between insulin concentrations and its resultant rates of glucose disposal in order to assess insulin sensitivity. It has been well documented that in healthy humans, insulin-mediated, whole-body glucose disposal was increased following an acute bout of exercise and this improved insulin action remained for as long as 48 hours (Mikines *et al.* 1988; Richter *et al.* 1989). The improved insulin action following an acute bout of exercise has also been demonstrated in patients with NIDDM and obesity (Devlin *et al.* 1985, 1987; Burstein *et al.* 1990). These findings clearly support the use of aerobic exercise in treating insulin resistance.

Some aforementioned studies have used the clamp technique combined with indirect calorimetry and the isotopic tracer technique in an effort to delineate the metabolic fate of enhanced glucose uptake (Devlin *et al.* 1985, 1987). It was found that the exercise-induced improvement in insulin-mediated glucose disposal can be largely explained by an increase in glucose storage as glycogen. A handful of studies of both animals and humans also revealed an increase in glycogen synthase and hexokinase during exercise recovery. Collectively, it is plausible to conclude that glycogen formation is an ultimate destination for the augmented glucose taken up by muscle after exercise.

The improved insulin action following exercise can also be ascribed to augmented glucose transport across the cell membrane. Glucose transport into tissues is achieved by the action of protein molecules called glucose transporters (GLUTs). As mentioned earlier, a number of different GLUT proteins have been identified and they are manifested in a variety of different tissues. GLUT 4 is the form of transporter found in skeletal muscle and adipose tissues. In a series of experiments using diabetic rats, Richter *et al.* (1982, 1984, 1985) found an acute increase in glucose transport following exercise. In these studies, glucose transport was determined by quantifying how much of the specially marked glucose (3-O-methylglucose) entered the muscle cell. Ren *et al.* (1994) also observed an increase in GLUT 4 expression along with improved insulin-stimulated glycogen storage in rat muscle following a prolonged swim. It appears that the improvement in insulin action occurs primarily in exercising muscle. Since exercise is also accompanied by major cardiovascular and hormonal changes, it is possible that insulin action is also affected in extramuscular tissues. Recent studies have shown that the liver, like muscle, becomes more insulin-sensitive after exercise (Pencek *et al.* 2003). Like muscle, a major portion of glucose taken up by the liver is also channeled into liver glycogen synthesis (Hamilton *et al.* 1996).

As glycogen synthesis represents a major metabolic fate for enhanced glucose uptake, it may be speculated that the greater the depletion of glycogen during prior exercise, the greater the improvement in insulin sensitivity following exercise. However, this hypothesis remains questionable. It has been found that improved insulin effect on glucose uptake can persist even when pre-exercise glycogen levels have been restored (Hamilton *et al.* 1996; Richter 1996). This finding suggests that this exercise benefit is not necessarily dependent on glycogen depletion. On the other hand, there are several lines of research that appear to support this hypothesis. For example, Bogardus *et al.* (1983) and Ivy *et al.* (1985) found an inverse relation between insulin sensitivity and muscle glycogen content after a single bout of exercise. Using a glucose tolerance test, Kang *et al.* (1996) also reported an improved insulin action following exercise at 70% but not 50% VO_2max. It may be safe to suggest that if an exercise program is prescribed with a goal of improving insulin sensitivity, then this program should entail at least some component of vigorous exercise.

Chronic changes resulting from physical training

Physical training consists of repeated bouts of acute exercise. With the use of an oral glucose tolerance test, studies that compared trained and untrained individuals have shown that trained individuals are better able to tolerate glucose and are more sensitive to insulin (King *et al.* 1987; Heath *et al.* 1983; Seals *et al.* 1984). This training-associated benefit was also found by studies using the hyperinsulinemic glucose clamp technique, although it is of interest that these studies demonstrated an improved insulin responsiveness, but not insulin sensitivity (Mikines *et al.* 1989a, 1989b). Remember, insulin responsiveness represents the maximal ability of the whole body to handle glucose, and as such an improvement in this measure may be due to wider changes that occur not only in skeletal muscle, but also in the liver, adipose tissue, and the pancreas. Exercise training has been found to increase muscle GLUT 4 expression (Lee *et al.* 2002; Terada *et al.* 2001). This increase in GLUT 4 may contribute to the enhanced capacity for insulin-stimulated glucose disposal in trained subjects. The training-induced improvement in insulin action may also be due to the fact that trained subjects are able to utilize more of their fat energy sources. This augmented fat utilization may be in part because of increased lipolysis. It was found that training can make adipose tissue more sensitive to adrenergic stimulation (Izawa *et al.* 1991). As discussed earlier, insulin resistance is linked to abdominal obesity as well as an excess accumulation of intramuscular triglyceride. It was also found that training would result in a decrease in the mRNA for proinsulin and glucokinase in the pancreas (Koranyi *et al.* 1991). A reduction in these protein molecules suggests decreased insulin secretion at a given level of blood glucose concentration.

SUMMARY

- In comparison with men, women are able to derive proportionally more of the total energy expended from fat oxidation during aerobic exercise. This gender difference may be attributed to the experimental observations that estrogen stimulates lypolysis and also inhibits carbohydrate utilization.
- Due to an increase in body mass, including fetal tissue, there is an increase in energy cost during weight-bearing activities – such as walking, jogging, and running – while pregnant. Energy costs during non-exercise periods also increases, primarily due to an increased resting metabolism.
- The inhibitive effect of estrogen and other placental hormones upon carbohydrate metabolism places pregnant women at higher risk of developing insulin resistance that can lead to gestational diabetes. As such, being physically active during pregnancy is of importance in preventing the occurrence of these metabolic disorders. This can be achieved by choosing primarily non-weight-bearing activities and exercising at low to moderate intensities.
- Both carbohydrate and fat utilization decrease as one ages, and these declines can impair the ability of the elderly to tolerate strenuous physical activities. The age-related reduction in substrate utilization is due to the loss of the size and/or oxidative capacity of metabolically active tissues such as skeletal muscle, as well as a decrease in tissues' sensitivity to insulin.
- Children and adolescents should not be regarded as miniature adults, because the age-related functional deficiency in children and adolescents is not always attributable to the fact that they are smaller in size. Many differences in function have been found to be either partially related to, or completely independent of, changes in size.
- In comparison with adults, children are inefficient metabolically due to the fact that they are less coordinated in performing physical activities. They are also less able to store and use carbohydrate as an energy source and this can limit their tolerance to a strenuous exercise for an extended period of time. However, children have a reduced oxygen deficit and are also able to derive proportionally more of their total energy from fat oxidation.

- Overweight or obesity can be simply attributed to the fact that energy intake is greater than energy expenditure. However, etiology of this chronic condition is rather complex, as either an increase in energy intake, a decrease in energy expenditure, or a combination of both can contribute to this positive energy balance. In addition, the impact of such energy imbalance can vary from individual to individual and can have different effects at different stages of development within the same individual.

- Both IDDM and NIDDM are characterized as metabolic disorders in that they impair the body's ability to regulate blood glucose concentration and are accompanied by a condition of hyperglycemia. However, the etiology to this impairment is different. IDDM is associated with a lack of insulin; NIDDM involves a reduced ability of tissues to respond to insulin.

- Many individuals with obesity and NIDDM are characterized as having insulin resistance. Insulin resistance is defined as a decreased ability of insulin to stimulate cellular glucose uptake and storage and to suppress hepatic glucose production. Insulin resistance is not only associated with an overall level of body fat, but also with body fat distribution. Those obese individuals with their body fat being distributed primarily in the abdominal region are most prone to the condition of insulin resistance.

- In healthy individuals, reduced muscle glycogen content will result in an increase in fat utilization. However, the skeletal muscle in those with insulin resistance is unable to make such a switch easily from carbohydrate to fat utilization. This condition is also known as metabolic inflexibility.

- Aerobic exercise helps reduce the blood glucose level of diabetic individuals. This blood glucose-lowering effect of exercise can occur with and without insulin. For both IDDM and NIDDM, the introduction of an exercise program should be accompanied by accordant modifications of diet, as well as medication or insulin so that an over-reduction in blood glucose concentration or hypoglycemia can be prevented.

- Insulin resistance is associated with an impaired utilization of intramuscular triglycerides. However, those with insulin resistance are not limited in their ability to use fat as an energy source during exercise. Therefore, exercise is considered ideal in that it can help stimulate greater fat utilization, which may alleviate insulin resistance.

- Research has shown that a greater reduction in muscle glycogen following an acute bout of exercise can produce a greater improvement in insulin sensitivity. It may be safe to suggest that if an exercise program is prescribed with a goal of improving insulin sensitivity, then this program should entail at least some component of vigorous exercise so that a greater utilization of muscle glycogen can provide a positive impact upon insulin sensitivity.

- Regular physical activity has proven beneficial in improving insulin sensitivity. The improved insulin action can be explained by cellular changes, including increased glucose transporters and activity of enzymes that are responsible for glycogen synthesis. The improved insulin sensitivity is not only observed in skeletal muscle, but also manifested in the liver and adipose tissue. These positive changes justify the use of exercise as part of therapy in treating insulin resistance associated with obesity and NIDDM.

CASE STUDY: EXERCISE TRAINING FOR TYPE-2 DIABETES

Julia is a 50-year-old sedentary, African-American woman who has had type-2 diabetes for five years. Julia's baseline weight is 160 lb, and her waist circumference is 99 cm. Her resting blood pressure is 144/68 mmHg. She is currently taking medication to lower her blood glucose concentration. As recommended by her family physician, Julia plans to enroll in a three-month supervised aerobic exercise program. She received clearance from her physician before beginning exercise training.

Baseline laboratory analyses reveal a fasting blood glucose of 246 mg per 100 ml, HbA1C of 9.9%, total cholesterol of 205 mg per 100 ml, LDL-cholesterol of 145 mg per 100 ml, and total cholesterol-to-HDL-cholesterol ratio of 4.8. Her VO_2max is 26.5 ml kg^{-1} min and her percentage body fat is 40%.

Questions

- How would you describe her results of baseline laboratory analyses in terms of blood chemistry, fitness, and body composition?
- What risk factors does Julia have that may contribute to cardiovascular diseases?
- How might regular exercise training benefit her diabetic condition?
- What type of exercise program should be prescribed for Julia?

REVIEW QUESTIONS

1 What is the experimental evidence that supports the observation that women are able oxidize proportionally more fat than men during exercise?

2 Why isn't the gender difference in substrate utilization always observed?

3 Provide a specific explanation as to why there is an impaired utilization of carbohydrate and fat in older individuals.

4 How are children different from adults in terms of their aerobic and anaerobic capacity?

5 How would you explain the notion that children should not be viewed as miniature adults?

6 What are the differences between insulin-dependent diabetes mellitus (IDDM) and non-insulin-dependent diabetes mellitus (NIDDM)?

7 Define the terms (1) insulin sensitivity and (2) insulin responsiveness. How can these parameters be determined?

8 Insulin resistance can be assessed by an oral glucose tolerance test or euglycemic, hyperinsulinemic glucose clamp test. Provide a brief explanation as to the procedure of each, and the advantages and disadvantages associated with each test.

9 Why is there a close relationship between insulin resistance and visceral adiposity?

10 Explain how a bout of exercise may cause a fall in blood glucose concentration in patients with IDDM and NIDDM. Why is such a glucose-lowering effect of exercise often more phenomenal in NIDDM than IDDM?

11 How does an improvement in insulin sensitivity come about following training?

SUGGESTED READING

Bar-Or, O. and Rowland, T.W. (2004) Physiologic and perceptual responses to exercise in the healthy child. In: *Pediatric Exercise Medicine*, Champaign, IL: Human Kinetics, pp. 3–59.

This section of the book presents comparative data demonstrating how the exercise-induced cardiorespiratory, metabolic, and perceptual responses differ between children and adults. It is clear that children are not miniature adults and some of the responses are specific to their biological age rather than their body size.

Elia, M., Ritz, P., and Stubbs, R.J. (2000) Total energy expenditure in the elderly. *European Journal of Clinical Nutrition*, 54: S92–S103.

Using doubly labeled water measurements and cross-sectional comparisons of different age groups, authors are able to examine how aging may affect the total energy expenditure and its sub-components, including resting metabolic rate and physical activity.

Goodpaster, B.H., Wolfe, R.R., and Kelley, D.E. (2002) Effect of obesity on substrate utilization during exercise. *Obesity Research*, 10: 575–584.

This original investigation provides strong evidence that characterizes substrate utilization patterns during exercise between obese and lean individuals.

Ivy, J.L. (1997) Role of exercise training in the prevention and treatment of insulin resistance and non-insulin-dependent diabetes mellitus. *Sports Medicine*, 24: 321–336.

This review provides readers with further understanding of how physical training helps in treating and preventing insulin resistance and non-insulin-dependent diabetes. The cellular mechanisms responsible for improved insulin-mediated glucose disposal are thoroughly discussed in this article.

Tarnopolsky, M.A. (2000) Gender differences in substrate metabolism during endurance exercise. *Canadian Journal of Applied Physiology*, 25: 312–327.

This article reviews both animal and human studies pertaining to gender differences in substrate utilization during endurance exercise. It also discusses the underlying mechanism for why such gender difference exists.

11 Measurement of energy consumption and output

KEY TERMS

- missing foods
- phantom foods
- food diary
- food frequency questionnaires
- diet history
- signs
- symptoms
- creatinine
- direct calorimetry
- indirect calorimetry

- respiratory quotient
- caloric equivalent of oxygen
- respiratory exchange ratio
- doubly labeled water
- isotope
- acceleration
- pedometer
- accelerometer
- algorithm
- metabolic equivalent

ASSESSING NUTRITIONAL HEALTH

Measurement of nutrient intake is probably the most widely used indirect indicator of nutritional status. It is used routinely in national nutrition surveys, epidemiologic or clinical studies, and various federal and state health and nutrition program evaluations. Assessing dietary status includes taking into account the types and amounts of foods consumed and the intake of the nutrients and other components contained in foods. When the food consumption data are combined with information on the nutrient composition of food, the intake of particular nutrients and other food components can be estimated. Various methods for collecting food consumption data are available. However, no single best method exists. For example, food consumption data can be obtained by observing all the food and drink consumed by the individual for a specific period of time or by asking the individual to record or recall their intake. Neither option is ideal in that being observed can affect an individual's intake, and recording and recalling intake can be erroneous because these methods rely on memory, reliability and cognitive level of the consumer.

It has been reported that a person who is attempting to lose weight may tend to report smaller portions that were actually eaten (Johansson *et al.* 1998). Despite these disadvantages, properly collected and analyzed dietary intake data have considerable value. For example, using dietary analysis has allowed us to assess adequacy of dietary intakes of individuals and groups, to monitor trends in food and nutrient consumption, to study the relationship between diet and health, to establish food and nutrition regulations, and to evaluate success and cost-effectiveness of nutrition and risk-reduction programs. The commonly used methods described below are the best tools available for evaluating dietary intake to predict nutrient deficiencies and excesses. Once this information on intake has been collected, the nutrient content of the foods consumed can be estimated using computer software or food composition tables.

24-hour recall

This method is the most commonly used for assessing dietary intake. It involves a registered dietitian or a trained interviewer asking people to recall exactly what they ate during the preceding 24-hour period. Occasionally, the time period is the previous eight hours, the past seven days, or, in rare instances, even the preceding month (Lee-Han *et al.* 1989). However, memories of intake may fade rather quickly beyond the most recent day or two, so that loss of accuracy may exceed gain in representativeness.

In this method, a detailed description of all food and drink, including descriptions of cooking methods and brand names of products, is recorded. The interviewer helps the respondent remember all that was consumed during a pre-determined period and assists the respondent in estimating portion sizes of foods consumed. The method typically begins by asking what the respondent first ate or drank on last awakening. The recall proceeds from the morning of the present day to the current moment. The interviewer then begins at the point exactly 24 hours in the past and works forward to the time of awakening. Some researchers use time periods from midnight to midnight of the previous day. Asking the respondent about his or her activities during the day and inquiring how they might have been associated with eating and drinking can help in recalling food intake. The recall can then be analyzed using a computerized diet analysis program.

The 24-hour recall method has several advantages (Table 11.1). It is inexpensive and quick to administer (30 min or less), and it can provide detailed information on specific foods, especially if brand names can be recalled (Block 1989). It requires only short-term memory. It is well received by respondents because they are not asked to keep records and their time and effort of involvement is relatively low. The method is also considered to be more objective than the methods of **food diary** and **food frequency questionnaire** because its administration does not alter the usual diet (Guenther 1994).

Recalls have several limitations (Table 11.1). The primary limitation of the method is that data on a single day, no matter how accurate, are a poor representation of an individual's usual nutrient intake because intra-individual variability. However, multiple 24-hour recalls performed on an individual and spaced over various seasons may provide a reasonable estimate of that person's usual nutrient intake (Block 1989; Lee-Han *et al.* 1989). In addition, respondents may withhold or alter information about what they ate because of poor memory, embarrassment, or trying to please or impress the interviewer. The items that respondents tend to under-report include binge eating, consumption of alcoholic beverages, and consumption of foods perceived as unhealthy. Respondents tend to over-report consumption of brand-name foods, expensive cuts of meat, and foods considered healthy (Feskanich *et al.* 1993). Foods eaten but not reported are known as **missing foods**, while foods not eaten but reported are known as **phantom foods**.

Errors associated with under-report and over-report can be minimized by using multiple-pass 24-hour recall methods in which the interviewer and respondent review the previous day's eating events several times to obtain detailed and accurate information about food intake. For example, a quick list of foods eaten in the previous 24-hours is initially compiled. In the second pass, a detailed description of foods on the quick list is obtained by asking respondents to clarify the description and preparation methods of foods on the quick list. In the third pass, the interviewer reviews the data collected, probes for additional eating occasions, and clarifies portion sizes using standardized methodology.

Food diary or food intake record

Although the method of 24-hour recall is relatively simple, its usefulness depends on a person's memory. To more accurately assess a person's diet, it is better to record foods

Table 11.1 Advantages and disadvantages of various methods assessing diet

METHOD	ADVANTAGES	DISADVANTAGES
24-hour recall	• Quick and easy to administer • Relatively inexpensive • Requires only short-term memory • Does not alter eating pattern • Low respondent burden • Can provide detailed information on the type of food consumed • Useful in clinical settings	• Not representative of individual diet • Subject to under- or over-reporting of serving size • May omit foods • Relies on memory • Results vary with season • Data entry can be labor intensive
Food diary or food intake record	• Provides detailed information on nutrient intake • Does not rely on memory • Multiple days of recording are more representative • Can provide data on eating habits	• Requires a high degree of cooperation from respondents • Respondents must be literate • May alter diets • Labor intensive and time consuming in data analysis
Food frequency questionnaire	• Can be self-administered • Results are machine readable • Modest demand on respondents • Relatively inexpensive for a large sample size • May better represent usual intake than several days of food record or 24-hour recall • Helps to reveal diet–disease relationship	• May not give a good estimate of quantity of foods consumed • Requires a good memory of diet over weeks and months • Data can be compromised when multiple foods are grouped within single listings
Dietary history	• Assesses usual eating habits, not just recent intake • Can detect seasonal changes • Data on all nutrients can be obtained • Can correlate well with biochemical measures	• Lengthy interview process • Requires highly skilled interviewers • Difficult to analyze • Requires cooperative respondent with ability to recall usual diet

and beverages while the person is consuming them. This method is known as a food diary or food intake record. In this method, the respondent records, at the time of consumption, the identity and amounts of all foods and beverages consumed for a period of time, usually ranging over 1–7 days and including at least one weekend day. Directions for conducting the food diary, as well as a sample recording sheet, can be found in Appendix G. In this method, food and beverage consumption are determined by estimating portion size, using household measures, or weighing the food or beverage on the scales. Portion sizes are often quantified by using household measures such as cups, tablespoons, and teaspoons, or measurements made with a ruler. Certain items, such as eggs, apples, bananas, bread, or 12-oz cans of soft drinks, may be simply counted as units. The record should be as complete as possible, including not only food and beverages, but also dressing, condiments, brand names, and preparation methods. This method allows portion sizes to be estimated. Ideally, food and leftovers should be weighed so that results of analysis can be more accurate. However, using weighed food records requires a great deal of effort and cooperation from respondents and can be cost-prohibitive in terms of measurement scales.

One of the most important advantages of this method is that the food record does not depend on memory because the respondent ideally records food and beverage consumption at the time of consumption (Table 11.1). In addition, data from a multiple-day food record are more representative of usual intake than that from a 24-hour recall or one-day food record. However, it is recommended that such a food record is derived from non-consecutive, random days (including weekends) and that multiple food records should be carried out to cover different seasons (Rebro *et al.* 1998; Macdiarmid and Blundell 1997). Via the use of dietary software, this method also provides more detailed nutrient intake information that can be specific to each day or each meal. Results of analysis typically include the consumption of total calories, calories from macronutrients, dietary fiber, sodium, cholesterol, saturated and unsaturated fat, all of which have been directly linked to health. Data outputs also include the consumption of essential nutrients, such as protein, vitamins, and most minerals, relative to the recommended daily allowance (RDA).

The food diary has several disadvantages (Table 11.1). The method requires respondents to be literate and cooperative, so that they are able and willing to spend the time and effort necessary to record their dietary intake for multiple days carefully. In fact, it has been suggested that those who can complete the method may not be representative of the general population (Rebro *et al.* 1998). In some cases, the tedious nature of this method can discourage the respondent from continuing or cause the respondent to change his or her intake rather than record certain items. When asked directly about recording their food intake, 30–50% of respondents have reported changing their eating habits while keeping a food record (Macdiarmid and Blundell 1997). Therefore, it is likely that this method can lead to an under-reporting of energy and nutrient intake (Sawaya *et al.* 1996).

Food frequency questionnaires

Food frequency questionnaires assess energy or nutrient intake by determining how frequently an individual consumes a limited checklist of foods that are major sources of nutrients or of a particular dietary component in question. For example: How often do you drink orange juice? How many times per week do you eat red meat? The foods included in the questionnaires are usually important contributors to the population's intake of energy and nutrients. They may focus on particular food groups, preparation methods, or nutrients. Respondents indicate how many times per day, week, month, or year they usually consume the foods in question.

Depending on the type of dietary information needed, the food frequency questionnaire can include questions about serving size. If only a general sense of serving size is needed, the respondent may be asked to indicate whether the servings were small, medium, or large relative to a standard serving. Some questionnaires may ask subjects to estimate serving sizes; these are called semi-quantitative. In terms of data analysis, a Scantron is often used so respondents can mark their answers on an answer sheet, which can then be optically scanned and scored on a computer. This saves the researchers considerable time and effort, and makes food frequency questionnaires a cost-effective approach for measuring diet in large epidemiologic studies. Both the 24-hour recall and the food frequency questionnaire are often regarded as retrospective dietary assessment methods, which require the person to recall what he or she ate in the past. An example of a food frequency questionnaire is shown in Figure 11.1.

Figure 11.1 An example of a food frequency questionnaire.

	Never	Once per week	2–4 per week	5–6 per week	Daily	Once per month	Once per 3 months	Once per year
Milk, yogurt, regular fat (1 cup)	○	○	○	○	○	○	○	○
Milk, yogurt, low-fat (1 cup)	○	○	○	○	○	○	○	○
Spinach, kale, other green leafy vegetables (1/2 cup)	○	○	○	○	○	○	○	○
Carrots (1 medium)	○	○	○	○	○	○	○	○
Beef (3 oz)	○	○	○	○	○	○	○	○
Rice, white (1 cup)	○	○	○	○	○	○	○	○
Rice, brown (1 cup)	○	○	○	○	○	○	○	○
Cookies (2–2" diameter)	○	○	○	○	○	○	○	○
Ice cream, regular fat (1/2 cup)	○	○	○	○	○	○	○	○

Food frequency questionnaires have several advantages (Table 11.1). They place a modest demand on the time and effort of respondents and generate estimates of food and nutrient intake that may be more representative of usual intake than a 24-hour recall or a few days of diet records. They are relatively easy and quick to administer, and the average time needed to complete the survey is usually less than 30 min. Although better data are obtained if an interviewer collects the data, the food frequency questionnaires can be self-administered and machine readable, and thus they are more cost-effective if used in studies that involve a large sample size. The questionnaires can be specific about foods or groups of foods to investigate the relationship between diet and disease. For example, if you are interested in the relationship between calcium intake and osteoporosis, you can assess calcium intake by using a food frequency questionnaire that asks only about foods high in calcium in subjects with and without osteoporosis.

There are some limitations associated with the food frequency questionnaire (Table 11.1). The most important limitation is that it requires the subject to have a good memory of intakes over weeks, months, or sometimes years (Willett *et al.* 1985; Feskanich *et al.* 1993). This method cannot be used to obtain the actual quantity of nutrient intakes; it only provides qualitative data on types and frequency of foods or food groups consumed over an extended period of time, although this limitation can be partially overcome by using a semi-quantitative questionnaire. In addition, the limited number of foods listed may cause the respondent to under-estimate consumption, or the respondent may eat things that are not listed on the questionnaire. The food list varies depending upon the questionnaire, but is usually limited to approximately 100–150 foods or drinks. Short questionnaires are faster and easier to administer, but are likely to be less representativeness. On the other hand, long questionnaires may do a better job of assessing nutrient intake patterns, but can be time-consuming and too tedious to complete.

Dietary history

Unlike the 24-hour recall and the food records, which assess recent intake for a short period, the **diet history** is used to assess an individual's usual dietary intake over an extended period of time, usually a month to a year. This historical approach was initially developed by Burke in the 1940s (Burke 1947). Burke's procedure involves four steps. The first is to collect general information about the respondent's health habits. The second is to collect information on usual eating habits, such as number of meals eaten per day, snacking pattern, likes and dislikes, and seasonal variation of eating patterns. This allows the interviewer to become acquainted with the respondent in ways that may be helpful in obtaining further information. At this step, a 24-hour recall is also collected.

The third step is to collect information about the frequency of consumption of specific foods. This step also serves as a check on the information given in the second step. For example, the respondent may have said that he or she drinks an 8-oz glass of milk in the morning. The interviewer then should inquire about the participant's milk-drinking habits to clarify the information given about the respondent's milk intake. The fourth and final step is to have the respondent complete a three-day food record, which serves as an additional means of checking the usual intake. This last step is sometimes omitted, since it is only another measure of recent intake and it adds to the cost and time needed to complete the analysis.

The dietary history has several advantages over other methods of dietary assessment. It gives an overall picture of eating habits and patterns including seasonal changes, not just an estimate of recent intake. The method is one of the preferred methods for obtaining estimates of usual nutrient intake (Block 1989; van Staveren *et al.* 1985). If researchers are only interested in a list of items that are typical of an individual's diet rather than a specific list of items eaten during a certain period of time, the diet history seems adequate to determine the typical diet. Most people are able to report what they typically eat, even if they cannot report exactly what they ate during a specific period of time.

This method also has some limitations. The data collection is a time-consuming process. The entire interview can take up to two hours to complete. Interviewers who use this method must be highly trained. Because of the large amount of information collected, the data is very difficult to analyze. In addition, the method also requires a cooperative respondent with the ability to recall his or her usual diet.

Next step: analyzing nutrient and energy content

Once information on food intake has been obtained, the next step is to determine nutrient and energy content, so that results can be compared to the dietary recommendations. To get a general picture of dietary intake, an individual's food record can be compared with a guide for diet planning, such as the MyPyramid (Chapter 8). For example, does the individual consume the recommended number of servings of milk per day? A more precise and detailed analysis of dietary intake can be done by determining the nutrient content of each food item. Information concerning the nutrient content of foods can be found on food labels, in published food composition tables, and in computer databases. Food labels provide information only for some nutrients, and they are not always available, especially for raw foods such as fruits, vegetables, fish, meats, and poultry. Food composition tables generated by government and industry laboratories can be found on the US Department of Agriculture (USDA) website: www.ars.usda.gov/nutrientdata.

Using food composition tables can be time consuming and tedious. Such a shortcoming can be overcome by using computer programs and their associated nutrient databases, which are readily accessible for both professional and home use. To analyze nutrient intake using a computer program, one must enter each food and exact quantity consumed into the program. If a food is not found in the database, an appropriate substitute can be used or the food can be broken down into its individual ingredients. For example, home-made vegetable soup could be entered as generic vegetable soup, or as vegetable broth, carrots, green beans, rice, and so on. If a new product has come on the market, the information from the food label can be added to the database. The advantage of using a computer program for dietary analysis is that it is fast and accurate. The program can calculate nutrient intake for each day or average it over several days. It can also compare nutrient intake to recommended amounts.

Computer analysis does have several limitations. It is not available to all people, and the lack of computer equipment and/or the cost may be prohibitive. Most computer programs have a limited number of foods in the database, which may cause a problem when a certain degree of accuracy is needed. This may be overcome by choosing a good program and having a good knowledge of foods, so that adequate substitutions can be made.

Other methods of assessing nutritional health

The other components of nutritional assessment include anthropometric measurement, laboratory measurements, and clinical assessments. Each of these assessment methods provides a unique perspective of an individual's nutritional status.

Height, weight, and body size

Anthropometric measurement assesses physical dimensions and body composition. Common examples of anthropometric measurements include health, weight, and body size. These measurements can be compared with population standards or used to monitor changes in an individual over time. If an individual's measurements differ significantly from standards, it could indicate health problems. For example, weight is often used to assess a person's risk for certain chronic diseases, such as heart disease and type-2 diabetes, because being overweight or obese can lead to many health problems, especially as an individual ages. Height and weight are also used to assess nutritional status in infants and children and are typically measured shortly after birth and throughout childhood. For example, children who are small for their age may have a nutritional deficiency, although this information should be evaluated only within the context of their personal and family history. Those who are smaller than the standard may simply have inherited their small body size or may be considered adequately nourished if they have never weighed more than their current weight and are otherwise healthy. Chapter 12 provides more detail with regard to assessment of body composition (e.g., percentage body fat).

Clinical assessments

Nutritional status can be evaluated using a clinical assessment. During this procedure, the clinician or researcher takes a medical history to obtain information about previous disease, weight loss and/or gain, surgeries, medications, and other relevant information such as family history. The health care provider asks the patient whether he or she has been experiencing any unusual symptoms such as a lack of energy, blurred vision, or reduced appetite. As these symptoms are not observable by others and can only be reported by the patient, symptoms are often overlooked.

The medical history is typically followed by a physical examination to determine whether there are visible signs of a health problem. In a physical exam, all areas of the body, including the mouth, skin, hair, eyes, and fingernails, are examined for indications of poor nutrition status. **Signs** of illness are different from **symptoms**, because signs can be seen by others, whereas symptoms are what the patient experiences, which may be more subjective. Some signs may indicate poor nutritional status, such as skin rashes, swollen ankles or edema, or bleeding gums. Physical examinations almost always include anthropometric measurements such as height and weight. Determining whether the signs or symptoms noted in a physical exam are due to malnutrition requires that they be evaluated in conjunction with the results of laboratory measurements and within the context of each individual's medical history.

Clinical assessment has advantages. For example, it is the only way health care providers can ask questions concerning symptoms of malnutrition. Furthermore, signs of some extreme forms of malnutrition are very distinct, and observation of them can make clinical diagnosis of a particular nutrient deficiency or toxicity quite accurate. However, because signs and symptoms of many nutrient deficiencies are not apparent until they become severe, clinical assessment may not be adequate when malnutrition is more moderate in nature.

Laboratory measurements

Measurement of nutrients or their metabolic by-products in the body, such as in the blood and urine, can be used to detect nutrient deficiencies or toxicities when compared with standard reference values (Table 11.2). They require laboratory analysis of biological samples taken from blood or urine. In some cases, the sample is analyzed for a specific nutrient. Blood concentrations of calcium and vitamin D are often measured to assess the risk for osteoporosis, especially in women who are approaching menopause. To assess whether the body maintains its muscle mass, one can determine urinary concentration of **creatinine**, a by-product of skeletal muscle metabolism. Because blood carries newly absorbed nutrients to the cells of the body, the amounts of some nutrients in the blood may reflect the amount in the current diet rather than the total body status of the nutrient. In this case, it may be necessary to analyze the cells in the blood or other tissues for indications of abnormal function, such as altered rates of chemical reactions. For example, vitamin B-6 is needed for chemical reactions involved in amino acid metabolism; when vitamin B-6 is deficient, the rate of these reactions will be slower than normal.

There are advantages and disadvantages to using laboratory measurements. On the positive side, laboratory measurements can help diagnose a specific nutrient deficiency or to detect a potential health problem, whereas other methods of nutritional analysis cannot. Laboratory data has been used a lot to evaluate risk for nutrition-related chronic diseases. For example, risks for cardiovascular diseases can be assessed by measuring levels of cholesterol in the blood. Measuring the amount of glucose in the blood can be used to diagnose diabetes. However, collection and analysis of biological samples requires technical expertise and is considered costly. In addition, many factors such as time of day, age, sex, activity pattern, and use of certain drugs can influence the level of nutrients or their biomarkers in blood or urine. These extraneous factors must be taken into consideration when interpreting results of laboratory measurements.

ASSESSING ENERGY EXPENDITURE AND SUBSTRATE UTILIZATION

Interest in monitoring energy expenditure and respiratory gas exchange during physical activity can be traced back almost 100 years, to when Haldane and Douglas, in preparation for the 1911 Anglo-American expedition to Pike's Peak in Colorado, developed the "Douglas bag" method, which measures oxygen consumption and carbon dioxide

Table 11.2 Normal blood values or reference range of nutritional relevance	
TEST VARIABLE	VALUE/REFERENCE RANGE
Acidity (pH)	7.35–7.45
Alcohol	0 mg dL
Ammonia	15–50 μg of nitrogen/dL
Bicarbonate	18–23 mEq L (carbon dioxide content)
Blood volume	8.5–9.1% of total body weight
Calcium	8.2–10.6 mg dL
Carotene	48–200 μg L
Chloride	100–108 mEq L
Copper	70–150 μg dL
Creatinine	0.6–1.2 mg dL
Folate	2–20 ng ml
Glucose (fasting)	70–110 mg dL
Hematocrit	Men: 45–55%
	Women: 37–48%
Hemoglobin	Men: 14–18 g dL
	Women: 12–16 g dL
Iron	Men: 75–175 μg dL
	Women: 65–165 μg dL
Lactate (lactic acid)	Venous: 4.5–20 mg dL
	Arterial: 4.5–14.4 mg dL
Lipids	Cholesterol: <200 mg dL
	LDL Cholesterol: <130 mg dL
	HDL Cholesterol: ≥40 mg dL
	Triglycerides: <150 mg dL
Magnesium	1.8–3.0 mEq L
Oxygen partial pressure	83–100 mmHg
Oxygen saturation (arterial)	96–100%
Phosphorous	3.5–5.4 mg dL
Platelet	150,000–350,000/mL
Potassium	3.5–5.0 mEq L
Proteins	Total: 6.0–8.4 g dL
	Albumin: 3.0–4.0 g dL
	Globulin: 2.3–3.5 g dL
Red blood cells	Men: 4.6–6.2 million per mm^3
	Women: 4.2–5.2 million per mm^3
Sodium	135–145 mEq L
Urea nitrogen (BUN)	7–18 mg dL
Vitamin A	30–70 μg dL
Vitamin B-12	200–800 pg dL
Vitamin C	0.6–2.0 mg dL
White blood cells	4500–10,000 per mm^3
Zinc	0.75–1.4 μg ml

production. Since then there has been a wide range of electronic and computer-assisted metabolic systems developed, which allow a precise quantification of energy expenditure during various physical activities. In today's world, where physical inactivity often plays a contributing role in the development of many chronic diseases, such technological advancement has enabled us not only to simply quantify energy cost of an activity, but also to assess the association between physical activity and health and the effectiveness of interventions aimed at increasing physical activity in order to treat or prevent diseases.

Laboratory approaches

In general, under the laboratory setting there are two techniques employed in the measurement of energy expenditure and substrate utilization: (1) direct calorimetry; and (2) indirect calorimetry.

Direct calorimetry

Energy metabolism can be defined as the rate of heat production. This definition recognizes the fact that, when the body uses energy to do work, heat is liberated. This production of heat occurs through cellular respiration and mechanical work. **Direct calorimetry** involves measurement of heat that is produced during metabolism. This technique works in a similar way as the bomb calorimeter mentioned in Chapter 6, except it is large enough to allow an individual to live and work inside for an extended period of time (Figure 11.2).

Energy expenditure during muscular exercise can be measured by installing an exercise device such as a treadmill, bicycle ergometer, etc. in the chamber. A thin copper sheet lines the interior wall of this insulated calorimeter, to which heat exchangers are attached. A known amount of water circulates through the heat exchanger, regularly absorbing the heat radiated from the subject in the chamber, which reflects the metabolic rate of that person. Insulation protects the entire chamber, so any change in water temperature relates directly to the individual's energy metabolism. The air is recirculated, and carbon dioxide and water are filtered out of the air before it re-enters the chamber, together with added oxygen. Direct measurement of heat production in humans is proven to be very precise based upon the well-established concept that one calorie is equivalent to the amount of heat needed to raise 1 g of water by 1°C. However, its application is limited because the technique requires considerable time, expense, and engineering expertise in operating and maintaining the equipment (Ravussin and Rising 1992). In addition, though to a much lesser extent, the results of this technique could be affected in that not all the heat produced is liberated to the environment and so cannot be captured; there will also be extra heat produced due to operation of exercise equipment itself, which does not result from metabolism.

Indirect calorimetry

Indirect calorimetry is the method by which measurement of whole-body respiratory gas exchange is used to estimate the amount of energy produced through the oxidative process. The rationale behind indirect calorimetry is that virtually all bioenergetic processes are oxygen dependent (Ravussin and Rising 1992). Indirect calorimetry

Figure 11.2 Direct calorimetry chamber.

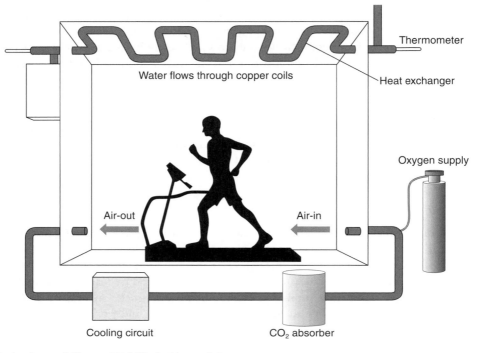

Source: Jeukendrup and Gleeson 2004. Used with permission.

differs from direct calorimetry in that it determines how much oxygen is required for biological combustion to be completed, whereas the latter measures directly the heat that is produced as a result of metabolism. The principle of indirect calorimetry can be explained by the following relationship:

$$\text{Substrate} + O_2 \rightarrow \text{Heat} + CO_2 + H_2O$$

In light of this direct relationship between oxygen consumption and the amount of heat produced, it makes sense that measuring the amount of oxygen consumed can be a logical replacement for measuring the heat produced as a result of biological oxidation.

In order to convert the amount of oxygen consumed into heat equivalents, it is necessary to know the type of energy substrates that are metabolized, e.g., carbohydrate and fat. The energy liberated when fat is the only substrate being oxidized is 19.7 kJ or 4.7 kcal per liter of oxygen used. However, the energy released when carbohydrate is the only fuel being oxidized is 21.1 kJ or 5.05 kcal per liter of oxygen used. Although less accurate, energy expenditure of exercise is often estimated by using 5 kcal or 21 kJ per liter of oxygen used. Therefore, a person exercising at oxygen consumption of $2.0\,l\,min^{-1}$ would expend approximately 42 kJ or 10 kcal of energy per minute. Despite being simplistic, this assumed value of energy equivalency should be used with caution because it implies that over 95% of the energy comes from oxidation of carbohydrate. In reality, this may not always be the case, as energy contribution from carbohydrate should be much lower when exercise is performed at low intensities. Consequently, it has been suggested that one should use an energy equivalent of 4.825 kcal per liter of oxygen for occasions such as resting or low-intensity, steady-state exercises, such as walking.

With indirect calorimetry, subjects are required to inhale ambient air with a constant composition of 20.93% oxygen, 0.03% carbon dioxide, and 79.04% nitrogen. VO_2 is determined from the change in volume of oxygen inhaled compared with volume of oxygen exhaled, shown as follows:

$$\text{Volume of } O_2 \text{ consumed} = \text{volume of } O_2 \text{ inhaled} - \text{volume of } O_2 \text{ exhaled}$$

The laboratory equipment used to measure oxygen consumption is illustrated in Figure 11.3. The volume of air inhaled and exhaled is measured with a gas meter that is attached to a subject through a flexible hose and a face-fitting mask. The exhaled gas from the subject is channeled to a mixing chamber to be analyzed for the fractions of

Figure 11.3 An open-circuit indirect calorimetry system.

Source: Medical Graphics Corporation.

oxygen and carbon dioxide by electronic gas analyzers. Results are then sent to a computer via an analog–digital converter; the computer is programmed to perform necessary calculations of VO_2 and other metabolic parameters.

The technology of indirect calorimetry has become increasingly sophisticated during recent years. A room-sized chamber has now been made available based on the principle of open-circuit indirect calorimetry. Such a chamber appears similar to the chamber used with direct calorimetry mentioned earlier, but without the coils to measure heat exchange. It has all the basic furniture and utility necessary to carry out various daily functions so that measurements can be made in a real-life situation. The chamber is equipped with the instruments that can measure oxygen uptake and carbon dioxide, as well as pulmonary ventilation, and these measurements can be performed continuously for an extended period of time. By using this large chamber, all components of daily energy expenditure can be assessed. These include the basal metabolic rate (BMR), sleeping metabolism, energy cost of arousal (basal metabolic rate minus sleeping metabolism), thermal effect of meals, and the energy cost of physical activities.

Although indirect calorimetry does not involve the direct measurement of heat production, this technique is still considered to provide a quite accurate reflection of energy metabolism and substrate oxidation. In fact, this technique has been extensively used as a criterion measurement in validation studies aimed to develop a new field-based method for quantifying energy expenditure. Compared to direct calorimetry, this indirect approach is relatively simpler to operate and less expensive to maintain and staff. With recent emergence of portable versions, this technique can also be used under many free-living conditions such as common household and garden tasks and physical leisure activities.

Measurement of substrate oxidation

In addition to quantifying energy expenditure, indirect calorimetry provides a means of estimating the composition of fuels oxidized. This is accomplished by determining the ratio of VCO_2 produced to VO_2 consumed, which is referred to as **respiratory quotient** (RQ). Due to structural differences in the composition of carbohydrates, lipids, and proteins, complete oxidation of each nutrient requires different amounts of oxygen and produces different amounts of carbon dioxide. For example, the oxidation of 1 g of glucose requires 0.746 l of oxygen and produces 0.743 l of carbon dioxide, and as a result, RQ is close to 1. On the other hand, the oxidation of 1 g of free fatty acid (palamitic acid) requires 2.009 l of oxygen and produces 1.414 l of carbon dioxide; as a result, RQ is close to 0.7. Differences in RQ caused by carbohydrate and fat oxidation are also illustrated in the following oxidative chemical reactions:

Glucose

$$C_6H_{12}O_6 + 6O_2 \rightarrow 6CO_2 + 6H_2O$$
$$RQ = 6CO_2 / 6O_2 = 1$$

Palmitic acid

$$C_{16}H_{32}O_2 + 23O_2 \rightarrow 16CO_2 + 16H_2O$$
$$RQ = 16CO_2 / 23O_2 = 0.7$$

RQ serves as a convenient measure to provide quantitative information on the relative contributions of energy nutrients to the total energy provision at rest and during steady-state exercise. If RQ is found to be equal to 1, then all energy is derived from oxidation of carbohydrate energy substrates. If RQ is found to be equal to 0.7, then all energy is derived from oxidation of fat energy substrates. It is, however, very unlikely that fat or carbohydrate would be the only fuel used in most circumstances. In fact, during rest and submaximal exercise, RQ is often found to be somewhere between 0.7 and 1.0. Table 11.3 lists a range of RQ values and corresponding percentages of energy derived from fat or carbohydrate oxidation, which was first published by American

Table 11.3 Thermal equivalents of oxygen for the non-protein respiratory quotient (RQ) and percentages of calories derived from carbohydrate and fat

NON-PROTEIN RQ	KCAL PER LITER OF O_2	% CARBOHYDRATE	% FAT
0.70	4.686	0.0	100.0
0.71	4.690	1.1	98.9
0.72	4.702	4.8	95.2
0.73	4.714	8.4	91.6
0.74	4.727	12.0	88.0
0.75	4.739	15.6	84.4
0.76	4.750	19.2	80.8
0.77	4.764	22.8	77.2
0.78	4.776	26.3	73.7
0.79	4.788	29.9	70.1
0.80	4.801	33.4	66.6
0.81	4.813	36.9	63.1
0.82	4.825	40.3	59.7
0.83	4.838	43.8	56.2
0.84	4.850	47.2	52.8
0.85	4.862	50.7	49.3
0.86	4.875	54.1	45.9
0.87	4.887	57.5	42.5
0.88	4.889	60.8	39.2
0.89	4.911	64.2	35.8
0.90	4.924	67.5	32.5
0.91	4.936	70.8	29.2
0.92	4.948	74.1	25.9
0.93	4.961	77.4	22.6
0.94	4.973	80.7	19.3
0.95	4.985	84.0	16.0
0.96	4.998	87.2	12.8
0.97	5.010	90.4	9.6
0.98	5.022	93.6	6.4
0.99	5.035	96.8	3.2
1.00	5.047	100.0	0.0

nutrition scientist Graham Lusk (1924). This table also illustrates the **caloric equivalent of oxygen** corresponding to each RQ value. The caloric equivalent of oxygen is defined as the amount of calories produced for each liter of oxygen used and, as shown in Table 11.3, this parameter is subject to the change in composition of carbohydrate and fat oxidation. In order to determine energy expenditure accurately, we need to know not only the level of oxygen uptake, but also the composition of energy fuels being utilized.

RQ represents gas exchange across the blood–cell barrier within an organ or tissue bed. Thus, direct measurement of this parameter can be difficult because it requires an invasive medical procedure. However, this methodological limitation is overcome by determining VO_2 and VCO_2 at the lungs using indirect calorimetry. To recognize the fact that VO_2 and VCO_2 were measured at the lungs, the **respiratory exchange ratio** (RER) is used instead of RQ. RQ and RER depict the same ratio, but RER can be interpreted as the ratio of VCO_2/VO_2 corresponding to metabolism of the overall body rather than a specific tissue bed, and is typically determined using indirect calorimetry.

Caution should be used in situations where one measures RER but uses the RQ table for quantifying substrate utilization. First, the RQ table was developed based upon the assumption that the amount of protein oxidized is small and negligible, or that it can be corrected for the oxidation of protein computed from nitrogen excretion in urine and sweat (Ferrannini 1988; Frayn 1983). Second, application of RQ assumes that oxygen and carbon dioxide exchange measured at the lungs reflects the actual gas exchange from macronutrient metabolism within tissues. In this regard, exercise intensity could be of concern. It has been found that this assumption works well during exercise of light to moderate intensity. During heavy exercise, however, VCO_2 measured at the lungs represents not only that produced during energy metabolism, but also that derived from buffering of metabolic acid, which increases at a greater rate during high-intensity exercise. Consequently, use of RER would no longer be accurate in reflecting the pattern of substrate utilization. Finally, use of RER may not be adequate for those with pulmonary disorders because the pattern of gas exchange at the lungs can be altered due to obstructed ventilation.

Field-based techniques

The traditional chamber and calorimetry technology can be inadequate for reasons such as cost, instruments, and time necessary to run the test. It is very difficult to use these sophisticated approaches to capture the complexity of activities in which people are engaged as they go about their daily lives. Consequently, there have been many attempts to develop relatively simple and more convenient methods to allow energy demands associated with free-living activities to be determined. Of those methods, the doubly labeled water technique, motion sensors, heart rate monitoring, and physical activity questionnaires or logs are perhaps the most common attempts for which the validity and reliability have been highly investigated. These field-based approaches enable us to track our physical activity participation in many free-living conditions. Although field-based techniques provide a convenient approach to the measurement of energy expenditure, they generally do not measure or distinguish metabolism of energy substrates such as carbohydrate and fat.

Doubly labeled water technique

The use of **doubly labeled water** for assessing energy expenditure in humans was first reported by Schoeller and Santen (1982). This technique requires the subject to consume a quantity of water containing a known concentration of the stable isotopes of hydrogen (^2H or deuterium) and oxygen (^{18}O or oxygen-18).

The term **isotope** means one of two or more species of the same chemical element that have different atomic weights (Shier *et al.* 1999). Isotopes have nuclei with the same number of protons but varying numbers of neutrons. Stable isotopes denote those whose nuclei will not emit radiation and thus are not radioactive. Both ^2H and ^{18}O are used as tracers as they are slightly heavier and can be measured within various body compartments. For example, through oxidative metabolism, labeled hydrogen is lost as ^2H_2O in sweat, urine, and water vapor during respiration, while labeled oxygen leaves as $H_2^{18}O$ in water and $C^{18}O_2$ in expired air. A mass spectrometer is then used to determine the difference in excretion rate between the two tracers and such difference then represents the rate of carbon dioxide production. Oxygen uptake is further estimated from VCO_2 as well as RQ, which is often assumed to be 0.85 (Black *et al.* 1986).

The primary advantage of using this technique to measure total energy expenditure is that it does not interfere with everyday life and thus can be used in a variety of free-living settings. The fact that this technique is not constrained by time would allow an measurement of the typical daily energy expenditure. To date, the technique has been used in circumstances such as bed rest and during prolonged activities like climbing Mount Everest, cycling the Tour de France, rowing, endurance running, and swimming (Hill and Davis 2002; Stroud *et al.* 1997; Mudambo *et al.* 1997). The potential drawbacks of this technique include the high cost of the ^{18}O and both the expense and specialized expertise required for the analysis of the isotope concentrations in body fluids by a mass spectrometer. As measurement is often taken to cover a long period of time, no information is obtained about brief periods of peak energy expenditure.

Motion sensors

Motion sensors are mechanical and electronic devices that capture motion or **acceleration** of a limb or trunk, depending on where the device is attached to the body. There are several different types of motion sensors that range in cost and complexity from the **pedometer** to the triaxial accelerometer. A pedometer is a relatively simple device primarily used to measure walking distance. It can be clipped to a belt or worn on the wrist or ankle. Pedometers do not operate on the vertical pendulum principle. Rather, they count the motion by responding to vertical acceleration. The early version of this instrument is mechanical in that it has a lever arm that is attached to a gear, which rotates each time the lever arm clicks. The horizontal spring-suspended lever arm moves vertically up and down as a result of each step being made. More sophisticated pedometers are now commercially available. They rely on the use of a micro-electromechanical system (MEMS) comprised of inertia sensors and computer software to detect steps. These pedometers are battery operated and have digital readouts that can display not only the total steps and distance, but also values in calories. Some of these can be adjusted for stride length so that walking distance can be more precisely calculated.

Modern pedometers are generally small, low-cost, and can be used in epidemiological studies that deal with large populations (Figure 11.4; Table 11.4). However, a pedometer has a number of limitations when used as a research tool. It is unable to distinguish vertical accelerations above a certain threshold, and thus cannot discriminate walking from running or different levels of exercise intensity (Bassett 2000). In terms of converting steps into energy expenditure, this device works on the assumption that a person expends a constant amount of energy per step. In Yamax pedometers, for example, this constant is assumed to be 0.55 cal per kg per step, regardless of how fast the person is moving (Hatano 1993). To date, most instruments have not been equipped with the ability to store data over a specific time period. As such, it generally cannot furnish information on the duration, frequency, and intensity of physical activity. It is also important to note that for activities that do not involve locomotion, such as cycling or upper body exercise, the unit may need to be attached to the body part that is moving.

Figure 11.4 Examples of digital pedometers.

Table 11.4 Advantages and disadvantages of various objective field methods for assessing physical activity and energy expenditure

METHOD	ADVANTAGES	DISADVANTAGES
Pedometers	• Small in size • Low cost • Suitable for epidemiological	• Unable to detect acceleration • Unable to quantify intensity, duration, and frequency • Unable to detect certain movements such as weight lifting, cycling and upper-body exercise
Accelerometers	• Small in size • Detect the rate of movement or acceleration • Able to provide information on intensity, duration, and frequency	• Questionable in converting motion data into energy expenditure • Unable to detect certain movements such as weight lifting, cycling, and upper-body exercise • Unable to discriminate walking/running performed on soft or graded terrain
Heart rate monitors	• Correlate closely with VO_2 • Measure all movements, including those that can't be detected by motion sensors • Able to provide information on intensity, duration, and frequency	• Weak relation with VO_2 in the low-intensity domain • Require individual calibration curves for accurate estimates of energy expenditure • Heart rate is subject to change according to stress, body posture, dehydration, environmental temperature
Combining motion and heart rate monitoring	• Overcomes major weakness associated with motion sensors and heart rate monitors in addition to the advantages mentioned above	• Time consuming in data analysis • Cost prohibitive • Necessary to validate the use of algorithm to estimate energy expenditure using large and heterogeneous samples and during various activities

Accelerometers are more sophisticated electronic devices that measure acceleration by body movement. Unlike pedometers, accelerometers are able to detect the rate of movement or the intensity of exercise, as acceleration is directly proportional to muscular force being exerted. Accelerometers can also measure acceleration in one (uniaxial) or three (triaxial) planes. Although a variety of different models are now commercially available, the Caltrac™ and Computer Science Applications™ (CSA) are the two commonly used uniaxial accelerometers, whereas the Tritrac R3D™ and Tracmor™ are the two commonly used triaxial accelerometers (Ainslie *et al.* 2003). Structurally, the accelerometer is equipped with a transducer that is made of piezoceramic material with a brass center layer. When the body accelerates, the transducer, which is mounted in a cantilever-beam position, bends, producing an electrical charge that is proportional to the force being exerted by the subject. This creates an acceleration–deceleration wave and the area under this wave is summed and converted to digital signals referred to as "counts." Results can be displayed on a screen as an accumulated total or downloaded as raw data to be further analyzed. Most current models also have the ability to display the level of accumulated energy expenditure for an extended period of time. This is done through a microprocessor that utilizes an activity-energy conversion factor as well as prediction equations for BMR based upon age, body size, and gender as independent variables (Washburn *et al.* 1989).

The most notable advantage of accelerometers is that they have the ability to detect the rate of movement and thus the intensity of exercise (Table 11.4). Together with the use of an internal clock, this intensity-discriminating feature helps to characterize the intensity and duration of physical activity being performed. In doing so, a dose–response relationship of physical activity to health and fitness outcome can be assessed. Other advantages are that they are small in size, can be worn without interfering with normal movement, and record data for extended periods of time. They also seem to be reliable instruments. By having a subject wear two Caltrac™ devices on the left and right sides of the body, Sallis *et al.* (1990) observed that the inter-instrument reliability reached 0.96.

A number of studies have reported significant correlations between energy expenditure estimated by accelerometers and by other proven, accurate methods, such as indirect calorimetry and the doubly labeled water technique. However, an equal amount of studies also found that this technique under-estimated energy expenditure. The high validity of this instrument for use in assessing physical activity appears to be demonstrated primarily in studies that employed level walking and running (Hendelman *et al.* 2000). Questions still remain as to whether accelerometers are able to accurately assess energy expenditure during leisure household and occupational activities, weight-bearing and static exercises such as cycling and load carriage, and during walking/running performed on soft or graded terrain (Hendelman *et al.* 2000; Sherman *et al.* 1998).

Given that a triaxial accelerometer combines three independent sensors to detect acceleration in three-dimensional space, it seems logical that it would be more accurate in capturing physical activities than a uniaxial accelerometer. However, validation results for supporting this contention are mixed. Welk and Corbin (1995) reported that both Caltrac™ (uniaxial) and Tritrac™ (triaxial) were similarly accurate in reflecting aspects of lifestyle activities. On the other hand, Bouten *et al.* (1996) and Eston *et al.* (1998) found that a three-dimensional monitor was better at predicting oxygen consumption during physical activities as compared to a uniaxial monitor. In light of advantages and concerns associated with accelerometers, there is still a need to continue improving not only the hardware in order to better track motion, but also the software so that converting activity counts into energy values can be more accurate.

Heart rate monitoring

Due to the difficulties encountered in measuring VO_2, and thus energy expenditure in the field, a steady interest has been devoted to developing less direct methods of recording physiological responses that are associated with VO_2. Among those physiological parameters investigated are heart rate (HR), pulmonary ventilation, and body temperature. However, monitoring HR appears to be the most popular technique. There is a fairly close and linear relationship between HR and VO_2 or energy expenditure during dynamic exercise involving large muscle groups; the greater the HR, the greater the VO_2. This is especially the case when HR ranges from 110 to 150 beats per minute (bpm), in which case the relationship between these two parameters is found to be linear. As such, it is reasonable to use HR as a physiological marker of VO_2 to assess physical activity and its associated energy expenditure. HR is relatively low-cost, non-invasive, and easy to measure. With today's technology, HR can be monitored and recorded with the use of a chest strap transmitter and a small receiver watch. A typical HR monitor transmits the R-R waves of electromyography (ECG) into a receiver in which ECG signals can be digitized and displayed (R-R waves refer to time intervals between two successive ventricular depolarizations). Many advanced models are also equipped with an internal clock that allows sampling over different time intervals, and also have the ability to store data over a period of days or weeks, thereby providing information on various components of physical activity, including intensity, duration, and frequency.

With the HR monitoring technique, the key issue lies in the precision of converting HR into energy expenditure. This is because HR in relation to energy cost can be affected by many factors other than physical activity per se. Factors such as age, fitness, and resting metabolism may be accounted for by using individualized HR–VO_2 curves, or by using measures of relative intensity – e.g., percentage of HR reserve – that adjusts for age and fitness. However, there are still some doubts deserving of attention (Table 11.4). For example, a question remains as to whether HR is a valid indicator of energy expenditure during low-intensity activities due to its weak relationship with VO_2 within the low-intensity domain (Freedson and Miller 2000). This is a pertinent question in that the intensities at which many daily activities are performed range from low to moderate. HR is also more susceptible to emotional stress that would result in a disproportional rise in HR for a given VO_2. In addition, HR can vary due to changes in stroke volume, and this latter parameter is influenced by body posture, exercise modes, and heat stress and dehydration.

Combining HR and motion monitoring

Both accelerometry and HR monitoring are the field-based methods that have been commonly chosen for assessing physical activity and energy expenditure. However, as discussed earlier, there are limitations associated with each method when used alone. It appears that limitations of HR monitoring are primarily due to biological variance. For example, as mentioned earlier, the HR–VO_2 relationship has been found to be affected by age, gender, fitness, stroke volume, and psychological stress. Responses of HR can also be influenced by ambient temperature, hydration status, and quantity of muscle mass involved in the activity (Haskell *et al.* 1993; Brage 2003). On the other hand, the limitations of accelerometry are mainly biomechanical in that the technique is generally unable to adequately detect increases in energy expenditure due to (1) movement up inclines; (2) an increase in resistance to movement; or (3) static exercise. In addition, a single sensor cannot identify movement that involves various parts of the body. As errors associated with the two methods are not inherently related, the combination of HR and accelerometry should in theory yield a more precise estimate of physical activity and energy expenditure as compared to either used independently.

Over the past decade or so, some studies have been attempted that have aimed to examine the validity of the simultaneous HR–motion sensor technique for measuring energy expenditure during exercise in the laboratory and a field setting (Haskell *et al.* 1993; Luke *et al.* 1997; Rennie *et al.* 2000; Strath *et al.* 2001a, 2001b; Brage *et al.* 2003). In general, these studies have found that measuring both HR and movement concurrently is a better approach in estimating oxygen consumption and energy expenditure. In these studies, both HR and movement counts were recorded at the same time. VO_2 was estimated by using data on HR, motion, and both HR and motion. Each estimated VO_2 was then compared against criterion VO_2 measured with a standard technique, e.g., indirect calorimetry. As information on both HR and motion were available, these studies were able to use the motion data to exclude HR that was increased due to non-exercise reasons or use the HR data to capture an increase in energy metabolism that was not detected by motion sensors. Some studies recorded HR and motion using two or more devices or sensors that were attached to different body parts, which may be problematic in terms of this technique being used in a field setting for a long period of time. To date, a single unit that detects HR and motion with only one sensor has been developed for use in tracking physical activity (Figure 11.5). This device uses a sophisticated **algorithm** to determine energy expenditure and can record data for as long as 11 days.

Apparently, this combined approach has many advantages. It overcomes the major weakness associated with HR monitoring and motion sensing alone (Table 11.4). For

Figure 11.5 Illustration of Actiheart™, which combines heart rate and motion monitoring to track physical activity and energy expenditure.

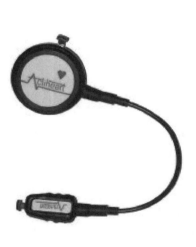
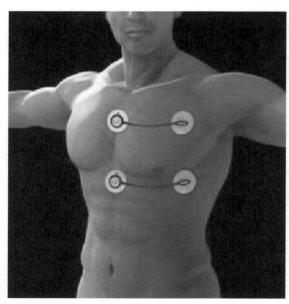

example, HR monitoring will not be subject to the error of movement sensing, such as detecting the level of activity during resistance exercise, swimming, and cycling. Likewise, movement sensing complements HR monitoring as it allows differentiation between increased HR caused by physical activity and that caused by other non-exercise-related influences. Nevertheless, this method still requires an individualized calibration curve to be established for both HR and movement counts, which could be time-consuming. Perhaps due to such technical difficulty in dealing with multiple sensing, most validation studies have used only a relatively homogenous and small sample size. Hopefully, future efforts will be directed to examine whether this combined approach will be sufficiently precise to preclude the need for individual calibration.

Multi-sensor monitoring system

Recently, a new device called SenseWare™ armband (SWA) was made available on the market to assess energy expenditure. This device is worn on the right upper arm over the tricep muscle. It has an ergonomic design that allows it to be easily slipped on and off and does not interfere with everyday activities or sleeping. The device contains multiple sensors that can measure various physiological and movement parameters simultaneously, including body surface temperature, skin vasodilation, rate of heat dissipation, and two-axis accelerometry. Data from these parameters, together with demographic information including gender, age, height, and weight, are then used to estimate energy expenditure using a generalized algorithm. The principle difference of this system compared to what was previously discussed is the inclusion of a heat flux sensor. This allows the system to detect a change in heat produced as a result of metabolism.

This system is still at an early stage of development and there is limited data available concerning its validity and reliability. It appears that the system is accurate in tracking resting energy expenditure (Fruin and Walberg Rankin 2004; Malavolti *et al.* 2007). However, it remains questionable with regard to its ability to detect energy expenditure during exercise. In comparing SWA with indirect calorimetry, Fruin and Walberg Rankin (2004) found that the SWA over-estimated energy expenditure of flat walking and under-estimated inclined walking, although the device was found to be reasonably accurate for exercise performed on a cycle ergometer. Jakicic *et al.* (2004) also observed differences between SWA and indirect calorimetry during treadmill walking, cycling, stair stepping, and arm cranking. Interestingly, in this latter study, no difference was observed when the authors adopted a series of mode-specific algorithms. This finding is appealing in the sense that to use multiple algorithms would require the development of a mechanism that allows the device to switch between algorithms to be developed so that minimal burden can be placed upon the user.

Subjective measures

There is a range of subjective methods available for the assessment of energy expenditure in humans. These methods may be classified into categories of (1) questionnaires and (2) activity log/diary. In general, questionnaires can be viewed as primarily recall-based; subjects are required to provide information on a pattern of daily activities that have already occurred. Questionnaires have been used to assess physical activities for the previous 24 hours, the previous week, and the previous year. Some questionnaires have been structured to specifically focus on occupational or leisure-time activities. Other questionnaires have been quite general in searching for information on activities that have occurred both on and off the job. Questionnaires can be further divided into those that are self-reported or those that are interviewer-administered.

Activity logs or diaries involve logging subjects' activities periodically; this can be done in a frequency ranging from every minute to every several hours. This method often includes a standardized form used to facilitate both compliance and quality. Some forms require subjects to record activities minute by minute, whereas others require subjects to record the change in activity. In these forms, names of activities are often abbreviated to make recording easier. In some activity logs, activities to be recorded could be specific enough to include not only the type, but also the intensity and duration. This method demands a greater level of attention from subjects in order to maintain their diary. To alleviate this burden upon subjects, some researchers have used a wristwatch that alerts the subject to record activities at specified times.

Both questionnaires and activity logs/diaries have been frequently used in large-population surveys and epidemiological studies. This is mainly because these techniques are relatively inexpensive compared to HR or motion monitoring. With these techniques, specific activities can be identified together with information on intensity, duration, and frequency. In addition, data can be collected by many subjects simultaneously. However, there are some limitations to each of these methods. Quantifying various aspects of physical activity such as type, intensity, duration, and frequency, which are important in estimating energy expenditure, is difficult. For example, with questionnaires, subjects may not necessarily recall all activities they have performed and/or they may over-estimate or under-estimate intensity and duration for some activities. These shortcomings could be particularly problematic when a questionnaire is administered to children, as they have lower cognitive functioning (Janz *et al.* 1995). There is also evidence that physical activity was underestimated in women when household chores were not included in the surveys (Ainsworth and Leon 1991). With the activity log/diary technique, recording activities frequently for a long period of time can be difficult. This technique can also influence the pattern of activity of the subject who records his/her on-going activities – the subject may purposely modify his/her behavior due to the survey.

Ainsworth *et al.* (1993, 2000) developed the Compendium of Physical Activity that provides a comprehensive list of physical activities and their associated level of exercise intensity (Appendix H). This standardized classification system is designed for use in constructing a questionnaire or activity log so results can be compared across different studies. The major strength of this system is that it furnishes information on intensity for each activity listed. The intensity associated with each activity is expressed as a **metabolic equivalent** (MET), which is a measure of energy cost of physical activities as multiples of the resting metabolic rate and is defined as the ratio of the metabolic rate (and therefore the rate of energy consumption) during a specific physical activity to a reference rate of metabolic rate at rest equivalent to 3.5 ml of oxygen per kilogram per minute, $1\,kcal\,kg^{-1}\,h^{-1}$, or $4.184\,kJ\,kg^{-1}\,h^{-1}$. In this context, energy expenditure can be calculated if necessary. By convention, 1 MET is considered as the resting metabolic rate obtained during quiet sitting, and MET values of physical activities range from 0.9 (sleeping) to 18 (running at 17.5 kmh or a 5:30 mile pace).

As shown in Appendix H, the first activity in the compendium is listed as follows:

TABLE C			
Code	MET	Category	Activity
01009	8.5	Bicycling	Bicycle, BMX or mountain

Column 1 shows a five-digit code that describes the class and type of activity. In this case, 01 refers to bicycling in general and 009 refers specifically to BMX (bicycle motocross) or mountain cycling. Column 2 shows the MET level. Columns 3 and 4 describe the category and activity. For example, regular resistance exercises are classified as 02050, and a competitive football game is classified as 15210. This particular method allows for the most detailed description of activities. However, the coding process is very time-consuming and impractical for people with poor reading and writing skills. For this reason, this methodology is used primarily in research settings.

SUMMARY

- Using dietary analysis allows us to assess the adequacy of dietary intakes of individuals and groups, to monitor trends in food and nutrient consumption, to study the relationship between diet and health, to establish food and nutrition regulations, and to evaluate success and cost-effectiveness of nutrition and risk-reduction programs.
- 24-hour recall, food intake records, and food frequency questionnaires are the three commonly used methods for assessing dietary status. No single best method exists for measuring dietary intake; each method has certain advantages and disadvantages. The method used depends on time constraints, subject characteristics, available resources, and whether the intent is to provide quantitative information or merely estimates of an individual's usual intake.
- The 24-hour recall method involves a registered dietitian or a trained interviewer asking people to recall exactly what they ate during the preceding 24-hour period. This method is quickly administered, does not alter eating patterns, and has a low respondent burden. However, it does not provide data representative of an individual's usual intake.
- A food diary or food intake record can assess an individual's diet more accurately than 24-hour recall. In this method, the respondent records, at the time of consumption, the identity and amounts of all foods and beverages consumed for a period of time, usually ranging 1–7 days and including at least one weekend day. This method does not rely on memory and can provide more detailed intake data, but requires a high degree of respondent cooperation and may result in alterations in the diet.

- Food frequency questionnaires assess energy or nutrient intake by determining how many times per day, week, month, or year an individual consumes a limited check-list of foods that are major sources of nutrients or of a particular dietary component in question. This method provides data that is more representative of usual intake and is useful in revealing diet–disease relationships. However, it may not give a good estimate of quantity of foods consumed and respondents must be able to recall and describe their diets correctly.

- Being able to accurately determine energy expenditure of daily living, including physical activity, is critical for a number of reasons: (1) to quantify the participation of physical activity; (2) to monitor compliance with physical activity guidelines; (3) to understand the dose–response relationship between physical activity and health; and (4) to assess the effectiveness of intervention programs designed to improve physical activity levels.

- There are two laboratory procedures commonly used to assess human energy expenditure: (1) direct calorimetry; and (2) indirect calorimetry. Direct calorimetry measures directly the heat produced as a result of metabolism, whereas indirect calorimetry measures the amount of oxygen being used during energy transformation.

- Respiratory quotient (RQ), which represents gas exchange across the blood–cell barrier within tissue beds, is a key parameter used in determining the composition of substrate utilization. This is based upon the fact that oxidizing carbohydrate and fat requires different amounts of oxygen and produces different amounts of carbon dioxide. Under most circumstances, this parameter can be estimated by measuring gas exchange occurring at the lungs; the ratio of VCO_2/VO_2 obtained is referred to as the respiratory exchange ratio (RER).

- Accelerometer and heart rate monitoring are by far the two most commonly chosen techniques for assessing energy expenditure. However, there are limitations associated with each method. The errors associated with accelerometry appear to be mechanical in that the technique is generally unable to detect increases in energy expenditure due to (1) movement up inclines; (2) an increase in resistance to movement; or (3) static exercise. Limitations of HR monitoring are primarily biological. For example, the $HR–VO_2$ relationship has been found to be affected by age, gender, fitness, stroke volume, psychological stress, and environmental temperature.

- As errors associated with accelerometer and HR monitoring are not inherently related, a simultaneous HR–motion sensor technique has been developed. This combined approach has proved to be a better approach in estimating energy expenditure than using either technique alone.

- Subjective measures of physical activity and energy expenditure are mainly accomplished by using (1) questionnaires and (2) activity logs/diaries. Questionnaires are often done in a recall fashion, and subjects are required to provide information on activities they have already performed. Activity logs or diaries are carried out by an individual periodically as their day goes by. These subjective methods provide the least expensive approach to tracking physical activity and energy expenditure and this advantage can be important when carrying out studies that use large sample sizes.

CASE STUDY: USING INDIRECT CALORIMETRY TO OPTIMIZE NUTRITION

Richard is on the college track team. His coach suggests that he participate in an off-season training program aimed to optimize his weight and body composition. He also needs to watch carefully what he eats in order to achieve a desirable energy balance. One of the major activities in the training program is a 60-min steady-state run at about 7 mph, three times per week. Richard wonders how many calories he will burn in this activity and how many of the calories expended come from fat. He also wants to know his resting metabolic rate in order to better plan his diet.

Richard was able to take part in a testing session in the human performance laboratory at his college, where the lab technician used indirect calorimetry to determine his resting metabolic rate and energy utilization during running at 7 mph. The following are the results of this test:

- oxygen uptake (VO_2) at rest = 0.28 liters per minute
- respiratory exchange ratio at rest = 0.75
- average VO_2 during running = 1.8 liters per min
- respiratory exchange ratio during running = 0.90.

Questions
- What is Richard's resting metabolic rate in (1) kcal min and (2) kcal day?
- How many calories did Richard expend during the 60 min run?
- What is the primary fuel that Richard used (1) at rest and (2) during running?
- How many calories of carbohydrate and fat did Richard use during running?

(Note: Table 11.3 is needed for solving these questions)

REVIEW QUESTIONS

1 What is the 24-hour recall method? What advantages and disadvantages are associated with it?

2 Describe how the food intake record method should be carried out? How does this method differ from the 24-hour recall method?

3 Why are food frequency questionnaires considered qualitative? How can this method be made semi-quantitative?

4 Both clinical assessments and laboratory measurements are used in an overall assessment of one's nutritional status. What are the advantages of incorporating these additional tools?

5 How is the term Calorie defined? What is the caloric equivalent of oxygen?

6 What is the difference between direct and indirect calorimetry?

7 What is the respiratory quotient? How is this parameter used to determine the pattern of substrate utilization?

8 If someone walks on a treadmill and is consuming oxygen at $1.5 \, l \, min^{-1}$ and expiring carbon dioxide at $1.2 \, l \, min^{-1}$, what is the rate of energy expenditure of this

individual? What are the percentages of carbohydrate and fat being used? How much of the energy expended comes from oxidizing carbohydrate and how much comes from oxidizing fat?

9 Describe how pedometer, accelerometer, and heart rate monitoring works in tracking energy expenditure of daily living.

10 What are the advantages and disadvantages associated with subjective measures of physical activity using questionnaires or activity logs?

SUGGESTED READING

Ferrannini, E. (1988) The theoretical basis of indirect calorimetry: a review. *Metabolism*, 37: 287–301.
This article offers further understanding of how indirect calorimetry works. It reviews the theoretical basis of and the advantages and limitations associated with this laboratory technique.

Freedson, P.S. and Miller, K. (2000) Objective monitoring of physical activity using motion sensors and heart rate. *Research Quarterly for Exercise & Sport*, 71: 21–29.
Motion sensor and heart rate monitoring are the two objective methods of tracking one's physical activity level. This article covers the theoretical basis of these two field-based techniques and also discusses advantages and limitations associated with each method. Validity and feasibility of developing a technique that uses both of these methods simultaneously is also discussed.

Sallis, J.F. and Saelens, B.E. (2000) Assessment of physical activity by self-report: status, limitations, and future directions. *Research Quarterly for Exercise & Sport*, 71: 1–14.
Self-report instruments are the most widely used type of physical activity measure. In this review, the authors summarize their findings on validity and reliability of this method, identify its strengths and limitations, and discuss areas for further improvement.

Schutz, Y., Weinsier, R.L., and Hunter, G.R. (2001) Assessment of free-living physical activity in humans: an overview of currently available and proposed new measures. *Obesity Research*, 9: 368–379.
This article is an excellent piece, which reviews not only the methods discussed in this textbook, but also some of the more recently developed measures and techniques, such as day-time physical activity level, activity-related time equivalent, day-time physical activity level heart rate, as well as using global-positioning-system measures to track human motion.

12 Body composition

KEY TERMS

- body mass index
- overweight
- obese
- body composition
- essential fat
- storage fat
- visceral fat
- subcutaneous fat
- cellulite
- resistance exercise
- endomorph

- mesomorph
- ectomorph
- densitometry
- hydrostatic weighing
- body density
- Archimedes' principle
- air displacement plethysmography
- dual-energy X-ray absorptiometry
- skinfold
- bioelectrical impedance analysis
- circumference

HEIGHT AND WEIGHT MEASUREMENTS

One of the most important measurements in nutritional assessment is body weight. Weight is an important variable in equations that predict energy expenditure and indices of body composition. Measurement of a person's weight in conjunction with his/her stature or height do provide some value in predicting health and risk of death. There is good evidence that overweight individuals tend to die sooner than average-weight individuals, especially those who are overweight at a younger age (Must *et al.* 1999; National Task Force on the Prevention and Treatment of Obesity 2000). A positive relationship has been demonstrated between **body mass index** (BMI – a weight–height index discussed later in this chapter) and mortality rate for cancer, heart disease, and diabetes mellitus, when BMI is above $25\,\text{kg m}^2$. Some researchers believe that the lowest mortality rates in the United States and many other Western nations are associated with body weights somewhat below the average weight of the population under consideration (National Task Force on the Prevention and Treatment of Obesity 2000). On the other hand, one should be cautioned against encouraging weight loss when it is not needed. A body weight that is too low is unhealthy and increases the risk of death. This can be seen in people suffering starvation or anorexia nervosa. Existing evidence suggests that as BMI falls below about 20, the risk of death increases.

Weight–height tables

Body weight can be better assessed through the use of weight–height tables. Weight–height tables are convenient, quick, and easy to use. They are designed to assess the extent of overweight based on sex and frame size, with the frame size being determining by measuring the width of the elbow. Weight and height can be easily measured, and most adults and many adolescents and children are able to understand and use the tables. The life insurance industry has been at the forefront of developing weight–height tables. The tables were developed by comparing the heights and weights of life insurance policyholders with statistical data on mortality rates and/or longevity of policyholders. The insurance industry has used the data from the tables to help screen applicants to avoid insuring people who have high risks. Table 12.1 is the 1983 Metropolitan Life weight–height table. This weight–height table has served as a benchmark based on the average ranges of body mass related to stature in which men and women aged 25 to 59 years have the lowest mortality rate. Based on the data used for developing this table, it was found that the lowest mortality rates occurred among non-smokers weighing 80–89% of the average weight (Manson *et al.* 1987). Therefore, the weights defined by the Metropolitan tables as recommended for a given stature were actually less than the average weights of the population under study.

Classifying subjects on the basis of frame size is a common feature of the Metropolitan weight–height table. The frame size of an individual can be determined objectively, although this parameter has often been estimated based on subject self-appraisal. Several approaches to determining frame size have been developed, including biacromial breadth (distance between the tips of the biacromial processes at the top of the shoulders) and bitrochanteric breadth (distance between the most lateral projections of the greater trochanter of the two femurs) (Katch *et al.* 1982), the ratio of stature to wrist circumference (Grant *et al.* 1981), and the width of knee, wrist, and elbow (Frisancho and Flegel 1983). Measuring the width of the elbow appears to be the most common and practical way of determining frame size and was taken into consideration when the 1983 Metropolitan weight–height table was developed. When measuring the width of the elbow, the subject should stand erect, with the arm extended forward perpendicular to the body. The subject then flexes the arm, forming a 90° angle at the elbow, with the palm facing the subject. The person who measures places the heads of a sliding caliper at the points that represent the widest bony width of the elbow, and pressure should be firm enough to compress the soft tissue outside the bony structure. The measurement should be read to 0.1 cm or $\frac{1}{8}$ in. Table 12.2 provides normative data allowing classification of frame size based on elbow breadth for men and women of different height.

There are several considerations to bear in mind when using the 1983 weight–height table. The table is formulated based on insurance industry data, which are derived from people who were able to purchase non-group insurance and who were 25–59 years of age. These people were predominately white, middle-class adults. Therefore, African-Americans, Asians, Native Americans, Hispanics, and low-income individuals are not proportionally represented. The weight–height table was based on the lowest mortality, and did not take into account the health problems often associated with obesity. These health problems include cardiovascular disease, cancer, hypertension, hyperlipidemia, and insulin resistance, which are more prevalent among the obese. In addition, no special consideration was given to the status of cigarette smoking, which is associated with lower weight and shorter life span. Including data on smokers in the table tends to make lower weights appear less healthy and higher weight more healthy (Willett *et al.* 1991). Finally, the weight–height table does not differentiate between fat mass and fat-free mass. What really matters is the

Table 12.1 1983 gender-specific weight–height tables proposed by Metropolitan Life Insurance Company

HEIGHT		SMALL FRAME		MEDIUM FRAME		LARGE FRAME	
INCHES	CENTIMETERS	POUNDS	KILOGRAMS	POUNDS	KILOGRAMS	POUNDS	KILOGRAMS
Men							
62	157	128–134	58–61	131–141	60–64	138–150	63–68
63	160	130–136	59–62	133–143	60–65	140–153	64–70
64	163	132–138	60–63	135–145	61–66	142–156	65–71
65	165	134–140	61–64	137–148	62–67	144–160	65–73
66	168	136–142	62–65	139–151	63–69	146–164	66–75
67	170	138–145	63–66	142–154	65–70	149–168	68–76
68	173	140–148	64–67	145–157	66–71	152–172	69–78
69	175	142–151	65–68	148–160	67–73	155–176	70–80
70	178	144–154	66–69	151–163	69–74	158–176	72–80
71	180	146–157	67–70	154–166	70–75	161–184	73–84
72	183	149–160	68–71	157–170	71–77	164–188	75–85
73	185	152–164	69–72	160–174	73–79	168–192	76–87
74	188	155–172	70–73	164–178	75–81	172–197	78–90
75	191	158–172	71–74	167–182	76–83	176–202	80–92
76	193	162–176	72–75	171–187	78–85	181–207	82–94
Women							
58	147	102–111	46–50	109–121	50–55	118–131	54–60
59	150	103–113	47–51	111–123	50–56	120–134	55–61
60	152	104–115	47–52	113–126	51–57	122–137	55–62
61	155	106–118	48–54	115–126	52–57	125–140	57–64
62	157	108–121	49–55	118–132	54–60	128–143	58–65
63	160	111–124	50–56	121–135	55–61	131–147	60–67
64	163	114–127	52–58	124–138	56–63	135–151	61–69
65	165	117–130	53–59	127–141	58–64	137–155	62–70
66	168	120–133	55–60	130–144	59–65	140–159	64–72
67	170	123–136	56–62	133–147	60–67	143–163	65–74
68	173	126–139	57–63	136–150	62–68	146–167	66–76
69	175	129–142	59–65	139–153	63–70	149–170	68–77
70	178	132–145	60–66	142–156	65–71	152–173	69–79
71	180	135–148	61–67	145–159	66–72	155–176	70–80
72	183	138–151	63–69	148–162	67–74	158–179	72–81

Note: Adapted by including values in centimeters and kilograms. Weights are obtained from men and women of 25–49 years of age with indoor clothing weighing 5 lb and 3 lb, respectively.

quality of the weight, not the quantity, and an "ideal weight" is not ideal for everyone at a given height. Many athletes weigh more than the average weight–height standards due to additional muscle mass. Being above the ideal weight based on the weight–height table should not necessarily dictate whether someone should lose weight.

Body mass index

Another measure of weight for a given height is the **body mass index** (BMI), also known as the Quetelet index. Such an index was an attempt by mathematician Lambert Adolphe Jacques Quetelet early in the nineteenth century to describe the relationship between body weight and stature in humans. BMI is calculated in the metric system by dividing weight by the square of height:

$$\text{BMI} = (\text{weight in kilograms}) / (\text{height in meters})^2$$

An individual who is 1.75 m (or 5' 9") and weighs 75 kg (165 lb) has a BMI of $75 / (1.75)^2 = 24.49 \, \text{kg m}^2$. The normal range is between $18.5 \, \text{kg m}^2$ and $25.0 \, \text{kg m}^2$. Individuals with a BMI higher than $25 \, \text{kg/m}^2$ are classified as **overweight** and individuals with a BMI higher than $30 \, \text{kg m}^2$ are classified as **obese**.

As reflected in the formula, the Quetelet index or BMI provides a height-free measure of obesity by adjusting body weight for height. In studies that involved a large sample size, a commonly used measure of obesity is the BMI. This is mainly because of its simplicity of measurement and calculation and its low cost. Studies

Table 12.2 Elbow breadth classifications for males and females of various statures

HEIGHT		SMALL FRAME		MEDIUM FRAME		LARGE FRAME	
Inches	Centimeters	Inches	Millimeters	Inches	Millimeters	Inches	Millimeters
Men							
61–62	155–158	$<2\frac{1}{2}$	<64	$2\frac{1}{2}-2\frac{7}{8}$	64–73	$>2\frac{7}{8}$	>73
63–66	159–168	$<2\frac{5}{8}$	<67	$2\frac{5}{8}-2\frac{7}{8}$	67–73	$>2\frac{7}{8}$	>73
67–70	169–178	$<2\frac{3}{4}$	<70	$2\frac{3}{4}-3$	70–76	>3	>76
71–74	179–188	$<2\frac{3}{4}$	<70	$2\frac{3}{4}-3\frac{1}{8}$	70–80	$>3\frac{1}{8}$	>79
≥75	≥189	$<2\frac{7}{8}$	<73	$2\frac{7}{8}-3\frac{1}{4}$	73–83	$>3\frac{1}{4}$	>83
Women							
57–58	145–148	$<2\frac{1}{4}$	<57	$2\frac{1}{4}-2\frac{1}{2}$	57–64	$>2\frac{1}{2}$	>64
59–62	149–158	$<2\frac{1}{4}$	<57	$2\frac{1}{4}-2\frac{1}{2}$	57–64	$>2\frac{1}{2}$	>64
63–66	159–168	$<2\frac{3}{8}$	<60	$2\frac{3}{8}-2\frac{5}{8}$	60–67	$>2\frac{5}{8}$	>67
67–70	169–178	$<2\frac{3}{8}$	<60	$2\frac{3}{8}-2\frac{5}{8}$	60–67	$>2\frac{5}{8}$	>67
≥71	≥179	$<2\frac{1}{2}$	<64	$2\frac{1}{2}-2\frac{3}{4}$	64–70	$>2\frac{3}{4}$	>70

have shown that the BMI correlates rather well ($r \approx 0.70$) with actual measurement of body fat from hydrostatic weighing. This index has also been found to correlate well with body composition estimated from total body water, total body potassium, and the skinfold technique. It is recommended by the National Institute of Health that physicians use the BMI in evaluating patients (National Task Force on the Prevention and Treatment of Obesity 2000). Many scientists also consider the BMI to be an appropriate way to assess body weight in children and adolescents (Dietz and Bellizzi 1999). BMI may be used in conjunction with skinfold measurements or waist circumference as an improved means of assessing increased risk in adults for heart disease, stroke, type-2 diabetes, and premature death (National Heart, Lung, and Blood Institute 1998). Table 12.3 shows classifications of overweight and obesity and associated disease risk based on BMI and waist circumference in adults.

The BMI may be a useful screening device for health problems. However, it reveals nothing about body composition and cannot distinguish between overweight resulting from obesity or from muscular development. Two individuals might have the same BMI but completely different body compositions. One could achieve his/her body weight mainly with muscle mass as a result of hard training, whereas the other could achieve his/her body weight by fat deposition as a result of a sedentary lifestyle. Without information about body composition, they both might be classified as obese. The possibility of misclassifying someone as overweight or obese by BMI applies particularly to large-size males, particularly field athletes, body builders, weight lifters, upper-weight class wrestlers, and American football players, such as linemen. It has been estimated that more than 50% of the National Football League were obese, with a BMI greater than 30, yet the average percentage body fat of linemen is about 18%. Thus, when assessing athletic populations, measuring body composition that includes fat mass and fat free mass is more appropriate.

Table 12.3 Classification of obesity and overweight and disease risk associated with BMI and waist circumference

CLASSIFICATION	OBESITY CLASS	BMI (KG/M²)	DISEASE RISK RELATIVE TO NORMAL WEIGHT AND WAIST CIRCUMFERENCE*	
			MEN ≤ 40 INCHES WOMEN ≤ 35 INCHES	MEN >40 INCHES WOMEN >35 INCHES
Underweight		<18.5	–	–
Normal		18.5–24.9	–	–
Overweight		25.0–29.9	Increased	High
Obesity	I	30.0–34.9	High	Very high
	II	35.0–39.9	Very high	Very high
Extreme obesity	III	≥40	Extremely high	Extremely high

Source: Adapted from National Heart, Lung, and Blood Institute 1998.

Note: *Diseases risk for type-2 diabetes, hypertension and cardiovascular disease.

BODY COMPOSITION ESSENTIALS

Body composition is defined as the ratio of fat to fat-free mass and frequently expressed as a percentage of body fat. The percentage of body fat is also regarded as relative fatness (% body fat). Table 12.4 presents recommended percentage body fat standards for young adults, the middle aged, and the elderly. The minimal averages and the threshold values for obesity vary with age and gender. For example, the average or median percentage body fat values for adult men and women (18–34 years) are 13% for men and 28% for women and the standards for obesity are > 22% for men and > 35% for women.

Essential and storage fat

Total body fat consists of both essential fat and storage fat. **Essential fat** is stored in the bone marrow, heart, lungs, liver, spleen, kidneys, intestines, muscles, and lipid-rich tissues of the central nervous system. This fat is necessary for proper functioning of certain body structures such as the brain, nerve tissue, bone marrow, heart tissue, and cell membranes. The essential fat in adult males represents 3–5% of the body weight. In females, essential fat includes additional sex-specific essential fat, which serves biologically important child-bearing and other reductive functions. This additional sex-specific fat gives adult females roughly 12–15% essential fat. Essential fat is the level below which physical and physiological health would be negatively affected.

Storage fat is simply a depot for excess fat. The major fat depot consists of fat accumulation in adipose tissue. This fat is also referred to as an energy reserve and contains about 83% pure fat in addition to 2% protein and 15% water within its

Table 12.4 Percentage body fat standards for healthy and physically active men and women

POPULATION	AGE (YEARS)	RECOMMENDED % BODY FAT LEVEL			
		LOW	MID	UPPER	OBESITY
Healthy males	18–34	8	13	22	>22
	35–55	10	18	25	>25
	55+	10	16	23	>23
Healthy females	18–34	20	28	35	>35
	35–55	25	32	38	>38
	55+	25	30	35	>35
Physically active males	18–34	5	10	15	
	35–55	7	11	18	
	55+	9	12	18	
Physically active females	18–34	16	23	28	
	35–55	20	27	33	
	55+	20	27	33	

Source: Adapted from Lohman *et al.* 1997.

supporting structures. Average males and females have storage fat of approximately 12% and 15% of the body weight, respectively. Storage fat includes the **visceral fat** that protects the various organs within the thoracic and abdominal cavities, but a much larger portion of the storage fat is found just beneath the skin's surface and is called **subcutaneous fat**. Subcutaneous fat accounts for over 50% of the total body fat. When this type of fat is separated by connective tissue into small compartments that extrude into the dermis, it gives a dimpled, quilt-like look to the skin; this is known as **cellulite**. Cellulite is primarily fat, but may contain high concentrations of glycoproteins, particles that can attract water and possibly give cellulite skin a waffle-like appearance. Such skin change is much more common in women than men.

The two-compartment body composition model

In order to make the most valid assessment of body composition, it is necessary to understand the underlying theoretical models used in the development of various body composition assessment techniques. Scientists have developed a variety of techniques to measure various body components, including fat, protein, bone mineral, and body water. Of these components, fat remains the one that generates the most interest due to its direct relationship with health and sports performance. A simplified, two-compartment model was developed by Brozek *et al.* (1963), and has been used as a theoretical basis for many of the body composition assessment techniques developed during the last several decades. In the two-compartment model, the body can be divided into fat mass and fat-free mass, or, according to an alternative approach, into adipose tissue and lean body mass (Figure 12.1). An individual who has 20% body fat has 80% fat-free mass. It has long been considered that the fat mass includes all the solvent-extractible lipids contained in body adipose and other tissues, and the fat-free mass is

Figure 12.1 The two-compartment model for body composition.

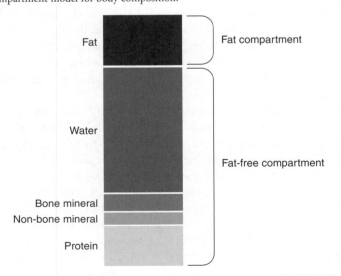

residual (Keys and Brozek 1953). Adipose tissue contains about 15% water, is nearly 100% free of the electrolyte potassium, and is assumed to have a density of $0.9\,g\,cm^3$. The fat-free mass is less homogenous and primarily composed of muscle, water, bone, and other tissues devoid of lipids. For example, solvent ether could be used to extract all the fat and lipid from minced animal carcass. That remaining after all the fat and lipid were extracted would be the fat-free mass. The fat-free mass has a density of $1.1\,g\,cm^3$, although this value may change depending upon age, ethnicity, nutritional status, degree of fitness, and the body's state of hydration. The fat-free mass is also referred to as lean body mass. The lean body mass is similar to the fat-free mass, except that lean body mass includes a small amount of lipids that the body must have – for example, lipid that serves as a structural component of cell membranes or lipid contained in the nervous system. The essential lipid constitutes about 1.5–3% of the lean body mass.

Computation of percentage body fat

Based upon cadaver analysis, which reveals that the density of body fat is $0.9\,g\,cm^3$ and the density of fat-free mass is $1.1\,g\,cm^3$, two equations were developed to estimate percentage body fat by incorporating the measured values of whole-body density (Brozek et al. 1963; Siri 1961):

Siri equation

% fat = [(4.95 / body density) − 4.50] × 100

Brozek equation

% fat = [(4.57 / body density) − 4.412] × 100

These formulas are based on the two-compartment model and yield similar percentage body fat estimates for body densities ranging from $1.03\,g\,cm^3$ to $1.09\,g\,cm^3$. For example, if an individual's measured whole-body density is $1.05\,g\,cm^3$, the percentage body fat, obtained by running this value through the Siri and Brozek equations, are 21.4% and 21.0%, respectively. In these formulas, body density is obtained from hydrostatic or underwater weighing, once considered the "gold standard" method. This method of measuring body composition is described in detail later in this chapter.

The generalized density values for fat $(0.9\,g\,cm^3)$ and fat-free $(1.1\,g\,cm^3)$ tissues represent averages for young and middle-aged adults. However, recent technological advances for measuring water (isotope dilution), mineral (dual-energy X-ray absorptiometry (DEXA)), and protein (neutron activation analysis) have shown that the fat-free mass varies widely among population subgroups because of age, sex, ethnicity, level of body fatness, training background, and nutritional status. For example, a significantly larger average density of fat-free mass was found in black $(1.113\,g\,cm^3)$ and Hispanics $(1.105\,g\,cm^3)$ people compared to white people $(1.100\,g\,cm^3)$ (Ortiz et al. 1992; Schutte et al. 1984; Stolarczyk et al. 1995). Consequently, using the existing equations established on the assumptions for whites to calculate body composition from whole-body density of blacks and Hispanics under-estimates percentage body fat. Applying such constant density values for fat and fat-free tissues to growing children or aging adults also introduces errors in predicting body composition (Lohman and Going 1993). For example, the water and mineral contents of the fat-free mass continually increases during the growth period and decreases during the process of aging. This will reduce the density of the fat-free tissue of children and the elderly below the assumed constant of $1.1\,g\,cm^3$, thus over-estimating percentage body fat. In addition, regular resistance training increases muscularity disproportionately to changes in bone mass, thereby reducing density of fat-free mass. Modlesky *et al.* (1996) found that the fat-free mass density of young white men with high musculoskeletal development due to regular resistance training was lower ($1.089\,g\,cm^3$) than the assumed value of $1.1\,g\,cm^3$, which caused an over-estimation of percentage body fat by use of the Siri equation.

Therefore, for certain population subgroups, scientists have applied multi-compartment models of body composition based on measured total body water and bone mineral values. With the multi-compartment approach, one can avoid systematic errors in estimating body fat by taking into account other influencing factors, such as the age, gender, ethnicity, nutritional status of the individuals. Table 12.5 provides selected fat-free mass densities measured for different population subgroups using the multi-compartment models (Heyward and Stolarczyk 1996).

BODY COMPOSITION, HEALTH, AND SPORTS PERFORMANCE

How much should I weigh or how much body fat should I have? This is a complex question, and the response depends on whether you are concerned primarily about your health, sports performance, or simply your physical appearance. The effect of body weight and fat on health has received considerable attention. Although being underweight may impair health, most of the focus has been on excess body weight

Table 12.5 Selected Population-Specific Fat-Free Mass Density

POPULATION	AGE (YEARS)	GENDER	FAT-FREE MASS DENSITY
White	20–80	Males	1.100
		Females	1.097
Black	18–32	Males	1.113
	24–79	Females	1.106
American Indian	18–60	Females	1.108
Hispanic	20–40	Females	1.105
Asian Natives	18–48	Males	1.099
		Females	1.111
Obese	17–62	Females	1.098
Anorexic	15–30	Females	1.087

Source: Adapted from Heyward V.H. and Stolarczyk L.M. (1996) Applied Body Composition Assessment. Human Kinetics, Champion, IL.

and fat, particularly the relationship of obesity to health. For athletes, on the other hand, extra body weight might prove to be an advantage, especially in American football, ice hockey, heavy weight or sumo wrestling, and other sports in which body contact may occur or in which maintaining body stability is important. However, caution must be exercised because the effect of extra weight can be neutralized or proven ineffective if the athlete loses speed. For physical appearance, the individual is the best judge of how he or she wishes to look – nevertheless, a distorted image can lead to serious health problems or impairment in sports performance.

Body composition and health

Most of the attention to body composition and its influence on health has focused on the proportion of body fat. The percentage can vary from 3–5% of body weight in excessively lean individuals to as much as over 50% of body weight in morbidly obese individuals. By definition, obesity is simply an accumulation of fat in the adipose tissue. As shown in Table 12.4, for adult males, a level of body fat above 25% is considered the low threshold for obesity, whereas adult females are considered obese when their body fat exceeds 38%. Obesity is associated with various risk factors for diseases, such as heart disease, hypertension, diabetes, some types of cancer, and osteoarthritis. Body fat can influence health in several ways; this is clearly demonstrated as when body fat increases over time, health conditions worsen in a parallel manner. However, when body fat levels fall too low, the resultant excessive leanness is also potentially problematic.

Extremely low levels of body fat are associated with some activities and sports. For example, some athletes, such as body builders or fitness competitors may reduce their body fat percentages as competition approaches or try to maintain lower levels of body fat throughout the year. Athletes participating in sports where body physique and leanness are of importance to success, such as gymnastics, figure skating, and diving, or sports involving weight classes, such as wrestling, weight lifting, boxing, and lightweight rowing, can become excessively lean during their competition seasons. Distance runners and cyclists also tend to have very low levels of body fat, usually as a result of the high energy demands of their sports. As mentioned earlier, body fat percentages of approximately 3–5% for men and 12–15% for women are considered the minimal level of adiposity necessary for maintaining health. Excessive leanness appears to affect female athletes more than their male counterparts. An extremely low body fat percentage has been considered a primary cause of reduced estrogen production and disruption of menstrual cycle in some female athletes. It becomes more evident that the condition of amenorrhea may be more directly related to a chronic reduction in energy consumption. Often, an excessively lean body reflects conditions known as eating disorders, and the female athlete triad discussed in Chapter 8. These conditions, if left untreated, can have long-term health and psychological consequences.

Fat-free mass is an important component of body composition for health. Higher levels of fat-free mass can predict more desirable bone density and skeletal muscle mass. Bakker *et al.* (2003) revealed that, compared with total body weight, standing height, BMI, waist circumference, waist–hip ratio, and skinfold thickness and fat mass, the fat-free mass was the most important determinant of ten-year longitudinal development of lumbar bone mineral density in adult men and women. As the fat-free mass is largely composed of skeletal muscle tissues, such a relationship between fat-free mass and bone mineral can be explained by mechanical stresses mediated through gravitational action and muscle contractions on bone. Greater bone density is associated with greater bone integrity, which in turn reduces the risk of bone fracture as well as osteoporosis in the future. Likewise, having more skeletal muscle is associated with increased muscular strength and decreased risk of physical injury. As muscle is more active metabolically than adipose tissue, increased skeletal muscle mass can be related to increased daily energy expenditure, which can in turn reduce the possibility of weight gain and obesity. In addition, having more skeletal muscle can improve glucose tolerance and decrease the risk of diabetes mellitus. Indeed, skeletal muscle is one of the major target organs responsible for taking up glucose upon the action of insulin.

Recently, a good deal of attention has been focused on the benefits that resistance exercise can have in aiding glucose regulation in people with type-2 diabetes. This is in part because **resistance exercise**, which is defined as performance of dynamic or static muscular contractions against external resistance of varying intensities, is most effective in improving or maintaining skeletal muscle mass and function. A number of studies have shown improved insulin sensitivity and enhanced glucose clearance following programs of resistance training (Holten *et al.* 2004; Ferrara *et al.* 2006). Resistance training has also been shown to be effective in lowering blood pressure (Fagard Cornelissen 2007). It is clear that a regular conditioning involving repeated muscular contractions against relatively high loads can be beneficial in reducing risks for many chronic diseases.

Body composition and sports performance

Many athletes believe that they must be big to be good at their sports because size traditionally has been associated with performance quality in certain sports, such as American football, basketball, and baseball. The bigger the athlete, the better the performance. But big doesn't always mean better. In certain other sports, smaller and lighter are considered to be more ideal, e.g., gymnastics, figure skating, and diving. Yet, this can be taken to extremes, compromising the health of the athlete. The following sections are intended to discuss further how performance can be affected by body types, weight/size, and composition.

Somatotype

Somatotype concerns physical types or physique of the body. A simple observation of the track-and-field events at the Olympic Games suggests that the physical characteristics of those successful in the shot put are different from those successful in the marathon. Indeed, there exist different body types, some of which may be considered desirable for particular sporting events. According to William Sheldon (1898–1977), an American psychologist who devoted his life to observing the variety of human bodies and temperaments, each person could be characterized as possessing a certain amount of the following three types of body form:

1 **Endomorph** – relative predominance of soft roundness and large digestive viscera.
2 **Mesomorph** – relative predominance of muscle, bone, and connective tissue ultimately derived from the mesodermal embryonic layer.
3 **Ectomorph** – relative predominance of linearity and fragility with a great surface-to-mass ratio giving sensory exposure to the environment.

Most people are a mixture of these three body types, with tendency toward one. Only 5% of the population are considered "pure" for each type. William Sheldon also examined 137 Olympic track-and-field athletes. He revealed that, although variation exists, the majority of athletes are considered to fall between mesomorph and ectomorph. Via further analysis by Heath and Carter (1967), it was reported that Olympic weight lifters and throwers are mostly mesomorphic, but slightly endomorphic. On the other hand, most distance runners and basketball players are considered mainly ectomorphic. Indeed, each athlete's build is a unique combination of these three components. Athletes in certain sports usually exhibit a predominance of one component over the other two. For example, body builders exhibit primarily mesomorph or muscularity, basketball centers exhibit primarily ectomorph or linearity, and sumo wrestlers exhibit primarily endomorph or fatness.

A predominantly endomorphic individual typically has short arms and legs and a large amount of mass on their frame. This hampers their ability to compete in sports requiring high levels of agility or speed. Sports of pure strength, like power lifting, are perfect for an endomorph. Their extra weight can make it difficult to perform sustained weight-bearing aerobic activities such as running. They can gain weight easily and lose condition quickly if training is ceased.

A predominantly mesomorphic individual excels in strength, agility, and speed. Their medium structure and height, along with their tendency to gain muscle and strength easily makes them a strong candidate for a top athlete in any sport. They respond well to cardiovascular and resistance training due to their adaptability and responsive physiology. They can sustain low body fat levels and find it easy to lose and gain weight.

A predominantly ectomorphic individual is long, slender, and thin, and therefore may not be suitable for power and strength sports. While they can easily get lean and hard, their lack of musculature severely limits their chances in sports requiring mass. Typically, ectomorphs dominate endurance sports, such as distance running and cycling. However, they can achieve low levels of body fat, which can be detrimental to health and for females in endurance sports it can result in a cessation of periods and iron deficiency.

Table 12.6 summarizes the physical characteristics and sports benefits of the three somatotypes.

Table 12.6 Physical characteristics of somatotypes and their suitability in sports

SOMATOTYPE	PHYSICAL CHARACTERISTICS	SPORTS BENEFITS
Endomorph	• A pear-shaped body • Wide hips and shoulders • Wider front to back rather than side to side • High fat mass and fat-free mass	• Size benefits sports such as football or rugby where bulk is useful, given that it can be moved powerfully • Often associated with larger muscle mass compared with ectomorphs • Tend to have large lung capacity which can make them suited to sports such as swimming or rowing
Mesomorph	• A wedge-shaped body • Wide, broad shoulders • Narrow hips • Narrow from front to back rather than side to side • Low or normal fat mass and high fat-free mass	• Respond well to cardiovascular and resistance training • They can easily gain or lose weight depending on the sport's needs • Often associated with low body fat and high fat-free mass, which is suitable for all sports events
Ectomorph	• Narrow shoulders and hips • A narrow chest and abdomen • Thin arms and legs • Low normal fat mass and low fat-free mass	• Light frame makes them suited for aerobic activity like long-distance running or gymnastics • A larger body surface area-to-mass ratio enhances heat tolerance

Body fatness and performance

Relative body fat or the percentage of body fat is a major concern for athletes. Adding more fat to the body just to increase the athlete's weight and overall size is generally detrimental to performance. Many studies have shown that the higher the percentage of body fat, the poorer the person's athletic performance. This is true of all activities in which the body weight must be moved through space, such as running and jumping. A negative association between level of fatness and sports performance has been demonstrated for a wide variety of sports events relating to speed, endurance, jumping ability, and balance and agility. In general, leaner athletes perform better.

Body composition can profoundly influence running performance in highly trained distance runners. Less fat generally leads to better performance. Studies have demonstrated that male and female long-distance runners of national and international caliber averaged 4.3% and 15.2% body fat, respectively (Pollock *et al.* 1977; Wilmore and Brown 1974). These values are similar to the more recent reports of elite male and female Kenyan endurance runners who averaged 6.6% and 16% body fat, respectively (Billat *et al.* 2003). In terms of gender comparisons, male runners normally have much less relative body fat than female runners; this is thought to be one of the most important reasons for the differences in running performance between elite male and female distance runners. This contention is supported by a study of male and female runners who, when matched by their 24-km road race time, did not differ in relative body fat (Pate *et al.* 1985). For body dimension and structure, distance runners generally have smaller girths and bone diameters than untrained counterparts. The best long-distance runners possess an overall smaller body frame, including both the stature and skeletal dimension. This ensures that much less weight needs to be carried over the long-distance run. It has been shown that runners with extra weight added to their trunk can increase the metabolic cost of submaximal exercise and reduce maximal aerobic power (Cureton and Sparling 1980).

Male and female competitive swimmers generally have more body fat than distance runners. Male swimmers who competed in the Tokyo and Mexico City Olympics were found to have an average body fat of 12% and 9%, respectively. It was speculated that because of a lower core temperature due to cool water of the training environment, swimmers tend to have an increased appetite, which often is not the case in athletes who are trained on land. In a study that compared collegiate swimmers and runners, Jang *et al.* (1987) found that female swimmers consumed a greater amount of energy (2490 kcal) than female runners (2040 kcal). However, they failed to demonstrate any difference in energy intake between male swimmers (3380 kcal) and male distance runners (3460 kcal).

Swimming is a weight-supported sport; although the athletes move their own body mass, they are supported by the buoyancy of the water, reducing the energy cost associated with movement. It is sometime argued that a certain level of body fat is useful for the swimmer because of enhanced buoyancy and body position or a reduced drag due to more rounded body surfaces. In analyzing the correlation between swimming performance and the physical characteristics of a large group of female swimmers aged 12–17, Stager *et al.* (1984) found that fat-free mass was a better predictor of swimming performance than body fat.

Resistance-trained athletes, particularly body builders and Olympic weight lifters, exhibit remarkable muscular development and relatively lean physiques. These athletes possess a large fat-free mass; percentages of body fat measured by hydrostatic weighing averaged 9.3% in body builders and 10.8% in Olympic weight lifters (Katch *et al.* 1980). However, using the weight–height tables, about 20% of these athletes would be classified as overweight. Interestingly, among weight lifters, adding extra fat weight is sometimes considered advantageous; some weight lifters add large amounts of fat weight just before the competition under the premise that the additional weight will lower their center of gravity and give them a greater mechanical advantage in lifting, although this practice has not yet been confirmed using a scientific approach. Sumo wrestlers are another notable exception to the notion that overall body fat may not be a major determinant of athletic success. In this sport, the larger individual has an advantage toward winning, but even so, the wrestler with the higher fat-free mass should have the best overall success.

Body weight and composition have always been intriguing issues among American football players because American football is a game that emphasizes the importance of the size of athletes in their success. This is especially true for football linemen. The results of body composition analysis for professional American football players carried out by Welham and Behnke in 1942 appear to be quite acceptable by today's standards. The players as a group had body fat content that averaged only 10.4% of body mass, while fat-free mass averaged 81 kg (179 lb). In addition, the heaviest lineman weighed about 118 kg (260 lb) and had 17.4% body fat, whereas the lineman with the most body fat (23.2%) weighed 115 kg (252 lb). During the last 70 years, however, there has been a steady increase in body weight, especially in defensive and offensive linemen. For example, according to the National Football League, the average body weight of linemen of the (NFL) reached 127 kg (280 lb) by 1995. Collegiate linemen are also increasing in size. In a large study evaluating the size of collegiate athletes, football linemen were found to have the largest increase in weight and BMI (Yamamoto *et al.* 2008). A recent study by Borchers *et al.* (2009) also showed an average body fat of 17.3% in division 1 collegiate football players, with linemen exhibiting the highest body fat percentages (25.6%). Both defensive and offensive linemen would have definite advantages from being big and strong in American football. However, this group of athletes has been associated with increased risks of cardiovascular diseases and high prevalence of metabolic syndrome and insulin resistance.

There are a number of sports in which competition is conducted with weight limits or weight classes, such as weight lifting, boxing, and wrestling. In these sports, weight classes are designed to promote competition between athletes of roughly equal size. In this case, body mass is considered a proxy for fat-free mass or muscle mass and, therefore, the athlete's strength and power. Despite such intention to promote fair and interesting competition by matching opponents of equal size and capability, the prevailing attitude in these sports is that the athletes will gain a performance advantage by competing against smaller and lighter opponents in a weight class that is lower than his natural training weight. Typically, therefore, the athlete aims to reduce his body mass to the lowest level possible, with much of this effort taking place in the days before competition. The rapid weight-loss tactics used by athletes to successfully weigh in at lower weight class are commonly referred to as "making weight." The medical and scientific communities have been concerned over problems in athletes associated with making weight. In a survey that involved 63 collegiate wrestlers and 368 high-school wrestlers, Steen and Brownell (1990) found that the athletes lost weight an average of 15 times during a normal season, and the average for the most weight lost at any one time was 7.2 kg (15.8 lb). A variety of aggressive methods have been used by these athletes to lose weight, including dehydration, food restriction, fasting, and, for a few, vomiting, laxatives, and diuretics. In addition, making weight was associated with fatigue, anger, and anxiety, and over one-third of the wrestlers, at both high-school and college level, reported being preoccupied with food and eating being out of control after a match.

Establishing an appropriate weight goal

In light of previous discussion, athletes could be pushed well below the optimal body weight range. Therefore, it is critically important to properly set weight standards. Body weight standards should be based on an athlete's body composition. Once body composition is known, the athlete's fat-free mass can then be determined and used to estimate what the athlete should weigh at a specific relative body fat. Table 12.7 illustrates how to determine a weight goal for a female athlete who weighs 75 kg and wants to reduce her body fat from 20% to 15%. We know her goal weight will consist of 15% fat mass and 85% fat-free mass, so to estimate her weight goal at 15% body fat, we divide her current fat-free mass by 85%, which is the fraction of her weight goal that is to be represented by her fat-free mass. The calculation yields a goal weight of 70.6 kg, which means she needs to lose 4.4 kg. This method allows a determination of weight goal based on the assumption that the fat-free mass stays unchanged despite a loss of body weight. Table 12.8 provides gender-specific ranges of body fat percentage for various sports, which can be used to determine an athlete's relative fat goal. In most cases, these values represent elite athletes in those sports.

BODY COMPOSITION ASSESSMENTS

It is clear that measurements of body composition are necessary and important because the percentages of body fat, as well as its placement, can have profound effects on health and performance. Estimating body fat content is a routine practice in assessing one's nutritional status. It is also part of a comprehensive health-, fitness-, and sports-related assessment. Desirable body fat is deemed important for the health and wellness of all individuals. As already noted, excessive body fat is related to increased risk of cardiovascular, metabolic, pulmonary, and neuromuscular diseases. Body composition is an important factor in most athletic events. A competitive edge may be gained by the athlete who can achieve the optimal balance between fat and fat-free mass for his or her particular sport. Too low a body fat percentage can adversely affect metabolism and health. Female athletes with overtly low body fat can suffer from conditions of amenorrhea and osteoporosis, which are often accompanied by an eating disorder such as anorexia nervosa. The major reasons why practitioners, clinicians, and researchers conduct body composition assessment are:

- to monitor nutrition;
- to assess risk factors for diseases;
- to evaluate overall health- and sports-related fitness;
- to track changes in body composition in response to an intervention program;
- to advance research regarding the impact of body composition on health and performance;
- to set safety standards.

Table 12.7 Example of computing a weight goal

PARAMETER	RESULTS
Current weight	75 kg
Percentage of body fat	20%
Fat weight	15 kg (weight × 20%)
Fat-free weight	60 kg (weight − fat weight)
Relative fat goal	15% (or 85% fat-free mass)
Weight goal	70.6 kg (fat-free mass / 85%)
Weight loss goal	4.4 kg (current weight − weight goal)

Table 12.8 Ranges of body fat percentages for male and female athletes of selected sports

SPORTS	MEN	WOMEN
Baseball/softball	8–14	12–18
Basketball	7–11	21–27
Bodybuilding	5–8	6–12
Cycling	5–11	10–16
American football (Backs)	9–12	–
American football (Linebackers)	11–14	–
American football (Linemen)	14–18	–
American football (Quarterbacks)	11–14	–
Gymnastics	5–12	8–16
Field hockey	8–16	12–16
Rowing	11–15	–
Rugby	6–16	–
Skiing (alpine)	7–14	18–24
Skiing (cross-country)	8–13	16–22
Soccer	7–13	10–18
Swimming	6–12	18–22
Tennis	8–16	18–22
Track and field (field events)	10–18	18–24
Track and field (sprint)	8–14	12–18
Track and field (long distance)	4–10	10–16
Track and Field (jump)	5–10	8–14
Triathlon	5–12	8–15
Volleyball	7–15	10–20
Olympic weightlifting	5–12	10–18
Wrestling	5–15	–

Source: Adapted from McArdle *et al.* 2009; Powers and Howley 2009; Wilmore and Costill 2004.

Dozens of methods have been developed within the past few decades and are currently used in clinical, fitness, and athletic settings. These methods vary considerably in their accuracy, required instruments, and practicality. Each method is associated with advantages and disadvantages, and in general there is a trade-off between accuracy and practicality. The following provides a review of methods commonly used for assessing body composition.

Laboratory method for assessing body composition

In many laboratory and clinical settings, **densitometry** and dual-energy X-ray absorptiometry (DEXA) are used to obtain reference measures of body composition. Due to high precision and reliability of these two techniques, they are also often used in research settings. For densitometric methods, total body density (Db) is estimated from the ratio of body mass (BM) to body volume (BV) (Db = BM / BV). Body volume can be determined by hydrostatic weighing or air replacement plethysmography. As mentioned earlier in this chapter, once body density is known, the percentage of body fat can be calculated using either the Brozek or the Siri equation.

Hydrostatic weighing

Hydrostatic weighing is a valid, reliable, and widely used technique for determining whole-body density (Wang *et al.* 2000). The technique determines body volume by measuring the volume of water displaced by the body. According to the Archimedes' Principle, weight loss under water is directly proportional to the weight of water displaced by the body (Figure 12.2). Such a measure of weight, however, must be converted to volume. As water density equals to 1 under normal circumstances (i.e., warm water temperature of 34–36°C), the weight of water is quantitatively the same as the volume of water. Therefore, by weighing a subject under the water, his or her body volume can be determined. **Body density** can be calculated from the following formula:

$$\text{Body density} = WA / BV = WA / [(WA - UWW) / DW - (RV + GV)]$$

where:

WA = body weight in air
BV = body volume
UWW = body weight submerged in water
DW = density of water
RV = residual lung volume
GV = volume of gas in the gastrointestinal tract.

This approach is based on the two-compartment model of body composition: fat and fat-free mass. It assumes a constant fat mass density of $0.90\,\text{g cm}^3$ and a density of the fat-free mass of $1.10\,\text{g cm}^3$. The densities of bone and muscle tissue are greater than

Figure 12.2 Illustration of the Archimedes' Principle.

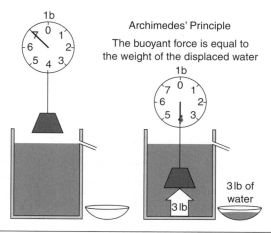

the density of water (density of distilled water is $1.00\,g\,cm^3$), whereas fat is less dense than water. Thus, a muscular subject having a low percentage of body fat tends to weigh more submerged in water than do subjects having higher percentages of body fat. Caution should be exercised with regard to these assumptions. As mentioned earlier, the density of fat-free mass varies among individuals. Athletes, for example, tend to have denser bone and muscle tissue, whereas older individuals tend to have less dense bones. It has been reported that hydrostatic weighing has a tendency to under-estimate body fat for athletes (possibly even giving negative body fat values) and over-estimate body fat for the elderly (Brodie 1988).

To use underwater weighing, one can use a tank, tub, or pool of sufficient size for total body submersion and a chair attached to a weighing scale (Figure 12.3) or platform attached to load cells (Figure 12.4). The water should be comfortably warm, filtered, chlorinated, and undisturbed by wind or other activity in the water during testing. It is important to note that the **Archimedes' Principle** applies to the condition where there is minimal movement in the body relative to the surrounding water. This may be more easily achieved in an isolated water tank; a greater amount of error may occur when underwater weighing is undertaken in a pool. When measuring in a pool, it has been suggested to use a wooden shell to reduce water movement, thereby enhancing precision.

Figure 12.3 Hydrostatic weighing using a scale and chair.

Source: Heyward 2002. Used with permission.

Figure 12.4 Hydrostatic weight using electronic load cells and platform.

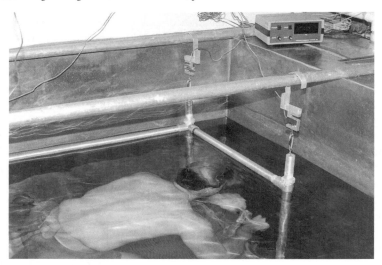

Source: Heyward 2002. Used with permission.

As shown in the formula above, net underwater weight is the difference between the UWW and the weight of the equipment, such as the chair or platform and its supporting device. The BV must be corrected for the volume of air remaining in the lungs after a maximal expiration, which is residual volume or RV, as well as the volume of air in the gastrointestinal tract (GV). The GV is assumed to be 100 ml (Heyward 2002). The RV is commonly measured using helium dilution, nitrogen washout, or oxygen dilution techniques.

Hydrostatic weighing has been considered the good standard against which other body composition estimates are validated. There are, however, distinct limitations associated with this method. It is inappropriate for settings that cannot house a 1000-gallon tank. The tank must be cleaned and disinfected regularly, and the tank water must be maintained within a range of acceptable temperatures, because the density of water varies with temperature. The method requires the determination of residual lung volume, which can be technically challenging. Because air density is zero, even a small error in the estimate of air volume can significantly affect the calculation of body density. Hydrostatic weighing may not be used with people who are afraid of total immersion in water, especially considering that the procedure requires forcible exhalation while submerged. Neither is it an appropriate method for persons with certain pulmonary disorders, such as asthma or emphysema. In addition, special adjustments must be made for morbidly obese people, who tend to float in water; they often have to wear weights to keep them under water.

Air replacement plethysmography

As an alternative to underwater weighing, body volume and, consequently, body density and percentage of body fat can be measured by a technique known as **air**

displacement plethysmography. Compared to hydrostatic weighing, this method is relatively expensive, requiring the use of a whole-body plethysmography (e.g., Bod Pod). The Bod Pod is a large, egg-shaped fiberglass shell (Figure 12.5). It uses air displacement and pressure–volume relationships, and allows for the derivation of an unknown volume by directly measuring pressure. The molded front seat separates the unit into front and rear chambers. The electronics, housed in the rear of the chamber, include pressure transducers, a breathing circuit, and an air circulation system. During measurement, small changes in volume occur inside the chamber, and the pressure response to these small volume changes is measured with the use of a pressure transducer. This is done by measuring the interior volume of the empty chamber, then measuring it again when the subject is inside. By subtraction, the subject's body volume is obtained. For example, the interior air volume of the empty chamber is 450 liters; if the volume of the chamber is reduced to 350 liters with the subject inside, the body volume of the subject is 100 liters.

In this method, the thoracic gas volume, which includes all air in the lungs and airway, is measured instead of residual lung volume. The measurement uses the same principle of pressure–volume relationships. During this measurement, the subject's nostrils are sealed with a nose clip, and the subject is instructed to breathe quietly through a single-use, disposable breathing tube that connects to a breathing circuit housed in the unit's rear chamber when the chamber remains closed. After several

Figure 12.5 Air displacement plethysmograph.

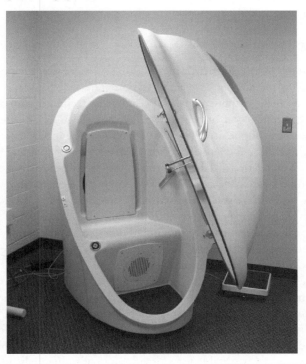

normal breaths, a valve in the breathing circuit momentarily closes at the midpoint of an exhalation. The subject compresses and then relaxes the diaphragm muscle. This produces small pressure fluctuations, which are measured and used to calculate the thoracic air volume.

Compared to hydrostatic weighing, this method is quick and usually takes 5–10 min to complete. It is also relatively simple to operate, and can accommodate those who are obese, elderly, disabled, as well as those who are afraid of total immersion in water. Limited demands are placed on the subjects as the procedure does not require maximal exhalation in order for the residual volume to be used for calculation. The Bod Pod is mobile, so it can be moved from one location to another. In addition, this method requires minimal training and equipment maintenance.

Despite the fact that air displacement plethysmography overcomes some of the methodological and technical constraints of hydrostatic weighing, more studies are necessary to continue validating this method. Some studies, especially when using a heterogeneous group of healthy adults, have reported good test–retest reliability and validity compared to hydrostatic weighing (McCrory et al. 1995; Vescovi et al. 2001). However, this method has been questioned for its validity in assessing body composition among athletes, children, and different ethnic subgroups. For example, the Bod Pod was found to under-estimate percentage body fat in children (Lockner et al. 2000). It was also reported that this method under-estimated percentage body fat in collegiate football players (Collins et al. 1999) and over-estimated percentage body fat in female collegiate track-and-field athletes (Bentzur et al. 2008). In addition, this method was found to over-estimate percentage body fat in African-American men (Wagner et al. 2000).

Dual-energy X-ray absorptiometry

Dual-energy X-ray absorptiometry (DEXA) is gaining in popularity as a standard technique for measuring body composition. This technique was initially developed for the measurement of bone density. Because this technique can also identify soft tissue and categorize it as fat or lean mass, there has been a great deal of interest in using DEXA for assessing relative body fat content. The DEXA system generates two X-ray beams with differing energy levels that can penetrate bone and soft tissue. The procedure involves a series of cross-sectional scans from head to toe at 1-cm intervals, and body composition is determined based upon differential photon absorption of bone minerals and soft tissue. An entire DEXA scan takes approximately 10–20 min, depending upon body size. Computer software reconstructs the attenuated X-ray beams to produce an image of the underlying tissue and quantify bone mineral density, fat content, and fat-free mass for the whole body, as well as for various sections of the body, including the head, trunk, hips, and limbs (Figure 12.6).

DEXA is precise, accurate, and reliable. Its measurements are based on a three-compartmental model – bone mineral, fat mass, and soft lean mass – rather than two compartments, as in most other methods. In this context, DEXA can provide accurate assessments of fat and fat-free mass in individuals with below- and above-average bone mineral density. DEXA requires little effort and cooperation from the subjects and, as such, the method is suitable for a broad range of individuals, including those who are

Figure 12.6 Dual-energy X-ray absorptiometer.

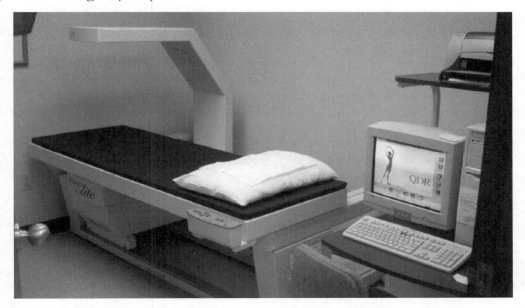

very young, very old, or diseased. A single scan produces regional and whole-body estimates of fat and fat-free mass. By comparing regional fat content in the trunk and
limbs, clinicians and researchers can also obtain information with regard to body fat
distribution of the subject. There is a high degree of agreement between percentage
body fat estimates obtained by hydrostatic weighing and by DEXA.

DEXA has some limitations. The equipment is expensive, and often requires
trained radiology personnel to operate. As differences in hydration of lean tissue can
produce errors in estimating body composition, concerns have been raised about the
appropriateness of using DEXA for assessing body composition in individuals with
acute or chronic changes in body water. In addition, DEXA measurements of body
composition are sensitive to the anterior–posterior thickness of the body or body parts,
so that results can simply be confounded by the size and shape of an individual
(Roubenoff *et al.* 1993).

Field methods for assessing body composition

While laboratory methods are commonly used to assess fat mass, field measures, such
as skinfold thickness and bioelectrical impedance, may be more practical for screening
large numbers of individuals within a short period of time. These methods are relatively inexpensive and simple to perform. In order to use these methods appropriately
and to produce results that are reasonably accurate and reliable, one should understand
the basic assumptions and rationale, as well as potential sources of error associated with
each method. Sufficient practice is also necessary in order to ensure measurement accuracy and reliability.

Skinfold method

The **skinfold** method is the most widely used in fitness and clinical settings. A skinfold is a double fold of skin and the immediate layer of subcutaneous fat. It is measured in millimeters using a special tool known as a caliper. The accuracy of skinfold measurements is affected by the type of caliper used. Three calipers are shown in Figure 12.7; these operate on the same principle as a micrometer, which measures the distance between two points. However, Lange or Harpenden calipers, which cost several hundred dollars, are considered the better choice compared with plastic calipers in terms of yielding reasonably reliable skinfold results. Across the entire range of skinfold thickness, these calipers exert constant pressure. On the other hand, plastic calipers tend to be most accurate through the middle range of percentage body fat; the measurements of very thin and very thick skinfolds are most likely to be inaccurate.

There are more than 100 equations available for estimating percentage body fat from skinfold thickness (Heyward 2002). These equations differ by number and location of skinfold sites. They also differ in terms of the populations used for developing the equation. Some of them are quite general, meaning that equations can be used for any healthy adult within the same gender. Others are more specific and were developed for relatively homogeneous subgroups of populations. The population-specific equations are assumed to be valid only for subjects who have similar characteristics, such as age, gender, ethnicity, or level of physical activity. For example, the equation derived from adults would not be valid for use among children and adolescents. Generally speaking, in a research setting, a seven-site procedure is most often adopted, with measurements taken by Harpenden or Lange calipers. In clinical or field settings, four-site, three-site, and even two-site procedures are more commonly chosen. In these settings, the loss of accuracy due to the use of fewer skinfold sites is largely compensated by the gain in practicality. It is also recommended to use equations that have skinfolds measured from a variety of sites, including both upper and lower body regions (Martin *et al.* 1985). Table 12.9 presents some of the commonly used generalized skinfold equations for predicting body density. These questions allow for calculation of body density, with the latter then being converted to percentage body fat using the appropriate population-specific conversion formula as previously shown in Table 12.5.

Figure 12.7 Common skinfold calipers. The Lange and Harpenden calipers provide constant tension at all jaw openings.

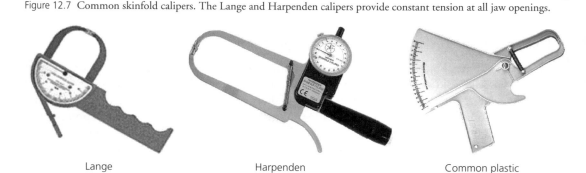

Lange   Harpenden   Common plastic

Table 12.9 Generalized equations for predicting body density (Db) for adult men and women

GENDER	SKINFOLD SITES	EQUATIONS
Men	Seven sites (chest, midaxillary, triceps, subscapula, abdomen, suprailiac, thigh)	$Db = 1.112 - 0.00043499 (\Sigma 7 \text{ skinfolds}) + 0.00000055$ $(\Sigma 7 \text{ skinfolds})^2 - 0.00028826 \text{ (age)}$
	Four sites (abdomen, suprailiac, triceps, thigh)	$Db = 0.29288 (\Sigma 4 \text{ skinfolds}) + 0.0005 (\Sigma 4 \text{ skinfolds})^2 + 0.15845$ $(\text{age}) - 5.76377$
	Three sites (chest, abdomen, thigh)	$Db = 1.10938 - 0.0008267 (\Sigma 3 \text{ skinfolds}) + 0.0000016$ $(\Sigma 3 \text{ skinfolds})^2 - 0.0002574 \text{ (age)}$
Women	Seven sites (chest, midaxillary, triceps, subscapula, abdomen, suprailiac, thigh)	$Db = 1.097 - 0.00046971 (\Sigma 7 \text{ skinfolds}) + 0.00000056$ $(\Sigma 7 \text{ skinfolds})^2 - 0.00012828 \text{ (age)}$
	Four sites (abdomen, suprailiac, triceps, thigh)	$Db = 0.29669 (\Sigma 4 \text{ skinfolds}) + 0.00043 (\Sigma 4 \text{ skinfolds})^2 + 0.02963$ $(\text{age}) - 1.4072$
	Three sites (triceps, suprailiac, thigh)	$Db = 1.0994921 - 0.0009929 (\Sigma 3 \text{ skinfolds}) + 0.0000023$ $(\Sigma 3 \text{ skinfolds})^2 - 0.0001392 \text{ (age)}$

Sources: Jackson *et al.* 1980; Jackson and Pollock 1978.

To reduce error, skinfold sites should be precisely determined and, in some cases, marked for enhancing repeatability. The following describes the eight most commonly used skinfold sites as outlined in the *Anthropometric Standardization Reference Manual* (Lohman 1988):

- Chest: measured with a diagonal skinfold at midway between the right nipple and anterior axillary line (imaginary line dropping straight down from front of armpit).
- Triceps: measured with a vertical skinfold at the posterior midline of the right upper arm over the triceps, midway between the lateral projection of the acromion process of the scapula and the inferior margin of the olecranon process of the ulna.
- Subscapular: measured at 1 cm below the lowest angle of the scapula on diagonal fold that runs from the vertebral border downward.
- Midaxillary: measured with a horizontal skinfold at the right midaxillary line (a vertical line extending from the middle of the axillar) level with the xiphisternal junction.
- Suprailliac: measured at just above the crest of the right ilium on a diagonal skinfold that runs from the anterior axillary line downward.
- Abdomen: measured with a horizontal fold at approximately 2 cm to the right of and 1 cm below the midpoint of the umbilicus.
- Thigh: measured with a vertical skinfold at the midline of the anterior aspect of the thigh midway between the junction of the midline and the inguinal crease and the proximal border of the patella.
- Medial calf: measured with a vertical skinfold at the point of maximal circumference and medial aspect of the calf.

The technique used to measure a skinfold is the same regardless of the particular equations being used to assess body composition. The accuracy and reliability of measurement depend very much on the skill of the examiner. Examiners can improve the accuracy of results by training with an experienced and skilled technician, practicing on clients of different body types, carefully identifying and marking skinfold sites, and following standardized procedures of measuring skinfold thickness. Additionally, because the conversion from skinfolds to body fatness is influenced by ethnicity, gender, and age, it is especially important to use an appropriate prediction equation for the individual being tested. The following are the general guidelines for taking the skinfold measurement:

- All skinfolds should be taken on the right side of the body.
- Grasp the skinfold firmly between the thumb and index finger of the left hand, with the thumb and index finger placed about 1 cm above the skinfold site. It is important to lift only skin and fat tissue, not muscle.
- Do not take measurements when the subject's skin is moist because there is a tendency to grasp extra skin, obtaining inaccurately large values. Also, do not take measurements immediately after exercise or when the subject being measured is over-heated because the shift of body fluid to the skin may expand normal skinfold thickness.
- The skinfold is lifted by placing the thumb and index finger on the skin, about 8 cm (3 in) apart and perpendicular to the desired skinfold. However, for individuals with extremely large skinfolds, the starting separation between the thumb and index finger should be greater than 8 cm in order to lift the fold.
- Keep the fold elevated when the caliper is being applied to the skinfold site. The caliper jaws should be placed perpendicular to the long axis of the skinfold and about midway between the base and crest of the skinfold, and jaw pressure is then released slowly.
- The measurement should be read within 2 s of the caliper pressure being applied, and should always be repeated two or three times so that an average skinfold value is recorded for each site.
- Upon completion, open the caliper jaws before remove the caliper from the skinfold site.

Bioelectrical impedance analysis

Bioelectrical impedance analysis (BIA) is a relatively new technique for assessing body composition. It is based on the understanding that lean tissue, because of its higher water content, is a much better conductor of electricity than is fat tissue, which contains considerably less water. In BIA, an electronic instrument generates a low-dose, harmless current (800 μA at 50 kHz), which is passed through the person being measured. BIA determines the electrical impedance, or opposition to the flow of this electric current. This information is then used to estimate total body water and fat-free mass. Figure 12.8 demonstrates two examples of bioelectrical impendence analyzers, which have also been marketed for home use. The Tanita analyzer measures lower body resistance between the two legs as the individual stands on the electrode plates. The Omron analyzer, which is hand-held, measures upper body resistance between the two arms.

Figure 12.8 Common bioelectrical impedance analyzers.

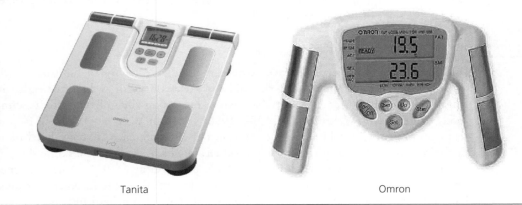

Tanita   Omron

There are various BIA equations available for predicting fat-free mass and percentage body fat. These prediction equations are based on either population-specific or generalized models, which can be found in Hayward and Stolarczyk's (1996) book. Early models of BIA instruments simply provide the operator with a value of resistance, which requires further calculation that uses an appropriate equation to determine the subject's percentage body fat. However, most of the current and more sophisticated BIA instruments now contain some prediction equations to be selected along with a computer and printer, which have made data analysis much easier. In general, it has been recommended not to use the fat-free mass and percentage body fat estimates obtained directly from the BIA instrument unless the equations are known to apply directly to the subjects being measured. In other words, one can obtain a value of resistance from the analyzer and then look for an appropriate equation for converting the resistance into percentage body fat. The prediction equations are valid only for subjects whose physical characteristics are similar to those specified in the equation.

BIA has the advantage of being safe, quick, portable, and easy to use. It requires little or no technical knowledge or skills of the operator or the client. It is less intrusive compared to hydrostatic weighing and skinfold measurement. When the proper measurement guidelines are followed and the appropriate BIA equation is used, results of BIA have been found to have about the same accuracy as the skinfold method.

The major disadvantage is that BIA assumes that subjects are normally hydrated, an assumption that is often not correct. The state of hydration can greatly affect the accuracy of BIA. This is because any change in bodily water can alter normal electrolyte concentrations, which in turn affect the flow of electrical current independent of a real change in body composition. For example, dehydration caused by insufficient water intake, excessive perspiration, heavy exercise, or caffeine or alcohol use will lead to an over-estimation of body fat content by increasing bioelectrical impedance. To prevent this, subjects should be advised to refrain from consuming caffeine and alcohol for 24 hours before testing and to avoid heavy exercise for 12 hours before testing. It is also recommended that BIA measurements be made in a room with a normal ambient temperature and that no testing be given to anyone who has used diuretic medications within 7 days of the test or female subjects who perceive that they are retaining water during the stage of their menstrual cycle.

Circumference measurements

Measuring girth or **circumference** represents another category of anthropometric methods in addition to measuring skinfold and skeletal breadth as discussed earlier in this chapter. Methods are available that involve measuring circumference for assessing body composition (Weltman *et al.* 1987, 1988; Tran and Weltman 1988). In fact, in comparison with skinfold technique, measures of circumference are considered more accurate and feasible for use in obese individuals in which conducting a skinfold test is often difficult and sometimes impossible (Seip and Weltman 1991). Circumference measurements are also useful in clinical settings. For example, measurements of waist circumference and the ratio of waist to hip circumference can help distinguish between pattern of fat distribution in the upper and lower body. They provide an important indication of disease risk. As shown in Table 12.3, a waist circumference of larger than 102 cm in men or 88 cm in women is considered a risk factor independent of obesity. Likewise, young adults with a waist–hip ratio larger than 0.94 for men and 0.82 for women are at high risk for adverse health consequences (Bray and Gray 1988). Waist circumference reflects abdominal fat accumulation. It is considered that fat stored in this region is more responsible for the pathological processes that cause insulin resistance, type-2 diabetes, and heart disease.

Using measures of circumference to assess body composition has its unique advantage in testing that involves obese individuals. When studying obese populations, it is often difficult to obtain accurate skinfold measurements because (1) the skinfold may exceed the maximum opening capacity of the caliper; (2) caliper tips may slide on large skinfolds; and (3) readings tend to decrease with subsequent measurements due to repeated compression of the subcutaneous fat. To overcome these shortcomings associated with skinfold method, Weltman *et al.* (1987, 1988) developed and cross-validated body composition prediction equations for obese men and women using height, weight, and measures of waist and abdomen circumference as predictor variables. These gender-specific equations are:

Obese men

$$\% \text{ fat} = 0.31457(\text{mean abdomen}) - 0.10969(\text{Wt}) + 10.8336$$

Obese women

$$\% \text{ fat} = 0.11077(\text{mean abdomen}) - 0.17666(\text{Ht}) + 0.14354(\text{Wt}) + 51.03301$$

In these equations, mean abdomen is the average of two circumferences measured at waist and abdomen, Wt is weight; and Ht is height. Waist circumference is determined by placing the tape in a horizontal plane at the level of the narrowest part of the torso, as seen from the anterior aspect. The abdomen circumference is determined by placing the tape at the level of the umbilicus. Both measurements are taken at the end of a normal expiration.

Circumferences should be measured by using an anthropometric tape, which is made from flexible material that does not stretch with use. Some anthropometric tapes have a spring-loaded handle that allows a constant tension to be applied during the measurement. The examiner should hold the zero end of the tape in the left hand, just above or below the remaining tape held by the right hand. The tape should be snug around the body part without indenting the skin or compressing subcutaneous adipose tissue. Duplicate measurements should be taken, and the circumference measurement should be recorded to the nearest 0.5 cm or ¼ in. Compared to the skinfold technique, skill is not a major source of measurement error. However, examiners should practice sufficiently and closely follow standardized testing procedures for locating measurement sites, positioning the anthropometric tape, and applying tension during the measurement.

SUMMARY

- Weight–height tables provide a rough estimate of ideal weight for a given height. The lowest mortality in the United States is associated with body weights that are somewhat below average for a given group based on sex and stature. However, weight–height tables reveal little about an individual's body composition. Studies of athletes clearly show that overweight does not coincide with excessive body fat.
- BMI relates more closely to body fat and health risk than does body mass and stature. BMI may be used in conjunction with skinfold measurements or waist circumference as an improved means of assessing increased risk in adults for heart disease, stroke, type-2 diabetes, and premature death. Still, BMI fails to consider the proportion of fat mass and fat-free mass.
- Total body fat consists of essential fat and storage fat. Essential fat contains fat present in bone marrow, nerve tissue, and organs. It serves as an important component for normal biological function. Storage fat represents the energy reserve that accumulates mainly as adipose tissue beneath the skin and in visceral depots.
- Athletes generally have physique characteristics unique to their specific sport. Field event athletes have relatively large fat-free body mass; distance runners have the lowest fat-free mass and fat mass. Competitive male and female swimmers generally have higher body fat levels than distance runners. American football players are among the heaviest of all athletes, yet maintain a relatively lean body composition.
- Densitometry involves measuring the density of the entire body, usually by hydrostatic weighing, with body density being later converted into percentage body fat. Hydrostatic weighing remains the laboratory standard, but the time, expense, and expertise needed is prohibitive for many clinical settings.
- Air displacement plethysmography also measures body density and thus body composition. Subjects better tolerate this method than hydrostatic weighing. This method requires less subject cooperation, and residual volume measurements are not needed. It is as accurate as hydrostatic weighing, but the equipment is relatively more complex and costly.
- Measurement of skinfolds is the most widely used method of indirectly estimating percentage body fat. The equipment is inexpensive and portable. Measurements

can be easily and quickly obtained, and they correlate well with body density measurements. However, proper measurement of skinfolds requires careful site selection and strict adherence to the guidelines.

- In comparison with the skinfold technique, measures of circumference are considered more accurate and feasible for use in obese individuals in which conducting a skinfold test is often difficult and sometimes impossible. In addition, measures of waist circumference and the ratio of waist to hip circumference can help distinguish between patterns of fat distribution in the upper and lower body.

CASE STUDY: CALCULATING IDEAL BODY WEIGHT

Michael, a 45-year-old business man, works for a consulting firm and has a frequent travel schedule. Lately, he has been experiencing a steady weight gain. He is 5′ 8″ tall and weighs 200 lb. His body mass index is 30.5, which is considered borderline obesity. He decides to see a personal trainer to solve his weight problem. His first meeting with the personal trainer included a body composition assessment using the bioelectrical impedance analysis. This initial assessment reveals that he has a body fat of 28%. He is unhappy with this result and decides to do something to lose weight. Via discussion with the trainer, he sets an initial goal to cut his body fat from 28% to 20%, while maintaining his lean body mass.

Questions

- What is Michael's current fat mass and lean body mass?
- How much should Michael weigh in order to reach his target of 20% body fat? In this case, how much fat does he need to lose?
- Considering that there are many other techniques available for assessing body composition, what are the unique advantages and disadvantages associated with the bioelectrical impedance analysis that was used by Michael's trainer?

REVIEW QUESTIONS

1. Why is it important to assess the body composition of your client or an athlete?
2. Explain the differences between essential and storage fat. How do they differ between males and females?
3. Define body density. How is this parameter computed? How is body density related to percentage body fat?
4. Why do competitive swimmers generally have more body fat than distance runners?
5. Explain how BMI, waist circumference, and waist–hip ratio may be used to identify clients at risk due to obesity.
6. A 20-year-old college male is 180 lb, with 28% fat. What is his target body weight to achieve 17% fat?
7. What are the advantages and disadvantages of using a weight–height table or body mass index (BMI) in assessing body composition? How is BMI calculated?

8 What is the principle of hydrostatic weighing? What are the potential errors that can cause this technique to be inaccurate?

9 Why should a different body density equation be used for children compared to adults?

10 Identify potential sources of measurement error for the skinfold method.

11 To obtain accurate estimates of body composition using the bioelectric impedance assessment method, your client must follow the pre-testing guidelines. Identify these client guidelines.

12 How would you rate the skinfold, bioelectric impedance assessment, and hydrostatic weighing techniques in terms of their suitability for very lean, very obese, and older individuals?

SUGGESTED READING

Ellis, K.J. (2001) Selected body composition methods can be used in field studies. *Journal of Nutrition*, 131: S1589–S1595.

This article provides an overview of the present status of in vivo body composition methodologies that have potential for use in field-based studies. The methods discussed in this article include four general categories: anthropometric indices and skinfold, body volume measurements, body water measurements including bioelectrical methods, and imaging techniques.

Wang, J., Thornton, J.C., Kolesnik, S., Pierson, R.N., Jr. (2000) Anthropometry in body composition: an overview. *Annals of the New York Academy of Sciences*, 904: 317–326.

Anthropometry is a simple, reliable method for quantifying body size and proportions by measuring body length, width, circumference, and skinfold thickness. This article is an excellent resource for those who are interested in using anthropometric measurements to predict body composition.

13 Energy balance and weight control

▊ KEY TERMS

- set point
- hypothalamus
- ghrelin
- positive energy balance
- negative energy balance
- resting metabolic rate
- oxygen deficit
- excess post-exercise oxygen consumption
- circuit weight training
- thermal effect of food
- obligatory thermogenesis

- facultative thermogenesis
- settling-point theory
- weight cycling
- physical activity
- exercise
- intensity
- duration
- frequency
- lactate threshold
- intermittent exercise

▊ HOW IS BODY WEIGHT REGULATED?

In most people, body fat and weight remain remarkably constant over long periods, despite fluctuations in food intake and activity level. It appears that when energy intake or activity level changes, the body compensates to prevent a significant change in body weight and fat. This is mainly because the body has the ability to balance energy intake and expenditure at a particular level or **set point**. For example, the body takes in an average of 2500 kcal per day, or nearly one million kilocalories per year. However, the average gain of 0.7 kg (1.5 lb) of fat each year represents an imbalance of only 5250 kcal between energy intake and expenditure (3500 kcal is equivalent of 0.45 kg (1 lb) of adipose tissue). This translates into a surplus of less than 15 kcal per day.

Set point

According to the set-point theory, there is a control system built into every person dictating how much fat he or she should carry – a kind of thermostat for body fat. The set point for body fatness is determined by genetics. Some individuals have a high setting, others have a low one. In an obese individual, body fat is set to remain at a higher level than it is in a lean individual. When people lose weight, regardless of whether they are lean or obese, metabolic signals are generated to decrease energy output and increase energy intake in order to return their weight to its set point. According to this theory, body fat percentage and body weight are matters of internal controls that are set differently in different people.

The set-point theory has been well proven in both animal and human studies. When animals are fed or starved for various periods of time, their weights respectively increase or decrease markedly. But when they go back to their normal eating patterns, they always return to their original weight or to the weight of control animals. Similar results have been found in humans, although the number of studies is limited. Subjects placed on semi-starvation diets have lost up to 25% of their body weight but regained that weight within months of returning to a normal diet. It is the existence of such a set point that makes weight loss a very difficult task, and most people who lose weight eventually regain all they have lost.

The set point that controls body weight is not always constant. Changes in physiological, psychological, and environmental circumstances do cause the level at which body weight is regulated to change, usually increasing over time. For example, body weight increases in most adults between 30 and 60 years of age, and after giving birth, most women return to a weight that is 1–2 lb higher than their pre-pregnancy weight.

Regulation of energy balance

The regulation of human energy balance is complex, involving numerous feedback loops to keep it controlled. The central nervous system, the brain in particular, is the center for appetite control, either creating a sensation of satiety or stimulating food-seeking behavior. However, its activity is dependent upon a complex array of afferent signals from various body systems. The interaction of the brain with these afferent signals helps regulate the appetite on a short-term (daily) or on a long-term basis in order to keep the body weight constant. It is believed that signals related to food intake affect hunger or satiety over a short period of time, whereas signals from the adipose tissue trigger the brain to adjust both food intake and energy expenditure for long-term regulation.

Short-term regulation

The short-term regulation of energy balance involves the control of food intake from meal to meal. We eat in response to hunger, which is the physiological drive to consume food. We stop eating when we experience satiety, the feeling of fullness and satisfaction that follows food intake. What, when, and how much we eat are also affected by appetite – the drive to eat specific foods that is not necessarily related to hunger. Signals to eat or stop eating can be external, originating from the environment, or they can be internal, originating from the gastrointestinal tract, circulating nutrients, or centers in the brain. The **hypothalamus**, which lies between the brain and brainstem, contains neural centers that help regulate appetite and hunger. It is believed that the hypothalamus contains a hunger center that stimulates eating behaviors, and a satiety center that, once stimulated, inhibits the hunger center. As a means of controlling energy intake, specific neural receptors within the hypothalamus monitor various afferent stimuli that may augment or inhibit food intake.

The external signals that motivate eating include the sight, taste, and smell of food, the time of day, cultural and social conventions, the appeal of the foods available, and ethnic and religious rituals (Friedman 1995). As discussed in Chapter 5, sensory input such as the sight of a meal being presented on a table may make your mouth become moist and your stomach begin to secrete digestive substances, and such responses can occur even when the body is not in need of food. Some people eat lunch at noon out of social convention, not because they are hungry. We eat turkey on Thanksgiving because it is a tradition, and we eat cookies and cinnamon rolls while walking through the mall because the smell entices us to buy them. Likewise, external factors such as religious dietary restrictions or negative experiences associated with certain foods can signal us to stop eating.

Internal signals that promote hunger and satiety are triggered after food is consumed and absorbed, thereby eliciting meal consumption or termination. The simplest type of signal about food intake comes from local nerves in the walls of the stomach and small intestine, which sense the volume or pressure of food and send a message to the brain to either start or stop food intake. When small amounts of food are consumed, the feeling of fullness associated with gastric stretching is barely noticeable. As the volume of food increases, stretch receptors are stimulated, relaying this information to the brain to cause the sensation of satiety.

The presence of food in the gastrointestinal tract also sends information directly to the brain and triggers the release of gastric hormones, the majority of which promote satiety (Woods *et al.* 1998). Of these satiety-promoting hormones, cholecystokinin is the best understood. It is released from intestinal cells, particularly in response to dietary fat and protein, signaling the brain to decrease food intake. The hormone that triggers hunger is **ghrelin**, produced mainly by the stomach. Ghrelin is released in response to a lack of food, and its circulating concentration decreases after food is consumed. There is evidence that over-production of ghrelin may contribute to obesity (Inui *et al.* 2004)

Absorbed nutrients may also send information to the brain to modulate food intake. Circulating levels of nutrients, including glucose, fatty acids, amino acids, and ketone, are monitored by the brain and may trigger signals to eat or not to eat. The liver may also be involved in signaling hunger and satiety by monitoring changes in fuel metabolism. Changes in liver metabolism, in particular the amount of ATP, are believed to modulate food intake (Friedman 1995). The pancreas is also involved in food intake regulation because it releases insulin, which may affect hunger and satiety by lowering the levels of circulating nutrients.

Long-term regulation

In addition to short-term regulation of food intake, the body also regulates energy intake on a long-term basis. Short-term regulators of energy balance affect the size and timing of individual meals. If a change in food intake is sustained over a long period, however, it can affect long-term energy balance and, hence, body weight and fatness. Long-term regulatory signals communicate the body's energy reserves to the brain, which in turn releases neuropeptides that influence energy intake and/or energy expenditure. If this long-term system functions effectively, body weight remains somewhat stable over time.

The mechanisms that regulate long-term energy balance are complex and not well understood. However, the hormones leptin and, to a lesser extent, insulin, appear important (Figure 13.1). Leptin is produced primarily by adipose tissue; when body fat increases, circulating leptin concentration increases as well. Likewise, when body fat decreases, leptin production decreases. Leptin travels in the blood to the hypothalamus, where it binds to proteins called leptin receptors in order for the brain to release catabolic neuropeptides. Catabolic neuropeptides help the body to resist further weight gain by decreasing energy intake and increasing energy expenditure. When fat stores shrink, less leptin is released. Low leptin levels trigger the brain to release anabolic neuropeptides, thereby decreasing energy expenditure and increasing food intake. This will then protect the body against further weight loss. In this context, leptin acts like a thermostat to keep body fatness from changing significantly.

Along with leptin, the hormone insulin is also important in communicating adiposity to the brain. Insulin is secreted from the pancreas when blood glucose levels rise; its circulating concentration is proportional to the amount of body fat. Insulin can affect food intake and body weight by sending signals to the brain and by affecting the amount of leptin produced and secreted (Schwartz *et al.* 1999). When insulin levels are high, there will be a reduced drive for food as well as increased energy expenditure. As you may recall, insulin secretion increases after a meal. Such an acute rise in insulin acts as an anabolic hormone, favoring energy storage in peripheral tissues. Unlike insulin, acute changes in leptin are not easily detected after meals.

Figure 13.1 Operation of leptin in maintaining body fat at a set-point level.

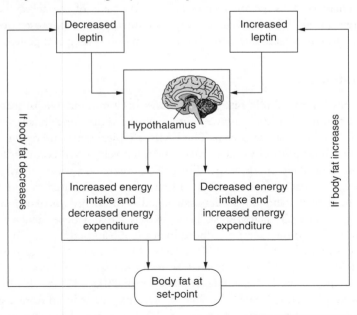

Overall, leptin and insulin are part of a long-term homeostatic system that helps prevent large shifts in body weight. Some researchers believe that defects in the leptin/insulin signaling system may lead to impaired body weight regulation in some people (Cancello *et al.* 2004). In fact, many obese people are found to have high levels of both leptin and insulin, suggesting that these individuals may have developed tissue resistance in their responses to these hormones. There is more to learn about this regulatory system. It seems clear that both leptin and insulin help protect the body during times of food scarcity. However, food regulation involving these two hormones can be altered or impaired during times of food surplus.

COMPONENTS OF DAILY ENERGY EXPENDITURE

A healthy weight can result from paying more attention to the important concept of energy balance. Think of energy balance as an equation: energy input = energy output. While energy input refers to calories from food intake, energy output is accomplished by metabolism, digestion, absorption, transport of nutrients, and physical activity. When energy input is greater than energy output, the result is **positive energy balance**. The excess calories consumed are stored, which results in weight gain. There are some situations in which positive energy balance is normal and healthy. For example, during pregnancy, a surplus of calories supports the developing fetus. Infants and children require a positive energy balance for growth and development. On the other hand, if energy input is less than energy output, a **negative energy balance** is said to occur. A negative energy balance is necessary for successful weight loss. It is important to realize that in instances of negative energy balance, weight loss involves a reduction in both lean and adipose tissue, not just "fat." So far, issues related to energy intake have been discussed in previous chapters. This section focuses on the other side of the energy balance equation – energy output. The body uses energy for three general purposes: resting metabolism, physical activity, and digestion, absorption, and processing of ingested nutrients.

Resting metabolic rate

The resting metabolic rate (RMR) represents a minimal rate of metabolism necessary to sustain life. It is the energy requirement of a variety of cellular events that are essential to the life of an organism. The RMR is typically measured 3–4 hours after a light meal without prior physical activity. Quantitatively, RMR accounts for about 60–75% of the total daily energy expenditure. For this reason, this energy component is the focus of a great deal of attention and has often been treated as a major player in contributing to metabolic disorders associated with energy imbalance. In many exercise-intervention studies, RMR has been used as a major dependent variable expected to increase due to the exercise-induced increase in lean body mass. On average, RMR has been estimated to be about $1680\,kcal\,day^{-1}$ for men and $1340\,kcal\,day^{-1}$ for women.

RMR can simply be estimated by using the factor of 1 kcal per kilogram of body weight per hour. For example, for a male of 70 kg (154 lb), his daily RMR will be 1680 kcal (1 kcal × 70 kg × 24 h). Such a factor may be reduced from 1 to $0.9\,kcal\,kg^{-1}\,hr^{-1}$ for use in women given that the average RMR is about 10% lower in women than men.

This method is convenient, but does not discriminate the age- or body composition-related differences. To eliminate this drawback, a revised equation was developed, which allows estimation of RMR from fat-free mass in kilograms (McArdle *et al.* 2001). The equation is expressed as:

$$RMR = 370 + 21.6 \times (\text{fat-free mass})$$

For example, a male who weighs 70 kg at 20% body fat has a fat-free mass of 56 kg. By using the equation, his estimated daily RMR will be about 1580 kcal ($370 + 21.6 \times 56 = 370 + 1209.6 = 1579.6$). This equation was developed based upon studies of mixed sample of males and females and therefore should apply uniformly to both genders. The major advantage with this equation is that it takes into account the impact of body composition on RMR and in doing so results can be more accurate in reflecting the gender- and age-related differences in metabolism, despite the use of a single equation. This estimation approach, however, requires body composition to be measured in the first place, which could be problematic for those who don't have access to the body composition measuring equipment.

RMR can be affected by body weight and composition. As shown in Table 13.1, this energy component can also be influenced by many other factors, such as age, climate, and hormones. Compared to adults, infants have a large proportion of metabolically active tissue. Hence, their mass-specific RMR is higher than that of adults. However, RMR declines through childhood, adolescence, and adulthood as full growth and maturation are achieved. Mass-specific RMR will show a continued decline in those who become elderly (reach or pass the age of 60). This is because aging has been associated with the loss of lean body mass, which includes muscle as well other metabolically active organs. Climate conditions, especially temperature changes, can also raise resting energy expenditure. For example, exposure to the cold may stimulate muscle shivering as well as the secretion of several thermogenic hormones such as epinephrine and thyroid hormone. Exposure to a warm environment will also provoke an increase in energy metabolism, although this increase may not be as great in magnitude as that induced in a cold environment. RMR is also subject to change in hormonal concentration. The two major hormones linked to RMR are epinephrine and thyroid hormone.

Physical activity

Physical activity is a powerful metabolic stressor. It stimulates chemical processes in which the potential energy stored in energy substrates is converted to the type of energy that cells can utilize – namely, ATP. As mentioned in Chapter 7, in an aerobic event, energy utilization associated with a particular exercise can be quantified by determining the amount of oxygen used, with oxygen being later converted to calories. Determination of energy cost for an anaerobic event is much more complex and may be achieved via measurement of production of muscular force and power, utilization of energy substrates (e.g., ATP, PCr, glucose), or energy utilization following the completion of exercise. Knowing the energy cost of exercise is important if a comparable nutritional requirement needs to be provided or if the efficiency of the body during performance of exercise is to be calculated.

FACTORS	EFFECT
Body size	An increase in body size increases RMR
Body composition	An increase in lean body mass increases RMR
Growth	Mass-specific RMR declines through childhood, adolescence, and adulthood
Age	Mass-specific RMR continues to decline from adulthood into being elderly
Ambient temperature	Both cold and warm exposure will stimulate RMR
Hormones	Both thyroid hormone and epinephrine will stimulate RMR
Aerobic fitness	A high level of aerobic fitness is linked to an increased RMR
Resistance training	Resistance training increases RMR or prevents a decline in RMR as one ages
Smoking	Nicotine increases RMR
Caffeine	Caffeine stimulates RMR
Sleeping	RMR decreases during the period of sleeping
Nutritional status	Underfeeding tends to reduce RMR, while overfeeding tends to increase RMR

Table 13.1 Factors that affect resting metabolic rate (RMR)

The total energy cost for a given activity should be assessed both during and following exercise. This is especially the case when exercise is performed at high intensities that require a longer recovery period. During exercise, there is an increase in VO_2 to support the increased energy need of the body. However, the body's ability to gauge such a demand for oxygen is not always perfected. At the onset of exercise, both respiration and circulation do not immediately supply the needed quantity of oxygen to the exercising muscle (Figure 13.2). Oxygen supply normally requires several minutes to reach the required level at which aerobic processes are fully functional. The difference in oxygen requirement and oxygen supply is regarded as the **oxygen deficit**. After exercise, VO_2 does not return to resting levels immediately, but does so rather gradually. This elevated oxygen consumption following exercise has been referred to as **excess post-exercise oxygen consumption** (EPOC). The current theory of EPOC reflects two factors: (1) the level of anaerobic metabolism in previous exercise; and (2) the exercise-induced adjustments in respiratory, circulatory, hormonal, and thermal function that still exert their influences during recovery.

Understanding the dynamics of EPOC enables us to more adequately quantify energy cost of an activity, especially when the activity performed is intense and brief, such as sprinting or resistance exercise. Due to their anaerobic nature, those strenuous and short-duration work bouts could drastically disturb the body's homeostasis throughout exercise that demands a greater level of oxygen consumption during recovery. As such, these types of exercise are often associated with a fairly large EPOC that can constitute a majority of the total energy associated with exercise. The phenomenon of EPOC has also been brought to attention in the area of weight management because of its role in facilitating energy expenditure. Overweight or obesity is the result of a positive energy balance and EPOC can contribute to the opposite when exercise is undertaken regularly. Nevertheless, it has been suggested that in order for EPOC to be effective, one would have to exercise at an intensity exceeding 70% VO_2max for more than an hour three times per week (Borsheim and Bahr 2003). This volume of exercise is often impractical for many overweight or obese individuals.

Figure 13.2 Response of oxygen uptake during steady-state exercise and recovery.

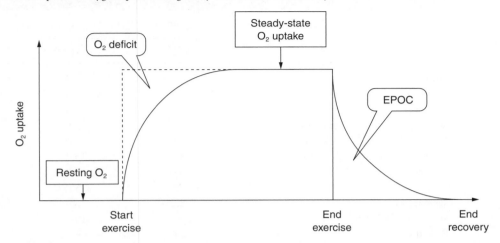

Walking and running are the two principle forms of human locomotion. From an energetic standpoint, walking is an energy-cheap activity, and its energy cost is generally no more than three times the resting metabolic rate. On the other hand, running can be very demanding metabolically as it engages more muscles that contract forcefully. On average, energy expenditure during running can reach as high as ten times the resting metabolic rate (Table 13.2).

Cycling is the major means of transportation in many countries. This form of exercise is also used as a recreational and competitive sport. During the last decade or so, off-road cycling has enjoyed an exponential growth in popularity. In the fitness and rehabilitation industry, cycling is often performed on a stationary ergometer, which allows exercise intensity to be regulated. Stationary cycling differs from outdoor cycling in that it provides a broad range of exercise intensities that can be adjusted by manipulating either pedal speed or brake resistance. Cycling uses less muscle mass compared to running and as a result expends less energy (Table 13.2). However, the non-weight-bearing feature of this activity makes cycling more tolerable for and commonly chosen by those who are extremely sedentary and/or obese or those who have difficulty performing weight-bearing activities.

Resistance exercises or weight lifting have been widely used in various sports training programs, as well as in health- and fitness-related exercise interventions. Every activity, including activities of daily living, requires a certain percentage of an individual's maximum strength and endurance. Regular resistance exercise can serve as a potent stimulus to the musculoskeletal system that is necessary to bring about gain in muscle size and function. It also helps in enhancing bone mass and the strength of connective tissue. A training routine that combines both aerobic and resistance exercises is highly recommended because the resulting improvement in cardiorespiratory and muscular function can allow individuals to not only reduce their risks for chronic diseases related to physical inactivity, but also be able to perform activities of daily living comfortably and safely.

Table 13.2 Energy expenditure during various physical activities

ACTIVITY	ENERGY EXPENDITURE (KCAL MIN⁻¹)	
	MEN	WOMEN
Sitting	1.7	1.3
Standing	1.8	1.4
Sleeping	1.2	0.9
Walking (3.5 mi h⁻¹, 5.6 km h⁻¹)	5.0	3.9
Running (7.5 mi h⁻¹, 12.0 km h⁻¹)	14.0	11.0
Running (10.0 mi h⁻¹, 16.0 km h⁻¹)	18.2	14.3
Cycling (7.0 mi h⁻¹, 11.2 km h⁻¹)	5.0	3.9
Cycling (10.0 mi h⁻¹, 16.0 km h⁻¹)	7.5	5.9
Weight lifting	8.2	6.4
Swimming (3.0 mi h⁻¹, 4.8 km h⁻¹)	20.0	15.7
Basketball	8.6	6.8
Handball	11.0	8.6
Tennis	7.1	5.5
Wrestling	13.1	10.3

Source: Adapted from Wilmore and Costill 2004.

Considerably less information is available concerning energy costs of resistance exercise. This may be due to the fact that this type of exercise is typically performed in an intermittent fashion, so the accumulated exercise time is relatively short. Although exertion can be quite strenuous at times during resistance exercise, these moments of strenuous activity are not usually sustained for more than 1 min. Table 13.3 provides results on net VO_2 from studies that have examined the energy cost of resistance exercises (Halton *et al.* 1999; Hunter *et al.* 1988, 1992; Olds and Abernethy 1993; Willoughby 1991; Wilmore *et al.* 1978). It appears that a greater energy cost is generated during a **circuit weight training** routine that contains a larger volume of exercise being performed. Energy cost during resistance exercise has also been reported using gross VO_2 for which the resting component is not subtracted from the total value (Burleson *et al.* 1998; Phillips and Ziuraitis 2003). With this approach, VO_2 was found to be near 1–1.5 l min⁻¹, a range that is generally higher than those in Table 13.3. When the total oxygen consumed is accumulated over an entire exercise session, it ranges over 30–45 l session⁻¹ or 150–225 kcal session⁻¹. This level of oxygen uptake is considered mild given that the gross caloric expenditure is capable of reaching 400–500 kcal during a typical endurance exercise of moderate intensity that lasts ~45 min.

Although it may not accumulate as much energy expended while exercising, resistance exercise can disturb the body's homeostasis to a greater extent than does aerobic exercise. The physiological strain it imposes can persist through a sustained period of recovery following exercise. As such, resistance exercise is often associated with a greater EPOC than aerobic exercise. For example, Burleson *et al.* (1998) found a greater EPOC following circuit weight training performed at 60% 1-RM (one-repetition

STUDIES	SUBJECTS	VOLUME/INTENSITY	VO_2 (L MIN^{-1})*
Willoughby et al. (1991)	10 men	Squat at 50% 1-RM, 7 reps	0.18
		Squat at 70% 1-RM, 6 reps	0.19
		Squat at 90% 1-RM, 5 reps	0.24
Halton et al. (1999)	7 men	1 circuit of 8 exercises at 75% 20-RM, 20 reps	0.29
Olds and Abernathy (1993)	7 men	2 circuits of 7 exercises at 60%, 15 reps	0.78
		2 circuits of 7 exercises at 75%, 12 reps	0.78
Hunter et al. (1988)	10 men	4 sets of bench press at 20% 1-RM, 30 reps	0.14
	7 women	4 sets of bench press at 40% 1-RM, 20 reps	0.23
		4 sets of bench press at 60% 1-RM, 10 reps	0.30
		4 sets of bench press at 80% 1-RM, 5 reps	0.51
Wilmore et al. (1978)	20 men	3 circuits of 10 exercises at 40% 1-RM,15–18 reps	1.00
	20 women		
Hunter et al. (1992)	14 men	4 sets of knee extension at 60% 1-RM, 10 reps	0.46
	8 women	4 sets of knee extension at 80% 1-RM, 5 reps	0.48
		4 sets of knee flexion at 60% 1-RM, 10 reps	0.58
		4 sets of knee flexion at 80% 1-RM, 5 reps	0.60

Note: * Values are the net VO_2 that is averaged over exercise and recovery periods. Net VO_2 is computed as exercise VO_2 + recovery VO_2 – resting VO_2 accumulated for the total period of exercise and recovery.

maximum – the maximum weight one can lift in a single repetition) as compared to treadmill exercise that was matched for the same VO_2 elicited during resistance exercise. The greater EPOC is attributable to the fact that a majority of the energy that supports the activity is derived from the use of anaerobic energy sources such as ATP and creatine phosphate, which requires oxygen for replenishment following exercise. It can also be due to a greater change in heart rate, as well as concentration of blood lactate and selected hormones imposed during exercise that will elevate oxygen utilization once exercise is completed. Many studies have examined the effect of resistance exercise on EPOC (Binzen et al. 2001; Kang et al. 2005b; Thornton et al. 2002; Melby et al. 1993; Melanson et al. 2002; Olds and Abernethy 1993; Schuenke et al. 2002). It is generally agreed that although the lifting mode and intensity (% 1-RM) can be influential, it is the volume of resistance training, which represents the total quantity of weights lifted, that serves as the most important factor that determines the magnitude of EPOC. Using the respiratory exchange ratio, several studies also revealed a greater fat utilization in addition to the greater EPOC following resistance exercise (Binzen et al. 2001; Melby et al. 1993).

Thermal effect of foods

Diet-induced thermogenesis – the **thermal effect of food** (TEF) – represents another important component of the total daily energy expenditure. This energy fraction is defined as the significant elevation of the metabolic rate that occurs after ingestion of a meal. Typically, this elevation reaches its peak within an hour and can last four hours after a meal. TEF is proportional to the amount of energy being consumed and is estimated at about 10% of energy intake. For example, an individual consuming 2000 kcal probably expends about 200 kcal on TEF.

TEF can be divided into two subcomponents: **obligatory thermogenesis** and **facultative thermogenesis**. The obligatory component of TEF is the energy cost associated with digestion, absorption, transport, and assimilation of nutrients, as well as the synthesis of protein, fat, and carbohydrate to be stored in the body. Facultative thermogenesis is thought to be mediated by the activation of the sympathetic nervous system, which functions to stimulate the metabolic rate. This classification in essence provides underlying mechanisms that explain the increment in thermogenesis following a meal.

TEF can be different depending upon whether protein, carbohydrate, or fat is being consumed. It is widely accepted that the TEF produced by protein is about 20–30% of the energy intake, whereas the TEF for carbohydrate and fat is approximately 5–10% and 0–5%, respectively. This relationship may be attributed to the differences in chemical structure of these nutrients, which dictate the amount of energy that is necessary for them to be digested, absorbed, assimilated, and stored. Consumption of protein is considered most thermogenic, partly because protein contains nitrogen that needs to be removed, which is expensive in terms of energy. In addition, most amino acids are absorbed by an energy-requiring process. Absorbed amino acids can also be used for protein synthesis. In this process, energy is mainly used for synthesizing peptide bonds.

The relatively large calorigenic effect of ingested protein has been used as evidence to promote a high-protein diet for weight loss. This is based on the belief that as a greater amount of energy is needed during the process in which protein is digested, absorbed, and assimilated, fewer calories will become available to the body for storage in comparison to a comparable meal consisting mainly of carbohydrate and fat. However, this notion needs to be considered with caution as it has been claimed that a high protein intake can lead to hypoglycemia and protein degradation, as well as harmful strain on kidneys and the liver in the long run.

ETIOLOGY OF OBESITY

Energy processes in the human body, like those of other machines, are governed by the laws of thermodynamics. If the human body consumes less energy in the form of food calories than it expends in metabolic processes, then a negative energy balance will occur and the individual will lose body weight. Conversely, a greater caloric intake in comparison to energy expenditure will result in a positive energy balance and a gain in body weight. In simple terms, obesity is caused by this latter condition of energy imbalance. Although a positive energy balance provides the basic answer to how we get fat, it does not provide any insight relative to the specific mechanisms. Claude Bouchard, a prominent international authority on obesity and weight control, noted that currently there is no common agreement on the specific determinants of obesity, emphasizing that numerous factors are correlated with body fat content (Bouchard 2008). In general, most leading scientists support a multi-causal theory involving the interaction of a number of genetic and environmental factors.

Heredity

Experimental studies on animals have linked obesity to hereditary (genetic) factors. Studies of humans by Dr. Albert Stunkard and his colleagues have shown a direct genetic influence on height, weight, and BMI (Stunkard *et al.* 1986, 1990). Perhaps a study from Laval University in Quebec provided the strangest evidence yet of a significant genetic component of obesity (Bouchard *et al.* 1990). The investigators took 12 pairs of young adult male monozygotic (identical) twins and housed them in a closed section of a dormitory under 24-hour observation for 120 consecutive days. The subjects' diets were monitored during the initial 14 days to determine their baseline caloric intake. Over the next 100 days, the subjects were fed 1000 kcal above their baseline consumption for six of every seven days. On the seventh day, the subjects were fed only their baseline diet. Thus, they were over-fed on 84 of 100 days. Activity levels were also tightly controlled. At the end of the study period, the actual weight gained varied widely from 4.3 kg to 13.3 kg (9.5 lb to 29.3 lb) despite an over-consumption of the same calories. However, the response of both twins in any given twin pair was quite similar. Similar results were found for gains in fat mass, percentage body fat, and subcutaneous fat. This intra-pair similarity in the adaptation to long-term overfeeding and for the variations in weight gain and fat distribution among the pairs of twins suggests that genetic factors are involved.

Research into the genetics of obesity has progressed at a rapid pace. To date, several obesity genes have been identified, which may explain why some individuals maintain an unhealthy set point. It is generally considered that individuals with a genetic susceptibility to obesity may be predisposed to abnormalities in neural function. These individuals establish neural circuits that are not easily abolished. In essence, obesity genes influence appetite to increase energy intake or affect metabolism to decrease energy expenditure. For example, genes in the hypothalamus may decrease the number of protein receptors for leptin, thus preventing leptin from inhibiting the appetite. Also, as noted in Chapter 6, a protein known as the uncoupling protein (UCP) is believed to activate thermogenesis and there are several forms of UCP that have been discovered from different tissues. For example, UCP1 activates thermogenesis in brown fat, whereas UCP2 activates thermogenesis in white fat and muscle tissue (Lowell and Spiegelman 2000). The brown fat tissue, which differs from the white fat that comprises most fat tissue in the body, is found in small amounts located near large blood vessels around the neck, back, and chest areas (Cannon and Nedergaard 2004). Defects in UCP genes may decrease resting energy expenditure. Over 300 genes may be involved in weight loss and weight maintenance. Genetic factors that have been implicated in the development of obesity include: (1) a predisposition to sweet and high-fat foods; (2) inability to control appetite; (3) impaired functions of hormones such as insulin and leptin; (4) reduced levels of human growth hormones; (5) decreased resting energy expenditure; (6) reduced rates of fat oxidation; and (7) enhanced efficiency in storing fat and conserving energy.

Environmental factors

Heredity may predispose one to obesity. However, environmental factors are also highly involved. Some would argue that body weight similarities between family members stem more from leaned behaviors than genetic similarities. Even married couples, who have no genetic link, may behave similarly toward food and eventually assume similar degrees of leanness or fatness. Proponents of the nurture view propose that environmental factors, such as high-calorie or -fat diet and inactivity, literally, shape us. This notion seems likely when considering that our gene pool has not changed much in the past 50 years, whereas according to the US Centers for Disease Control the prevalence of obesity has grown to epidemic proportions for the same time period (Mokdad *et al.* 2001).

Although excess calories, or over-feeding, may lead to weight gain and obesity, researchers suggest that the main "culprit" in the diet that leads to obesity is dietary fat. Researchers have postulated several reasons for why dietary fat plays a major role in causing weight gain and obesity. Dietary fat is highly palatable to most individuals, encouraging over-consumption. Dietary fat contains more calories per gram, and may not provide the same satiety as carbohydrate and protein. It has been demonstrated that high-fat foods give rise to higher energy intake during a meal than do carbohydrates and proteins, and calorie for calorie are less effective in suppressing subsequent food intake (Green and Blundell 1996). Shah and Garg (1996) also reported that spontaneous energy intake is higher in an unrestricted high-fat diet compared to a high-carbohydrate diet. Dietary fat may be stored as fat more efficiently compared to carbohydrate and protein. It takes some energy to synthesize fat and store it in adipose tissue, but in comparison to dietary fat, it may cost up to 3–4 times more energy to convert carbohydrate or protein to body fat. In addition, it has been found that chronic intake of a high-fat diet will produce resistance in the hypothalamus to various factors that normally suppress appetite, such as leptin, resulting in increased energy intake and body fat deposition (Tso and Liu 2004).

Excess dietary fat, for the reasons discussed above, may play a major role in the etiology of overweight and obesity. However, researchers have also observed that the prevalence of obesity in the United States has increased dramatically over the past two decades, even though the energy derived from dietary fat has actually decreased, both in absolute terms and as a percentage of total energy intake. This apparent inconsistency can be attributed to an increased daily caloric intake. Our society promotes increased food intake. Supermarkets, fast-food restaurants, and all-night convenience stores provide ready access to food throughout the day and night. Appetizing low-fat, but high-calorie food is everywhere, and in supersize proportions which significantly increase caloric content. Bigger is marketed as better in terms of portion size. People buy the large sizes and combinations, perceiving them to be good value, but then they eat more than they need. What was once small is now large. For example, the average soft drink is over 50% larger compared to what it was ten years ago, some of which may contain over 500 calories. The trend toward large portion sizes parallels the prevalence of overweight and obesity in the United States, beginning in the 1970s, increasing sharply in the 1980s, and continuing today (Young and Nestle 2002).

With excess dietary calories, it is possible for an individual to become obese even on a low-fat diet, because the body rapidly adjusts to oxidizing excess dietary carbohydrate and protein to meet its energy needs, sparing the use of body fat stores. In addition, some of the excess dietary carbohydrate may also be used to generate body fat. Recent research by Ma *et al.* (2005) indicates that a high glycemic index diet is associated with increased body weight. This evidence suggests that weight gain can occur without a necessary increase in fat consumption.

As noted, the prevalence of obesity is increasing despite a decrease in consumption of dietary fat. Other than the explanation that this observation is attributable to an increased consumption of low-fat, high-calorie foods, it also may be explained by concomitantly decreasing levels of physical activity (Astrup *et al.* 2002). Modern technology is helping to make our lives more comfortable and enjoyable in numerous ways. However, technology may also exert a negative effect on our health as the development of television, computers, and other labor-saving devices may decrease the level of physical activity. A number of cross-sectional studies have reported an inverse relationship between physical activity and body weight; that is, those who were physically inactive weighed more. Physical activity and exercise is an important factor in the prevention of overweight or obesity. Once an individual becomes obese, physical activity decreases, creating a vicious cycle of increasing body weight and less physical activity.

Interactions of heredity and environment

It is unlikely that the increasing incidence of obesity in the United States is due solely to genetics because it takes many generations to change the genes present in a population. Therefore, non-genetic factors such as increased energy intake and decreased physical activity are thought to be major contributors to our increasing body weights. When genetically susceptible individuals find themselves in an environment where food is appealing and plentiful and physical activity is easily avoided, obesity is a likely outcome. An example of human obesity that is due to the interaction of a genetic predisposition and an environment that is conducive to obesity is the Pima Indian tribe living in Arizona. More than 75% of this population is obese. Several genes have been identified and considered responsible for this group's tendency to store more fat (Norman *et al.* 1997). Typically, Pimas have low energy requirements per unit of fat-free mass, which, coupled with a low level of physical activity and a high intake of an energy-dense diet, has caused body fat to be maintained at a high level. A group of Pima Indians living in Mexico is genetically the same as those in the United States, but they are farmers who consume the food they grow and have a high level of physical activity (Esparza *et al.* 2000). They still have higher rates of obesity than would be predicted from their diet and exercise patterns, suggesting genes that favor high body weight. However, they are significantly less obese than the Arizona Pima Indians.

The interaction of heredity and environment may also explain why some people attain and maintain body weights higher than their set point. It is now believed that an individual's set-point is adjustable, possibly shifting with changes in the hypothalamus. Some scientists have proposed a new theory, the "settling-point" theory. The **settling-point theory** suggests that the set point may be modified, and in the case of weight gain, set at a higher level. In other words, whatever genes we have that make us susceptible to obesity may settle into a happy equilibrium with our environment. For example, a chronic high-fat diet may modify our genes, possibly increasing leptin resistance, and our body weight then rises to a new level. On the other hand, a weight-loss diet may lower the set point to help maintain body weight at a lower level. Weinsier *et al.* (2000) found that caloric restriction in obese women induced a transient decrease in resting energy expenditure, which would be counterproductive to weight loss. However, metabolism returned to normal at the completion of a weight-loss program when energy intake was then adequate to maintain the reduced body weight, suggesting that the set point may settle to a lower level over time. Clearly, genetics and environment interact to influence body weight and composition.

DIETARY THERAPIES DESIGNED TO REDUCE ENERGY INTAKE

Although the weight-loss industry would like us to think otherwise, the truth is clear: there is no quick and easy way to lose weight. Treatment of overweight and obesity should be long term, similar to that for any chronic disease. It requires a firm commitment to lifestyle changes, rather than a quick fix as promoted by many popular diet books. We often view a diet as something one goes on temporarily, only to resume prior (typically poor) habits once satisfactory results have been achieved. This is why so many people regain lost weight. Instead, an emphasis on healthy, active living with acceptable dietary modifications will promote weight loss and later weight maintenance.

Weight loss guidelines and approaches

Although there are many reasons why people want to lose weight, the most important one is to improve health. Rather than focusing on weight loss, which is unlikely to be successful in the long term, a weight problem should be viewed in terms of weight management. One goal of weight management is to prevent excess body weight gain. Healthy eating habits and active lives that promote the maintenance of a healthy weight should be developed in childhood and maintained throughout life. Just as people adopt dietary and lifestyle changes in response to a family history of heart disease or an increase in blood cholesterol, similar actions should be taken to maintain a healthy weight if someone is exposed to a family history of obesity or an increase in body weight. For those who are already overweight, the goal of weight management is to reduce body weight and body fat to a healthy level that can be maintained over their lifetime.

Achieving and maintaining weight loss requires making lasting lifestyle changes, including what we choose to eat and how much physical activity we engage in. Most health experts suggest that people focus less on weight loss and more on eating healthily and on overall fitness. Misguided efforts of weight loss at any cost unfortunately present food as the enemy, rather than as a means to good health. Most people who successfully lose weight and keep it off do so by eating a balanced diet of nutrient-dense foods and by maintaining a moderately high level of physical activity (Wing and Phelan 2005). A healthy weight-loss and weight-maintenance program consists of four components: (1) setting reasonable goals; (2) choosing foods sensibly; (3) increasing physical activity; and (4) modifying behavior. These are described below.

Setting reasonable goals

Setting reasonable and attainable goals is an important component of any successful weight-management program. For most people, a loss of 5–15% of their body weight, most of which is fat, will significantly reduce disease risk. Therefore, it has been suggested that the initial goal of weight loss should be to reduce body weight by approximately 10% over a period of about six months (National Heart, Lung, and Blood Institute 1998). A slow loss of 10% of body weight is considered achievable for most individuals and is easier to maintain than larger weight losses. Most people who lose large amounts of weight or lose weight rapidly eventually regain all that they have lost. Repeated cycles of weight loss and regain, often referred to as **weight cycling**, may increase the proportion of body fat with each successive weight regain and cause a decrease in RMR, making subsequent weight loss more difficult (Seagle *et al.* 2009).

When it comes to weight loss, slow and steady is the way to go, and weight loss should not exceed 1–2 lb per week. This will promote the loss of fat and not lean tissue. It is estimated that 1 lb of fat provides 3500 kcal. Therefore, to lose a pound of fat, one must decrease energy intake and/or increase energy output by this amount. To lose 1 lb per week, one must shift energy balance by 500 kcal per day ($3500\,kcal/7\ days = 500\,kcal\,day^{-1}$). Please note that this is the predicted average weight loss at this energy deficit. The actual amount of weight loss per week may vary over time. Rather than making dramatic dietary changes, small changes such as reducing portion sizes and cutting back on energy-dense snacks makes a big difference in overall energy intake. Once the body weight stabilizes and the new lower weight is maintained for a few months, a decision can be made whether additional weight loss is needed.

Table 13.4 outlines the recommendations of a weight-loss diet made by the National Institutes of Health: Obesity Education Initiative (2000). What is reflected in the recommendations is that energy intake should provide nutritional adequacy without excess – that is, somewhere between deprivation and completed freedom to eat. A reasonable suggestion is that an adult needs to increase activity level and to

Table 13.4 General recommendations for a weight-loss diet

NUTRIENT	RECOMMENDED INTAKE
kcal (BMI ≥ 35)	500–1000 kcal per day reduction from usual intake
kcal (BMI between 27 and 35)	300–500 kcal per day reduction from usual intake
Total fat	30% or less of total kcal
Saturated fatty acids	8–10% of total kcal
Monounsaturated fatty acids	Up to 15% of total kcal
Polyunsaturated fatty acids	Up to 10% of total kcal
Cholesterol	300 mg or less per day
Protein[a]	Approximately 15% of total kcal
Carbohydrate[b]	55% or more of total kcal
Sodium chloride (salt)	No more than 2400 mg of sodium or approximately 6 g of sodium chloride per day
Calcium	1000–1500 mg per day
Fiber	20–30 g per day

Source: National Institutes of Health: Obesity Education Initiative 2000, p. 27.

Notes:
a Protein should be derived from plant sources and lean sources of animal protein.
b Carbohydrate and fiber should be derived from vegetables, fruits and whole grains.

reduce food intake enough to create a deficit of 500 kcal per day. As mentioned earlier, such a deficit produces a weight loss of about 1 lb per week, a rate that supports the loss of fat efficiently, while retaining lean tissue. In general, weight-loss diets provide 1200–1600 kcal per day.

Choosing foods sensibly

Contrary to the claims of fad diets, no single food plan is magical and no specific food must be included or avoided in a weight-management program. In fact, weight-loss plans that drastically reduce calories and offer limited food choices leave people feeling hungry and dissatisfied. Thus, weight-loss diets that encourage people to eat foods that are healthy and appealing tend to have greater success.

The main characteristic of a weight-loss diet is that it provides less energy than the person needs to maintain their present body weight. Reducing energy intake is best achieved by cutting back on energy-dense foods that have little nutritional value, such as soft drinks, potato chips, cookies, and cakes. Aside from their low energy densities, foods such as whole grains, legumes, nuts, fruits, and vegetables offer many health benefits such as micronutrients and fiber. In addition, these foods tend to have greater volume compared to more energy-dense foods, thus helping people feel full more easily.

Most people still believe that to lose weight, one should avoid foods that contain fat. However, this is not the case! As discussed in Chapter 8, the Dietary Guidelines for Americans now recommend that we choose our fats as carefully as we choose our carbohydrates. In general, it is best to limit intake of foods containing trans and saturated fatty acids. Foods containing relatively more polyunsaturated and monounsaturated fatty acids are healthier. Another common misconception is that dairy products and meat are high-fat foods that should be avoided when trying to lose weight. Again, the key to good nutrition is moderation and choosing wisely. For example, switching from whole to reduced-fat milk is one way to lower caloric intake without losing many vitamins and minerals from dairy products. Likewise, how meat is prepared and what type of meats are consumed can greatly affect the amount of calories consumed. Lean meats prepared by broiling or grilling are both nutritious and satisfying.

Healthy eating also requires people to pay attention to hunger and satiety cues. Rather, the amount of food served or packaged often determines how much we eat. That is, visual cues, rather than internal cues, have a greater influence on the quantity of food consumed. For example, some commercially made muffins are extremely large, containing as many calories as eight slices of bread. Therefore, learning to choose reasonable portions of food is a critical component to successful weight management. In fact, reducing portion sizes by as little as 10–15% could reduce our daily caloric intake by as much as 300 kcal. One way to limit serving size is to consider sharing large and supersize meals the next time you eat at a restaurant.

Table 13.5 presents some examples of how to save kilocalories via simple substitutions. By choosing foods wisely, a significant reduction in fat and total calories can be achieved. As you should realize, it is best to consider eating healthily and making lifestyle changes, rather than a weight-loss plan. A person who adopts a lifelong "eating plan for good health" rather than a "diet for weight loss" will be more likely to keep

Table 13.5 Selected food substitutes for reducing fat and caloric intake

FOOD CLASS	HIGHER-FAT FOODS	LOW-FAT ALTERNATIVE
Dairy products	Whole milk	Fat-free (skim), low-fat (1%) or reduced-fat (2%) milk
	Whipping cream	Whipped cream made with fat-free milk
	Cream cheese	Light or fat-free cream cheese
	Cheese (cheddar, Swiss, American)	Low-calorie, reduced-fat or fat-free cheese
Cereals, grains, and pastas	Ramen noodles	Rice or spaghetti noodles
	Pasta with white source (alfredo)	Pasta with red/marinara source
	Pasta with cheese	Pasta with vegetables
	Granola	Bran flakes, oatmeal or reduced granola
Meat, fish, and poultry	Regular hot dogs	Low-fat hot dogs
	Bacon or sausage	Canadian bacon or lean ham
	Regular ground beef	Ground turkey or extra lean ground beef
	Chicken or turkey with skin	Chicken or turkey without skin
	Oil-packed tuna	Water-packed tuna
	Beef (chuck, rib, brisket)	Beef (round, loin) trimmed of external fat
	Pork (spare-ribs, untrimmed loin)	Pork tenderloin, trimmed or lean smoked ham
	Frozen breaded or fried fish	Fresh or unbreaded fish
	Whole eggs	Egg whites or egg substitutes
Baked goods	Croissants, brioches, etc.	French rolls or soft brown rolls
	Donuts, sweet rolls, muffins or pastries	English muffins, muffins with reduced fat or bagels
	Party crackers	Low-fat crackers or saltine or soda crackers
	Cake (pound, chocolate, yellow)	Cake (angel food, white, gingerbread)
	Cookies	Reduced-fat or fat-free cookies (graham crackers, ginger snaps, fig bars)
Snack and sweets	Nuts	Popcorn, fruits, vegetables
	Ice cream	Sorbet, sherbet, low-fat or fat-free yogurt or ice cream
	Custards or puddings made with whole milk	Puddings made with skim milk
Fats, oils, and salad dressings	Butter or margarine	Light spread or diet margarine, jelly, jam or honey
	Mayonnaise	Light mayonnaise or mustard
	Salad dressings	Reduced-calorie or fat-free dressings, lemon juice or plain, herb flavored or wine vinegar
	Oils, shortenings or lard	Use non-stick cooking spray for stir-frying or applesauce or prune puree for baked goods

the lost weight off. Keep in mind that well-balanced diets that emphasize fruits, vegetables, whole grains, lean meats or meat alternatives, and low-fat milk products offer many health benefits even when they don't result in weight loss.

Increasing physical activity

Physical activity is an important component of any well-designed weight-management program. Exercise promotes fat loss and weight maintenance. It increases energy expenditure, so if the intake remains the same, energy stored as fat is used as fuel. In theory, an increase in activity of 200 kcal five times per week will result in the loss of a pound of fat in about 3.5 weeks. Exercise also promotes muscle development. This is important during weight loss because muscle is metabolically active tissue. Increases in muscle mass help prevent the decrease in resting metabolic rate that occurs as body weight decreases. Weight loss is also better maintained when physical activity is included in the weight-management program (Wilmore 1996).

It is important to realize that people can be physically fit while being overweight. Normal blood pressure and healthy blood glucose concentration and lipid profile are important indicators of physical fitness. Studies show that obese individuals who are physically fit have fewer health problems than do average-weight individuals who are unfit (Lovejoy *et al.* 2003). An added benefit of including physical activity in a weight-reduction program is maintenance of bone health. Regular participation in physical activity also helps relieve boredom and stress and promotes a positive self-image. Physical activity is also an effective strategy for preventing unhealthy weight gain in normal, overweight, and obese individuals. An expert panel assembled by the International Association for the Study of Obesity concluded that 45–60 min of daily exercise can help prevent normal weight individuals from becoming overweight, overweight people from becoming obese, and already obese individuals from worsening their condition.

The latest advice for adults from the 2005 Dietary Guidelines for Americans is 60 min of physical activity per day to maintain body weight and prevent weight gain and 60–90 min per day for maintenance of weight loss. Duration and regular performance, rather than intensity, are the keys to success. One should search for activities that can be continued over time. In this regard, walking vigorously three miles per day can be as helpful as aerobic dancing or jogging if it is maintained. Moreover, activities of lighter intensity are less likely to lead to injuries. Some resistance exercises or weight training should be added to increase lean body mass and, in turn, fat use. As lean muscle mass increases, so does the basal metabolic rate. Various exercise strategies to maximize energy expenditure are discussed in detail in later sections of this chapter.

Modifying behavior

Behavior and attitude play an important role in supporting efforts to achieve and maintain appropriate body weight. In order to keep weight at a new lower level, food intake and exercise patterns must be changed for life. However, changing the hundreds of small behaviors of over-eating and under-exercising that lead to obesity requires time and effort. A person must commit to taking action.

Changing behaviors requires identifying the old patterns that led to weight gain and replacing them with new ones to maintain weight loss. This can be accomplished through a process called behavior modification, which is based on the theory that behaviors involve (1) antecedents or cues that lead to the behavior; (2) the behavior itself; and (3) consequences of the behavior. For example, sitting in front of the television and consuming a large bag of potato chips may leave you feeling bad because you consumed the extra calories. In this case, the antecedent is watching television, the behavior is eating the chips, and the consequence is feeling remorse and gaining weight. The key to modifying this behavior is to recognize the antecedent, change the behavior, and replace the negative consequence with a positive one.

Therefore, to solve a problem one must first identify all the behaviors that created the problem. Keeping a record will help to identify eating and exercise behaviors that may need changing. It will also establish a baseline against which to measure future progress. With so many possible behavior changes, a person can choose where to begin. Start simple and don't try to master them all at once. Attempting too many changes at one time can be overwhelming. Pick one trouble area that is manageable and start from there. Practice desired behavior until it becomes routine. Then select another trouble area to work on, and so on.

Options for reducing energy intake

An ideal weight-management program should provide a reduction in energy intake along with education about meeting nutrient needs, increasing energy expenditure, and changing lifestyle patterns that led to the weight gain. Fad diets, such as those that emphasize eating primarily a single food, special foods, or specific combinations of foods, may promote weight loss over the short term, but since they are not nutritionally sound, they cannot be consumed safely for long periods. These diets do not encourage exercise or promote the changes in eating habits that affect body weight over the long term. If the program's approach is not one that can be followed for a lifetime, it is unlikely to promote successful weight management. The following sections discuss some of the more common methods for reducing energy intakes.

Low-calorie diets

Weight-loss plans based simply on reducing energy intake are the most common choices. Some of the common plans include (1) fixed meal plans; (2) free-choice diets; (3) liquid-formula diets; and (4) very-low-calorie diets. Some recommended energy reduction without restricting the type of foods selected, some use exchange systems to plan energy and nutrient intake, and others provide low-calorie meals and formulas.

Fixed-meal plans. These are diet plans in which you either choose from a limited list of food options or in which the entire menu is decided for you. For example, a fixed meal plan might specify a cup of corn flakes with a banana for breakfast, tuna salad and an apple for lunch, and grilled chicken with broccoli for dinner. These diets are easy to follow and may make losing weight easier, but can be boring and thus are not practical for the long term. In addition, they don't teach food selection skills because the meals are pre-determined.

Free-choice diets. These diets allow dieters to choose the foods they eat as long as the total caloric intake is reduced. They offer flexibility and variety, and can suit different preferences. These types of diet often use food-guide pyramids to construct a balanced low-calorie diet. For example, a diet with as few as 1200 kcal can be planned by using the low end of the range of suggested servings and making low-calorie choices. These diets may not meet nutrient needs unless dieters are literate regarding basic nutritional principles and diets are based on sound food selection guidelines.

Liquid-formula diets. These diets recommend a combination of food and formula to provide a daily energy intake of about 800–1200 kcal. They normally consist of two liquid meal replacements and a normal meal. The dieter can also snack on fresh fruit or vegetables. Liquid diets can allow rapid weight loss, but this effect is purely short term. They can be very difficult to maintain due to the few calories that are permitted for consumption. The liquid diets do not teach you how to eat in order to stay slim for the long term. They need to be accompanied by a behavior-alteration course that deals with and prevents the reasons for over-eating. Otherwise, it is easy to put back on any weight lost on a liquid diet.

Very-low-calorie diets. These diets are defined as those containing fewer than 800 kcal per day. They become popular in response to a desire for rapid weight loss. These diets provide little energy and a high proportion of protein. The protein in the diet is used to meet the body's protein needs and will, therefore, prevent excessive loss of the body's protein. Often, very-low-calorie diets are offered as a liquid formula. These formulae provide 300–800 kcal and 50–100 g of protein per day. They also contain the recommended daily requirements for vitamins, minerals, trace elements, and fatty acids. The initial weight loss is fairly rapid, about 3–5 lb per week. However, in most cases, about 75% of this initial weight loss comes from water loss. Once initial weight loss ends, weight loss slows in part because the dieter's resting metabolism decreases to conserve energy, and physical activity decreases because the dieter often does not have the energy to continue their typical level of physical activity. Very-low-calorie diets are no more effective than other methods in the long term, and they carry more risks. At these low energy intakes, body protein is broken down and potassium is excreted. Depletion of potassium can result in irregular heart beats, which can be fatal. Studies have also shown that in about one in four individuals following a very-low-calorie diet for a few months, gallstones develop. Gallstone formation is facilitated by the more concentrated bile fluid and reduced flow as a result of the diet. Other side effects include fatigue, nausea, light-headedness, constipation, anemia, hair loss, dry skin, and menstrual irregularities. It is recommended that very-low-calorie diets be carried out in conjunction with medical supervision.

Diets that modify macronutrient intake

Rather than focusing on counting calories as a way of reducing body weight, some diets concentrate on modifying the proportion of energy-containing nutrients. Currently, one of the biggest controversies is the role of dietary carbohydrate versus dietary fat in promoting weight loss. Weight-loss diets that are low in fat and high in carbohydrates have long been considered the most effective in terms of weight management. In fact, as discussed in Chapter 8, the Dietary Guidelines for Americans advocate low-fat food choices with an emphasis on whole grains, fruits, and vegetables. They suggest that we consume 45–65% of energy from carbohydrate, 10–35% from protein, and

20–30% from fat. However, Dr. Robert Atkins, one of the first pioneers of the low-carbohydrate diet shocked the nutritional world in 1972 when he proposed that too much carbohydrate, rather than too much fat, may actually cause people to gain weight. Since then, there have been numerous diets developed that consist of an altered macronutrient distribution favoring a low carbohydrate intake. Caution is needed in using these diets as they generally lack supporting evidence on safety and long-term efficacy. These diets share some common characteristics:

- They promote quick weight loss.
- They limit food selections.
- They use testimonials from famous people and tie the diet to well-known cities.
- They claim to work for everyone.
- They make no attempt to change eating habits permanently.

Low-fat, high-carbohydrate diets. These diets contain approximately 10% of calories as fat and are very high in carbohydrates. The most notable are the "Pritikin Diet" and Dr. Dean Ornish's "Eat More, Weigh Less" diet plans. The diets advise dieters to avoid meat, dairy, oils, and olives; low-fat meat and dairy can be eaten in moderation. With an emphasis on fruits, vegetables, and whole grains, the diets provide about 65–75% of total calories from carbohydrates, with proteins making up the difference.

There are several reasons why advocates of low-fat diets believe such diets help promote weight loss. Gram for gram, fat has twice as many calories as carbohydrate and protein. Therefore, it is reasonable to assume that consuming less fat may lead to lower energy intake which in turn results in weight loss. Fat can also make foods more palatable, contributing to over-consumption. In addition, excess calories from fat are more readily stored by the body compared to those from carbohydrate and protein. This is due to energy costs associated with excess glucose and amino acids being turned into fatty acids prior to storage. Recent studies suggest that low-fat diets benefit overall health by lowering total and LDL-cholesterol concentrations, increasing HDL-cholesterol concentrations, and improving blood glucose regulation (Lovejoy *et al.* 2003).

It must be borne in mind that low-fat diets are not always low in energy; even a diet low in fat will result in weight gain if energy intake exceeds energy output. This is illustrated by the fact that the percentage of calories from fat in the typical American diet has decreased, but the number of people who are overweight continues to increase. Although this dietary approach is not harmful, it is difficult to follow. People are quickly bored with this type of diet because they cannot eat many of their favorite foods. The diets emphasize consumption of grains, fruits, and vegetables, which most people cannot stick to for very long.

Low-carbohydrate, high-fat diets. These diets are at the opposite end of the weight-loss spectrum. Some of these diets, such as the induction phase of the Atkins™ diet, severely restrict carbohydrate intake by prohibiting nearly all carbohydrate foods and allowing unlimited quantities of meat and high-fat foods that are low in carbohydrate. Others, such as the Zone™ and South Beach™ diets, take a more moderate approach by allowing fruits, vegetables, and whole grains. Low-carbohydrate diets have become popular in recent years, with almost 15% of Americans reporting to be on some type of low-carbohydrate diet to lose weight. These diets are all based on the premise that a high carbohydrate intake causes an increase in insulin levels, which promotes storage of body fat.

Weight loss produced by this dietary approach appears to be short term and is largely attributed to water loss. As there are 3 g of water stored per gram of glycogen formed, the low carbohydrate intake leads to less glycogen synthesis and less water in the body. A very low carbohydrate intake also forces the liver to produce needed glucose. The source of carbons for this glucose is mostly protein from tissues such as muscle. Therefore, such production of glucose will result in a loss of lean body mass. Essential ions such as potassium are also lost in the urine. With the loss of glycogen stores, lean tissue, and water, dieters lose weight very rapidly. However, when a normal diet is resumed, weight is regained.

Once glycogen stores are depleted, the body begins to use triglycerides for energy. However, limited glucose availability allows fatty acids to only be partially oxidized. As a result, ketone formation increases, indicating that the body is using stored fat as a major energy source. Limiting carbohydrate intake causes ketosis, and for this reason, these diets are also called ketogenic diets. Excretion of ketones causes additional water loss. In addition, ketosis has been found to suppress appetite, thereby making weight loss easier.

Studies comparing weight loss associated with low-carbohydrate diets and low-fat diets showed that at six months, a greater weight loss is achieved on low-carbohydrate diets. However, differences disappeared by 12 months (Klein 2004). There is no compelling evidence that low-carbohydrate diets are more effective than other types of diet in helping people lose weight. Weight loss associated with low-carbohydrate diets may not necessarily be caused by an altered macronutrient distribution, but rather by a reduction in caloric intake as a result of limited food choice and increased ketosis (Bravata *et al.* 2003). Perhaps one of the biggest concerns regarding low-carbohydrate diets is that diets contain too much total fat, saturated fat, and cholesterol, and may lack essential macronutrients, dietary fiber, and phytochemicals rich in antioxidants. This dietary approach still needs more investigation concerning its long-term efficacy and health consequences.

EXERCISE STRATEGIES IN MAXIMIZING ENERGY EXPENDITURE

Energy expenditure via **physical activity** or **exercise** typically accounts for about 30% of the total daily energy expenditure. Physical activity and exercise have been used interchangeably in the past, but more recently exercise has been referred to as physical activity that is planned, structured, and repetitive, and purposive in the sense that improvement or maintenance of one or more components of physical fitness is the objective (Caspersen *et al.* 1985). Pursuing exercise and physical activity regularly has long been part of the recommendations made by various healthy organizations and authorities. This is because exercise and physical activity enable individuals to increase their energy expenditure or create an energy deficit, while gaining cardiorespiratory fitness. It has been recommended that exercise in conjunction with dietary modification is the most effective behavioral approach for weight loss (National Health, Lung, and Blood Institute 1998).

Enhancing energy expenditure through physical activity

An exercise program typically used in a weight-management program consists of continuous, large-muscle activities with moderate to high caloric cost, such as walking, running, cycling, swimming, rowing, and stair stepping. This approach will increase daily energy expenditure and help tip the caloric equation so that energy output is greater than energy input. Most training studies that have demonstrated exercise-induced weight loss have adopted exercise programs that elicit weekly energy expenditure of 1500–2000 kcal. This suggests that energy expenditure of a minimal of 300 kcal should be achieved during each exercise session and exercise should be performed no less than five times per week. This amount of energy expenditure generally occurs with 30 min of moderate-to-vigorous running, swimming, or cycling or 60 min of brisk walking.

Energy expenditure during exercise can be influenced by **intensity**, **duration** and modes of exercise. Unlike most fitness programs in which intensity of exercise plays an important part of exercise prescription, any program aimed at weight loss should ultimately be guided by the measure of total energy expenditure and its relation to energy consumption. In order to burn the most calories, exercise duration is considered more important, and exercise intensity must be adapted for the amount of time one would like to exercise (American College of Sports Medicine 2006). It has been suggested that exercise intensity should be tailored to allow a minimum of 300 kcal to be expended during each exercise session. Previous studies that include dietary modification have suggested that overweight women who exercise for a longer duration each week are able to lose more weight during an 18-month intervention (Jakicic *et al.* 1999). In this study, individuals reporting >200 min of exercise per week also reported >2000 kcal wk^{-1} of leisure-time physical activity as measured by a questionnaire.

The higher the exercise intensity is, the more the energy expenditure during exercise per unit of time. Stated alternatively, it costs you more energy to move your body weight or a given resistance at a faster pace. However, it is often difficult to maintain vigorous exercise for a sufficient period of time. This occurs especially in overweight or obese individuals who have low exercise tolerance in general. Some of them may also have the risk factors that contradict vigorous exercise. Consequently, those who need to maximize their energy expenditure may have to depend more on the expansion of exercise duration in order to achieve this goal. The American College of Sports Medicine (ACSM) position stand suggests that in order to be effective in maximizing energy expenditure over the long term, exercise intensity should not exceed 70% of an individual's maximal heart rate, which corresponds to ~60% VO$_2$max or heart rate reserve (HRR) (American College of Sports Medicine 2001). This intensity allows the attainment of recommended exercise duration of ~45 minutes. The concept of mild exercise is also considered ideal for those who are beginning an exercise program. This will allow adequate time for individuals to adapt to their exercise routine and to progressively increase exercise intensity over time. It should be made clear that exercise at moderate to high intensities is proven effective in augmenting aerobic fitness. However, the level of exercise intensity necessary to improve fitness is generally higher than the level of exercise intensity necessary to facilitate energy expenditure and thus weight loss.

Exercise **frequency** complements duration and intensity. Frequency of exercise refers to how often each week one should perform exercise and is often determined based upon how vigorous each exercise session is. It has been recommended that a training program consisting of exercise of moderate intensity (50–70% VO$_2$max) be carried out no less than three times per week in order to develop and maintain cardio-respiratory fitness and improve body composition (American College of Sports Medicine 2006). More recent studies also suggest that energy expenditure necessary to bring about weight loss should be no less than 1500 kcal week^{-1}. In order to achieve this threshold, energy expended during each exercise session should reach at least 500 kcal if exercise is performed three times per week or 300 kcal if exercise is performed five times per week. Obviously, this latter exercise arrangement (≥5 times per week) is more tolerable for sedentary and overweight individuals who often have difficulty in sustaining vigorous exercise. The more frequent exercise regimen will also help in maximizing fat utilization and minimizing exercise-related injury (Wallace 1997).

In order to meet the caloric threshold necessary to bring about weight reduction, the ACSM recommends a target range of 150–400 kcal of energy expenditure per day in physical activity and/or exercise (American College of Sports Medicine 2001). The lower end of this range is considered a minimal threshold and an initial goal for the previously sedentary and/or overweight individuals. If such light activity can be performed daily, it will help in achieving a total of ~1000 kcal of energy expenditure each week. The caloric expenditure of ~1000 kcal week^{-1} as suggested by the ACSM may be adequate for an overweight individual to begin an exercise intervention. This amount, however, has been found to be insufficient for long-term weight maintenance and should be increased progressively. Jakicic *et al.* (2001) observed that those who reported >2000 kcal week^{-1} physical activity showed no weight regain from 6 months to 18 months of treatment, whereas there was a significant weight regain observed in individuals who had energy expenditure below this threshold. Table 13.6 provides a summary of exercise guidelines as well as a sample prescription plan for maximizing energy expenditure and long-term weight control.

Exercise intensity and fat utilization

Will mild exercise be superior in facilitating fat utilization? As discussed in Chapter 7, exercise at lower intensities is associated with a greater percentage of energy derived from fat oxidation. This observation has resulted in a myth that in order to burn fat, you must exercise at a lower percentage of your maximal oxygen uptake (VO$_2$max). It may also explain the growth of "fat burner" classes, which assume that low-intensity exercise will lead to more weight loss. It must be clear that at higher intensities, though the percentage of the total calories derived from fat is lower, the total energy expenditure resulting from exercise will be higher, especially if accompanied by sufficient exercise duration. Consequently, the absolute quantity of calories burned due to fat oxidation will also be higher. We made a comparison of fat metabolism during exercise between 50% and 70% VO$_2$max using data collected from our laboratory. As shown in Table 13.7, those who exercise at 50% VO$_2$max consume oxygen at 1 l min^{-1} and respiratory exchange ratio is 0.86. At this ratio, according to Table 11.3, a total of 4.875 calories per minute (1 l min$^{-1} \times 4.875$ kcal l^{-1}) are expended and 2.24 or 46% of these are fat calories. On the

COMPONENT	GUIDELINES	SAMPLE PRESCRIPTION
Mode	Exercises include low-impact and non-weight-bearing activities involving large muscle groups. Other non-conventional modes such as yoga, weight lifting, and household activities may also be considered	Walking, cycling, swimming, low-impact group exercise such as water aerobics
Frequency	Exercise should be performed daily or at least five times per week. If necessary, a single session can be split into two or more mini-sessions	Daily or 5–7 times per week
Duration	Exercise duration should be maximized within the limit of tolerance and can be determined by time or caloric expenditure	40–60 minutes per day, 20–30 minutes twice per day or 150–400 kcal per day
Intensity	Intensity should generally stay at the lower end of the target range and be determined in accordance with amount of time or calories one would like to accomplish	40–70% VO$_2$max, 40–70% HRreserve or 60–80% age-predicted HRmax
Progression	In order to achieve successful weight loss and maintenance, there should be a progressive increase in the volume of exercise as intervention continues.	Proper increment in exercise volume should be made gradually as one's tolerance increases. Both intensity and duration may eventually reach the upper end of their respective range.

Table 13.6 Exercise guidelines and sample prescription plan for maximizing energy expenditure and long-term weight control

other hand, for those who exercise at 70% VO$_2$max, oxygen consumption and respiratory exchange ratio are $1.5 \, l\,min^{-1}$ and 0.88, respectively. As a result, a total of 7.349 calories per minute ($1.5 \, l\,min^{-1} \times 4.899 \, kcal\,l^{-1}$) are expended and 2.87 or 39% of these are fat calories. Although this is a lower percentage of fat calories, it is a higher total number of fat calories. From this example, it is clear that even though a greater percentage of fat is elicited at a lower level of exercise intensity, this does not necessarily mean that a greater quantity of fat is burned.

One may always try to pursue an exercise routine of sufficient intensity that will bring about a decent level of total energy expenditure. However, exercise intensity should be carefully chosen to prevent premature fatigue and to allow an exercise program to continue. Exercise duration could decrease drastically if intensity is set too high or exceeds **lactate threshold**, the intensity at which blood lactic acid begins to accumulate drastically. Based upon the current literature, exercise at 60–65% VO$_2$max would help in eliciting the maximal rate of fat oxidation (Achten *et al.* 2002). For most individuals, this moderate-level intensity can allow them to achieve the recommended exercise duration (~45 min). As such, this intensity appears to be effective in not only facilitating fat utilization, but also maximizing the total energy expenditure.

Table 13.7 Comparisons of fat and total calories expended during stationary cycling at 50% and 70% VO_2max

METABOLIC VARIABLES	EXERCISE INTENSITY	
	50% VO_2MAX	70% VO_2MAX
Oxygen uptake ($l\,min^{-1}$)	1.0	1.5
Respirator exchange ratio	0.86	0.88
Caloric equivalent ($kcal^{-1}$)	4.875	4.899
Energy output ($kcal\,min^{-1}$)	4.875	7.349
Relative fat contribution (%)	46	39
Fat calories ($kcal\,min^{-1}$)	2.24	2.87

Performing more vigorous exercise has also been related to a greater reduction in fat from the abdominal areas. This is shown by a number of epidemiological studies that involve middle-aged and elderly individuals (Buemann and Tremblay 1996; Tremblay *et al.* 1990, 1994; Visser *et al.* 1997). For example, Visser *et al.* (1997) found that intense exercise such as playing sports was negatively associated with abdominal fat. Tremblay *et al.* (1990) also observed a preferential reduction in abdominal fat in those who performed more intense exercise. This association that favors the use of more vigorous exercises may be explained by the fact that in contrast to gluteal fat, the fat stored in the abdominal area is more responsive to lipolysis, which is intensity dependent. It has been found that lipolysis is subject to the influence of epinephrine, which increases with increased exercise intensity (Wahrenberg *et al.* 1991).

Other exercise strategies

Exercise intensity and duration are the two prescription indices that are often used to develop an effective but safe exercise program. An easy application of such prescription is to have a steady-state exercise performed at the target intensity and for the duration desired. In reality, however, many exercise sessions are conducted in a more complex fashion. For example, a single exercise session can be divided into two smaller sessions performed at different times of the day, so that the target caloric expenditure can be achieved via accumulation. In many cases, despite the provision of target intensity, an exercise is performed with fluctuating intensity, such as in Spinning®, Treading®, and many other forms of group activities involving changing workload during different stages of the work-bout. Additionally, recent evidence suggests that resistance training should be included in a comprehensive weight-loss program to maximize its effectiveness.

Intermittent exercise

A few studies have examined the efficacy of adopting **intermittent exercise** for weight management (Donnelly *et al.* 2000; Jakicic *et al.* 1995, 1999). This direction of research was driven by a question as to whether the same metabolic and weight-loss benefits can be achieved by exercising in multiple sessions of shorter duration throughout the day. The intermittent exercise is typically defined as accumulation of

30–40 min of exercise each day through participation in multiple 10–15-min exercise sessions daily. This exercise strategy is considered advantageous for those who dislike or are unable to tolerate continuous exercise or have a daily schedule that prohibits a typical workout session. Intermittent exercise has long been proven effective in improving exercise compliance and enhancing cardiorespiratory fitness and improving risk factors for cardiovascular diseases. Its direct impact upon energy metabolism and weight loss has also been recognized. Jakicic *et al.* (1995) reported that exercising in multiple short bouts per day was just as effective as a single long exercise session in producing weight loss while gaining cardiorespiratory fitness over a 20-week intervention period that included dietary modification. In this same study, the program of multiple short bouts of exercise was also found to increase exercise adherence, which implies that this exercise strategy has the potential to facilitate the long-term adoption of a weight-loss program and thus to prevent weight regain.

Whether performing multiple short sessions of exercise instead of a long exercise bout will elicit more energy expenditure is another intriguing question that emerged recently, but remained largely unanswered. The total energy expenditure of a single exercise session includes the energy expended during the actual exercise period, as well as that during the recovery period following exercise. The latter is referred to as excessive post-exercise oxygen consumption (EPOC). As mentioned earlier, EPOC can be viewed as a compensatory response resulting from a disruption to homeostasis caused by the preceding exercise. In this context, it may be speculated that exercise of shorter duration performed more than once daily would be associated with a greater EPOC due to multiple occurrences of recovery. Almuzaini *et al.* (1998) have found that splitting a 30-min session into two 15-min sessions elicited a greater overall post-exercise VO_2. Although the long-term impact of this exercise arrangement on energy metabolism and weight loss remains to be elucidated, exercise for 20–30 min twice daily has been adopted by the ACSM as an alternative approach to exercise programming to combat overweight and obesity (Wallace 1997).

Variable intensity protocols

In line with the concept of EPOC, we recently examined whether an aerobic exercise performed at variable intensities, such as Spinning®, would produce a greater energy expenditure (Kang *et al.* 2005a). This type of exercise has gained popularity within recent years because it replicates the experience of outdoor cycling during which intensity often varies and is considered more effective in engaging exercise participants, especially when conducted to the accompaniment of music and/or visualization (Francis *et al.* 1999). In this study, we found a greater EPOC following the variable intensity – Spinning® – exercise as compared to the constant intensity exercise, despite the fact that the average intensity was kept the same. We attributed this greater EPOC to the fact that intensity was fluctuated during variable-intensity exercise, which may have disturbed homeostasis to a greater extent. This exercise arrangement was also found to be associated with a greater accumulation of blood lactic acid. However, the level of exertion during the entire workout was not perceived to be harder than the constant-intensity exercise, which makes this variable-intensity protocol more attractive to those who seek to maximize energy expenditure while participating in an exercise program. It should be noted that a variable-intensity exercise regimen differs from intermittent exercise protocols in that the former is the exercise in which there is no rest period and intensity fluctuates in a repeating pattern.

Resistance exercise

Resistance training is a potent stimulus to increasing fat-free mass, which may help in preserving lean body mass while reducing body weight and body fat. An increase in fat-free mass will also help maintain or augment the resting energy rate, which accounts for up to 60% of the total daily energy expenditure (Poehlman and Melby 1998). The resting metabolic rate is primarily related to the amount of fat-free mass. It is known that the resting metabolic rate decreases with advancing age at a rate of 2–3% per decade, and this decrease is primarily attributed to the loss of fat-free mass. Incorporating resistance training is especially important as people age, and can help counteract the age-related decrease in neuromuscular function and resting energy rate.

Resistance training also improves muscular strength and power. This adaptation is necessary in terms of athletics. It also allows an ordinary individual to be more capable of performing daily tasks such as carrying, lifting, and changing body posture. The improvement in muscular strength may not impact energy balance directly. However, it will facilitate adoption of a more active lifestyle in sedentary overweight and obese individuals (American College of Sports Medicine 2001). Some other metabolic benefits of resistance training may include improvement in blood lipid profile and increases in fat oxidation (Van Etten *et al.* 1995; Treuth *et al.* 1995).

As with other forms of exercise, resistance training increases energy expenditure during both exercise and recovery. However, resistance exercise differs from aerobic exercise in that its energetic contribution to the daily total energy expenditure is very small. Energy expenditure during a typical weight-lifting workout ranges from 100 kcal to 200 kcal, a figure that is less than half as much as what is normally achieved during a single session of aerobic exercise. On the other hand, resistance training can trigger a profound increase in post-exercise oxygen consumption due to the fact that the exercise is performed intermittently at a very high intensity. In fact, a number of studies have found that the average oxygen consumption following resistance exercise is even greater than that following aerobic exercise when both types of exercise were calculated for the total energy expenditure (Gillette *et al.* 1994; Burleson *et al.* 1998).

Similar to cardiorespiratory fitness, the resistance-training prescription should be made based on the health and fitness status and the specific goals of the individual. For weight-management purposes, the major goal of the resistance-training program is to develop sufficient muscular strength so that an individual is able to sustain a regular training routine and at the same time to live a physically independent lifestyle. In order to maximize energy expenditure, a circuit-training program may be introduced in order to allow individuals to work on multiple muscle groups (e.g., gluteals, quadriceps, hamstrings, pectorals, latissimus dorsi, deltoids, and abdominals) in one session that can last for as long as 60 min. The metabolic advantage is that this training format involves multiple sets of low intensity (e.g., 40% 1-RM) and high repetition (e.g., 10–15 repetitions) on each muscle group, coupled with a relatively shorter rest interval between sets (e.g., ~15 s). In addition to more energy being expended during the workout, this type of training also elicits greater EPOC as compared to regular strength-training programs. It is recommended that resistance training be performed at least twice per week, with at least 48 hours of rest between sessions to allow proper recuperation (American College of Sports Medicine 2006).

Limitations of exercise alone in weight management

Despite the ability of physical activity and exercise to create a negative caloric balance, the actual impact of exercise alone on weight loss is often found to be minimal (Garrow 1995; Wilmore 1995; Saris 1993). Although there is a negative association between the level of physical activity and the prevalence of obesity, and those who are physically active tend to be leaner, it remains unclear to what extent physical activity or exercise can do, even combined with dietary restriction, in treating those who are already overweight and obese. In a meta-analysis of 493 studies over a 25-year period, Miller *et al.* (1997) reported that exercise alone has a relatively minor effect on weight loss and does not add much to the weight-loss effect of a reduced diet. There are some studies that have demonstrated positive effects of exercise on weight loss (Ross *et al.* 2000). However, the exercise intervention adopted was rather vigorous and resulted in an energy deficit of 700 kcal per day, a caloric value twice as much as what is normally recommended during an exercise session. It appears that in order for physical activity and exercise to have a major impact on body weight reduction, exercise prescription should entail exercise daily with each exercise being performed at moderate to high intensity for more than an hour. Obviously, most obese individuals cannot tolerate and sustain this exercise dosage.

The weak effect of exercise alone on weight loss may be explained by the fact that when people begin exercise training, they tend to rest more after each exercise session, which negates the calories expended during exercise. One should also realize that the amount of energy expended during exercise that is suitable to sedentary or obese individuals is actually rather small. For example, the net energy cost of a three-mile brisk walk for a 70-kg (154-lb) obese woman is only about 150–160 kcal (215 total calories minus 50 calories for the BMR). Given that 1 lb of fat contains 3500 calories, it would take nearly a month of daily walking of three miles to lose 1 lb of fat if all else stays the same. Exercise by itself being ineffective in weight loss should not be attributed to the claim that people will eat more when they become more active. In fact, in those studies that failed to demonstrate the effect of added exercise, the exercise intervention was implemented under controlled dietary conditions in which all subjects, whether treated with exercise or not, were fed similarly.

It appears that in terms of exercise alone, the key question is "What is the minimal exercise dose necessary to achieve a desirable weight loss?" While this issue remains to be investigated, future studies should also be able to adequately control energy intake as well as all other components of daily energy expenditure when a planned exercise intervention is introduced. From a public health standpoint, it seems prudent to conclude that any exercise is better than none and, within the range of tolerance, more is probably better. Still, regular physical activity has proven beneficial in many aspects, irrespective of body weight control, such as improving cardiorespiratory fitness, augmenting overall feeling of well being, and reducing risk factors for developing chronic conditions such as coronary heart disease, hypertension, diabetes, and osteoporosis.

SUMMARY

- According to the set-point theory, there is a control system built into every person dictating how much fat he/she should carry – a kind of thermostat for body fat. The set point for body fatness is determined by genetics. Some individuals have a high setting, others have a low one. In an obese individual, body fat is set to remain at a higher level than it is in a lean individual.

- Body fatness is regulated through the interaction of the brain with an array of afferent signals from various body systems. Signals from the gastrointestinal tract, hormones, and circulating nutrients regulate short-term hunger and satiety. Signals such as the release of leptin from fat cells regulate long-term energy intake and expenditure.

- Leptin functions to help with body weight regulation by decreasing energy intake and increasing energy expenditure. Defects in the leptin signaling system cause the brain to assess adipose tissue status improperly and thus impairs body weight regulation. Many obese people are found to have high levels of leptin, suggesting that these individuals may have developed tissue resistance to respond to these hormones.

- The total energy expenditure on any given day can be further divided into (1) resting energy expenditure (RMR); (2) thermal effect of food; and (3) energy expenditure during physical activities. Quantitatively, RMR accounts for about 60–75% of the total daily energy expenditure. It is for this large fraction that RMR has drawn a great deal of attention with regard to its role in mediating weight gain and obesity.

- Upon completion of an exercise, VO_2 does not return to resting levels immediately, but does so rather gradually. This elevated oxygen consumption following exercise has been referred to as excess post-exercise oxygen consumption (EPOC). The EPOC represents the energy necessary to restore homeostasis, which is disrupted during the preceding exercise; its quantity is proportional to the level of exercise intensity.

- In contrast to aerobic exercise, resistance exercise is often performed in an intermittent fashion. Exertion can be very strenuous, but is usually sustained for no more than 1 min. The energy cost of an entire session of resistance training is usually lower than most aerobic exercises. However, such training can produce significant EPOC. Circuit weight training is by far the best choice for promoting energy expenditure through resistance exercise.

- The thermic effect of feeding can be divided into two subcomponents: obligatory and facultative thermogenesis. The obligatory component is the energy used for digestion, absorption, assimilation, and storage; the facultative component is additional increase in energy expenditure caused by the activation of the sympathetic nervous system.

- Of three macronutrients, consumption of protein is the most thermogenic – its energy cost can be 20–30% of the total energy intake. This is followed by carbohydrate (5–10%) and fat (0–5%). The greatest thermic effect associated with protein can be mostly ascribed to the extra energy needed for absorption by the gastrointestinal tract, as well as deamination of amino acids and synthesis of protein in the liver.

- Heredity may predispose individuals to obesity. However, environmental factors are also highly involved. When genetically susceptible individuals find themselves in an environment where food is appealing and plentiful and physical activity is easily avoided, obesity is a likely outcome. Among those environmental factors are: high calorie and fat intake; reduced energy expenditure due to technology advancement; and personal choices concerning the amount and type of food consumed.

- Treatment of overweight and obesity should be long term, similar to that for any chronic disease. It requires a firm commitment to lifestyle changes, rather than a quick fix as promoted by many popular diet books. An ideal weight-management program should provide for a reduction in energy intake, along with education about meeting nutrient needs, increasing energy expenditure, and changing the lifestyle patterns that led to weight gain.

- Studies comparing weight loss associated with low-carbohydrate diets and low-fat diets showed that at six months, a greater weight loss is achieved on low-carbohydrate diets. However, differences disappeared by 12 months. There is no compelling evidence that low-carbohydrate diets are more effective than other types of diet in helping people lose weight.

- Specific weight-management strategies include: (1) increasing physical activity to use more energy than is consumed in food; (2) make nutritional adequacy a priority and emphasize foods with a low energy density and a high nutrient density; (3) limit high-fat foods – make legumes, whole grain, vegetables, and fruits central to your diet plan; (4) eat small portions; (5) limit concentrated sweets and alcoholic beverages; (6) keep a record of diet and exercise habits; and (7) learn alternative ways to deal with emotions and stresses.

- Aerobic training is proven to be effective in weight control and management. It helps increase energy expenditure via repeated muscle contractions. It also augments fat utilization, which can be explained in part by an improved transport of oxygen and fatty acids due to increased capillary density of muscle tissue being trained.

- Exercise intensity and duration are the two important measures of exercise prescription. Unlike most fitness programs in which intensity of exercise plays an important part of exercise prescription, any program aimed at weight loss should ultimately be guided by the measure of total energy expenditure and its relation to energy consumption. In order to burn the most calories, exercise duration is considered more important, and exercise intensity must be adapted to the amount of time or calories one would like to accomplish.

- Contrary to the common myth, it appears that a relatively more vigorous exercise program is necessary in order to facilitate fat utilization. Exercise intensity that provides the maximal rate of fat oxidation was found to be about 60–65% VO_2max. This finding may be explained in part by the fact that lipolysis is intensity dependent.

- Circuit weight training allows individuals to work on 8–10 major muscle groups in one session that can last for as long as 60 min. It involves multiple sets of low intensity and high repetitions, coupled with a relatively shorter rest interval between sets. It can stimulate both muscular and cardiorespiratory systems simultaneously. Hence, it may be the choice for those who are interested in losing fat tissue while gaining muscle mass.

CASE STUDY: BALANCING ENERGY INTAKE AND ENERGY OUTPUT

Joe is unhappy about the 20 lb he gained during his first two years at college. He is 21 years old, 5' 7" inches tall, and weighs 180 lb. He would like to weigh around 160 lb. In analyzing why he gained weight, Joe realizes that he has a hectic schedule. While studying full-time, Joe also works 30 hours per week at a warehouse distribution center filling orders. During his leisure time, Joe likes to watch sports on TV, spend time with friends, and study. Joe has little time to think about what he eats. On a typical day, he stops for coffee and a pastry on his way to class in the morning, has a burger and pizza for lunch in a quick-service restaurant, and picks up fried chicken or fish at the drive-through on his way to work. He gets less exercise than he used to and often eats cookies or candy bars while studying late at night. By recording and analyzing his food intake for three days, Joe found that he takes in about 3200 kcal per day. He also calculated his estimated energy requirement just to see how this compares to his recommended intake. By keeping an activity log, he estimated that his typical day includes 30 minutes of low to moderate activity, such as brisk walking, which puts his activity level in the "low active" category.

Questions

- What is Joe's estimated energy requirement (EER)? How does his EER compare to his caloric intake? (Note: See EER equations in Chapter 8.)
- What changes could Joe make in his diet that would promote weight loss?
- What aspects of Joe's lifestyle (other than diet) are causing his weight gain? How should he change them in order to promote weight loss and maintenance?

REVIEW QUESTIONS

1. What is the set point? How does the set point function in regulating body weight?
2. Define the terms: (1) hunger; (2) satiety; (3) appetite; and (4) hypothalamus.
3. Discuss the role leptin plays in regulating body weight.
4. What is the resting metabolic rate average for men and women? List three factors that can influence resting metabolic rate and explain how each one works.
5. What is the thermic effect of feeding? How does it differ between consumption of protein, carbohydrate, and fat?
6. Explain each of the following low-calorie diets: (1) fixed-meal plan; (2) free-choice diets; (3) liquid-formula diets; and (4) very-low-calorie diets.
7. How is weight loss created by (1) high-carbohydrate, low-fat diets; and (2) low-carbohydrate, high-fat diets? Which one is more effective?
8. Explain the differences between the terms "physical activity" and "exercise."
9. What is a negative energy balance? Why is it important to achieve this caloric balance and how?
10. Define the terms: (1) intensity; (2) duration; (3) frequency; and (4) progression. How would you use these terms in establishing an exercise program for people who want to lose excess body fat?

11 How is circuit weight training carried out? What are the advantages of using this type of resistance-training program?

12 What is the best approach to weight management? Why?

SUGGESTED READING

Baile, C.A., Della-Fera, M.A., and Martin, R.J. (2000) Regulation of metabolism and body fat mass by leptin. *Annual Review of Nutrition*, 20: 105–127.

   This article discusses the role leptin plays in maintaining a stable body weight over the long term, despite the fact that a variety of environmental conditions alter short-term energy intake.

Donnelly, J.E., Blair, S.N., Jakicic, J.M., Manore, M.M., Rankin, J.W., Smith, B.K., and the American College of Sports Medicine (2009) American College of Sports Medicine position stand: appropriate physical activity intervention strategies for weight loss and prevention of weight regain for adults. *Medicine & Science in Sports & Exercise*, 41: 459–471.

   Overweight and obesity affect more than two-thirds of the adult population and are associated with a variety of chronic diseases. This position stand provides evidence-based guidelines as to how exercise prescription should be made in order to facilitate weight loss and prevent weight regain.

Hill, J.O. (2006) Understanding and addressing the epidemic of obesity: an energy balance perspective. *Endocrine Reviews*, 27: 750–761.

   This article helps readers further understand the etiology of obesity. It focuses on how behavioral and environmental factors interact to produce the positive energy balance that results in weight gain.

Paddon-Jones, D., Westman, E., Mattes, R.D., Wolfe, R.R., Astrup, A., and Westerterp-Plantenga, M. (2008) Protein, weight management, and satiety. *American Journal of Clinical Nutrition*, 87: S1558–S1561.

   Diets high in protein are popular in treating overweight and obesity. This article provides additional evidence to support the potential benefits associated with a moderately elevated protein intake.

14 Thermoregulation and fluid balance

■ KEY TERMS

- non-shivering thermogenesis
- radiation
- conduction
- convection
- evaporation
- relative humidity
- dehydration
- osmolality

- glucose–electrolyte solutions
- glucose polymer solutions
- maltodextrin
- gastric emptying
- hyperhydration
- rhabdomyolysis
- heat acclimatization

■ THERMOREGULATION AT REST AND DURING EXERCISE

In humans, normal body temperature is approximately 37°C (98.6°F). This value refers to the internal or core temperature, which is commonly measured orally and rectally. On the other hand, shell temperature, which represents the temperature of the skin and the tissues directly under it, varies considerably depending upon the surrounding environmental temperature. At rest, rectal temperature is normally 0.5–1°F higher than oral temperature; however, it was reported that following a road race, rectal temperature was 5.5°F higher than oral temperature, suggesting that an oral reading may not be an accurate reflection of the true body temperature. The body is able to maintain its core temperature by controlling the rate of heat production and the rate of heat loss. As shown in Figure 14.1, body temperature reflects a careful balance between heat production and heat loss. When out of balance, the body either gains or loses heat. The temperature control center, which is located in the hypothalamus, works like a thermostat; it can initiate an increase in heat production when body temperature falls and an increase in heat loss when body temperature rises.

Heat production

The body produces internal heat due to normal metabolic processes. At rest or during sleep, metabolic heat production is low. However, during intense exercise, heat

Figure 14.1 Factors that contribute to body temperature homeostasis.

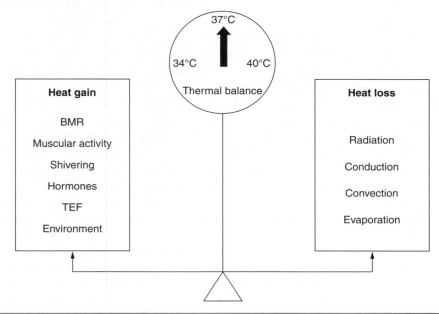

production is high. Heat production can be classified as voluntary and involuntary. Voluntary heat production is brought about by exercise, whereas involuntary heat production results from shivering or the secretion of hormones, such as thyroxine and catecholamines. Increased muscular activity during exercise causes an increase in heat production in the body because of the inefficiency of the metabolic reactions that provide energy for contraction. The body is, at most, 20–30% efficient. Therefore 70–80% of the energy expended during exercise appears as heat. The total amount of heat produced in the body depends on the intensity and duration of the exercise. A more intense exercise will produce heat at a faster rate; the longer the exercise lasts, the more total heat is produced.

During heavy exercise, this can result in a large heat load. As discussed in Chapter 11, each liter of oxygen used is equivalent to a production of 5 kcal. Therefore, it can be discerned that for every liter of oxygen consumed during exercise, approximately 16 kJ (4 kcal) of heat is produced and only about 4 kJ (1 kcal) is actually used to perform mechanical work. Therefore, for athletes consuming oxygen at a rate of $4 l min^{-1}$ during exercise, the rate of heat production in the body is about $16 kcal min^{-1}$ or $960 kcal h^{-1}$. It is estimated that at an intensity equivalent to about 80–90% $VO_2 max$, the amount of heat produced in a fit individual could potentially increase body temperature by 1°C (1.8°F) every 4–5 minutes if no changes occur in the body's heat dissipation mechanism.

A normal body temperature at rest ranges across 36–38°C (97–100°F) and may rise to 38–40°C (100–104°F) during exercise. Further increases are commonly associated with heat exhaustion and subsequently with heat stroke (both of which will be discussed later in this chapter). The elevated body temperature during exercise is not caused by resetting of the hypothalamic set point. Rather, it is caused by the temporary imbalance between the rates of heat production and heat dissipation during early stages of exercise. A more rapid response of heat dissipation mechanisms can attenuate or delay the exercise-induced rise in body temperature. This can then allow an athlete to compete at a relatively lower body temperature in a hot/warm environment. One of the heat acclimatization responses is that athletes will be able to respond to the heat more quickly upon start of exercise.

Involuntary heat production by shivering is the primary means of increasing heat production during exposure to cold. Maximal shivering can increase the body's heat production by approximately five times the resting values. In addition, release of thyroxine from the thyroid gland can also increase heat production. Thyroxine acts by increasing the metabolic rate of all cells in the body. Finally, an increase in the blood level of catecholamines (epinephrine and nor-epinephrine) can cause an increase in the rate of cellular metabolism. The increase in heat production due to the combined influences of thyroxine and catecholamines is also referred to as **non-shivering thermogenesis**.

Environmental factors, such as air temperature, relative humidity, and air movement, will also affect heat stress imposed on an active individual. Caution should be used when the air temperature is 27°C (80°F) or higher. However, if the relative humidity and solar radiation are high, lower air temperature – even 21°C (70°F) – may still pose a risk of heat stress during exercise. As the water content in air increases, the relative humidity rises. The increased humidity impairs the ability of sweat to evaporate and thus may restrict the effectiveness of the body's main cooling system when exercising. With humidity levels close to 90–100%, heat loss via evaporation nears zero. Extra caution should be used when the relative humidity exceeds 50–60%, especially when accompanied by warmer temperatures. Thermal balance can also be affected by air movement. Still air hinders heat from being carried away by convection. Even a gentle breeze may help keep body temperature near normal by moving heat away from the skin surface.

Heat dissipation

For the body to transfer heat to the environment, the heat produced in the body must have access to the outside world. The heat from deep in the body (core) is moved by the blood to the skin (the shell). Once heat nears the skin, it can be transferred to the environment by any of the four mechanisms: conduction, convection, radiation, and evaporation (Figure 14.1).

Radiation

Radiation is heat loss in the form of infrared rays. This involves the transfer of heat from the surface of one object to the surface of another, with no physical contact between them. At rest, radiation is the primary means for dissipating the body's excess heat. At normal room temperature (70–77°F/21–25°C), the nude body loses about 60% of its excess heat by radiation. This is possible because skin temperature is greater than the temperature of surrounding objects, e.g., walls, floors, furniture, etc., and a net loss of body heat occurs due to the thermal gradient. If the temperature of the surrounding objects is greater than that of the skin surface, the body will experience a net heat gain via radiation. A tremendous amount of heat is received via radiation from exposure to the sun.

Conduction

Conduction is defined as the transfer of heat from the body into the molecules of cooler object in contact with its surface. In general, the body loses only small amounts of heat due to this process. For example, heat generated deep in your body can be conducted through adjacent tissues until it reaches your body's surface. It can then be conducted to clothing or a chair that you are sitting on as long as the chair is cooler than the body surface it is in contact with. Conversely, when a hot object is pressed again the skin, heat from the object will be conducted to the skin to warm it.

Convection

Convection is a form of conductive heat loss in which heat is transmitted to either air or water molecules in contact with the body. In other words, it involves moving heat from one place to another by the motion of gas or liquid across the heated surface. Although we are not always aware of it, the air around us is in constant motion. As it circulates around us, passing over the skin, it sweeps away the air molecules that have been warmed by their contact with the skin. The greater the movement of the air or liquid, such as water, the greater the rate of heat dissipation by convection. For example, cycling at high speeds would improve convective cooling when compared to cycling at slow speeds or running. Swimming in cool water also results in convective heat loss. In fact, water's effectiveness in cooling is about 25 times greater than that of air at the same temperature.

Evaporation

In **evaporation**, heat is transferred from the body to water on the surface of the skin. When this water gains sufficient heat, it is then converted to a gas or water vapor, taking the heat away from the body. Evaporation accounts for about 80% of the total heat loss during exercise, but for only 20–25% of body heat loss at rest. Some evaporation occurs without our awareness. This is referred to as insensible water loss and happens wherever body fluid is brought into contact with the external environment, such as in the lungs, the lining of the mouth, and the skin. Insensible water loss removes about 10% of the total metabolic heat produced by the body. This heat-loss mechanism is relatively constant, so when the body needs to lose more heat, such as during exercise, it becomes insufficient. Instead, as body temperature increases, sweat production increases. As sweat reaches the skin, it is converted from a liquid to a vapor by heat from the skin. Therefore, sweat evaporation becomes increasingly important as body temperature increases. Each liter of sweat that evaporates transfers 2400 kJ (580 kcal) of heat energy from the body to the environment.

Evaporation is our major defense against over-heating during exercise; this is especially the case when the environmental temperature rises. For example, evaporation of sweat can account for as much as 70% of total heat loss when exercising in an ambient temperature of 30°C (86°F). During exercise, when body temperature rises above normal, the nervous system stimulates sweat glands to secrete sweat onto the surface of the skin. As sweat evaporates, heat is lost to the environment, which in turn lowers skin temperature. It has been estimated that the sweat rate during exercise has to be at least $1.6 \, l \, h^{-1}$ if all the heat produced is to be dissipated by evaporative heat loss alone. In fact, a much higher sweat rate ($\sim 2 \, l \, h^{-1}$) may be required because some of the sweat rolls off the skin, which has virtually no cooling effect. If we assume that $2 \, l$ of sweat are lost, then this individual would have lost $2 \, kg$ ($4.4 \, lb$) of body fluids during the 1-hour run; $1 \, l$ of sweat weighs $1 \, kg$ ($2.2 \, lb$).

Sweat rate can vary considerably between individuals. Some of the most important non-hereditary factors that determine maximal sweat rates include age, sport, climate, acclimatization, hydration status, and perhaps body fat content. A common estimate of sweating during athletic competition is $1–2 \, l \, h^{-1}$, but a rate $2–3 \, l \, h^{-1}$ or more can also occur. Also, larger individuals have higher maximal sweat rates during a given exercise task than smaller individuals.

Evaporation of sweat from the skin is dependent on three factors: (1) the temperature and relative humidity; (2) the convective current around the body; and (3) the amount of skin surface exposed to the environment (Nadel 1979). At high environmental temperatures, relative humidity is the most important factor in determining the rate of evaporative heat loss. **Relative humidity** refers to the percentage of water in ambient air at a particular temperature compared with the total quantity of moisture that air could carry. High relative humidity reduces the rate of evaporation. This is because evaporation occurs due to a vapor pressure gradient. That is, in order for evaporative cooling to occur during exercise, the vapor pressure on the skin must be greater than the vapor pressure in the air. At any given temperature, a rise in relative humidity results in increased vapor pressure. Practically speaking, this means that less evaporative cooling occurs during exercise on a hot/humid day. When this occurs, a greater sweat rate is induced, and individuals will dehydrate more rapidly. This dehydration poses further problems for the athlete because progressive dehydration impairs the ability to sweat and, consequently, to thermo-regulate. Sweat itself does not cool the skin; rather, skin cooling occurs only when sweat evaporates.

The amount of skin surface exposed to the environment is another important consideration. This factor may be best explained by using the example of American football. The football uniform and equipment present a considerable barrier to heat dissipation. This can blunt not only evaporative heat loss, but also heat loss through radiation and convection. Even with loose-fitting, porous jerseys, the wrappings, padding, helmet, and other objects of armor effectively seal off nearly 50% of the body's surface from the benefits of evaporative cooling, not to mention the heat stress from a hot artificial playing surface. Heat load can increase even further for those that are larger in size, such as offensive and defensive linemen. This is because they possess a smaller body surface area to body mass ratio, and a higher percentage of body fat than players of smaller size.

The relative contributions of evaporative heat loss during exercise of varying intensities are summarized in Table 14.1. The values are simple averages, because individual metabolic heat production varies with body size, body composition, and environmental conditions, such as air current, humidity, and sun exposure.

Table 14.1 Illustration of heat production and heat loss at rest and during exercise of varying intensities

	REST	LOW-INTENSITY EXERCISE	MODERATE-INTENSITY EXERCISE	HIGH-INTENSITY EXERCISE
Heat production (kcal hr^{-1})	80	200	350	650
Total heat loss (kcal hr)	80	200	300	450
Heat balance (kcal hr^{-1})	0	0	+50	+200
Evaporative heat loss (kcal hr^{-1})	25	90	170	350
Evaporative heat loss (%)	31	45	57	78

Role of circulation in heat dissipation

The circulatory system serves as the driving force to maintain thermal balance. In hot weather, heart rate and blood flow from the heart increase, while superficial arterial and venous blood vessels dilate to divert warm blood to the body's shell. This manifests as a flushed or reddened face on a hot day or during vigorous exercise. With extreme heat stress, as much as 25% of the cardiac output passes through the skin, raising skin temperature, which can facilitate heat loss via radiation. The down side of this circulatory adjustment is that it may hamper blood and oxygen supply to the working muscle as well as possibly reducing stroke volume and increasing heart rate. Stated simply, when exercising in hot conditions, the skin and working muscle will compete for blood.

EFFECT OF DEHYDRATION ON EXERCISE PERFORMANCE

Dehydration (or hypohydration) is defined as the excessive loss of body fluid. It reflects an imbalance in that fluid intake does not replenish water lost from the normally hydrated state. Dehydration is one of the most significant nutritional factors that can reduce physical performance. All too often endurance athletes and athletes participating in sports associated with heavy sweating, such as soccer, field hockey, lacrosse, tennis, and American football, experience at least some level of dehydration during training and competition. Other athletes such as wrestlers, boxers, body builders, and fitness competitors sometimes deliberately restrict their fluid consumption prior to competition to qualify for a lower weight category or to enhance aesthetic presentation. This latter approach is referred to as voluntary dehydration, and is generally discouraged due to its potential to cause health problem (Oppliger *et al.* 1996).

Voluntary dehydration

Athletes participating in sports such as wrestling, light-weight crew, judo, and boxing, often attempt to dehydrate in order to "make weight" and compete within a lower weight class. The term "hypohydration" is often used to describe voluntary efforts to reduce body water levels. Some of the commonly used techniques for voluntary dehydration include exercise-induced sweating, thermal-induced sweating such as the use of saunas, use of diuretics to increase urine losses, and reduced intake of fluids and foods. As mentioned earlier, reduction of body water by 1 l is equivalent to 1 kg (2.2 lb). These athletes then attempt to rehydrate between the weigh-in qualification and actual competition in order to "size up."

Much of the research with voluntary dehydration has been conducted with wrestlers. Evaluation criteria have emphasized factors such as muscular strength, power, and endurance and anaerobic activities designed to mimic wrestling. Controversies seem to exist as to the impact of voluntary dehydration on muscular strength and endurance and ultra short-term performance. Many studies conducted in this regard suggest that hypohydration, even up to levels of 8% of body weight, will not affect physical performance in events involving brief, intense muscular effort. For example, Greiew *et al.* (1998) reported that a 4% reduction in body weight had no effect on isometric muscle strength or endurance. On the other hand, Schoffstall *et al.* (2001) reported that dehydration resulting in approximately 1.5% loss of body mass adversely affected bench press 1-RM performance, but these adverse effects seem to disappear following a two-hour rest period and water consumption. The adverse effects on strength are not consistent, but anaerobic tasks lasting longer than 20 seconds have been impaired when subjects were hypohydrated. For example, Montain *et al.* (1998) observed a 15% reduction in time of repeated knee-extension to exhaustion following a 4% decrease water body weight. This reduced performance by hypohydration was attributed to a loss of potassium from muscle and higher muscle temperature during exercise.

Research studies are more consistent in demonstrating a negative impact of voluntary dehydration on aerobic, endurance performance. Dehydration induced by hypohydration practices may have a different influence on physical performance than involuntary or exercise-induced dehydration in that the effects of dehydration may be experienced at the onset of physical activity. This can result in reduced performance even in shorter-duration sports. It was reported that when runners performed 1.5-, 5-, and 10-km runs in either a well-hydrated state or a partially hypohydrated state (2% reduction in body weight as water), their running speed was significantly lower at distances of 5 km and 10 km, and a similar trend was observed during a 1.5-km run (Sawka and Pandolf 1990). Maxwell *et al.* (1999) also reported a reduction in performance in 20-s sprinting bouts separated by 100 s of rest.

In addition to a potential to reduce physical performance, hypohydration can influence the general health of individuals. As with involuntary dehydration, hypohydration-induced fluid loss can compromise thermal regulation during exercise in hot environments and causes cardiovascular complications. Deaths linked to hypohydration have been reported in wrestlers who experienced kidney and heart failure while working out in hot environments while dehydrated. Coaches and athletes should be well-educated on dangers and symptoms of severe dehydration and electrolyte imbalances. Readers are referred to the American College of Sports Medicine position stand in 1996 with regard to weight loss approach in wrestlers (Oppliger *et al.* 1996).

Involuntary or exercise-induced dehydration

Body water loss is rapid during exercise in the heat and is often not matched by an athlete's fluid consumption. This involuntary dehydration is most common during prolonged physical activity, particularly under warm, humid environmental conditions. Exercise performance is impaired when an individual is dehydrated by as little as 2% of body weight. Losses in excess of 5% of body weight can decrease the capacity for work by about 30% (Maughan 1991; Sawka and Pandolf 1990). Even in cool laboratory conditions, maximal aerobic power (VO_2max) decreases by about 5% when subjects experience fluid losses equivalent to 3% of body mass or more (Pinchan *et al.* 1988). In hot conditions, similar water deficits can cause a greater decrease in VO_2max. The dehydration-induced impairment in exercise performance is much more in hot environments than in cool conditions, which implies that altered thermoregulation due to heat is an important factor responsible for the reduced exercise performance associated with dehydration.

The negative impact of dehydration on endurance performance seems to also occur when exercise is performed at lower intensities. A study investigated the endurance capacity of eight subjects while performing treadmill walking at 25% VO_2max for 140 min in very hot and dry conditions (49°C/120°F at 20% relative humidity) (Sawka *et al.* 1985). All subjects were able to complete 140 min of walking when euhydrated and 3% dehydrated. But when dehydrated by 7%, six subjects stopped walking after an average of only 64 min. Therefore, even for relatively mild exercise, dehydration can bring about early fatigue from heat stress. In another study, this same research group had subjects walk to exhaustion at 47% VO_2max in the same environmental conditions as their previous study (Sawka *et al.* 1992). Comparisons were made between subjects who were euhydrated and those who were dehydrated to a loss of 8% of total body water. Dehydration reduced endurance time from 121 min to 55 min. In this study, it was also found that rectal temperature at exhaustion was about 0.4°C (0.7°F) lower in the dehydrated state, which suggests that dehydration may reduce the core temperature a person can tolerate.

The reasons dehydration has an adverse effect on exercise performance can be attributed mainly to factors that are cardiovascular and metabolic in origin. Fluid flux from the plasma during the early phase of endurance exercise is important to support efficient sweating. If sweating is mild or if fluid consumption is adequate, plasma volume can be stabilized during exercise. However, when sweating is heavy and is uncompensated by fluid consumption, plasma volume continues to decrease. As a result, less blood is returned to the heart, which then reduces stroke volume – the amount of the blood ejected by the heart per beat. In an attempt to maintain cardiac output, heart rate is accelerated. When plasma is further reduced, heart rate maximizes and cardiac output peaks. Cardiac output begins to decrease with further reductions in plasma volume.

Hence, a reduced maximal cardiac output (i.e., the highest pumping capacity of the heart that can be achieved during exercise) is the most likely physiological mechanism by which dehydration decreases a person's VO_2max and impairs work capacity in fatiguing exercise of an incremental nature. This is because reduced cardiac output decreases athletic performance by delivering less blood and nutrients to working muscle. Also, when blood flow to working muscle is decreased, there is reduced ability to transport heat and waste products such as lactic acid away from the muscle. In addition to its negative impact on stroke volume, decreased plasma volume also increases blood thickness (viscosity), which is a measure of internal resistance to blood flow. Also, during exercise in the heat, the opening up of the skin blood vessels reduces the proportion of the cardiac output available to the working muscles.

Dehydration can reduce blood flow to working muscle and influence substrate utilization. This may lead to an increased use of carbohydrate and reduced use of fat. The larger rise in core temperature during exercise in the dehydrated state is associated with a bigger catecholamine response, which may lead to increased rates of glycogen breakdown in the exercising muscle. Together, these responses will in turn increase the production of lactic acid and at the same time exhaust glycogen stores more quickly. Gonzalez-Alonso *et al.* (1999) evaluated cyclists riding at 60% VO_2max on two different occasions; once while they were provided adequate fluid to prevent dehydration, and another time when they were dehydrated to a reduction in body weight by 3.9%. When the cyclists were not provided fluids, blood flow to their legs was lower and glycogen breakdown and lactate content were greater during later stages of exercise in comparison to when they were provided sufficient fluids.

Other factors associated with dehydration may also contribute to a decrement in exercise performance. Disturbed fluid and electrolyte balance in the muscle cells may affect neuromuscular function and energy transformation processes, while adverse effects of hyperthermia upon mental processes may contribute to central fatigue. It has been reported that 1–2% dehydration can significantly impair cognitive function (Armstrong and Epstein 1999). Other dehydration-related gastrointestinal symptoms include nausea, vomiting, bloating, cramps, diarrhea, and bleeding, many of which could impair performance if severe enough.

REHYDRATION STRATEGIES

Adequate fluid replacement during exercise sustains the potential for evaporative cooling and helps maintain or restore plasma volume to near pre-exercise levels. As compared to dehydration, adequate hydration will help decrease fluid loss, reduce cardiovascular strain, enhance performance, and prevent some heat-related illnesses. For decades, hydration status was not recognized as important during activity. Some coaches and athletes still believe water consumption hinders performance. Today, hydration status is viewed as a crucial component of successful performance.

General rehydration guidelines

Athletes must be fully hydrated before they train or compete, because the body cannot adapt to dehydration. An adequately hydrated state can be assured by a high fluid intake in the last few days before competition. A useful check is to observe the volume and color of the urine. Voiding small volumes of dark yellow urine with a strong odor provides a qualitative indication of inadequate hydration. Well-hydrated individuals typically produce urine in large volumes, light in color, and without a strong smell. Observation of urine samples cannot be reliably used if the athlete is taking vitamin supplements, as some excreted water-soluble B vitamins add a yellowish hue to the urine. A more definite indication of hydration status is obtained by measuring urine **osmolality**. Osmolality is the measure of solute concentration of a solution. A urine osmolality of over 900 mOsmol kg indicates that the athlete is relatively dehydrated, whereas values of 100–300 mOsmol kg indicate that the athlete is well hydrated. Measuring the athlete's body weight after rising and voiding each morning may also prove useful. Some coaches require athletes to weigh-in before and after practice to monitor fluid balance. Remember, 1 kg (2.2 lb) of weight loss represents 1 l (35 fl. oz) of dehydration. A sudden drop in body mass on any given day is likely to indicate dehydration.

Athletes should be urged to rehydrate themselves because depending on their own thirst sensation to trigger water intake is not reliable. Ideally, athletes should consume enough fluids to ensure that body weight remains constant before and after exercise. Older individuals generally require a longer time to achieve rehydration after dehydration (Kenney and Chiu 2001). If rehydration were left entirely to a person's thirst, it could take several days after severe dehydration to re-establish fluid balance. Therefore, athletes should become accustomed to consuming fluid at regular intervals with and without thirst during training sessions so they do not develop discomfort during competition. American and Canadian Dietetic Associations recommend that approximately 500 ml of fluid be consumed two hours before exercise, followed by another 500 ml about 15 min before prolonged exercise. In a hot and humid environment, frequent consumption (every 15–20 min) of small amounts (120–180 ml) of fluid are recommended during exercise.

Until recently, athletes were generally encouraged to consume a volume of fluid equivalent to their sweat loss incurred during exercise to adequately rehydrate in the post-exercise recovery period. In other words, they were to consume about 1 l of fluid for every kilogram lost during exercise. However, this amount is considered insufficient because it does not take into account the urine loss that is associated with beverage consumption. In this context, it is suggested that ingestion of 150% or more of weight lost (i.e., consume 1.5 l of fluid for every kilogram lost) during recovery would be adequate to achieve a desirable hydration status following exercise (Sawka *et al.* 2007; Shirreffs and Maughan 2000). This 50% "extra" water accounts for that portion of ingested water lost in urine.

Composition of replacement fluid

In the 1960s, Robert Code, a scientist/physician working at the University of Florida, developed an oral fluid replacement that was designed for athletes to restore some of the nutrients lost in sweat. This product was eventually marketed as Gatorade and was the first of many **glucose–electrolyte solutions**. Later, **glucose polymer solutions** began to appear as sports drinks on the market. The glucose–electrolyte solutions were the first commercial fluid replacement preparations designed to replace fluid and carbohydrate. Common brands today include All-Sport, Gatorade, and PowerAde. Other than water, the major ingredients in these solutions are carbohydrates, usually in various combinations of glucose, glucose polymers, sucrose, or fructose, and some of the major electrolytes. The sugar content ranges from 5% to 10%, depending on the brand. The major electrolytes include sodium, chloride, potassium, and phosphorous. Some brands may also include other substances such as B vitamins, vitamin C, calcium, magnesium, branched-chain amino acids, caffeine, as well as artificial coloring and flavoring.

Obviously, one of the goals of researchers has been to develop a fluid that will help replace carbohydrate during exercise in the heat without sacrificing water absorption. In general, as osmolarity or solute concentration of a solution increases, fluid absorption decreases. Glucose polymer solutions are designed to provide carbohydrate while decreasing the osmotic concentration of the solution, thus helping with fluid absorption. **Maltodextrin** is a glucose polymer that exerts lesser osmotic effects compared with glucose, and thus is used in a variety of sports drinks as the source of carbohydrate. It is the main ingredient in a few commercial brands, such as Ultima. Other sports drinks combine maltodextrins with glucose, sucrose, and fructose. It is believed that sports drinks that contain maltodextrins will facilitate fluid absorption while maintaining an adequate carbohydrate supply. Table 14.2 compares the compositions of several commercially available sports drinks that are commonly chosen by athletes during training and competition.

How much carbohydrate a replacement fluid has can affect **gastric emptying**, which is another important consideration for effective rehydration. Fluid ingestion during exercise supplies exogenous fuel substrate, such as carbohydrate, as well as helping to maintain plasma volume and preventing dehydration. However, the availability of ingested fluids may be limited by the rate of gastric emptying or intestinal absorption. Gastric emptying refers to the process by which food leaves the stomach and enters the duodenum. High gastric emptying rates are advisable for sports drinks in order to maximize fluid absorption. Gastric emptying of fluids is slowed by the addition of carbohydrate or other macronutrients that increase the osmolarity of the solution ingested. Therefore, when glucose concentration in the fluid ingested

Table 14.2 Composition of commonly used carbohydrate beverages						
BEVERAGE	ENERGY (KCAL)	CHO (G)	CARBOHYDRATE SOURCES	SODIUM (MG)	POTASSIUM (MG)	CAFFEINE (MG)
Gatorade (Thirst Quencher)	50	14	Sucrose, glucose, fructose	110	25	0
Gatorade (Endurance Formula)	50	14	Sucrose, glucose, fructose	200	90	0
All Sports	60	16	HFCS	55	50	0
PowerAde	64	17	HFCS, maltodextrin	53	32	0
Accelerade (PacificHealth Laboratories)	80	14	Sucrose, fructose, maltodextrin	130	43	0
Vitamin Water	50	13	Fructose	0	0	0
Coca Cola	97	27	HFCS	33	0	23
Pepsi	100	27	HFCS, glucose	25	10	25
Mountain Dew	110	31	HFCS, orange juice concentrate	50	0	37
Orange Juice	110	27	Sucrose, fructose, glucose	15	450	0
Proper (Fitness Water)	10	3	Sucrose syrup	35	0	0

Notes: Data are gathered from product labels and sources provided by manufacturers. HFCS = high-fructose corn syrup.

increases, the rate of fluid volume delivery to the small intestine decreases, despite the fact that glucose delivery may increase.

Due to the fact that adding carbohydrate to a replacement fluid has the potential to hamper gastric emptying and thus fluid absorption, one should exercise caution in choosing a sports drink. In cool environments, where substrate provision to maintain endurance performance is more important, a concentrated solution containing large amounts of glucose is recommended. To avoid the limitation imposed by the rate of gastric emptying, the osmolarity of the beverage should be minimized by using glucose in the form of glucose polymers. The current evidence suggests the use of a 5–10% solution of multiple forms of carbohydrate, including glucose polymers and maltodextrin. When dehydration or hyperthermia is the major threat to performance, water replacement is the primary consideration. In this case, a more diluted solution that contains glucose in the form of maltodextrin is a more appropriate choice. In very prolonged exercise in the heat, which produces heavy sweat losses, such as ultra-marathons, electrolyte replacement may be essential to prevent heat injury.

Hyperhydration

Hyperhydration refers to an attempt to begin an exercise bout with a slight surplus of body water. The advantage of beginning an exercise hyperhydrated is that it will delay or eliminate the onset of hypohydration if one fails to completely replace sweat losses during exercise. This allows for a greater volume of sweat loss prior to a reduction in performance. In addition, beginning exercise with a slightly expanded plasma volume might provide a necessary buffer against detrimental reduction in plasma volume typically experienced during sustained vigorous activity. Hyperhydration may benefit athletes during prolonged exercise in a hot environment. Athletes competing in intermittently high-intensity sports for a couple of hours, with less opportunity to consume adequate fluids, might also consider this approach. The sports usually associated with hyperhydration include distance running, triathlon, soccer, and tennis. Disadvantages of hyperhydration include increased body weight, urine production, and incidence of gastrointestinal discomfort. Therefore, the decision to go forward with the hyperhydration procedure must be made based on whether the advantages of hyperhydration are considered superior to its temporary disadvantages.

Research generally supports the notion that hyperhydration reduces thermal and cardiovascular strain during exercise (Lamb and Shehata 1999). However, there is insufficient evidence to support the claim that pre-exercise hyperhydration improves exercise performance (Lamb and Shehata 1999; Sawka *et al.* 2001). Hyperhydration may increase water content in the body, but the effect was reported to be short lived. Much of the fluid overload is rapidly excreted. This could raise the question of whether water storage is still significantly higher at the onset of competition. Clearly, more research is needed to substantiate this hydration approach. Pre-exercise hyperhydration does not replace the need to continually replace fluid during exercise because during intense endurance exercise in heat, fluid loss usually outpaces fluid intake. On the other hand, hyperhydration is unnecessary when euhydration is maintained during exercise. Obviously, the hyperhydration procedure to be followed before key races must first have been thoroughly tested during specific training or low-key races.

Traditionally, the fluid of choice for hyperhydration has been water. But in the last 5–6 years, an increasing number of athletes are choosing to hyperhydrate exclusively with a glycerol solution. As discussed in previous chapters, glycerol is a three-carbon molecule formed mainly from carbohydrate and fat metabolism. Why glycerol? The kidneys are extremely efficient at rapidly excreting the excess water in water-induced hyperhydration, so the increase in total body water is typically short-lived. But when an athlete hyperhydrates with glycerol, its osmotic (soaking) property significantly reduces urine production and therefore increases water storages. It has been reported that using a glycerol solution can make the hyperhydration period last roughly twice as long as using water. Proponents of glycerol argue that because its use allows an athlete to maintain an enhanced fluid reservoir for a longer period of time, cardiovascular and thermoregulatory functions will be better preserved during exercise, thereby improving performance.

HEAT INJURIES

One of the most serious threats to the performance and health of physically active individuals is heat injuries, or heat illness. An early report indicates that heat-related injuries and illness cause 240 deaths annually, often in athletes (Barrow and Clark 1998). Heat injury is most common during exhaustive exercise in a hot, humid environment, particularly if the athlete is dehydrated. These problems affect not only highly trained athletes, but also less well-trained sports participants. In fact, those who are less well trained or overweight and poorly conditioned are more prone to heat injuries because they have less effective thermoregulation, work less economically, and use more carbohydrate for muscular work. From the perspective of health and safety, it is far easier to prevent heat injury then to remedy it. However, if one fails to pay attention to the normal signs of heat stress, such as thirst, tiredness, lightheadedness, and visual disturbances, the body's internal organs, especially the cardiovascular system, can lose their ability to further compensate, thereby triggering a series of disabling complications, some of which can have long-term consequences or be fatal.

In general, heat injuries result from chronic exposure to the combination of external heat stress and the inability to dissipate metabolically generated heat. Heat injuries, in order of increasing severity, include heat cramps, heat exhaustion, and exertional heat stroke. Though the following sections attempt to discuss each level of heat injury separately, readers must be aware that no clear-cut demarcation exists between these disorders because symptoms usually overlap. As shown in Figure 14.2, it is often the cumulative effects of multiple adverse stimuli that produce heat-related injuries.

Heat cramps

Heat cramps, the least serious of the three heat disorders, are characterized by severe involuntarily muscle spasms that occur during and after intense physical activity. They involve primarily the muscles that are most heavily used during exercise, although cramps can also occur in the muscles of the abdomen. This disorder is most likely brought on by the electrolyte losses and dehydration that accompany high rates of

Figure 14.2 Flow chart showing the causes and progression of heat injuries.

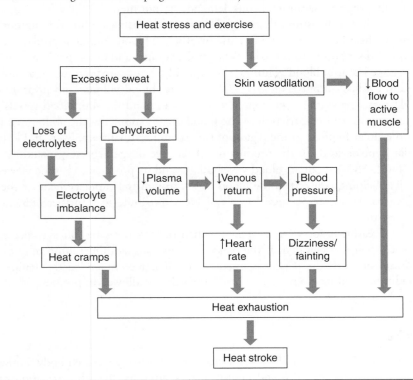

sweating (Figure 14.2). Those who often experience cramps tend to have high sweat rates and/or high sweat sodium concentrations. With heat cramps, body temperature does not necessarily increase. Heat cramps are treated by moving the stricken individual to a cooler location and administering fluids or a saline solution. One can prevent this heat-related disorder by consuming a relatively large quantity of fluids rich in electrolytes and/or increasing daily salt intake moderately (e.g., adding salt to foods at meal times) several days before heat stress is likely.

Heat exhaustion

Heat exhaustion is the most common heat illness among physically active individuals, and usually develops in those who are dehydrated, untrained, and unacclimatized. It is often seen during the first summer heat wave or first hard training session on a hot day. Heat exhaustion is typically accompanied by such symptoms as extreme fatigue, breathlessness, dizziness, vomiting, fainting, cold and clammy or hot and dry skin, low blood pressure, and a weak, rapid pulse. This more serious condition is caused by the cardiovascular system's inability to adjust, compounded by the depletion of extracellular fluid, including plasma, from excessive sweating. Blood pools in the dilated peripheral vessels. This drastically reduces the central blood volume required to maintain cardiac output. During exercise in heat, the active muscles and skin, through which excess heat is loss, compete for a share of the total blood volume. Heat exhaustion results when these simultaneous demands are not met.

In heat exhaustion, thermoregulatory mechanisms are functioning but cannot dissipate heat quickly enough because insufficient blood volume is available to allow adequate distribution to the skin. Both stroke volume and blood pressure also fall due to a reduced central blood volume. Consequently, heart rate increases in an effort to maintain cardiac output (Figure 14.2). As central blood volume and pressure continue to decrease, sympathetic nervous activity increases and the skin blood vessels constrict. A more powerful constriction of the blood vessels supplying the abdominal organs leads to cellular hypoxia in the region of the gastrointestinal tract, liver, and kidneys. Cellular hypoxia leads to the production of reactive oxygen species, including nitric oxide (NO). NO is a potent blood vessel dilator and its production can be viewed as protective, helping conserve some blood flow through the capillary beds of the abdominal organs. However, increased levels of NO contribute to a further reduction in blood pressure.

Treatment for victims of heat exhaustion involves rest in a cooler environment with their feet elevated. If the person is conscious, administration of salt water is recommended. If the person is unconscious, it is recommended that saline solution be administered intravenously. If heat exhaustion is allowed to progress, it can deteriorate to heat stroke.

Heat stroke

Heat stroke is a form of hyperthermia, an abnormally elevated body temperature with accompanying physical and neurological symptoms. It is the escalation of two other heat-related health problems: heat cramps and heat exhaustion (Figure 14.2). Unlike heat cramps and heat exhaustion, which are less severe, heat stroke is a life-threatening heat disorder that requires immediate medical attention. Heat stroke is caused by failure of the body's thermoregulatory mechanisms. With thermoregulatory failure, sweating usually ceases, the skin becomes hot and dry, body temperature rises to 40.5°C (105°F) or higher, and the circulatory system becomes excessively strained. Heat stroke is accompanied by rapid pulse and breathing, muscle cramps or weakness, and confusion or unconsciousness.

Sometimes a person experiences symptoms of heat exhaustion before progressing to heat strokes. As mentioned earlier, symptoms of heat exhaustion may include nausea, vomiting, fatigue, weakness, headache, muscle cramps and aches, and dizziness. These signs and symptoms are milder than those of heat stroke, and you can prevent heatstroke if you receive medical attention or take self-care steps as soon as you notice problems. If these symptoms are left unnoticed or untreated, these relatively milder symptoms can progress to result in heat stroke, which can be fatal from circulatory collapse, oxidative damage, systemic inflammatory response, and damage to the central nervous system. Heat stroke may also lead to **rhabdomyolysis**, a condition in which damaged tissue leaks its contents into the blood, eventually leading to kidney damage and possible death.

The populations most susceptible to heat strokes are infants, the elderly (often with associated heart diseases, lung diseases, or kidney diseases), athletes, and outdoor workers physically exerting themselves under the sun for an extended period of time. For athletes, heat stroke is a problem associated not only with extreme environmental conditions. Earlier studies have reported rectal temperatures above 40.5°C (105°F) in marathon runners who successfully competed in races conducted under moderate thermal conditions (e.g., 21°C/70°F and 30% relative humidity). Athletes who are on weight-loss supplements, such as ephedrine, are more prone to developing this more severe heat injury, as most weight-loss products impose extra strain on the cardiovascular system.

Treatment involves rapid mechanical cooling along with standard resuscitation measures. The body temperature must be lowered immediately. The victim should be moved to a cool area (e.g., indoors, or at least in the shade) and placed into the recovery position to ensure that the airway remains open. Active cooling methods may be used, and these methods include applying cool or tepid water to the skin, fanning the victim to promote evaporation of the water, and placing ice packs under armpits and the groin. Immersion in very cold water is counterproductive as it causes vasoconstriction in the skin and thereby prevents heat from escaping the body core. Hydration is of paramount importance in cooling the victim. This is achieved by drinking water if the victim is conscious. However, if the victim is unconscious or unable to tolerate oral fluids, intravenous hydration is necessary. These cooling efforts should be continued until the body temperature drops to ~38°C (100°F). Always notify emergency services immediately.

FACTORS INFLUENCING HEAT TOLERANCE

Who might be more subject to heat injuries? How can we prepare for prolonged activity in the heat? Does training in the heat make us more tolerant of thermal stress? Many studies have investigated these questions. To date, a number of predisposing factors have been identified to be associated with heat injury. These include gender, age, percentage body fat, level of fitness, previous history of heat injury, and degree of acclimatization.

Level of fitness

One of the major factors contributing to heat injuries is poor physical fitness. Using male Marine Corps recruits as subjects, Gardner *et al.* (1996) observed that the risk for developing heat illness increased with increase in time to complete a 1.5-mile run conducted during the first week of basic training. For a physically fit individual, a given physical task will represent a lower percentage of his or her VO_2max. In addition, those who are physically fit usually have augmented circulatory function that can allow a more effective heat transfer from the core to the skin. Therefore, the better the physical fitness, the better the heat tolerance to a given heat stress. However, it is a mistake to believe that athletes are immune to any heat injury. There have been reports of heat-related deaths in some elite athletes. These incidents were often related to the unsafe training practices or attempts to reduce body weight for competition through sweating in a hot environment. For example, one of the NCAA Division 1 wrestlers who died was wearing a rubber suit while riding a stationary bicycle in a steam-filled shower.

Gender

A distinct sex difference in thermoregulation exists for sweating. Women possess more heat-activated sweat glands per unit of skin area than men, yet they sweat less. Women begin to sweat at higher skin and core temperatures, which means they do not sweat as early as men during heat exposure or exercise. Despite a lower sweat output, women show a similar heat tolerance to men of equal aerobic fitness. For example, comparing male and female runners of comparable level of fitness, Arngrímsson *et al.* (2004) found that heat-induced reduction in VO_2max and physical performance were identical between men and women. This similarity may be in part attributable to the fact that, although women sweat comparatively less, they have a greater evaporative efficiency than men (Kaciuba-Uscilko and Crucza 2001). This may be because women possess a relatively large body surface area to body mass ratio, a favorable dimensional characteristic for heat dissipation. In other words, under identical conditions of heat exposure, women cool at a faster rate than men through a smaller body mass across a relatively larger surface area. The similarity in heat tolerance between men and women of equal aerobic fitness can also be ascribed to the fact that women rely more on circulatory cooling, whereas greater evaporative cooling occurs in men. This gender difference may impose a greater circulatory strain (i.e., a greater increase in heart rate) on women than men. Nevertheless, less sweat production to maintain thermal balance protects women from dehydration during exercise in the heat.

Age

Does aging impair one's ability to thermoregulate and exercise in the heat? Some of the earlier evidence suggests that the ability to exercise in the heat decreases with advancing age (Dill and Consolazio 1962). However, more recent studies reported that older and young men with a similar level of training show little difference in thermoregulation during exercise (Pandolf *et al.* 1988; Thomas *et al.* 1999). It appears that heat tolerance may not be compromised by age in healthy and physically active older individuals. A decrease in heat tolerance with age reported in some of the earlier studies seems to be due to de-conditioning and/or a lack of heat acclimatization in older subjects.

As more and more people become and remain physically active throughout middle age and advanced years, it can be expected that older individuals tolerate exercise in the heat just as well as younger adults. However, caution should still be exercised in that with advancing age, people lose their sweat glands progressively over time. In addition, Kenney and Chiu (2001) found that during exercise in a warm environment, older adults exhibit reduced thirst sensation and thus decreased voluntary fluid intake. Therefore, older individuals need to be especially diligent in following recommended hydration strategies before, during, and after exercise. Voluntary fluid consumption in the elderly can be enhanced by using electrolyte–carbohydrate solutions. Baker *et al.* (2005) found that older adults drank enough to maintain fluid balance when palatable fluid, such as carbohydrate–electrolyte solution, was readily available.

When compared to adolescents and adults, children may produce more metabolic heat during exercise relative to their body size. This is caused by the fact that children are not as metabolically efficient in performing physical activity as adults. Children don't sweat as much, although they have a greater number of heat-activated sweat glands per unit of skin area than adults (Falk *et al.* 1992). In addition, children have a reduced capacity to convey heat from the core to the skin and take a longer time to acclimatize to heat than adolescents and adults. These age differences suggest that, when exposed to environmental heat stress, children should exercise at reduced intensity and receive more time to adapt to the environment.

Level of body fat

Excess body fat negatively affects exercise performance in hot environments. Extra fat directly adds to the metabolic cost of weight-bearing activity. Extra fat also increases insulation of the body shell and thus retards heat dissipation from the body to the surroundings. Therefore, obese individuals not only generate more heat during exercise, but also have high amounts of body fat to deter heat losses. In the case of American footballers, the additional demands of equipment weight (i.e., football gear), intense competition, and a hot, humid environment compound these effects. Thus, overweight athletes, especially linemen who generally have more body fat, can experience considerable difficulty in temperature regulation and exercise performance. In the Marine Corps study cited previously, another major predictor of exertional heat illness was a high body mass index, which correlates with percentage body fat in most cases.

Heat acclimatization

Another important factor in determining an individual's response to exercise in the heat is their degree of **heat acclimatization**. Heat acclimatization refers to the process by which regular exercise in a hot environment results in a series of physiological adjustments designed to minimize disturbances in homeostasis due to heat stress. The primary adaptations that occur during heat acclimatization are an increased plasma volume, earlier onset of sweating, higher sweat rate, reduced sodium loss in sweat, reduced skin blood flow, and reduced use of muscle glycogen. These adaptations take place rather rapidly, with almost complete acclimatization achieved by 7–14 days after the first exposure to the heat. Although partial heat acclimatization can occur by training in a cool environment, athletes must exercise in a hot environment to obtain maximal heat acclimatization (Armstrong and Maresh 1991).

As stated previously, heat acclimatization results in an earlier onset of sweating. This means that sweating begins rapidly after the commencement of exercise, which translates to lower heat storage at the beginning of exercise and a lower core temperature. In addition, heat acclimatization also increases sweat capacity almost three-fold above the rate achieved prior to heat acclimatization (Yanagimoto *et al.* 2002). Therefore, much more evaporative cooling is possible. As a result, skin temperatures are lower. This then increases the temperature gradient from deep in the body to the skin and the environment. Because heat loss is facilitated via sweat, less blood must flow to the skin for body heat transfer, so more blood is available for the active muscles. In addition, the sweat produced is more dilute following training in heat, so the body's mineral stores are conserved despite an increase in sweat activity.

Heat acclimatization can result in a 10% gain in plasma volume (Gisolfi and Cohen 1979). The increase in plasma volume is due to an increase in plasma proteins and sodium content, which function to draw more fluids into the blood via osmosis. This increased plasma volume maintains central blood volume, venous return, stroke volume, and sweating capacity, and allows the body to store more heat with smaller rises in body temperature. The increased plasma volume coupled with reduced blood flow to the skin causes heart rate to increase less in response to a submaximal exercise. This smaller increase in heart rate contributes to a greater increase in stroke volume.

SUMMARY

- The body's thermostat is located in the hypothalamus. An increase in core temperature results in the hypothalamus initiating a series of physiological actions aimed at increasing heat loss. These actions include sweating and skin vasodilation.
- The body produces heat due to normal metabolic processes. At rest or during sleep, metabolic heat production is small. However, during intense exercise, heat production is large. Heat production can be classified as voluntary or involuntary. Voluntary heat production is brought about by exercise, whereas involuntary heat production results from shivering or the secretion of hormones, such as thyroxine and catecholamines.
- The heat produced in the body must have access to the outside world. The heat from deep in the body (core) is moved by the blood to the skin (the shell). Once heat nears the skin, it can be transferred to the environment by any of the four mechanisms: conduction, convection, radiation, and evaporation. Radiation is the primary means for dissipating the body's excess heat at rest, whereas evaporation functions as a major defense against over-heating during exercise.
- Evaporation of sweat from the skin is dependent on three factors: (1) the temperature and relative humidity; (2) the convective current around the body; and (3) the amount of skin surface. Warm, humid environments dramatically decrease the effectiveness of evaporative heat loss. This increases one's vulnerability to a dangerous state of dehydration.
- The circulatory system serves as the "driving force" to maintain thermal balance. In hot weather, heart rate and blood flow from the heart increases, while superficial arterial and venous blood vessels dilate to divert warm blood to the body's shell. This circulatory adjustment may hamper blood and oxygen supply to the working muscle.
- Fluid loss in excess of 2% of body mass impedes heat dissipation, compromises cardiovascular function, and diminishes exercise capacity in a hot environment. The reasons dehydration has an adverse effect on exercise performance can be attributed mainly to factors that are cardiovascular and metabolic in origin, and these factors include reduced cardiac output, increased heart rate, accelerated glycogen use, altered electrolyte balance, and impaired cognitive function.
- Rehydration based upon thirst sensation is not a reliable approach. Athletes should become accustomed to consuming fluid at regular intervals with and without thirst during training sessions. The amount to be rehydrated should match sweat loss incurred during exercise. Athletes are encouraged to ingest more than their sweat loss in order to account for urine loss associated with beverage consumption.
- Adding carbohydrate to a replacement fluid has the potential to hamper gastric emptying and thus fluid absorption. Thus, caution should be used in choosing a sports

drink. In cool environments, where substrate provision to maintain endurance performance is more important, a concentrated solution containing large amounts of glucose is recommended. When dehydration or hyperthermia is the major threat to performance, water replacement is the primary consideration. In a very prolonged exercise in the heat with heavy sweat losses, such as ultra-marathons, electrolyte replacement may be essential to prevent heat injury.

- Heat cramps, heat exhaustion, and heat stroke are the major forms of heat illness. Heat stroke represents the most serious and complex of these heat-related disorders.
- Repeated heat stress initiates thermoregulatory adjustments that improve exercise capacity. The primary adaptations that occur during heat acclimatization are an increased plasma volume, earlier onset of sweating, higher sweat rate, reduced sodium loss in sweat, reduced skin blood flow, and reduced use of muscle glycogen.
- When controlling for fitness and acclimatization levels, women and men show equal thermoregulatory efficiency during exercise. Women produce less sweat than men do and rely more on circulatory cooling. This gender difference may impose a greater circulatory strain on women than men. However, less sweat production to maintain thermal balance can protect women from dehydration during exercise in the heat.
- Aging may impair thermoregulatory function. However, this impairment appears to be fitness-related. Older and young men with a similar level of training show little difference in thermoregulation during exercise or in the ability to acclimatize to heat stress.

CASE STUDY: FIGHTING AGAINST HEAT CRAMPS

Devon is a talented 17-year-old tennis player. He trained hard, competed regularly with the best, and attained a respectable national ranking. He was accustomed to doing well in the early rounds of most tournaments. Unfortunately, after winning several matches, particularly in events held in hot weather, Devon often faced an unyielding opponent that many of his counterparts seemed to avoid: heat cramps. These debilitating muscle cramps, which primarily affected Devon's legs, occurred despite his efforts to hydrate and eat well between matches. Devon has talked to a number of trainers and physicians and tried a variety of "remedies," but none of them have worked. It wasn't until recently that Devon's physician noticed that Devon's family regularly consumed a diet fairly low in salt due to the fact that his father had high blood pressure. Subsequent laboratory testing also revealed that Devon had high rates of fluid and sodium loss via sweating during competitions. Because of these findings, Devon was urged by his physician to increase daily salt intake and to pay more attention to appropriate fluid intake in order to prevent heat cramps.

Questions

- How would you define the condition of heat cramp? What causes heat cramps?
- How is heat cramp different from other heat illnesses, such as heat exhaustion or heat stroke?
- What would you recommend Devon do to maintain an adequate sodium concentration, especially during tournaments?

REVIEW QUESTIONS

1 Discuss the role of the hypothalamus in temperature regulation.
2 List and define the four heat-dissipating mechanisms. Which of these heat-loss avenues plays the most important part during exercise in a hot/dry environment?
3 How does an increase in exercise intensity affect total heat production, evaporative heat loss, convective heat loss, and radiant heat loss?
4 Explain the role of circulation in heat dissipation. Why does heart rate usually increase to a greater extent during exercise in heat as compared to a thermo-neutral condition?
5 Your friend is going to run a marathon. The projected weather forecast is sunny, warm, and humid. What advice would you offer regarding consumption of fluid, including carbohydrate and electrolytes, before and during the race?
6 What are the physiological mechanisms responsible for dehydration reducing sports performance?
7 How do men and women differ in responding to heat stress?
8 Define heat cramp, heat exhaustion, and heat stroke. What are the major symptoms associated with each of these heat-related injuries?
9 How would you define heat acclimatization? What are the physiological adaptations that occur during heat acclimatization?

SUGGESTED READING

American College of Sports Medicine, Sawka, M.N., Burke, L.M., Eichner, E.R., Maughan, R.J., Montain, S.J., and Stachenfeld, N.S. (2007) American College of Sports Medicine position stand: exercise and fluid replacement. *Medicine & Science in Sports & Exercise*, 39: 377–390.
This position stand provides the most current and evidence-based guidance on fluid replacement to sustain appropriate hydration of individuals performing physical activity. It represents an official view of the American College of Sports Medicine toward thermoregulation and fluid replacement.

Maughan, R.J. and Shirreffs, S.M. (2010) Development of hydration strategies to optimize performance for athletes in high-intensity sports and in sports with repeated intense efforts. *Scandinavian Journal of Medicine and Science in Sports*, 20 (Suppl 2): 59–69.
Athletes should assess their hydration status and develop a personalized hydration strategy that takes account of exercise, environment, and individual needs. This review provides a comprehensive yet practical guide that can be used by competitive athletes to minimize the strain imposed by heat and dehydration.

Wendt, D., van Loon, L.J., and Lichtenbelt, W.D. (2007) Thermoregulation during exercise in the heat: strategies for maintaining health and performance. *Sports Medicine*, 37: 669–682.
During exercise, several powerful physiological mechanisms of heat loss are activated to prevent an excessive rise in body core temperature. However, a hot and humid environment can significantly add to the challenge that physical exercise imposes on the body. This article provides a thorough review of thermoregulation and hydration strategies for maintaining health and performance.

Appendix A

Metric units, English-metric conversions, and cooking measurement equivalents

▌DRY MEASURES

1 pint = 0.55 liters
1 quart = 1.10 liters
1 peck = 8.81 liters
1 bushel = 35.24 liters

▌LIQUID MEASURES

1 pint = 0.47 liters
1 quart = 0.946 liters
1 gallon = 3.79 liters

▌WEIGHT

1 ounce = 28.35 grams
1 pound = 0.45 kilograms
1 short ton = 0.91 metric ton
1 long ton = 1.02 metric tons
1000 milligrams = 1 gram
1000 grams = 1 kilogram
1000 kilograms = 1 metric ton

▌LENGTH

1 inch = 2.54 centimeters
1 yard = 0.91 meters
1 mile = 1.61 kilometers
1 kilometer = 0.6 miles
10 millimeters = 1 centimeter
10 centimeters = 1 decimeter
10 decimeters = 1 meter
1000 meters = 1 kilometer

AREA/SQUARE MEASURES

1 square inch = 6.4516 square centimeters
1 square foot = 9.29034 square decimeters
1 square yard = 0.836131 square meters
1 acre = 4046.85642 square meters
1 square mile = 2.59 square kilometers
100 square millimeters = 1 square centimeter
100 square centimeters = 1 square decimeter
100 square decimeters = 1 square meter

VOLUME/CUBIC MEASURES

1 cubic inch = 16.34 cubic centimeters
1 cubic foot = 0.28 cubic meters
1 cubic yard = 0.76 cubic meters
10 milliliters = 1 centiliter
10 centiliters = 1 deciliter
10 deciliters = 1 liter

COOKING MEASUREMENT EQUIVALENTS

16 tablespoons = 1 cup

8 tablespoons = $\frac{1}{2}$ cup

4 tablespoons = $\frac{1}{4}$ cup

2 tablespoons = $\frac{1}{8}$ cup

1 tablespoons = $\frac{1}{16}$ cup

1 tablespoons = 3 teaspoons
8 fluid ounces = 1 cup
1 pint = 2 cups
1 quart = 2 pints
4 cups = 1 quart
1 gallon = 4 quarts
16 ounces = 1 pound

Appendix B

Chemical structure of amino acids

Histidine (His)
(essential)

Tryptophan (Trp)
(essential)

Glycine (Gly)

Methionine (Met)
(essential)

Leucine (Leu)
(essential)

Alanine (Ala)

Arginine (Arg)
(essential in infancy)

Lysine (Lys)
(essential)

Proline (Pro)

Glutamic acid (Glu)

Aspartic acid (Asp)

Serine (Ser)

Phenylalanine (Phe)
(essential)

Isoleucine (Ile)
(essential)

Tyrosine (Tyr)

Glutamine (Gln)

Asparagine (Asn)

Threonine (Thr)
(essential)

Valine (Val)
(essential)

Cysteine (Cys)

Appendix C

Dietary reference intakes for energy, macronutrients, and micronutrients

The dietary reference intakes (DRIs) includes two sets of values that serve as goals for nutrient intake: recommended dietary allowance (RDA) and adequate intake (AI). The RDA reflects the average daily amount of a nutrient considered adequate to meet the needs of most healthy people. If there is insufficient evidence to determine an RDA, an AI is set.

APPENDIX C1

Dietary reference intakes (DRIs): recommended dietary allowances and adequate intakes, total water, and macro-nutrients

LIFE STAGE GROUP	TOTAL WATER[a] (L/D)	CARBOHYDRATE (G/D)	TOTAL FIBER (G/D)	FAT (G/D)	LINOLEIC ACID (G/D)	A-LINOLENIC ACID (G/D)	PROTEIN[b] (G/D)
Infants							
0–6 months	0.7*	60*	ND	31*	4.4*	0.5*	9.1*
6–12 months	0.8*	95*	ND	30*	4.6*	0.5*	11.0
Children							
1–3 years	1.3*	130	19*	ND[c]	7*	0.7*	13
4–8 years	1.7*	130	25*	ND	10*	0.9*	19
Males							
9–13 years	2.4*	130	31*	ND	12*	1.2*	34
14–18 years	3.3*	130	38*	ND	16*	1.6*	52
19–30 years	3.7*	130	38*	ND	17*	1.6*	56
31–50 years	3.7*	130	38*	ND	17*	1.6*	56
51–70 years	3.7*	130	30*	ND	14*	1.6*	56
>70 years	3.7*	130	30*	ND	14*	1.6*	56
Females							
9–13 years	2.1*	130	26*	ND	10*	1.0*	34
14–18 years	2.3*	130	26*	ND	11*	1.1*	46
19–30 years	2.7*	130	25*	ND	12*	1.1*	46
31–50 years	2.7*	130	25*	ND	12*	1.1*	46

LIFE STAGE GROUP	TOTAL WATER[a] (L/D)	CARBOHYDRATE (G/D)	TOTAL FIBER (G/D)	FAT (G/D)	LINOLEIC ACID (G/D)	A-LINOLENIC ACID (G/D)	PROTEIN[b] (G/D)
Females cont.							
51–70 years	2.7*	**130**	21*	ND	11*	1.1*	**46**
>70 years	2.7*	**130**	21*	ND	11*	1.1*	**46**
Pregnancy							
14–18 years	3.0*	**175**	28*	ND	13*	1.4*	**71**
19–30 years	3.0*	**175**	28*	ND	13*	1.4*	**71**
31–50 years	3.0*	**175**	28*	ND	13*	1.4*	**71**
Lactation							
14–18 years	3.8*	**210**	29*	ND	13*	1.3*	**71**
19–30 years	3.8*	**210**	29*	ND	13*	1.3*	**71**
31–50 years	3.8*	**210**	29*	ND	13*	1.3*	**71**

Source: *Dietary Reference Intakes for Energy, Carbohydrate, Fiber, Fat, Fatty Acids, Cholesterol, Protein, and Amino Acids* (2002/2005) and *Dietary Reference Intakes for Water, Potassium, Sodium, Chloride, and Sulfate* (2005). The report may be accessed via www.nap.edu.

Note: This table (taken from the DRI reports, see www.nap.edu) presents recommended dietary allowances (RDA) in **bold type** and adequate intakes (AI) in ordinary type followed by an asterisk (*). An RDA is the average daily dietary intake level; sufficient to meet the nutrient requirements of nearly all (97–98%) healthy individuals in a group. It is calculated from an estimated average requirement (EAR). If sufficient scientific evidence is not available to establish an EAR, and thus calculate an RDA, an AI is usually developed. For healthy breastfed infants, an AI is the mean intake. The AI for other life stage and gender groups is believed to cover the needs of all healthy individuals in the groups, but lack of data or uncertainty in the data prevent being able to specify with confidence the percentage of individuals covered by this intake.

a Total water includes all water contained in food, beverages, and drinking water.

b Based on grams of protein per kilogram of body weight for the reference body weight, e.g., for adults 0.8 g kg body weight for the reference body weight.

c Not determined.

APPENDIX C2

Dietary reference intakes (DRIs): recommended dietary allowances and adequate intakes, vitamins

LIFE STAGE GROUP	VITAMIN A (µg/d)[a]	VITAMIN C (mg/d)	VITAMIN D (µg/d)[b,c]	VITAMIN E (µg/d)[d]	VITAMIN K (µg/d)	THIAMIN (mg/d)	RIBOFLAVIN (mg/d)	NIACIN (mg/d)[e]	VITAMIN B-6 (mg/d)	FOLATE (µg/d)[f]	VITAMIN B-12 (µg/d)	PANTOTHENIC ACID (mg/d)	BIOTIN (µg/d)	CHOLINE (mg/d)[g]
Infants														
0–6 months	400*	40*	10	4*	2.0*	0.2*	0.3*	2*	0.1*	65*	0.4*	1.7*	5*	125*
6–12 months	500*	50*	10	5*	2.5*	0.3*	0.4*	4*	0.3*	80*	0.5*	1.8*	6*	150*
Children														
1–3 years	300	15	15	6	30*	0.5	0.5	6	0.5	150	0.9	2*	8*	200*
4–8 years	400	25	15	7	55*	0.6	0.6	8	0.6	200	1.2	3*	12*	250*
Males														
9–13 years	600	45	15	11	60*	0.9	0.9	12	1.0	300	1.8	4*	20*	375*
14–18 years	900	75	15	15	75*	1.2	1.3	16	1.3	400	2.4	5*	25*	550*
19–30 years	900	90	15	15	120*	1.2	1.3	16	1.3	400	2.4	5*	30*	550*
31–50 years	900	90	15	15	120*	1.2	1.3	16	1.3	400	2.4	5*	30*	550*
51–70 years	900	90	15	15	120*	1.2	1.3	16	1.3	400	2.4[h]	5*	30*	550*
>70 years	900	90	20	15	120*	1.2	1.3	16	1.3	400	2.4[h]	5*	30*	550*
Females														
9–13 years	600	45	15	11	60*	0.9	0.9	12	1.0	300	1.8	4*	20*	375*
14–18 years	700	65	15	15	75*	1.0	1.0	14	1.2	400[i]	2.4	5*	25*	400*
19–30 years	700	75	15	15	90*	1.1	1.1	14	1.3	400[i]	2.4	5*	30*	425*
31–50 years	700	75	15	15	90*	1.1	1.1	14	1.3	400[i]	2.4	5*	30*	425*
51–70 years	700	75	15	15	90*	1.1	1.1	14	1.3	400	2.4[h]	5*	30*	425*
>70 years	700	75	20	15	90*	1.1	1.1	14	1.3	400	2.4[h]	5*	30*	425*
Pregnancy														
14–18 years	750	80	15	15	75*	1.4	1.4	18	1.9	600	2.6	6*	30*	450*
19–30 years	770	85	15	15	90*	1.4	1.4	18	1.9	600	2.6	6*	30*	450*
31–50 years	770	85	15	15	90*	1.4	1.4	18	1.9	600	2.6	6*	30*	450*
Lactation														
14–18 years	1200	115	15	19	75*	1.4	1.6	17	2.0	500	2.8	7*	35*	550*
19–30 years	1300	120	15	19	90*	1.4	1.6	17	2.0	500	2.8	7*	35*	550*
31–50 years	1300	120	15	19	90*	1.4	1.6	17	2.0	500	2.8	7*	35*	550*

Source: Dietary Reference Intakes for Calcium, Phosphorous, Magnesium, Vitamin D, and Fluoride (1997); Dietary Reference Intakes for Thiamin, Riboflavin, Niacin, Vitamin B$_6$, Folate, Vitamin B$_{12}$, Pantothenic Acid, Biotin, and Choline (1998); Dietary Reference Intakes for Vitamin C, Vitamin E, Selenium, and Carotenoids (2000); Dietary Reference Intakes for Vitamin A, Vitamin K, Arsenic, Boron, Chromium, Copper, Iodine, Iron, Manganese, Molybdenum, Nickel, Silicon, Vanadium, and Zinc (2001); Dietary Reference Intakes for Water, Potassium, Sodium, Chloride, and Sulfate (2005); and Dietary Reference Intakes for Calcium and Vitamin B (2011). These reports may be accessed via www.nap.edu.

Note: This table (taken from the DRI reports, see www.nap.edu) presents recommended dietary allowances (RDA) in **bold type** and adequate intakes (AI) in ordinary type followed by an asterisk (*). An RDA is the average daily dietary intake level; sufficient to meet the nutrient requirements of nearly all (97–98%) healthy individuals in a group. It is calculated from an estimated average requirement (EAR). If sufficient scientific evidence is not available to establish an EAR, and thus calculate an RDA, an AI is usually developed. For healthy breast-fed infants, an AI is the mean intake. The AI for other life stage and gender groups is believed to cover the needs of all healthy individuals in the groups, but lack of data or uncertainty in the data prevent being able to specify with confidence the percentage of individuals covered by this intake.

a As retinol activity equivalents (RAEs). 1 RAE = 1 μg retinol, 12 μg β-carotene, 24 μg α-carotene or 24 μg β-cryptoxanthin. The RAE for dietary provitamin A carotenoids is two-fold greater than retinol equivalents (RE), whereas the RAE for preformed vitamin A is the same as RE.

b As cholecalciferol. 1 μg cholecalciferol = 40 IU vitamin D.

c Under the assumption of minimal sunlight.

d As α-tocopherol. α-tocopherol includes *RRR*-α-tocopherol, the only form of α-tocopherol that occurs naturally in foods, and the 2R-stereoisomeric forms of α-tocopherol (*RRR*-, *RSR*-, *RRS*-, and *RSS*-α-tocopherol) that occur in fortified foods and supplements. It does not include the 2S-stereoisomeric forms of α-tocopherol (*SRR*-, *SSR*-, *SRS*-, and *SSS*-α-tocopherol), also found in fortified foods and supplements.

e As niacin equivalents (NE). 1 mg of niacin = 60 mg of tryptophan; 0–6 months = preformed niacin (not NE).

f As dietary folate equivalents (DFE). DFE = 1 μg food folate = 0.6 μg of folic acid from fortified food or as a supplement consumed with food = 0.5 μg of a supplement taken on an empty stomach.

g Although AIs have been set for choline, there are few data to assess whether a dietary supply of choline is needed at all stages of the life cycle, and it may be that the choline requirement can be met by endogenous synthesis at some of these stages.

h Because 10–30% of older people may malabsorb food-bound B-12, it is advisable for those older than 50 years to meet their RDA mainly by consuming foods fortified with B-12, or a supplement containing B-12.

i In view of evidence linking folate intake with neural tube defects in the fetus, it is recommended that all women capable of becoming pregnant consume 400 μg from supplements or fortified foods in addition to intake of food folate from a varied diet.

j It is assumed that women will continue consuming 400 μg from supplements or fortified food until their pregnancy is confirmed and they enter prenatal care, which ordinarily occurs after the end of the periconceptional period – the critical time for formation of the neural tube.

APPENDIX C3

Dietary reference intakes (DRIs): recommended dietary allowances and adequate intakes, elements

LIFE STAGE GROUP	CALCIUM (mg/d)	CHROMIUM (μg/d)	COPPER (μg/d)	FLUORIDE (mg/d)	IODINE (μg/d)	IRON (mg/d)	MAGNESIUM (mg/d)	MANGANESE (mg/d)	MOLYBDENUM (μg/d)	PHOSPHORUS (mg/d)	SELENIUM (μg/d)	ZINC (mg/d)	POTASSIUM (g/d)	SODIUM (g/d)	CHLORIDE (g/d)
Infants															
0–6 months	200*	0.2*	200*	0.01*	110*	0.27*	30*	0.003*	2*	100*	15*	2*	0.4*	0.12*	0.18*
6–12 months	260*	5.5*	220*	0.5*	130*	11	75*	0.6*	3*	275*	20*	3	0.7*	0.37*	0.57*
Children															
1–3 years	700	11*	340	0.7*	90	7	80	1.2*	17	460	20	3	3.0*	1.0*	1.5*
4–8 years	1000	15*	440	1*	90	10	130	1.5*	22	500	30	5	3.8*	1.2*	1.9*
Males															
9–13 years	1300	25*	700	2*	120	8	240	1.9*	34	1250	40	8	4.5*	1.5*	2.3*
14–18 years	1300	35*	890	3*	150	11	410	2.2*	43	1250	55	11	4.7*	1.5*	2.3*
19–30 years	1000	35*	900	4*	150	8	400	2.3*	45	700	55	11	4.7*	1.5*	2.3*
31–50 years	1000	35*	900	4*	150	8	400	2.3*	45	700	55	11	4.7*	1.5*	2.3*
51–70 years	1000	30*	900	4*	150	8	400	2.3*	45	700	55	11	4.7*	1.3*	2.0*
>70 years	1200	30*	900	4*	150	8	400	2.3*	45	700	55	11	4.7*	1.2*	1.8*
Females															
9–13 years	1300	21*	700	2*	120	8	240	1.6*	34	1250	40	8	4.5*	1.5*	2.3*
14–18 years	1300	24*	890	3*	150	15	360	1.6*	43	1250	55	9	4.7*	1.5*	2.3*
19–30 years	1000	25*	900	3*	150	18	310	1.8*	45	700	55	8	4.7*	1.5*	2.3*
31–50 years	1000	25*	900	3*	150	18	320	1.8*	45	700	55	8	4.7*	1.5*	2.3*
51–70 years	1200	20*	900	3*	150	8	320	1.8*	45	700	55	8	4.7*	1.3*	2.0*
>70 years	1200	20*	900	3*	150	8	320	1.8*	45	700	55	8	4.7*	1.2*	1.8*
Pregnancy															
14–18 years	1300	29*	1000	3*	220	27	400	2.0*	50	1250	60	12	4.7*	1.5*	2.3*
19–30 years	1000	30*	1000	3*	220	27	350	2.0*	50	700	60	11	4.7*	1.5*	2.3*
31–50 years	1000	30*	1000	3*	220	27	360	2.0*	50	700	60	11	4.7*	1.5*	2.3*
Lactation															
14–18 years	1300	44*	1300	3*	290	10	360	2.6*	50	1250	70	13	5.1*	1.5*	2.3*
19–30 years	1300	45*	1300	3*	290	9	310	2.6*	50	700	70	12	5.1*	1.5*	2.3*
31–50 years	1300	45*	1300	3*	290	9	320	2.6*	50	700	70	12	5.1*	1.5*	2.3*

Source: Dietary Reference Intakes for Calcium, Phosphorous, Magnesium, Vitamin D, and Fluoride (1997); Dietary Reference Intakes for Thiamin, Riboflavin, Niacin, Vitamin B_6, Folate, Vitamin B_{12}, Pantothenic Acid, Biotin, and Choline (1998); Dietary Reference Intakes for Vitamin C, Vitamin E, Selenium, and Carotenoids (2000); and Dietary Reference Intakes for Vitamin A, Vitamin K, Arsenic, Boron, Chromium, Copper, Iodine, Iron, Manganese, Molybdenum, Nickel, Silicon, Vanadium, and Zinc (2001); Dietary Reference Intakes for Water, Potassium, Sodium, Chloride, and Sulfate (2005); and Dietary Reference Intakes for Calcium and Vitamin D (2011). These reports may be accessed via www.nap.edu.

Note: This table (taken from the DRI reports, see www.nap.edu) presents recommended dietary allowances (RDA) in **bold type** and adequate intakes (AI) in ordinary type followed by an asterisk (*). An RDA is the average daily dietary intake level; sufficient to meet the nutrient requirements of nearly all (97–98%) healthy individuals in a group. It is calculated from an estimated average requirement (EAR). If sufficient scientific evidence is not available to establish an EAR, and thus calculate an RDA, an AI is usually developed. For healthy breast-fed infants, an AI is the mean intake. The AI for other life stage and gender groups is believed to cover the needs of all healthy individuals in the groups, but lack of data or uncertainty in the data prevent being able to specify with confidence the percentage of individuals covered by this intake.

Appendix D

Estimated energy requirement calculations and physical activity values

APPENDIX D1

Estimated energy requirement calculations	
AGE GROUP	EQUATIONS FOR ESTIMATED ENERGY REQUIREMENTS (EER; kcal/d)[a]
0–3 months	$[89 \times \text{weight (kg)} - 100] + 175$
4–6 months	$[89 \times \text{weight (kg)} - 100] + 56$
7–12 months	$[89 \times \text{weight (kg)} - 100] + 22$
13–36 months	$[89 \times \text{weight (kg)} - 100] + 20$
3–8 years (male)	$88.5 - [61.9 \times \text{age (y)}] + \text{PA} \times [26.7 \times \text{weight (kg)} + 903 \times \text{height (m)}] + 20$
3–8 years (female)	$135.3 - [30.8 \times \text{age (y)}] + \text{PA} \times [10.0 \times \text{weight (kg)} + 934 \times \text{height (m)}] + 20$
9–18 years (male)	$88.5 - [61.9 \times \text{age (y)}] + \text{PA} \times [26.7 \times \text{weight (kg)} + 903 \times \text{height (m)}] + 25$
9–18 years (female)	$135.3 - [30.8 \times \text{age (y)}] + \text{PA} \times [10.0 \times \text{weight (kg)} + 934 \times \text{height (m)}] + 25$
19+ years (male)	$662 - [9.53 \times \text{age (y)}] + \text{PA} \times [15.91 \times \text{weight (kg)} + 539.6 \times \text{height (m)}]$
19+ years (female)	$354 - [6/91 \times \text{age (y)}] + \text{PA} \times [9.36 \times \text{weight (kg)} + 726 \times \text{height (m)}]$
Pregnancy	
14–18 years	
1st trimester	Adolescent EER + 0
2nd trimester	Adolescent EER + 340
3rd trimester	Adolescent EER + 452
19–50 years	
1st trimester	Adult EER + 0
2nd trimester	Adult EER + 340
3rd trimester	Adult EER + 452
Lactation	
14–18 years	
1st six months	Adolescent EER + 330
2nd six months	Adolescent EER + 400
19–50 years	
1st six months	Adult EER + 330
2nd six months	Adult EER + 400
Overweight or obese[b]	
3–18 years (male)	$114 - [50.9 \times \text{age (y)}] + \text{PA} \times [19.5 \times \text{weight (kg)} + 1161.4 \times \text{height (m)}]$
3–18 years (female)	$389 - [41.2 \times \text{age (y)}] + \text{PA} \times [15.0 \times \text{weight (kg)} + 701.6 \times \text{height (m)}]$
19+ years (male)	$1086 - [5010.1 \times \text{age (y)}] + \text{PA} \times [13.7 \times \text{weight (kg)} + 416 \times \text{height (m)}]$
19+ years (female)	$448 - [7.95 \times \text{age (y)}] + \text{PA} \times [11.4 \times \text{weight (kg)} + 619 \times \text{height (m)}]$

Notes:

a "PA" stands for "physical activity" value appropriate for the age and physiological stage. These can be found in the next table.

b Body mass index (BMI) ≥ 25 kg m²; values represent estimated total energy expenditure (TEE; kcal/d) for weight maintenance; weight loss can be achieved by a reduction in energy intake and/or an increase in energy expenditure.

APPENDIX D2

Physical activity values		
AGE GROUP (SEX)	PHYSICAL ACTIVITY LEVEL[a]	PHYSICAL ACTIVITY (PA) VALUE
3–8 years (male)	Sedentary	1.00
	Low active	1.13
	Active	1.26
	Very active	1.42
3–8 years (female)	Sedentary	1.00
	Low active	1.16
	Active	1.31
	Very active	1.56
3–18 years (overweight male)[b]	Sedentary	1.00
	Low active	1.12
	Active	1.24
	Very active	1.45
3–18 years (overweight female)[b]	Sedentary	1.00
	Low active	1.18
	Active	1.35
	Very active	1.60
9–18 years (male)	Sedentary	1.00
	Low active	1.13
	Active	1.26
	Very active	1.42
9–18 years (female)	Sedentary	1.00
	Low active	1.16
	Active	1.31
	Very active	1.56
19+ years (male)	Sedentary	1.00
	Low active	1.11
	Active	1.25
	Very active	1.48
19+ years (female)	Sedentary	1.00
	Low active	1.12
	Active	1.27
	Very active	1.45
19+ years (overweight/obese male)[b]	Sedentary	1.00
	Low active	1.12
	Active	1.29
	Very active	1.59
19+ years (overweight/obese female)[b]	Sedentary	1.00
	low active	1.16
	Active	1.27
	Very active	1.44

Source: Institute of Medicine 2005.

Notes:

a *Sedentary* activity level is characterized by no physical activity aside from that needed for independent living. *Low active* level is characterized by walking 1.5–3 miles per day at 2–4 mph (or equivalent) in addition to the light activity associated with typical day-to-day life. People who are *active* walk 3–10 miles per day at 2–4 mph (or equivalent) in addition to the light activity associated with typical day-to-day life. *Very active* individuals walk 10+ miles per day at 2–4 mph (or equivalent) in addition to the light activity associated with typical day-to-day life.

b Body mass index (BMI) $\geq 25\,kg\,m^2$.

Appendix E

Daily values used in food labels with a comparison to RDAs

APPENDIX E1

DIETARY CONSTITUENT	UNIT OF MEASUREMENT	DAILY VALUES FOR PEOPLE OVER FOUR YEARS OF AGE*	RDA FOR MALES	RDA FOR FEMALES
Total fat	g	<65	–	–
Saturated fatty acids	g	<20	–	–
Protein	g	50	56	46
Cholesterol	mg	<300	–	–
Carbohydrate	g	~300	130	130
Fiber	g	25	38	25
Vitamin A	μg	1000	900	700
Vitamin D	IU	400	200	200
Vitamin E	IU	30	22–33	22–33
Vitamin K	μg	80	120	90
Vitamin C	mg	60	90	75
Folate	μg	400	400	400
Thiamin	mg	1.5	1.2	1.1
Riboflavin	mg	1.7	1.3	1.1
Niacin	mg	20	16	14
Vitamin B-6	mg	2	1.3	1.3
Vitamin B-12	μg	6	2.4	2.4
Biotin	mg	0.3	0.03	0.03
Pantothenic acid	mg	10	5	5
Calcium	mg	1000	1000	1000
Phosphorus	mg	1000	700	700
Iodide	μg	150	150	150
Iron	mg	18	8	18
Magnesium	mg	400	400	310
Copper	mg	2	0.9	0.9
Zinc	mg	15	11	8
Sodium	mg	<2400	1500	1500
Potassium	mg	3500	4700	4700
Chloride	mg	3400	2300	2300
Manganese	mg	2	2.3	1.8
Selenium	μg	70	55	55
Chromium	μg	120	35	25
Molybdenum	μg	75	45	45

Notes: * Daily values are based on a 2000 kcal diet, with a caloric distribution of 30% from fat (one-third of this total from saturated fat), 60% from carbohydrate and 10% from protein. They are generally set at the highest nutrient recommendation in a specific age and gender category, and thus many daily values exceed current nutrient standards. Note: g = gram; mg = milligram; μg = microgram; IU = International Unit.

Appendix F

Substances and procedures banned by the IOC and NCAA

The NCAA bans the following classes of drugs:

- stimulants
- anabolic agents
- alcohol and beta blockers (banned for rifle competition only)
- diuretics and other masking agents
- street drugs
- peptide hormones and analogues
- anti-estrogens
- beta-2 agonists

DRUGS AND PROCEDURES SUBJECT TO RESTRICTIONS

- blood doping
- local anesthetics (under some conditions)
- manipulation of urine samples
- beta-2 agonists permitted only by prescription and inhalation
- caffeine if concentrations in urine exceed 15 μg ml.

STIMULANTS

Amphetamine (Adderall); caffeine (Guarana); cocaine; ephedrine; fenfluramine (Fen); methamphetamine; methylphenidate (Ritalin); phentermine (Phen); synephrine (Bitter orange).

Exceptions: phenylephrine and pseudoephedrine are not banned.

ANABOLIC AGENTS

Boldenone; clenbuterol; DHEA; nandrolone; stanozolol; testosterone; methasterone; androstenedione; norandrostenedione; methandienone; etiocholanolone; trenbolone.

These are sometimes listed as a chemical formula, such as 3,6,17-androstenetrione.

ALCOHOL AND BETA BLOCKERS (BANNED FOR RIFLE ONLY)

Alcohol; atenolol; metoprolol; nadolol; pindolol; propranolol; timolol.

DIURETICS (WATER PILLS) AND OTHER MASKING AGENTS

Bumetanide; chlorothiazide; furosemide; hydrochlorothiazide; probenecid; spironolactone (canrenone); triameterene; trichlormethiazide.

STREET DRUGS

Heroin; marijuana; tetrahydrocannabinol (THC).

PEPTIDE HORMONES AND ANALOGUES

Growth hormone (hGH); human chorionic gonadotropin (hcg); erythropoietin (epo).

ANTI-ESTROGENS

Anastrozole; tamoxifen; formestane; 3,17-dioxo-etiochol-1, 4,6-triene(ATD), etc.

BETA-2 AGONISTS

Bambuterol; formoterol; salbutamol; salmeterol.

Appendix G

Direction for conducting three-day dietary analyses

STEP 1: COMPLETING FOOD RECORDS/DIARY

1   Select one weekend day and two weekdays that are most reflective of your usual eating patterns.
2   Record one food item per line in the recording sheet (see below).
3   Include brand names whenever possible.
4   Record amounts in household measures, such as ounce, tablespoon, cups, slices, or units.
5   Read food labels and, if possible, use standard measuring tools, such as a plastic ruler, measuring cup, measuring spoons, and weighing scale.
6   Include methods that were used to prepare food items, such as fresh, frozen, stewed, fried, baked, canned, broiled, raw, or braised.
7   Record the amounts of visible fats you eat or use in cooking, such as oils, butter, salad dressing, margarine.
8   Keep your food diary current, and list foods immediately after they are eaten.
9   Do not alter your normal diet during the period you keep this diary.

STEP 2: ANALYZING NUTRIENT INTAKES USING THE COMPUTER PROGRAM "NUTRITIONCALC PLUS"

1   Getting started:
   • Upon installation, select NutritionCalc Plus from your program files.
   • Be sure to enter your name and course information in the Student ID Info box. This information will be printed at the top of each of your report pages.
   • To begin your personal diet analysis, click on the Profiles tab. Simply complete each section and move to the next tab as indicated in your assignment. Note that each tabbed section will have a drop-down information box on the top right of screen. Click on the triangle to use the drop-down to learn about the purpose of each tabbed section.
2   Creating a new profile
   • On the Profiles tab, click New.
   • Enter the following information:
      • Name – Enter name here.
      • Birth date – Enter birthday as month/day/year. For example: 12/08/1987

- Gender – Select gender from the drop-down list
- Height – Enter height in English or metric units
- Weight – Enter current weight here in pounds or kilograms.

3 Calculating weight gain/loss
- To calculate the calorie change per week or duration needed to reach your goal, click *Weight Gain/Loss*.
- The current weight you entered will be shown. Enter your goal weight. Select Calorie Change (per day) or Duration (in weeks) and enter in a value. Click the Calculate button and the program will fill in the remaining values. A safe weight change per week is less than 2 lb.

   Example: Current weight = 165. Goal weight = 145. Select Duration and enter 20 (weeks). Click Calculate. The program calculates the weight change value as 20 and the Calorie Change per day as –375.

4 Activity level
- Select Activity Level from the drop-down menu that best matches your individual activity level. For a description of each level, click the link. The Activity Level is a way to estimate the Calories needed every day to maintain the current weight.

ACTIVITY LEVEL DESCRIPTIONS

Sedentary

The sedentary activity level includes activities of daily living, without additional exercise. These activities include housework, grocery shopping, walking the dog, walking to the bus, mowing the lawn, and gardening. Unless you do at least 30 minutes per day of intentional exercise, this level is for you!

Low active

The low active activity level includes activities of daily living plus exercise that is equal to walking for 30 minutes at 4 mph every day. For an adult of average weight, this amount of exercise will burn about 120 additional Calories. Exercises with less intensity (METS*) will need to be performed longer to burn the same amount of Calories. Likewise, exercises with greater intensity can be performed for less time to achieve the same goal.

Active

The active activity level includes activities of daily living plus exercise that is equal to walking for 1 hour 45 minutes at 4 mph every day. For an adult of average weight, this amount of exercise will burn about 410 additional Calories.

Very active

The very active activity level includes activities of daily living plus exercise that is equal to walking for 4 hours 15 minutes at 4 mph every day. For an adult of average weight, this amount of exercise will burn about 1000 additional Calories.

- After entering all required information, click Save Profile.
- Note that the active profile name will appear in the Profile box in the upper right of the screen.

5 Intakes

- Intakes are made up of the food consumed in one day. Each food item will be part of a meal: breakfast, dinner, lunch, or snack. You can track intakes for as many days as you like; there is no limit.
- When the Intake page opens for the first time it will open to today's date and you can begin entering food items consumed on that day. If you wish to enter foods for a different day, simply click on the date. That day will be the active day and be highlighted in orange. After you enter more than one day's worth of intakes, the previous days with intakes will be shaded in green.
- To create a new intake click the Intake tab. Follow the instructions on the screen to search for foods to add to your intake list. To create an intake for a different date, use the 365-day calendar to select the appropriate date.
- To search for and enter a food item, type the search term or terms in the search box and click find. The results will show each item available that meets the search criteria.

SEARCH TIPS

Single-word search

Enter one word to find all items that have that word in the name.

Multiple-word search

Enter two or more words to find all items that have all the words in the name. The words can appear anywhere in the name and in any order.

Grouping

Enclosing the words in quotation marks will find all items that contain that exact phrase. For example, entering "Fried Chicken" will find all items that contain "Fried Chicken," but NOT "Chicken, Fried" or "Fried Potatoes and Chicken."

Wildcard

Type in a partial word followed by an asterisk. The results will include any item beginning with the partial word. For example, if you type in "mil*" you will get millet, milk, etc.

NOT or – (minus) search

The search will eliminate items that contain the term after NOT or –. For example, searching for Chocolate NOT milk will return items with chocolate but not milk in the title.

OR search

Typing OR between search terms will give you a search list with either of those terms in the title. For example, typing chocolate OR milk will give you all items with chocolate in their name and all items with milk in their name.

- Select the food from the search results list and click the plus (+) button to add the food to your intake.

- Click the item name link to preview the item. To add the item to your intake from the preview page, click the plus sign at the top.
- When the item has been added to your intake, select the meal and serving amount in the drop-down menu.
- Click *Save Intake* after each entry.
- Continue adding the remaining items in the same manner.
- To assist in searching, check out the Search Tips and list of Common Abbreviations used in the food database. The Fast Entry function automatically saves your last 20 food entries. Use this function to quickly compile your food list using foods you eat or drink on a regular basis. Click on Fast Entry and place check marks in the box to the left of the foods you wish to add to your intake list.

RECORDING SHEET

FOOD/METHOD OF PREPARATION	QUANTITY	MEAL	NOTE

Appendix H

Compendium of physical activities

CODE	METS	SPECIFIC ACTIVITY	EXAMPLES
01009	8.5	Bicycling	Bicycling, BMX or mountain
01010	4.0	Bicycling	Bicycling, <10 mph, leisure, to work or for pleasure (Taylor Code 115)
01015	8.0	Bicycling	Bicycling, general
01020	6.0	Bicycling	Bicycling, 10–11.9 mph, leisure, slow, light effort
01030	8.0	Bicycling	Bicycling, 12–13.9 mph, leisure, moderate effort
01040	10.0	Bicycling	Bicycling, 14–15.9 mph, racing or leisure, fast, vigorous effort
01050	12.0	Bicycling	Bicycling, 16–19 mph, racing/not drafting or >19 mph drafting, very fast, racing general
01060	16.0	Bicycling	Bicycling, >20 mph, racing, not drafting
01070	5.0	Bicycling	Unicycling
02010	7.0	Conditioning exercise	Bicycling, stationary, general
02011	3.0	Conditioning exercise	Bicycling, stationary, 50 watts, very light effort
02012	5.5	Conditioning exercise	Bicycling, stationary, 100 watts, light effort
02013	7.0	Conditioning exercise	Bicycling, stationary, 150 watts, moderate effort
02014	10.5	Conditioning exercise	Bicycling, stationary, 200 watts, vigorous effort
02015	12.5	Conditioning exercise	Bicycling, stationary, 250 watts, very vigorous effort
02020	8.0	Conditioning exercise	Calisthenics (e.g. push-ups, sit-ups, pull-ups, jumping jacks), heavy, vigorous effort
02030	3.5	Conditioning exercise	Calisthenics, home exercise, light or moderate effort, general (example: back exercises), going up and down from the floor (Taylor Code 150)
02040	8.0	Conditioning exercise	Circuit training, including some aerobic movement with minimal rest, general
02050	6.0	Conditioning exercise	Weight lifting (free weight, nautilus or universal-type), power lifting or body building, vigorous effort (Taylor Code 210)
02060	5.5	Conditioning exercise	Health-club exercise, general (Taylor Code 160)
02065	9.0	Conditioning exercise	Stair/treadmill ergometer, general
02070	7.0	Conditioning exercise	Rowing, stationary ergometer, general
02071	3.5	Conditioning exercise	Rowing, stationary, 50 watts, light effort
02072	7.0	Conditioning exercise	Rowing, stationary, 100 watts, moderate effort
02073	8.5	Conditioning exercise	Rowing, stationary, 150 watts, vigorous effort
02074	12.0	Conditioning exercise	Rowing, stationary, 200 watts, very vigorous effort

CODE	METS	SPECIFIC ACTIVITY	EXAMPLES
02080	7.0	Conditioning exercise	Ski machine, general
02090	6.0	Conditioning exercise	Slimnastics, jazzercise
02100	2.5	Conditioning exercise	Stretching, hatha yoga
02101	2.5	Conditioning exercise	Mild stretching
02110	6.0	Conditioning exercise	Teaching aerobic exercise class
02120	4.0	Conditioning exercise	Water aerobics, water calisthenics
02130	3.0	Conditioning exercise	Weight lifting (free, nautilus or universal-type), light or moderate effort, light workout, general
02135	1.0	Conditioning exercise	Whirlpool, sitting
03010	4.8	Dancing	Ballet or modern, twist, jazz, tap, jitterbug
03015	6.5	Dancing	Aerobic, general
03016	8.5	Dancing	Aerobic, step, with 6–8 inch step
03017	10.0	Dancing	Aerobic, step, with 10–12 inch step
03020	5.0	Dancing	Aerobic, low impact
03021	7.0	Dancing	Aerobic, high impact
03025	4.5	Dancing	General, Greek, Middle Eastern, hula, flamenco, belly, swing
03030	5.5	Dancing	Ballroom, fast (Taylor Code 125)
03031	4.5	Dancing	Ballroom, fast (disco, folk, square), line dancing, Irish step dancing, polka, contra, country
03040	3.0	Dancing	Ballroom, slow (e.g. waltz, foxtrot, slow dancing), samba, tango, nineteenth century, mambo, chacha
03050	5.5	Dancing	Anishinaabe jingle dancing or other traditional Native American dancing
04001	3.0	Fishing and hunting	Fishing, general
04010	4.0	Fishing and hunting	Digging worms, with shovel
04020	4.0	Fishing and hunting	Fishing from river bank and walking
04030	2.5	Fishing and hunting	Fishing from boat, sitting
04040	3.5	Fishing and hunting	Fishing from river bank, standing (Taylor Code 660)
04050	6.0	Fishing and hunting	Fishing in stream, in waders (Taylor Code 670)
04060	2.0	Fishing and hunting	Fishing, ice, sitting
04070	2.5	Fishing and hunting	Hunting, bow and arrow or crossbow
04080	6.0	Fishing and hunting	Hunting, deer, elk, large game (Taylor Code 170)
04090	2.5	Fishing and hunting	Hunting, duck, wading
04100	5.0	Fishing and hunting	Hunting, general
04110	6.0	Fishing and hunting	Hunting, pheasants or grouse (Taylor Code 680)
04120	5.0	Fishing and hunting	Hunting, rabbit, squirrel, prairie chick, raccoon, small game (Taylor Code 690)
04130	2.5	Fishing and hunting	Pistol shooting or trap shooting, standing
05010	3.3	Home activities	Carpet sweeping, sweeping floors
05020	3.0	Home activities	Cleaning, heavy or major (e.g., wash car, wash windows, clean garage), vigorous effort
05021	3.5	Home activities	Mopping
05025	2.5	Home activities	Multiple household tasks all at once, light effort
05026	3.5	Home activities	Multiple household tasks all at once, moderate effort
05027	4.0	Home activities	Multiple household tasks all at once, vigorous effort

continued

CODE	METS	SPECIFIC ACTIVITY	EXAMPLES
05030	3.0	Home activities	Cleaning, house or cabin, general
05040	2.5	Home activities	Cleaning, light (dusting, straightening up, changing linen, carrying out trash)
05041	2.3	Home activities	Wash dishes – standing or in general (not broken into stand/walk components)
05042	2.5	Home activities	Wash dishes; clearing dishes from table – walking
05043	3.5	Home activities	Vacuuming
05045	6.0	Home activities	Butchering animals
05050	2.0	Home activities	Cooking or food preparation – standing or sitting or in general (not broken into stand/walk components), manual appliances
05051	2.5	Home activities	Serving food, setting table – implied walking or standing
05052	2.5	Home activities	Cooking or food preparation – walking
05053	2.5	Home activities	Feeding animals
05055	2.5	Home activities	Putting away groceries (e.g., carrying groceries, shopping without a grocery cart), carrying packages
05056	7.5	Home activities	Carrying groceries upstairs
05057	3.0	Home activities	Cooking Indian bread on an outside stove
05060	2.3	Home activities	Food shopping with or without a grocery cart, standing or walking
05065	2.3	Home activities	Non-food shopping, standing or walking
05070	2.3	Home activities	Ironing
05080	1.5	Home activities	Sitting – knitting, sewing, wrapping (presents)
05090	2.0	Home activities	Implied standing – laundry, fold or hang clothes, put clothes in washer or dryer, packing suitcase
05095	2.3	Home activities	Implied walking – putting away clothes, gathering clothes to pack, putting away laundry
05100	2.0	Home activities	Making bed
05110	5.0	Home activities	Maple syruping/sugar bushing (including carrying buckets, carrying wood)
05120	6.0	Home activities	Moving furniture, household items, carrying boxes
05130	3.8	Home activities	Scrubbing floors, on hands and knees, scrubbing bathroom, bathtub
05140	4.0	Home activities	Sweeping garage, sidewalk or outside of house
05146	3.5	Home activities	Standing – packing/unpacking boxes, occasional lifting of household items light–moderate effort
05147	3.0	Home activities	Implied walking – putting away household items – moderate effort
05148	2.5	Home activities	Watering plants
05149	2.5	Home activities	Building a fire inside
05150	9.0	Home activities	Moving household items upstairs, carrying boxes or furniture
05160	2.0	Home activities	Standing – light (pump gas, change light bulb, etc.)
05165	3.0	Home activities	Walking – light, non-cleaning (readying to leave, shut/lock doors, close windows, etc.)
05170	2.5	Home activities	Sitting – playing with child(ren) – light, only active periods
05171	2.8	Home activities	Standing – playing with child(ren) – light, only active periods
05175	4.0	Home activities	Walk/run – playing with child(ren) – moderate, only active periods
05180	5.0	Home activities	Walk/run – playing with child(ren) – vigorous, only active periods
05181	3.0	Home activities	Carrying small children
05185	2.5	Home activities	Child care: sitting/kneeling – dressing, bathing, grooming, feeding, occasional lifting of child – light effort, general

CODE	METS	SPECIFIC ACTIVITY	EXAMPLES
05186	3.0	Home activities	Child care: standing – dressing, bathing, grooming, feeding, occasional lifting of child – light effort
05187	4.0	Home activities	Elder care, disabled adult, only active periods
05188	1.5	Home activities	Reclining with baby
05190	2.5	Home activities	Sit, playing with animals – light, only active periods
05191	2.8	Home activities	Stand, playing with animals – light, only active periods
05192	2.8	Home activities	Walk/run, playing with animals – light, only active periods
05193	4.0	Home activities	Walk/run, playing with animals – moderate, only active periods
05194	5.0	Home activities	Walk/run, playing with animals – vigorous, only active periods
05195	3.5	Home activities	Standing – bathing dog
06010	3.0	Home repair	Airplane repair
06020	4.0	Home repair	Automobile body work
06030	3.0	Home repair	Automobile repair
06040	3.0	Home repair	Carpentry, general, workshop (Taylor Code 620)
06050	6.0	Home repair	Carpentry, outside house, installing rain gutters, building a fence (Taylor Code 640)
06060	4.5	Home repair	Carpentry, finishing or refinishing cabinets or furniture
06070	7.5	Home repair	Carpentry, sawing hardwood
06080	5.0	Home repair	Caulking, chinking log cabin
06090	4.5	Home repair	Caulking, except log cabin
06100	5.0	Home repair	Cleaning gutters
06110	5.0	Home repair	Excavating garage
06120	5.0	Home repair	Hanging storm windows
06130	4.5	Home repair	Laying or removing carpet
06140	4.5	Home repair	Laying tile or linoleum, repairing appliances
06150	5.0	Home repair	Painting, outside home (Taylor Code 650)
06160	3.0	Home repair	Painting, papering, plastering, scraping, inside house, hanging sheet rock, remodeling
06165	4.5	Home repair	Painting (Taylor Code 630)
06170	3.0	Home repair	Put on and removal of tarp – sailboat
06180	6.0	Home repair	Roofing
06190	4.5	Home repair	Sanding floors with a power sander
06200	4.5	Home repair	Scraping and painting sailboat or powerboat
06210	5.0	Home repair	Spreading dirt with a shovel
06220	4.5	Home repair	Washing and waxing hull of sailboat, car, powerboat, airplane
06230	4.5	Home repair	Washing fence, painting fence
06240	3.0	Home repair	Wiring, plumbing
07010	1.0	Inactivity quiet	Lying quietly and watching television
07011	1.0	Inactivity quiet	Lying quietly, doing nothing, lying in bed awake, listening to music (not talking or reading)
07020	1.0	Inactivity quiet	Sitting quietly and watching television
07021	1.0	Inactivity quiet	Sitting quietly, sitting smoking, listening to music (not talking or reading), watching a movie in a theater
07030	0.9	Inactivity quiet	Sleeping

continued

CODE	METS	SPECIFIC ACTIVITY	EXAMPLES
07040	1.2	Inactivity quiet	Standing quietly (standing in a line)
07050	1.0	Inactivity light	Reclining – writing
07060	1.0	Inactivity light	Reclining – talking or talking on phone
07070	1.0	Inactivity light	Reclining – reading
07075	1.0	Inactivity light	Meditating
08010	5.0	Lawn and garden	Carrying, loading or stacking wood, loading/unloading or carrying lumber
08020	6.0	Lawn and garden	Chopping wood, splitting logs
08030	5.0	Lawn and garden	Clearing land, hauling branches, wheelbarrow chores
08040	5.0	Lawn and garden	Digging sandbox
08050	5.0	Lawn and garden	Digging, spading, filling garden, composting (Taylor Code 590)
08060	6.0	Lawn and garden	Gardening with heavy power tools, tilling a garden, chainsaw
08080	5.0	Lawn and garden	Laying crushed rock
08090	5.0	Lawn and garden	Laying sod
08095	5.5	Lawn and garden	Mowing lawn, general
08100	2.5	Lawn and garden	Mowing lawn, riding mower (Taylor Code 550)
08110	6.0	Lawn and garden	Mowing lawn, walk, hand mower (Taylor Code 570)
08120	5.5	Lawn and garden	Mowing lawn, walk, power mower
08125	4.5	Lawn and garden	Mowing lawn, power mower (Taylor Code 590)
08130	4.5	Lawn and garden	Operating snow blower, walking
08140	4.5	Lawn and garden	Planting seedlings, shrubs
08150	4.5	Lawn and garden	Planting trees
08160	4.3	Lawn and garden	Raking lawn
08165	4.0	Lawn and garden	Raking lawn (Taylor Code 600)
08170	4.0	Lawn and garden	Raking roof with snow rake
08180	3.0	Lawn and garden	Riding snow blower
08190	4.0	Lawn and garden	Sacking grass, leaves
08200	6.0	Lawn and garden	Shoveling snow, by hand (Taylor Code 610)
08210	4.5	Lawn and garden	Trimming shrubs or trees, manual cutter
08215	3.5	Lawn and garden	Trimming shrubs or trees, power cutter, using leaf blower, edger
08220	2.5	Lawn and garden	Walking, applying fertilizer or seeding a lawn
08230	1.5	Lawn and garden	Watering lawn or garden, standing or walking
08240	4.5	Lawn and garden	Weeding, cultivating garden (Taylor Code 580)
08245	4.0	Lawn and garden	Gardening, general
08246	3.0	Lawn and garden	Picking fruit off trees, picking fruits/vegetables, moderate effort
08250	3.0	Lawn and garden	Implied walking/standing – picking up yard, light, picking flowers or vegetables
08251	3.0	Lawn and garden	Walking, gathering gardening tools
09010	1.5	Miscellaneous	Sitting – card playing, playing board games
09020	2.3	Miscellaneous	Standing – drawing (writing), casino gambling, photocopying
09030	1.3	Miscellaneous	Sitting – reading, book, newspaper, etc.
09040	1.8	Miscellaneous	Sitting – writing, desk work, typing
09050	1.8	Miscellaneous	Standing – talking or talking on the phone
09055	1.5	Miscellaneous	Sitting – talking or talking on the phone
09060	1.8	Miscellaneous	Sitting – studying, general, including reading and/or writing
09065	1.8	Miscellaneous	Sitting – in class, general, including note-taking or class discussion

CODE	METS	SPECIFIC ACTIVITY	EXAMPLES
09070	1.8	Miscellaneous	Standing – reading
09071	2.0	Miscellaneous	Standing – miscellaneous
09075	1.5	Miscellaneous	Sitting – arts and crafts, light effort
09080	2.0	Miscellaneous	Sitting – arts and crafts, moderate effort
09085	1.8	Miscellaneous	Standing – arts and crafts, light effort
09090	3.0	Miscellaneous	Standing – arts and crafts, moderate effort
09095	3.5	Miscellaneous	Standing – arts and crafts, vigorous effort
09100	1.5	Miscellaneous	Retreat/family reunion activities involving sitting, relaxing, talking, eating
09105	2.0	Miscellaneous	Touring/traveling/vacation involving walking and riding
09110	2.5	Miscellaneous	Camping involving standing, walking, sitting, light–moderate effort
09115	1.5	Miscellaneous	Sitting at a sporting event, spectator
10010	1.8	Music playing	Accordion
10020	2.0	Music playing	Cello
10030	2.5	Music playing	Conducting
10040	4.0	Music playing	Drums
10050	2.0	Music playing	Flute (sitting)
10060	2.0	Music playing	Horn
10070	2.5	Music playing	Piano or organ
10080	3.5	Music playing	Trombone
10090	2.5	Music playing	Trumpet
10100	2.5	Music playing	Violin
10110	2.0	Music playing	Woodwind
10120	2.0	Music playing	Guitar, classical, folk (sitting)
10125	3.0	Music playing	Guitar, rock and roll band (standing)
10130	4.0	Music playing	Marching band, playing an instrument, baton twirling (walking)
10135	3.5	Music playing	Marching band, drum major (walking)
11010	4.0	Occupation	Bakery, general, moderate effort
11015	2.5	Occupation	Bakery, light effort
11020	2.3	Occupation	Bookbinding
11030	6.0	Occupation	Building road (including hauling debris, driving heavy machinery)
11035	2.0	Occupation	Building road, directing traffic (standing)
11040	3.5	Occupation	Carpentry, general
11050	8.0	Occupation	Carrying heavy loads, such as bricks
11060	8.0	Occupation	Carrying moderate loads up stairs, moving boxes (16–40 pounds)
11070	2.5	Occupation	Chambermaid, making bed (nursing)
11080	6.5	Occupation	Coal mining, drilling coal, rock
11090	6.5	Occupation	Coal mining, erecting supports
11100	6.0	Occupation	Coal mining, general
11110	7.0	Occupation	Coal mining, shoveling coal
11120	5.5	Occupation	Construction, outside, remodeling
11121	3.0	Occupation	Custodial work – buffing the floor with electric buffer
11122	2.5	Occupation	Custodial work – cleaning sink and toilet, light effort
11123	2.5	Occupation	Custodial work – dusting, light effort
11124	4.0	Occupation	Custodial work – feathering arena floor, moderate effort

continued

413

CODE	METS	SPECIFIC ACTIVITY	EXAMPLES
11125	3.5	Occupation	Custodial work – general cleaning, moderate effort
11126	3.5	Occupation	Custodial work – mopping, moderate effort
11127	3.0	Occupation	Custodial work – take out trash, moderate effort
11128	2.5	Occupation	Custodial work – vacuuming, light effort
11129	3.0	Occupation	Custodial work – vacuuming, moderate effort
11130	3.5	Occupation	Electrical work, plumbing
11140	8.0	Occupation	Farming, baling hay, cleaning barn, poultry work, vigorous effort
11150	3.5	Occupation	Farming, chasing cattle, non-strenuous (walking), moderate effort
11151	4.0	Occupation	Farming, chasing cattle or other livestock on horseback, moderate effort
11152	2.0	Occupation	Farming, chasing cattle or other livestock, driving, light effort
11160	2.5	Occupation	Farming, driving harvester, cutting hay, irrigation work
11170	2.5	Occupation	Farming, driving tractor
11180	4.0	Occupation	Farming, feeding small animals
11190	4.5	Occupation	Farming, feeding cattle, horses
11191	4.5	Occupation	Farming, hauling water for animals, general hauling water
11192	6.0	Occupation	Farming, taking care of animals (grooming, brushing, shearing sheep, assisting with birthing, medical care, branding)
11200	8.0	Occupation	Farming, forking straw bales, cleaning corral or barn, vigorous effort
11210	3.0	Occupation	Farming, milking by hand, moderate effort
11220	1.5	Occupation	Farming, milking by machine, light effort
11230	5.5	Occupation	Farming, shoveling grain, moderate effort
11240	12.0	Occupation	Fire fighter, general
11245	11.0	Occupation	Fire fighter, climbing ladder with full gear
11246	8.0	Occupation	Fire fighter, hauling hoses on ground
11250	17.0	Occupation	Forestry, axe chopping, fast
11260	5.0	Occupation	Forestry, axe chopping, slow
11270	7.0	Occupation	Forestry, barking trees
11280	11.0	Occupation	Forestry, carrying logs
11290	8.0	Occupation	Forestry, felling trees
11300	8.0	Occupation	Forestry, general
11310	5.0	Occupation	Forestry, hoeing
11320	6.0	Occupation	Forestry, planting, by hand
11330	7.0	Occupation	Forestry, sawing, by hand
11340	4.5	Occupation	Forestry, sawing, power
11350	9.0	Occupation	Forestry, trimming trees
11360	4.0	Occupation	Forestry, weeding
11370	4.5	Occupation	Furriery
11380	6.0	Occupation	Horse grooming
11390	8.0	Occupation	Horse racing, galloping
11400	6.5	Occupation	Horse racing, trotting
11410	2.6	Occupation	Horse racing, walking
11420	3.5	Occupation	Locksmith
11430	2.5	Occupation	Machine tooling, machining, working sheet metal
11440	3.0	Occupation	Machine tooling, operating lathe
11450	5.0	Occupation	Machine tooling, operating punch press

CODE	METS	SPECIFIC ACTIVITY	EXAMPLES
11460	4.0	Occupation	Machine tooling, tapping and drilling
11470	3.0	Occupation	Machine tooling, welding
11480	7.0	Occupation	Masonry, concrete
11485	4.0	Occupation	Masseur, masseuse (standing)
11490	7.5	Occupation	Moving, pushing heavy objects, 75 lb or more (desks, moving van work)
11495	12.0	Occupation	Skindiving or SCUBA diving as a frogman (Navy Seal)
11500	2.5	Occupation	Operating heavy duty equipment/automated, not driving
11510	4.5	Occupation	Orange grove work
11520	2.3	Occupation	Printing (standing)
11525	2.5	Occupation	Police, directing traffic (standing)
11526	2.0	Occupation	Police, driving a squad car (sitting)
11527	1.3	Occupation	Police, riding in a squad car (sitting)
11528	4.0	Occupation	Police, making an arrest (standing)
11530	2.5	Occupation	Shoe repair, general
11540	8.5	Occupation	Shoveling, digging ditches
11550	9.0	Occupation	Shoveling, heavy (more than 16 lb per minute)
11560	6.0	Occupation	Shoveling, light (less than 10 lb per minute)
11570	7.0	Occupation	Shoveling, moderate (10–15 lb per minute)
11580	1.5	Occupation	Sitting – light office work, general (chemistry lab work, light use of hand tools, watch repair or micro-assembly, light assembly/repair), sitting, reading, driving at work
11585	1.5	Occupation	Sitting – meetings, general, and/or with talking involved, eating at a business meeting
11590	2.5	Occupation	Sitting; moderate (heavy levers, riding mower/forklift, crane operation), teaching stretching or yoga
11600	2.3	Occupation	Standing; light (bartending, store clerk, assembling, filing, duplicating, putting up a Christmas tree), standing and talking at work, changing clothes when teaching physical education
11610	3.0	Occupation	Standing; light/moderate (assemble/repair heavy parts, welding, stocking, auto repair, pack boxes for moving, etc.), patient care (as in nursing)
11615	4.0	Occupation	Lifting items continuously, 10–20 lb, with limited walking or resting
11620	3.5	Occupation	Standing; moderate (assembling at fast rate, intermittent, lifting 50 lb, hitch/twisting ropes)
11630	4.0	Occupation	Standing; moderate/heavy (lifting more than 50 lb, masonry, painting, paper hanging)
11640	5.0	Occupation	Steel mill, fettling
11650	5.5	Occupation	Steel mill, forging
11660	8.0	Occupation	Steel mill, hand rolling
11670	8.0	Occupation	Steel mill, merchant mill rolling
11680	11.0	Occupation	Steel mill, removing slag
11690	7.5	Occupation	Steel mill, tending furnace
11700	5.5	Occupation	Steel mill, tipping molds
11710	8.0	Occupation	Steel mill, working in general
11720	2.5	Occupation	Tailoring, cutting

continued

CODE	METS	SPECIFIC ACTIVITY	EXAMPLES
11730	2.5	Occupation	Tailoring, general
11740	2.0	Occupation	Tailoring, hand sewing
11750	2.5	Occupation	Tailoring, machine sewing
11760	4.0	Occupation	Tailoring, pressing
11765	3.5	Occupation	Tailoring, weaving
11766	6.5	Occupation	Truck driving, loading and unloading truck (standing)
11770	1.5	Occupation	Typing, electric, manual or computer
11780	6.0	Occupation	Using heavy power tools such as pneumatic tools (jackhammers, drills, etc.)
11790	8.0	Occupation	Using heavy tools (not power) such as shovel, pick, tunnel bar, spade
11791	2.0	Occupation	Walking on job, less than 2.0 mph (in office or lab area), very slow
11792	3.3	Occupation	Walking on job, 3.0 mph, in office, moderate speed, not carrying anything
11793	3.8	Occupation	Walking on job, 3.5 mph, in office, brisk speed, not carrying anything
11795	3.0	Occupation	Walking, 2.5 mph, slowly and carrying light objects less than 25 lb
11796	3.0	Occupation	Walking, gathering things at work, ready to leave
11800	4.0	Occupation	Walking, 3.0 mph, moderately and carrying light objects less than 25 lb
11805	4.0	Occupation	Walking, pushing a wheelchair
11810	4.5	Occupation	Walking, 3.5 mph, briskly and carrying objects less than 25 lb
11820	5.0	Occupation	Walking or walk downstairs or standing, carrying objects about 25–49 lb
11830	6.5	Occupation	Walking or walk downstairs or standing, carrying objects about 50–74 lb
11840	7.5	Occupation	Walking or walk downstairs or standing, carrying objects about 75–99 lb
11850	8.5	Occupation	Walking or walk downstairs or standing, carrying objects about 100 lb or over
11870	3.0	Occupation	Working in scene shop, theater actor, backstage employee
11875	4.0	Occupation	Teaching physical education, exercise, sports classes (non-sport play)
11876	6.5	Occupation	Teaching physical education, exercise, sports classes (participate in the class)
12010	6.0	Running	Jog/walk combination (jogging component of less than 10 minutes) (Taylor Code 180)
12020	7.0	Running	Jogging, general
12025	8.0	Running	Jogging, in place
12027	4.5	Running	Jogging on a mini-tramp
12030	8.0	Running	Running, 5 mph (12 min/mile)
12040	9.0	Running	Running, 5.2 mph (11.5 min/mile)
12050	10.0	Running	Running, 6 mph (10 min/mile)
12060	11.0	Running	Running, 6.7 mph (9 min/mile)
12070	11.5	Running	Running, 7 mph (8.5 min/mile)
12080	12.5	Running	Running, 7.5 mph (8 min/mile)
12090	13.5	Running	Running, 8 mph (7.5 min/mile)
12100	14.0	Running	Running, 8.6 mph (7 min/mile)
12110	15.0	Running	Running, 9 mph (6.5 min/mile)
12120	16.0	Running	Running, 10 mph (6 min/mile)
12130	18.0	Running	Running, 10.9 mph (5.5 min/mile)
12140	9.0	Running	Running, cross country
12150	8.0	Running	Running (Taylor Code 200)
12170	15.0	Running	Running, up stairs
12180	10.0	Running	Running, on a track, team practice

CODE	METS	SPECIFIC ACTIVITY	EXAMPLES
12190	8.0	Running	Running, training, pushing a wheelchair
13000	2.0	Self care	Standing – getting ready for bed, in general
13009	1.0	Self care	Sitting on toilet
13010	1.5	Self care	Bathing (sitting)
13020	2.0	Self care	Dressing, undressing (standing or sitting)
13030	1.5	Self care	Eating (sitting)
13035	2.0	Self care	Talking and eating or eating only (standing)
13036	1.0	Self care	Taking medication, sitting or standing
13040	2.0	Self care	Grooming (washing, shaving, brushing teeth, urinating, washing hands, putting on make-up), sitting or standing
13045	2.5	Self care	Hairstyling
13046	1.0	Self care	Having hair or nails done by someone else, sitting
13050	2.0	Self care	Showering, toweling off (standing)
14010	1.5	Sexual activity	Active, vigorous effort
14020	1.3	Sexual activity	General, moderate effort
14030	1.0	Sexual activity	Passive, light effort, kissing, hugging
15010	3.5	Sports	Archery (non-hunting)
15020	7.0	Sports	Badminton, competitive (Taylor Code 450)
15030	4.5	Sports	Badminton, social singles and doubles, general
15040	8.0	Sports	Basketball, game (Taylor Code 490)
15050	6.0	Sports	Basketball, non-game, general (Taylor Code 480)
15060	7.0	Sports	Basketball, officiating (Taylor Code 500)
15070	4.5	Sports	Basketball, shooting baskets
15075	6.5	Sports	Basketball, wheelchair
15080	2.5	Sports	Billiards
15090	3.0	Sports	Bowling (Taylor Code 390)
15100	12.0	Sports	Boxing, in ring, general
15110	6.0	Sports	Boxing, punching bag
15120	9.0	Sports	Boxing, sparring
15130	7.0	Sports	Broomball
15135	5.0	Sports	Children's games (hopscotch, 4-square, dodge ball, playground apparatus, t-ball, tetherball, marbles, jacks, acrace games)
15140	4.0	Sports	Coaching: football, soccer, basketball, baseball, swimming, etc.
15150	5.0	Sports	Cricket (batting, bowling)
15160	2.5	Sports	Croquet
15170	4.0	Sports	Curling
15180	2.5	Sports	Darts, wall or lawn
15190	6.0	Sports	Drag racing, pushing or driving a car
15200	6.0	Sports	Fencing
15210	9.0	Sports	Football, competitive
15230	8.0	Sports	Football, touch, flag, general (Taylor Code 510)
15235	2.5	Sports	Football or baseball, playing catch
15240	3.0	Sports	Frisbee playing, general
15250	8.0	Sports	Frisbee, ultimate

continued

CODE	METS	SPECIFIC ACTIVITY	EXAMPLES
15255	4.5	Sports	Golf, general
15265	4.5	Sports	Golf, walking and carrying clubs (see footnote at end of Ainsworth *et al.* 2000)
15270	3.0	Sports	Golf, miniature, driving range
15285	4.3	Sports	Golf, walking and pulling clubs (see footnote at end of Ainsworth *et al.* 2000)
15290	3.5	Sports	Golf, using power cart (Taylor Code 070)
15300	4.0	Sports	Gymnastics, general
15310	4.0	Sports	Hacky sack
15320	12.0	Sports	Handball, general (Taylor Code 520)
15330	8.0	Sports	Handball, team
15340	3.5	Sports	Hand gliding
15350	8.0	Sports	Hockey, field
15360	8.0	Sports	Hockey, ice
15370	4.0	Sports	Horseback riding, general
15380	3.5	Sports	Horseback riding, saddling horse, grooming horse
15390	6.5	Sports	Horseback riding, trotting
15400	2.5	Sports	Horseback riding, walking
15410	3.0	Sports	Horseshoe pitching, quoits
15420	12.0	Sports	Jai alai
15430	10.0	Sports	Judo, jujitsu, karate, kick boxing, tae kwan do
15440	4.0	Sports	Juggling
15450	7.0	Sports	Kickball
15460	8.0	Sports	Lacrosse
15470	4.0	Sports	Motor-cross
15480	9.0	Sports	Orienteering
15490	10.0	Sports	Paddleball, competitive
15500	6.0	Sports	Paddleball, casual, general (Taylor Code 460)
15510	8.0	Sports	Polo
15520	10.0	Sports	Racquetball, competitive
15530	7.0	Sports	Racquetball, casual, general (Taylor Code 470)
15535	11.0	Sports	Rock climbing, ascending rock
15540	8.0	Sports	Rock climbing, rappelling
15550	12.0	Sports	Rope jumping, fast
15551	10.0	Sports	Rope jumping, moderate, general
15552	8.0	Sports	Rope jumping, slow
15560	10.0	Sports	Rugby
15570	3.0	Sports	Shuffleboard, lawn bowling
15580	5.0	Sports	Skateboarding
15590	7.0	Sports	Skating, roller (Taylor Code 360)
15591	12.5	Sports	Roller blading (in-line skating)
15600	3.5	Sports	Sky diving
15605	10.0	Sports	Soccer, competitive
15610	7.0	Sports	Soccer, casual, general (Taylor Code 540)
15620	5.0	Sports	Softball or baseball, fast or slow pitch, general (Taylor Code 440)

CODE	METS	SPECIFIC ACTIVITY	EXAMPLES
15630	4.0	Sports	Softball, officiating
15640	6.0	Sports	Softball, pitching
15650	12.0	Sports	Squash (Taylor Code 530)
15660	4.0	Sports	Table tennis, ping pong (Taylor Code 410)
15670	4.0	Sports	Tai chi
15675	7.0	Sports	Tennis, general
15680	6.0	Sports	Tennis, doubles (Taylor Code 430)
15685	5.0	Sports	Tennis, doubles
15690	8.0	Sports	Tennis, singles (Taylor Code 420)
15700	3.5	Sports	Trampoline
15710	4.0	Sports	Volleyball (Taylor Code 400)
15711	8.0	Sports	Volleyball, competitive, in gymnasium
15720	3.0	Sports	Volleyball, non-competitive, 6–9 member team, general
15725	8.0	Sports	Volleyball, beach
15730	6.0	Sports	Wrestling (one match = 5 minutes)
15731	7.0	Sports	Wallyball, general
15732	4.0	Sports	Track and field (shot, discus, hammer throw)
15733	6.0	Sports	Track and field (high jump, long jump, triple jump, javelin, pole vault)
15734	10.0	Sports	Track and field (steeplechase, hurdles)
16010	2.0	Transportation	Automobile or light truck (not a semi) driving
16015	1.0	Transportation	Riding in a car or truck
16016	1.0	Transportation	Riding in a bus
16020	2.0	Transportation	Flying airplane
16030	2.5	Transportation	Motor scooter, motorcycle
16040	6.0	Transportation	Pushing plane in and out of hangar
16050	3.0	Transportation	Driving heavy truck, tractor, bus
17010	7.0	Walking	Backpacking (Taylor Code 050)
17020	3.5	Walking	Carrying an infant or 15 lb load (e.g. suitcase), level ground or downstairs
17025	9.0	Walking	Carrying load upstairs, general
17026	5.0	Walking	Carrying 1–15 lb load, upstairs
17027	6.0	Walking	Carrying 16–24 lb load, upstairs
17028	8.0	Walking	Carrying 25–49 lb load, upstairs
17029	10.0	Walking	Carrying 50–74 lb load, upstairs
17030	12.0	Walking	Carrying 74+ lb load, upstairs
17031	3.0	Walking	Loading/unloading a car
17035	7.0	Walking	Climbing hills with 0–9 lb pound load
17040	7.5	Walking	Climbing hills with 10–20 lb load
17050	8.0	Walking	Climbing hills with 21–42 lb load
17060	9.0	Walking	Climbing hills with 42+ lb load
17070	3.0	Walking	Downstairs
17080	6.0	Walking	Hiking, cross country (Taylor Code 040)
17085	2.5	Walking	Bird watching
17090	6.5	Walking	Marching, rapidly, military
17100	2.5	Walking	Pushing or pulling stroller with child or walking with children

continued

CODE	METS	SPECIFIC ACTIVITY	EXAMPLES
17105	4.0	Walking	Pushing a wheelchair, non-occupational setting
17110	6.5	Walking	Race walking
17120	8.0	Walking	Rock or mountain climbing (Taylor Code 060)
17130	8.0	Walking	Up stairs, using or climbing up ladder (Taylor Code 030)
17140	5.0	Walking	Using crutches
17150	2.0	Walking	Walking, household
17151	2.0	Walking	Walking, less than 2.0 mph, level ground, strolling, very slow
17152	2.5	Walking	Walking, 2.0 mph, level, slow pace, firm surface
17160	3.5	Walking	Walking for pleasure (Taylor Code 010)
17161	2.5	Walking	Walking from house to car or bus, from car or bus to go places, from car or bus to and from the workplace
17162	2.5	Walking	Walking to neighbor's house or family's house for social reasons
17165	3.0	Walking	Walking the dog
17170	3.0	Walking	Walking, 2.5 mph, firm surface
17180	2.8	Walking	Walking, 2.5 mph, downhill
17190	3.3	Walking	Walking, 3.0 mph, level, moderate pace, firm surface
17200	3.8	Walking	Walking, 3.5 mph, level, brisk, firm surface, walking for exercise
17210	6.0	Walking	Walking, 3.5 mph, uphill
17220	5.0	Walking	Walking, 4.0 mph, level, firm surface, very brisk pace
17230	6.3	Walking	Walking, 4.5 mph, level, firm surface, very, very brisk
17231	8.0	Walking	Walking, 5.0 mph
17250	3.5	Walking	Walking, for pleasure, work break
17260	5.0	Walking	Walking, grass track
17270	4.0	Walking	Walking, to work or class (Taylor Code 015)
17280	2.5	Walking	Walking to and from an outhouse
18010	2.5	Water activities	Boating, power
18020	4.0	Water activities	Canoeing, on camping trip (Taylor Code 270)
18025	3.3	Water activities	Canoeing, harvesting wild rice, knocking rice off the stalks
18030	7.0	Water activities	Canoeing, portaging
18040	3.0	Water activities	Canoeing, rowing, 2.0–3.9 mph, light effort
18050	7.0	Water activities	Canoeing, rowing, 4.0–5.9 mph, moderate effort
18060	12.0	Water activities	Canoeing, rowing, >6 mph, vigorous effort
18070	3.5	Water activities	Canoeing, rowing, for pleasure, general (Taylor Code 250)
18080	12.0	Water activities	Canoeing, rowing, in competition, or crew or sculling (Taylor Code 260)
18090	3.0	Water activities	Diving, springboard or platform
18100	5.0	Water activities	Kayaking
18110	4.0	Water activities	Paddle boat
18120	3.0	Water activities	Sailing, boat and board sailing, windsurfing, ice sailing, general (Taylor Code 235)
18130	5.0	Water activities	Sailing, in competition
18140	3.0	Water activities	Sailing, Sunfish/Laser/Hobby Cat, Keel boats, ocean sailing, yachting
18150	6.0	Water activities	Skiing, water (Taylor Code 220)
18160	7.0	Water activities	Skimobiling
18180	16.0	Water activities	Skindiving, fast
18190	12.5	Water activities	Skindiving, moderate

CODE	METS	SPECIFIC ACTIVITY	EXAMPLES
18200	7.0	Water activities	Skindiving, scuba diving, general (Taylor Code 310)
18210	5.0	Water activities	Snorkeling (Taylor Code 320)
18220	3.0	Water activities	Surfing, body or board
18230	10.0	Water activities	Swimming laps, freestyle, fast, vigorous effort
18240	7.0	Water activities	Swimming laps, freestyle, slow, moderate or light effort
18250	7.0	Water activities	Swimming, backstroke, general
18260	10.0	Water activities	Swimming, breaststroke, general
18270	11.0	Water activities	Swimming, butterfly, general
18280	11.0	Water activities	Swimming, crawl, fast (75 yards/minute), vigorous effort
18290	8.0	Water activities	Swimming, crawl, slow (50 yards/minute), moderate or light effort
18300	6.0	Water activities	Swimming, lake, ocean, river (Taylor Codes 280, 295)
18310	6.0	Water activities	Swimming, leisurely, not lap swimming, general
18320	8.0	Water activities	Swimming, sidestroke, general
18330	8.0	Water activities	Swimming, synchronized
18340	10.0	Water activities	Swimming, treading water, fast vigorous effort
18350	4.0	Water activities	Swimming, treading water, moderate effort, general
18355	4.0	Water activities	Water aerobics, water calisthenics
18360	10.0	Water activities	Water polo
18365	3.0	Water activities	Water volleyball
18366	8.0	Water activities	Water jogging
18370	5.0	Water activities	Whitewater rafting, kayaking, or canoeing
19010	6.0	Winter activities	Moving ice house (set up/drill holes, etc.)
19020	5.5	Winter activities	Skating, ice, 9 mph or less
19030	7.0	Winter activities	Skating, ice, general (Taylor Code 360)
19040	9.0	Winter activities	Skating, ice, rapidly, more than 9 mph
19050	15.0	Winter activities	Skating, speed, competitive
19060	7.0	Winter activities	Ski jumping (climb up carrying skis)
19075	7.0	Winter activities	Skiing, general
19080	7.0	Winter activities	Skiing, cross country, 2.5 mph, slow or light effort, ski walking
19090	8.0	Winter activities	Skiing, cross country, 4.0–4.9 mph, moderate speed and effort, general
19100	9.0	Winter activities	Skiing, cross country, 5.0–7.9 mph, brisk speed, vigorous effort
19110	14.0	Winter activities	Skiing, cross country, >8.0 mph, racing
19130	16.5	Winter activities	Skiing, cross country, hard snow, uphill, maximum, snow mountaineering
19150	5.0	Winter activities	Skiing, downhill, light effort
19160	6.0	Winter activities	Skiing, downhill, moderate effort, general
19170	8.0	Winter activities	Skiing, downhill, vigorous effort, racing
19180	7.0	Winter activities	Sledding, tobogganing, bobsledding, luge (Taylor Code 370)
19190	8.0	Winter activities	Snow shoeing
19200	3.5	Winter activities	Snowmobiling
20000	1.0	Religious activities	Sitting in church, in service, attending a ceremony, sitting quietly
20001	2.5	Religious activities	Sitting, playing an instrument at church
20005	1.5	Religious activities	Sitting in church, talking or singing, attending a ceremony, sitting, active participation
20010	1.3	Religious activities	Sitting, reading religious materials at home

continued

CODE	METS	SPECIFIC ACTIVITY	EXAMPLES
20015	1.2	Religious activities	Standing in church (quietly), attending a ceremony, standing quietly
20020	2.0	Religious activities	Standing, singing in church, attending a ceremony, standing, active participation
20025	1.0	Religious activities	Kneeling in church/at home (praying)
20030	1.8	Religious activities	Standing, talking in church
20035	2.0	Religious activities	Walking in church
20036	2.0	Religious activities	Walking, less than 2.0 mph – very slow
20037	3.3	Religious activities	Walking 3.0 mph, moderate speed, not carrying anything
20038	3.8	Religious activities	Walking, 3.5 mph, brisk speed, not carrying anything
20039	2.0	Religious activities	Walk/stand combination for religious purposes, usher
20040	5.0	Religious activities	Praise with dance or run, spiritual dancing in church
20045	2.5	Religious activities	Serving food at church
20046	2.0	Religious activities	Preparing food at church
20047	2.3	Religious activities	Washing dishes/cleaning kitchen at church
20050	1.5	Religious activities	Eating at church
20055	2.0	Religious activities	Eating/talking at church or standing eating, Native American feast days
20060	3.0	Religious activities	Cleaning church
20061	5.0	Religious activities	General yard work at church
20065	2.5	Religious activities	Standing – moderate (lifting 50 lb., assembling at fast rate)
20095	4.0	Religious activities	Standing – moderate/heavy work
20100	1.5	Religious activities	Typing, electric, manual, or computer
21000	1.5	Volunteer activities	Sitting – meeting, general, and/or with talking involved
21005	1.5	Volunteer activities	Sitting – light office work, in general
21010	2.5	Volunteer activities	Sitting – moderate work
21015	2.3	Volunteer activities	Standing – light work (filing, talking, assembling)
21016	2.5	Volunteer activities	Sitting, child care, only active periods
21017	3.0	Volunteer activities	Standing, child care, only active periods
21018	4.0	Volunteer activities	Walk/run play with children, moderate, only active periods
21019	5.0	Volunteer activities	Walk/run play with children, vigorous, only active periods
21020	3.0	Volunteer activities	Standing – light/moderate work (pack boxes, assemble/repair, set up chairs/furniture)
21025	3.5	Volunteer activities	Standing – moderate (lifting 50 lb., assembling at fast rate)
21030	4.0	Volunteer activities	Standing – moderate/heavy work
21035	1.5	Volunteer activities	Typing, electric, manual, or computer
21040	2.0	Volunteer activities	Walking, less than 2.0 mph, very slow
21045	3.3	Volunteer activities	Walking, 3.0 mph, moderate speed, not carrying anything
21050	3.8	Volunteer activities	Walking, 3.5 mph, brisk speed, not carrying anything
21055	3.0	Volunteer activities	Walking, 2.5 mph slowly and carrying objects less than 25 lb
21060	4.0	Volunteer activities	Walking, 3.0 mph moderately and carrying objects less than 25 lb
21065	4.5	Volunteer activities	Walking, 3.5 mph, briskly and carrying objects less than 25 lb
21070	3.0	Volunteer activities	Walk/stand combination, for volunteer purposes

Sources: Data are derived from Ainsworth *et al.* 2000.

Glossary

Appetite A neurological drive that influences one's desire to eat and is controlled by external factors such as sight, smell, hearing, and social functions.

Control group A group of subjects who serve as a standard of comparison for the treatment being tested in a research experiment.

Double-blind study A research design or setup in which neither the subjects nor the investigators know who is in which group until the results have been analyzed.

Energy-yielding nutrients Referred to as carbohydrates, lipids, and proteins, which provide energy to the body.

Epidemiological research Research that involves studying large populations in order to suggest relationships between two or more variables.

Essential nutrients Also referred to as indispensable, these are nutrients that are necessary to support life but must be supplied in the diet because the body either cannot produce them or cannot produce them in a large enough quantity to meet needs.

Experimental research Research that actively intervenes in the lives of individuals and usually involves studying a small group of subjects who receive either a treatment or placebo under either tightly controlled or free-living conditions.

Hunger A neurological drive that influences one's desire to eat and is controlled by internal body mechanisms, such as activities of the stomach and small intestine.

Hypothesis A proposed scientific explanation for a phenomenon.

Inorganic nutrients Nutrients that contain no carbon, such as minerals and water.

Macronutrients Nutrients required by the body in relatively large amounts, often measured in grams, such as carbohydrates, lipids, and proteins.

Malnutrition Often interpreted as under-nutrition only, but also includes a condition of over-nutrition.

Micronutrients Nutrients required by the body in relatively small amounts, often measured in milligrams or micrograms, such as vitamins and minerals.

Morbidity A diseased state, disability, or poor health.

Non-essential nutrients Also referred to as dispensable, which describes those nutrients required by the body, but which can be produced in sufficient amounts to meet needs.

Nutrients Substances contained in food which are necessary to support growth, maintenance, and repair of the body tissues.

Nutrition A science that links foods to health and diseases.

Obesity A condition attributable to a positive energy balance (i.e., energy brought in via foods being greater than energy expended via physical activities).

Organic compounds Nutrients that contain carbon in addition to hydrogen and oxygen, such as carbohydrates, lipids, proteins, and vitamins.

Over-nutrition A form of malnutrition that occurs when food is consumed in excess of energy requirement.

Placebo A sham or simulated treatment that is identical (i.e., in appearance and taste) to the actual treatment but has no therapeutic value.

Risk factor A health behavior or pre-existing condition that has been associated with a particular disease, such as cigarette smoking, physical inactivity, stress, insulin resistance, hyperlipidemia, etc.

Satiety A feeling of fullness and satisfaction that halts one's desire to continue eating.

Single-blind study A research design or set-up in which subjects do not know which treatment they are receiving.

Sports nutrition The study and practice of nutrition and diet as it relates to athletic performance.

Under-nutrition A form of malnutrition that occurs due to reduced intake of energy and nutrients, increased requirements, or an inability to absorb or use nutrients.

2 Macronutrients: carbohydrates, lipids, and proteins

Amylase An enzyme that breaks down starch during digestion.

Amylopectin A highly branched chain of glucose units that makes up 80% of digestible starches.

Amylose A long, straight chain of glucose units that makes up about 20% of digestible starches.

Atherosclerosis A condition in which an artery wall thickens as the result of a build-up of fatty materials such as cholesterol.

Cellulose A form of polysaccharide found in plants; it cannot be digested by human enzymes.

Complementary proteins Proteins that can be combined to compensate for deficiencies in essential amino acid content in each protein.

Complete proteins Proteins that contain all nine essential amino acids.

Deamination A process in which the nitrogen-containing group of an amino acid is removed and converted to ammonia in the liver.

Denaturation A process in which proteins or nucleic acids unfold to lose their tertiary and secondary structures.

Disaccharides The combination of two monosaccharides; also referred to as double sugar.

Edema Swelling of tissue caused by fluid accumulation in the interstitial space.

Essential or indispensable amino acids Amino acids that cannot be made by the body and must be consumed in the diet.

Fatty acids A long chain of carbons bonded together and flanked by hydrogen; one end of the molecule is an acid group (COOH) and the other is a methyl group (CH_3).

Fermentation A process during which the yeast cells convert sugars to alcohol or ethanol and carbon dioxide.

Fiber A type of carbohydrate that the body cannot digest.

Fructose A common monosaccharide; also called fruit sugar.

Galactose A part of lactose, the disaccharide in milk.

Gluconeogenesis The process of producing new glucose using non-glucose molecules such as amino acids.

Glucose The simplest form of carbohydrate, found primarily in the blood and used by the cells for energy.

Glycogen The stored form of carbohydrate, found primarily in the muscle and liver.

Incomplete proteins Proteins that lack adequate amounts of one or more essential amino acids.

Insoluble fiber A type of fiber that does not dissolve in water and cannot be broken down by bacteria in the large intestine.

Lactase An enzyme that splits lactose into glucose and galactose during digestion.

Lactose A disaccharide consisting of glucose and galactose; also referred to as milk sugar.

Limiting amino acids The amino acids that are missing or in a low quantity.

Lipids A broad group of naturally occurring molecules, which includes fats, waxes, sterols, and phospholipids.

Lipogenesis Formation of triglycerides.

Lipolysis A process in which triglycerides are broken down into glycerol and fatty acids.

Maltose A disaccharide consisting of two molecules of glucose and also referred to as malt sugar.

Monosaccharides The basic unit of carbohydrate; it has the formula $C_6H_{12}O_6$.

Monounsaturated fatty acids The fatty acids that contain one double carbon–carbon bond.

Non-essential or dispensable amino acids Amino acids that can be made by the human body and are not required in the diet.

Oligosaccharides A saccharide polymer containing a small number (typically 3–10) of component sugars.

Osmosis The movement of water molecules across a partially permeable membrane from an area of high water potential (low solute concentration) to an area of low water potential (high solute concentration).

Peptide bond The chemical bond formed between the acid group of one amino acid and the nitrogen atom of the next amino acid.

Phospholipids Lipids which differ from triglycerides in that at least one fatty acid is replaced with a compound containing phosphorus.

Polysaccharides Complex carbohydrates containing many sugar units linked together.

Polyunsaturated fatty acids The fatty acids that contain two or more carbon–carbon double bonds.

Saturated fatty acids The fatty acids that contain all single carbon–carbon bonds.

Soluble fiber A type of fiber that can form viscous solutions when placed in water and can be digested by bacteria in the large intestine.

Starch A long, branched or unbranched chain of hundreds or thousands of glucose molecules linked together.

Sterols A group of lipids that consist of a multiple-ring structure, such as cholesterol.

Sucrose A disaccharide consisting of glucose and fructose; also referred to as cane sugar or table sugar.

Trans fatty acid The unsaturated fatty acids formed during certain types of food process in which some hydrogens are transferred to opposite sides of the carbon–carbon double bond that results in a straight shape.

Transamination A process in which the amino group from one amino acid is transported to a carbon-containing molecule to form a different amino acid.

Triglycerides Molecules in which three fatty acids are attached to a backbone of the three-carbon molecule glycerol.

Unsaturated fatty acids The fatty acids that contain some carbons that are not saturated with hydrogen.

3   Micronutrients: vitamins

Bioavailability A measure of how well a nutrient can be absorbed and used by the body.

Chylomicrons Large lipoprotein particles that consist primarily of triglycerides.

Collagen A fibrous protein found mainly in skin, bone, cartilage, tendons, teeth, and blood vessels.

Enrichment A type of fortification in which nutrients are added for the purpose of restoring those lost in processing to the same or a higher level than originally present.

Fortification The process of adding specific nutrients to foods.

Free radical Atoms or groups of atoms with an odd or unpaired number of electrons.

Macrocytic anemia A condition in which red blood cells grow bigger but are immature and have limited oxygen-carrying capacity.

Microcytic hypochromic anemia A condition in which blood cells are small in size and light in color due to lower concentrations of hemoglobin.

Osteoclast A type of bone cell that removes bone tissue.

Osteomalacia The softening of bone tissue due to vitamin D deficiency, an adult version of rickets.

Osteoporosis A chronic condition characterized by demineralization of bone and a decrease in bone density and strength.

Pernicious anemia A decrease in red blood cells that occurs when the body cannot properly absorb vitamin B-12 from the gastrointestinal tract.

Provitamins Vitamins that are available from foods in inactive forms, but once inside the body will be converted to active forms.

Retinoids Provitamin A compounds that include retinol, retinoic acid, and retinal.

Rickets Inadequate bone mineralization due to vitamin D deficiency in infants and children who are in active stages of growth.

Vitamins Organic compounds that are essential in the diet in small amounts to promote and regulate body functions necessary for growth, reproduction, and maintenance of the body.

4   Micronutrients: minerals and water

Aldosterone   The hormone released from the adrenal gland; it signals the kidneys to retain more sodium.

Anti-diuretic hormone   The hormone released from the pituitary gland; it causes the kidneys to conserve water.

Cretinism   A condition characterized by symptoms such as mental retardation, deaf mutism, and growth failure due to deficiency in thyroid hormone.

Cupric   An oxidized form of copper (Cu^{2+}).

Cuprous   A reduced form of copper (Cu^{+}).

Cytochromes   Heme-containing protein complexes that function in the electron transport chain.

Dehydration   A condition associated with excessive loss of body fluid.

Extracellular fluid   Body water found outside cells.

Ferric iron   An oxidized form of iron (Fe^{3+}).

Ferrous iron   A reduced form of iron (Fe^{2+}).

Fluorosis   A health condition caused by a person receiving too much fluoride during tooth development.

Goiter   A condition associated with an enlarged thyroid gland.

Hemoglobin   The iron-containing protein found in red blood cells, which function to transport oxygen.

Hypertension   A blood pressure of 140/90 mmHg or greater.

Hypokalemia   A condition of lower than normal blood potassium concentrations.

Hyponatremia   A condition in which sodium (salt) concentration is low in the body fluids outside the cells.

Interstitial fluid   The body water found between cells.

Intracellular fluid   Body water found inside cells.

Major minerals   Minerals that are needed in the diet in amounts greater than 100 mg per day, or present in the body in amounts greater than 0.01% of the body weight.

Microcytic hypochromic anemia   A condition in which blood cells are small in size and light in color due to lower concentrations of hemoglobin.

Myoglobin   The iron-containing protein found in muscle cells; it functions to transport oxygen.

Osmosis   The movement of water molecules across a partially permeable membrane from an area of high water potential (low solute concentration) to an area of low water potential (high solute concentration).

Osteoblast   A type of bone cell responsible for bone formation.

Osteoclast   A type of bone cell that removes bone tissue.

Osteopenia   A moderate loss of bone mass that can lead to osteoporosis.

Specific heat   The amount of energy it takes to increase the temperature of 1 g of a substance by 1°C.

Tetany   A condition in which muscles tighten and become unable to relax.

Thirst   The desire to drink and an essential mechanism involved in fluid balance.

Trace minerals   Minerals that are required by the body in an amount of 100 mg or less per day, or present in the body in an amount of 0.01% or less of body weight.

5   Digestion and absorption

Appendix   A blind-ended tube connected to the cecum.

Arteries   The blood vessels that transport the blood away from the heart.

Arterioles   The smallest arteries, which branch to form capillaries.

Bolus   The soft, moist mass of food formed from chewing, grinding, and mixing with saliva.

Cecum   The first portion of the large intestine.

Cephalic phase   The time period before food enters the stomach.

Chemoreceptors   Sensory receptors that detect changes in the chemical composition of the luminal contents.

Cholecystokinin   The hormone produced from the small intestine that signals the pancreas to secrete enzymes and causes the gallbladder to contract and empty its contents into the duodenum; also referred to as CCK.

Chyme   A semi-liquid food mass found in the stomach before passing into the duodenum.

Coenzymes   Non-protein substances such as ions and/or smaller organic molecules that facilitate enzyme action.

Colon   The largest section of the large intestine, consisting of the ascending colon, the transverse colon, and the descending colon.

Condensation   An anabolic process during which individual components of nutrients bind together to form more complex molecules.

Duodenum   The first 12 inches of the small intestine.

Energy of activation   The energy required to initiate chemical reactions.

Enteric endocrine system   Hormone-producing cells located within the gastrointestinal tract.

Enteric nervous system   The local nervous system located within the gastrointestinal tract.

Enzymes   A group of proteins that function to regulate the speed at which the reaction takes place.

Epiglottis   A flap of tissue near the pharynx, which prevents the bolus of swallowed food from entering the trachea.

Gastric inhibitory protein   A hormone released from the small intestine, which slows the rate of gastric emptying.

Gastric phase   The time period that begins when food enters the stomach.

Gastrin   The hormone secreted from the upper portion of the stomach that triggers the release of gastric juice, such as hydrochloric acid.

Gastrointestinal tract   A hollow tube or alimentary canal that runs from the mouth to the anus.

Hepatic portal circulation   The circulatory arrangement in the abdominal cavity that drains blood from the gastrointestinal tract and spleen to capillary beds in the liver.

Hydrolysis   Chemical reactions that digest or break down complex molecules into simpler forms.

Ileum   The last section of the small intestine, which is 10 feet long and leads to the large intestine.

Intestinal phase The time period that begins when chyme is passed into the small intestine.

Jejunum The 8 feet of the small intestine between the duodenum and ileum.

Lacteals Lymphatic capillaries that absorb large particles such as the products of fat digestion.

Lower esophageal sphincter The sphincter located between the esophagus and the stomach; it prevents foods from moving back out of the stomach.

Lumen The inside of the gastrointestinal tract.

Mechanoreceptors Sensory receptors that detect stretching or distension in the walls of the gastrointestinal tract.

Microvilli Tiny, hair-like folds in the plasma membrane, which extend from the surface of epithelial cells of the intestinal wall.

Pepsin The enzyme that breaks protein into shorter chains of amino acids or polypeptides.

Peptic ulcers Erosion of the lining of the stomach or the first part of the small intestine, called the duodenum.

Peristalsis A type of gastrointestinal movement that involves rhythmic, wave-like muscle contractions that propel food along the entire length of the gastrointestinal tract.

Probiotic Living microorganisms that take up residence in the large intestine and provide some health benefits.

Pyloric sphincter The sphincter located at the base of the stomach, which controls the rate at which the chyme is released into the small intestine.

Rectum The last section of the large intestine, which follows the descending colon.

Secretin The hormone produced from the small intestine; it signals the pancreas to secrete bicarbonate ions and stimulates the liver to secrete bile.

Segmentation A type of gastrointestinal movement in which circular muscles in the small intestine move the food mass back and forth.

Sphincter A muscle that encircles the tube of digestive tract and acts as a valve.

Substrates Substances or chemical compounds that are acted upon by enzymes.

Veins The blood vessels that transport the blood toward the heart.

Venules The smallest veins, which receive blood flow from the capillaries and converge to form larger veins for return to the heart.

Villi Tiny, finger-like projections that protrude from the epithelial lining of the intestinal wall.

6 Energy and energy-yielding metabolic pathways

Acetylcholine Neurotransmitters released by neurons of parasympathetic division of the autonomic nervous system.

Acetyl-CoA A molecule that functions to convey the carbon atoms to the Krebs cycle to be oxidized for energy production.

Adenosine triphosphate A high-energy compound used for a variety of biological work, including muscle contraction, synthesizing molecules, and transporting substances.

Bioenergetics A field of biochemistry that concerns chemical pathways responsible for converting energy-containing nutrients into a biologically usable form of energy.

Biosynthesis A process in which energy in one substance is transferred into other substances so their potential energy increases.

Catabolism A process in which more complex substances are broken down into simpler ones.

Digestive efficiency Referred to as the coefficient of digestibility that represents the percentage of ingested food digested and absorbed to serve the body's metabolic needs.

Energy The ability to produce change; it is measured by the amount of work performed during a given change.

Flavin adenine dinucleotide A coenzyme found in all living cells; it functions as a hydrogen carrier.

Glycogenolysis A process in which glycogen is broken down into glucose molecules.

Glycolysis See glycolytic system.

Glycolytic system A system that uses strictly the energy stored in carbohydrate molecules for regenerating ATP; also referred to as glycolysis.

Homeostasis The maintenance of a constant or unchanging internal environment.

Kinetic energy The energy possessed by an object due to its motion.

Lipolysis A process in which triglycerides are broken down into fatty acids and glycerol.

Mechanical energy A form of energy possessed by an object due to its motion or its position or internal structure.

Neurotransmitters Endogenous chemicals that transmit signals from a neuron to a target cell across a synapse.

Nicotinamide adenine dinucleotide A coenzyme found in all living cells that functions as a hydrogen carrier.

Nor-epinephrine Neurotransmitters released by neurons of sympathetic division of the autonomic nervous system.

Oxidative phosphorylation A metabolic pathway that uses energy released by the oxidation of nutrients to produce ATP.

Parasympathetic A branch of the autonomic nervous system that slows the heart rate and stimulates digestion.

Phosphagen system A system that serves as the immediate source of energy for regenerating ATP; also referred to as the ATP–PCr system.

Phosphocreatine A phosphorylated creatine molecule that serves as a rapidly mobilizable reserve of high-energy compounds; also referred to as PCr.

Phosphorylation A process in which energy transfers from energy substrate to ADP via phosphate.

Potential energy The energy possessed by an object due to its position or internal structure.

Second messengers Intracellular molecules or ions that are regulated by neurotransmitters or hormones and function to activate another set of enzymes.

Steady state A steady physiological environment in which energy demand is met by energy supply.

Sympathetic A branch of the autonomic nervous system that promotes fight-or-flight responses, including increases in heart rate and breathing rate and decreases in digestion.

Uncoupling proteins Mitochondrial inner membrane proteins that function to disrupt the connection between food breakdown and energy production.

7 Nutrients metabolism during exercise

β-oxidation A sequence of reactions that reduce a long-chain fatty acid into multiple two-carbon units in the form of acetyl-CoA.

Branched-chain amino acids Amino acids that have side chains with a branch (a carbon atom bound to more than two other carbon atoms), such as leucine, isoleucine, or valine.

Carnitine palmitoyl transferase (CPT) An enzyme that facilitates the action of carnitine.

Carnitine A carrier protein that helps transport long-chain fatty acids from cytoplasm into mitochondria.

Endogenous Produced or growing within an organism, tissue, or cell.

Gluconeogenesis A metabolic pathway that involves the use of non-glucose molecules such as amino acids or lactate for the production of glucose in the liver.

Glycogen phosphorylase A key enzyme that regulates glycogenolysis or glycogen degradation.

Hepatic glucose output Glucose released from liver glycogen degradation.

Hyperglycemia A condition that occurs when blood glucose concentration is too high.

Hypoglycemia A condition that occurs when blood glucose concentration is too low.

Insulin-like growth factors See somatomedins.

Isocitrate dehydrogenase A rating-limiting enzyme that regulates the Krebs cycle.

Lactate threshold An point above which the production of lactate will increase sharply.

Lipase An enzyme responsible for the breakdown of triglycerides; it is found in the liver, adipose tissue, muscle, as well as blood vessels.

Malonyl-CoA An intermediate in fatty acid synthesis that regulates fat utilization.

Nitrogen balance A measure that assesses the relationship between the dietary intake of protein and protein that is degraded and excreted.

Phosphofructokinase A rate-limiting enzyme that regulates glycolysis.

Pyruvate dehydrogenase An enzyme that catalyzes the breakdown of pyruvate into acetyl-CoA.

Somatomedins A group of hormones that promote cell growth and division.

8 Guidelines for designing a healthy and competitive diet

Acceptable macronutrient distribution A standard that provides a recommended distribution of the macronutrients in terms of energy consumption.

Adequacy The diet provides sufficient energy and enough of all the nutrients to meet the needs of healthy people.

Adequate intake levels Standard values that represent the average daily amount of a nutrient that appears sufficient to maintain a specific criterion; often used when DRI and RDA are not available.

Balance Consuming enough, but not too much, of each type of food.

Daily value Information found in food labels that provide a guide to the nutrients in one serving of food and is based on a 2000 kilocalorie diet.

Dietary Guidelines for Americans Dietary recommendations issued by the US Department of Health and Human Services to provide specific advice about how good dietary habits can promote health and reduce risk for major chronic disease.

Dietary reference intakes A set of nutritional standards used for assessing the adequacy of a person's diet.

Discretionary calories The calories allowed from food choices rich in added sugars and solid fat.

Estimated average requirements Standard values that represent the average daily amount necessary to maintain a specific biochemical or physiological function in half the healthy people of a given age and gender group.

Estimated energy requirement A measure used to assess whether one's energy intake is sufficient.

Female athlete triad A condition that includes disordered eating, amenorrhea, and osteoporosis, and that is often found in competitive female athletes.

Glycogen supercompensation A diet and exercise regimen aimed to increase muscle glycogen stores to a level that is greater than would be achieved on a typical diet.

Health claims Information found in food labels that describe a relationship between a nutrient or a food and the risk of a disease or health-related condition.

Moderation Not consuming too much of a particular food.

Nutrient content claims Information found in food labels describing the content level of a nutrient in a food.

Nutrient density The amount of nutrients that are in a food relative to its energy content.

Recommended dietary allowances Standard values that represent the average daily amount of a nutrient considered adequate to meet the known nutrient needs of nearly all healthy people of a given age and gender group.

Tolerable upper intake levels Standard values that represent the maximum daily amount of a nutrient that appears safe for most healthy people.

Variety Consuming different foods within each food group.

9   Ergogenic aids and supplements

Androstenedione A steroid hormone that functions as a precursor between DHEA and testosterone.

β-hydroxy-β-methylbutyrate A metabolite of the essential amino acid leucine; it is used to prevent or reduce muscle damage and to suppress protein degradation associated with intense physical effort.

Bicarbonate A salt of carbonic acid containing the ion HCO_3^{-2}, which helps delay the onset of acidosis and thus fatigue.

Boron An essential trace mineral involved in bone mineral metabolism, steroid hormone metabolism, and membrane functions.

Caffeine Naturally occurring substance found in a variety of beverages and foods, including coffee, tea, and chocolate.

Chromium A trace mineral that potentiates insulin action; insulin stimulates the glucose and amino acid uptake by muscle cells.

Coenzyme Q10 An integral component of the mitochondrion's electron transport system; also referred to as ubiquinone.

Creatine A nitrogen-containing molecule used by the body to form the high-energy compound phosphocreatine (PCr).

Dehydroepiandrosterone A steroid hormone that functions as a precursor to androstenedione and testosterone.

Doping The use of drugs to enhance performance in sports.

Ephedrine A sympathomimetic amine commonly used as a stimulant, appetite suppressant, concentration aid, and decongestant.

Ergogenic Increasing work or potential to do work.

Ergolytic Having a negative effect on muscle capacity.

Glutamine A non-essential amino acid that assists in nitrogen transport between tissues, acid–base regulation, and production of antioxidant glutathione.

Glycerol A component of the triglyceride molecule; it is used for gluconeogenesis and water retention.

Inosine A purine ribonucleoside used for ATP synthesis.

L-carnitine A substance that functions as a carrier protein to transport long-chain fatty acids into the mitochondrial matrix.

Phosphate loading Ingesting phosphate and phosphorus prior to strenuous exercise for the purpose of enhancing ATP synthesis and oxygen delivery to muscle cells.

Sports supplements Various nutritional and pharmacological ergogenic aids.

10 Nutrient metabolism in special cases

Adenylate cyclase An enzyme that catalyzes the conversion of ATP to cyclic AMP.

Adolescence The period in which a child matures into an adult.

Childhood The phase of development in humans between infancy and adolescence.

Co-morbidities The presence of one or more disorders in addition to a primary disease.

Corpus luteum A yellow, progesterone-secreting mass of cells that forms from an ovarian follicle after the release of an ovum.

Estrogen A major female sex hormones produced primarily by the ovarian follicles and responsible for developing and maintaining secondary female sex characteristics and preparing the uterus for the reception of a fertilized egg.

Euglycemic A condition in which blood glucose concentrations are within a normal range.

Exogenous Arising from outside an organism.

Follicular   The phase during which follicles are formed during the menstrual cycle or the time period from onset of menstruation to ovulation.

Gestational diabetes   A form of diabetes that occurs only during pregnancy.

Glucose tolerance   The test that determines how quickly ingested glucose is cleared from the blood.

Glucose transporters   A family of membrane proteins found in most mammalian cells that function to transport glucose molecules across cell membrane.

Hyperinsulinemic glucose clamp   A method that assesses how sensitive the tissue is to insulin; it requires maintaining a high insulin level by infusion with insulin.

Hyperlipidemia   A condition in which there is an abnormal elevation of blood lipids, including cholesterol and triglycerides.

Indirect calorimetry   A method that calculates heat that living organisms produce from their consumption of oxygen.

Infancy   A stage of human development lasting from birth to approximately two years.

Insulin resistance   A condition in which normal amounts of insulin are inadequate to produce a normal insulin response from fat, muscle, and liver cells.

Insulin responsiveness   A measure of a tissue's ability to respond to insulin; it is determined as the peak rate of insulin-mediated glucose disposal using the euglyemic hyperinsulinemic clamp technique.

Insulin sensitivity   A measure of a tissue's ability to respond to insulin; it can be determined as the insulin concentration that produces half of the maximal response in insulin-mediated glucose disposal using the euglyemic hyperinsulinemic clamp technique.

Lipoprotein lipase   An enzyme that cleaves one fatty acid from a triglyceride.

Luteal   The phase during which the corpus luteum secretes progesterone, which prepares the uterus for the implantation of an embryo or the second half of the menstrual cycle from ovulation to the beginning of the next menstrual flow.

Metabolic inflexibility   The inability to switch the utilization of lipids and carbohydrates in the peripheral tissue (i.e., muscle) based upon substrate availability.

Oxygen deficit   A lag of oxygen consumption at the onset of exercise; it is computed as the difference between oxygen uptake during the early stages of exercise and during a similar duration in a steady state of exercise.

Placenta   A membranous vascular organ in the uterus developed during pregnancy; it provides oxygen and nutrients to, and transfers wastes from, the developing fetus.

Placental lactogen   A polypeptide hormone that has similar structure and function as human growth hormone and is important in facilitating the energy supply of the fetus.

Portal circulation   See hepatic portal circulation, Chapter 5.

Post-absorptive state   The time period when the gastrointestinal tract is empty and energy comes from the breakdown of the body's reserves, such as glycogen and triglycerides.

Progesterone   A steroid hormone that prepares the uterus for the fertilized ovum and maintains pregnancy.

Prolactin   A peptide hormone secreted by the pituitary gland and primarily associated with lactation.

Puberty   The time period when children begin to mature biologically and become an adult capable of reproduction.

Respiratory exchange ratio   A qualitative indicator of which fuel (carbohydrate or fat) is being metabolized to supply the body with energy.

Subcutaneous   Beneath or under the skin.

Testosterone   A steroid hormone secreted by the testes that stimulates the development of male sex organs, secondary sexual traits, and sperm.

Thermogenesis   The process by which the body generates heat or energy through metabolism.

Vastus lateralis   The largest part of the quadriceps muscle.

Visceral   Internal organs of the body, specifically those within the chest such as the heart and lungs or within the abdomen such as the liver, pancreas, or intestine.

11   Measurement of energy consumption and output

Acceleration   The rate of change of velocity; expressed in meters/second^2.

Accelerometer   A device that measures the acceleration of a person's center of gravity.

Algorithm   A procedure or formula for solving a problem; it often has steps that repeat or require decisions such as logic or comparison.

Caloric equivalent of oxygen   Amount of energy yielded from 1 liter of oxygen used.

Creatinine   A breakdown product of creatine, which is an important part of muscle.

Diet history   A method of dietary analysis that assesses an individual's usual dietary intake over an extended period of time, usually a month to a year.

Direct calorimetry   A laboratory procedure that directly measures the heat produced during metabolism.

Doubly labeled water   Water in which both the hydrogen and the oxygen are partly or completely labeled with an uncommon isotope of these elements for the purpose of tracing metabolic rate.

Food diary   A method of dietary analysis that requires a person to record foods and beverages while consuming them.

Food frequency questionnaires   A method of dietary analysis that assesses energy or nutrient intake by determining how frequently an individual consumes a limited checklist of foods that are major sources of nutrients or of a particular dietary component.

Indirect calorimetry   A method by which measurement of whole-body respiratory gas exchange is used to estimate the amount of energy produced though the oxidative process.

Isotope   Any of the several different forms of an element, each having a different atomic mass.

Metabolic equivalent   A measure of the amount of oxygen consumed by an organism per minute in an activity relative to the resting metabolic rate. A single unit of metabolic equivalent (1 MET) equals to resting oxygen consumption, which is approximately 3.5 ml per kilogram of body weight per minute.

Missing foods Foods eaten but not reported in dietary analysis.

Pedometer A device, usually portable and electronic or electromechanical, that counts each step a person takes by detecting the motion of their hips.

Phantom foods Foods not eaten but reported in dietary analysis.

Respiratory exchange ratio Similar to respiratory quotient, except that both oxygen consumption and carbon dioxide production are measured at the lungs rather than tissue beds.

Respiratory quotient A ratio of oxygen consumption to carbon dioxide production, which is influenced by utilization of carbohydrate and fat as energy fuels.

Signs Objective evidence of disease that can be seen by others.

Symptoms Subjective evidence of disease; often referred to as a sensation people other than the patient cannot see.

12 Body composition

Air displacement plethysmography A body composition technique that is considered an alternative to hydrostatic weighing and uses air rather than water to determine body volume and thus body fat level.

Archimedes' principle A principle that relates buoyancy to displacement and states that weight loss under water is directly proportional to the weight of water displaced by the body.

Bioelectrical impedance analysis A body composition technique that determines body fat levels by measuring impedance or opposition to the flow of an electric current.

Body composition The ratio of fat to fat-free mass; frequently expressed as a percentage of body fat.

Body density A measurement that expresses total body mass or weight relative to body volume or the amount of space or area that your body occupies.

Body mass index Calculated by dividing weight in kilograms by the square of height in meters; also known as the Quetelet index.

Cellulite A dimpled, quilt-like skin caused by fat being separated by connective tissue into small compartments that extrude into the dermis.

Circumference A body composition technique that determines body fat levels or distribution based upon the circumference of selected body parts such as arms, legs, abdomen, and hips.

Densitometry The quantitative measurement of body composition that involves the determination of body density.

Dual-energy X-ray absorptiometry A body composition technique that use a series of cross-sectional scans from head to toe using photon beams to determine body fat levels.

Ectomorph Relative predominance of linearity and fragility, with a greater surface-to-mass ratio giving sensory exposure to the environment.

Endomorph Relative predominance of soft roundness and large digestive viscera.

Essential fat The body fat stored in the bone marrow, heart, lungs, liver, spleen, kidneys, intestines, muscles, and lipid-rich tissues of the central nervous system.

Hydrostatic weighing A body composition technique that determines body volume by measuring the volume of water displaced by the body based on the Archimedes' principle that weight loss under water is directly proportional to the weight of water displaced by the body.

Mesomorph Relative predominance of muscle, bone, and connective tissue ultimately derived from the mesodermal embryonic layer.

Obese A condition in which the BMI of a individual is higher than $30\,kg/m^2$.

Overweight A condition in which the BMI of a individual is higher than $25\,kg/m^2$.

Resistance exercise Performance of dynamic or static muscular contractions against external resistance of varying intensities.

Skinfold A body composition technique that determines body fat levels based upon the thickness of a double-fold of skin and the immediate layer of subcutaneous fat.

Storage fat The body fat stored in adipose tissue; often referred to as a depot for excess fat.

Subcutaneous fat A portion of the storage fat that is found just beneath the skin's surface.

Visceral fat A portion of storage fat that protects the various organs within the thoracic and abdominal cavities.

13 Energy balance and weight control

Circuit weight training A weight-training routine consisting of 10–12 resistance exercises for both the upper and lower body, executed in a planned fashion with limited rest between exercises.

Duration The length of time physical activity continues; it can also be expressed in terms of the total number of calories expended.

Excess post-exercise oxygen consumption The amount of extra oxygen required by the body during recovery from prior exercise.

Exercise Physical activity that is planned, structured, repetitive, and purposive.

Facultative thermogenesis The diet-induced thermogenesis mediated by the activation of the sympathetic nervous system.

Frequency The rate of occurrence within a given period of time or a number of exercise sessions completed every week.

Ghrelin A hormone produced by the stomach in response to a lack of food, triggering hunger.

Hypothalamus A portion of the brain that contains neural centers, helping regulate appetite and hunger.

Intensity The state of exertion or quality of being intense; it can be expressed using heart rate, oxygen consumption, or blood lactate concentration.

Intermittent exercise An exercise regimen during which exercise stops and resumes in an alternate fashion.

Lactate threshold The intensity at which blood lactic acid begins to accumulate drastically.

Negative energy balance A condition in which energy input is lower than energy output.

Obligatory thermogenesis The diet-induced thermogenesis due to digestion and absorption, as well as the synthesis of protein, fat, and carbohydrate to be stored in the body.

Oxygen deficit A lag of oxygen consumption at the onset of exercise; it is computed as the difference between oxygen uptake during the early stages of exercise and during a similar duration in a steady state of exercise.

Physical activity Muscular contraction that increases energy expenditure.

Positive energy balance A condition when energy input is greater than energy output.

Resting metabolic rate A minimal rate of metabolism necessary to sustain life.

Set point A control system built into every person, dictating how much fat he or she should carry.

Settling-point theory The idea that the set point may be modified, and in the case of weight gain, is re-established at a higher level.

Thermal effect of food An increase in energy expenditure associated with consumption of foods.

Weight cycling Repeated cycles of weight loss and regain that may increase the proportion of body fat with each successive weight regain and cause a decrease in RMR, making subsequent weight loss more difficult.

14 Thermoregulation and fluid balance

Conduction A process of heat loss via the transfer of heat from the body into the molecules of cooler objects in contact with the body's surface.

Convection A process of heat loss through movement of air or water molecules in contact with the body.

Dehydration A condition in which there is an excessive loss of body fluid.

Evaporation A process of heat loss in which water is converted from its liquid form to its vapor form.

Gastric emptying The process by which food leaves the stomach and enters the duodenum.

Glucose polymer solutions Fluid replacement preparations designed to provide carbohydrate while decreasing the osmotic concentration of the solution, thus helping with fluid absorption.

Glucose–electrolyte solutions Commercial fluid replacement preparations designed for athletes to replace fluid and carbohydrate lost during training and competition.

Heat acclimatization The process in which regular exercise in a hot environment results in a series of physiological adjustments designed to minimize disturbances in homeostasis due to heat stress.

Hyperhydration An attempt to begin an exercise bout with a slight surplus of body water.

Maltodextrin A glucose polymer that exerts lesser osmotic effects compared with glucose and is thus used in a variety of sports drinks as the source of carbohydrate.

Non-shivering thermogenesis An increase in heat production due to the combined influences of thyroxine and catecholamines.

Osmolality   A measure of solute concentration of a solution.

Radiation   A process of heat loss in the form of infrared rays.

Relative humidity   The percentage of water in ambient air at a particular temperature compared with the total quantity of moisture that air could carry.

Rhabdomyolysis   A condition in which damaged tissue leaks its contents into the blood, eventually leading to kidney damage and possible death.

Bibliography

Achten, J. and Jeukendrup, A.E. (2004) Relation between plasma lactate concentration and fat oxidation rates over a wide range of exercise intensities. *International Journal of Sports Medicine*, 25: 32–37.

Achten, J., Gleeson, M., and Jeukendrup, A.E. (2002) Determination of the exercise intensity that elicits maximal fat oxidation. *Medicine & Science in Sports & Exercise*, 34: 92–97.

Ainslie, P.N., Reilly, T., and Westerterp, K.R. (2003) Estimating human energy expenditure: a review of techniques with particular reference to doubly labelled water. *Sports Medicine*, 33: 683–698.

Ainsworth, B.E. and Leon, A.S. (1991) Gender differences in self-reported physical activity. *Medicine & Science in Sports & Exercise*, 23: S105.

Ainsworth, B.E., Haskell, W.L., Leon, A.S., Jacobs, D.R. Jr., Montoye, H.J., Sallis, J.F., and Paffenbarger, R.S. Jr. (1993) Compendium of physical activities: classification of energy costs of human physical activities, *Medicine & Science in Sports & Exercise*, 25: 71–80.

Ainsworth, B.E., Haskell, W.L., Whitt, M.C., Irwin, M.L., Swartz, A.M., Strath, S.J., O'Brien, W.L., Bassett, D.R., Jr., Schmitz, K.H., Emplaincourt, P.O., Jacobs, D.R. Jr. and Leon, A.S. (2000) Compendium of physical activities: an update of activity codes and MET intensities, *Medicine & Science in Sports & Exercise*, 32: S498–S516.

Alghamdi, A.A., Al-Radi, O.O., and Latter, D.A. (2005) Intravenous magnesium for prevention of atrial fibrillation after coronary artery bypass surgery: a systematic review and meta-analysis. *Journal of Cardiac Surgery*, 20: 293–299.

Almuzaini, K.S., Potteiger, J.A., and Green, S.B. (1998) Effects of split exercise sessions on excess post-exercise oxygen consumption and resting metabolic rate. *Canadian Journal of Applied Physiology*, 23: 433–443.

American College of Obstetricians and Gynecologists (1994) Exercise during pregnancy and post-partum period. *American College of Obstetricians and Gynecologists Technical Bulletin*, 189: 2–7.

American College of Sports Medicine (2001) Appropriate intervention strategies for weight loss and prevention of weight regain for adults. *Medicine & Science in Sports & Exercise*, 33: 2145–2156.

American College of Sports Medicine (2006) *ACSM's Guidelines for Exercise Testing and Prescription*, 7th edn., Baltimore, MD: Lippincott Williams and Wilkins.

Andersen, R.E., Crespo, C.J., Bartlett, S.J., Cheskin, L.J., and Pratt, M. (1998) Relationship of physical activity and television watching with body weight and level of fatness among children. *Journal of the American Medical Association*, 279: 938–942.

Anderson, T. (1996) Biomechanics and running economy. *Sports Medicine*, 22: 76–89.

Andresen, V. and Camilleri, M. (2006) Irritable bowel syndrome: recent and novel therapeutic approaches. *Drugs*, 66: 1073–1088.

Armellini, R., Zamboni, M., Mino, A., Bissoli, L., Micciolo, R., and Bosello, O. (2000) Postabsorptive resting metabolic rate and thermic effect of food in relation to body composition and adipose tissue distribution. *Metabolism*, 49: 6–10.

Armon, Y., Cooper, D.M., Flores, R., Zanconato, S., and Barstow, T.J. (1991) Oxygen uptake dynamics during high-intensity exercise in children and adults. *Journal of Applied Physiology*, 70: 841–848.

Armstrong, L.E. and Epstein, Y. (1999) Fluid–electrolyte balance during labor and exercise: concepts and misconceptions. *International Journal of Sport Nutrition*, 9: 1–12.

Armstrong, L.E. and Maresh, C.M. (1991) The induction and decay of heat acclimatisation in trained athletes. *Sports Medicine*, 12: 302–312.

Arngrímsson, S.A., Petitt, D.S., Borrani, F., Skinner, K.A., and Cureton, K.J. (2004) Hyperthermia and maximal oxygen uptake in men and women. *European Journal of Applied Physiology*, 92: 524–532.

Åstrand, I. (1960) Aerobic work capacity in men and women with a special reference to age. *Acta Physiologica Scandinavica*, 49 (Suppl. 169): 1–92.

Astrup, A., Lundsgaard, C., Madsen, J., and Christensen, N.J. (1985) Enhanced thermogenic responsiveness during chronic ephedrine treatment in man. *The American Journal of Clinical Nutrition*, 42: 83–94.

Astrup, A., Buemann, B., Flint, A., and Raben, A. (2002) Low-fat diets and energy balance: how does the evidence stand in 2002? *Proceedings of the Nutrition Society*, 61: 299–309.

Astrup, A., Meinert Larsen, T., and Harper, A. (2004) Atkins and other low-carbohydrate diets: hoax or an effective tool for weight loss? *The Lancet*, 364: 897–899.

Atwater, W.O. and Benddict, F.G. (1903) *Experiments on the Metabolism of Matter and Energy in the Human Body, 1900–1902*, US Department of Agriculture Office of Experiment Stations, Bulletin 136, US Government Printing Office, Washington, DC.

Babij, P., Matthews, S.M., and Rennie, M.J. (1983) Changes in blood ammonia, lactate, and amino acids in relation to workload during bicycle ergometer exercise in man. *European Journal of Applied Physiology*, 50: 405–411.

Baecke, J.A.H., van Staveren, W.A., and Burema, J. (1983) Food consumption, habitual physical activity, and body fatness in Dutch adults. *The American Journal of Clinical Nutrition*, 37: 278–286.

Baker, L.B., Munce, T.A., and Kenney, W.L. (2005) Sex differences in voluntary fluid intake by older adults during exercise. *Medicine & Science in Sports & Exercise*, 37: 789–796.

Bakker, I., Twisk, J.W., van Mechelen, W., and Kemper, H.C. (2003) Fat-free body mass is the most important body composition determinant of 10-yr longitudinal development of lumbar bone in adult men and women. *Journal of Clinical Endocrinology & Metabolism*, 88: 2607–2613.

Balsom, P.D., Wood, K., Olsson, P., and Ekblom, B. (1999) Carbohydrate intake and multiple sprint sports: with special reference to football (soccer). *International Journal of Sports Medicine*, 20: 48–52.

Banerji, M., Chaiken, R., Gordon, D., and Lebowitz, H. (1995) Does intra-abdominal adipose tissue in black men determine whether NIDDM is insulin-resistant or insulin sensitive? *Diabetes*, 44: 141–146.

Barnett, C., Costill, D.L., Vukovich, M.D., Cole, K.J., Goodpaster, B.H., Trappe, S.W., and Fink, W.J. (1994) Effect of L-carnitine supplementation on muscle and blood carnitine content and lactate accumulation during high-intensity sprint cycling. *International Journal of Sport Nutrition*, 4: 280–288.

Barone, J.J. and Roberts, H.R. (1996) Caffeine consumption. *Food and Chemical Toxicology*, 34: 119–129.

Bar-Or, O. and Rowland, T.W. (2004) Physiologic and perceptual responses to exercise in healthy child, In: *Pediatric Exercise Medicine*, Champaign, IL: Human Kinetics, pp. 3–59.

Barrow, M.W. and Clark, K.A. (1998) Heat-related illnesses. *American Family Physician*, 58: 749–756.

Bassett, D., Jr. (2000) Validity and reliability issues in objective monitoring of physical activity. *Research Quarterly for Exercise & Sport*, 71: 30–36.

Bell, D.G., Jacobs, I., and Ellerington, K. (2001) Effect of caffeine and ephedrine ingestion on anaerobic exercise performance. *Medicine & Science in Sports & Exercise*, 33: 1399–1403.

Bell, D.G., McLellan, T.M., and Sabiston, C.M. (2002) Effect of ingesting caffeine and ephedrine on 10-km run performance. *Medicine & Science in Sports & Exercise*, 34: 344–349.

Bell, R.D., MacDougall, J.D., Billeter, R., and Howald, H. (1980) Muscle fibers types and morphometric analysis of skeletal muscle in six years old children. *Medicine & Science in Sports & Exercise*, 12: 28–31.

Bentzur, K.M., Kravitz, L., and Lockner, D.W. (2008) Evaluation of the BOD POD for estimating percent body fat in collegiate track and field female athletes: a comparison of four methods. *Journal of Strength & Conditioning Research*, 22: 1985–1991.

Bergström, J. and Hultman, E. (1967) Synthesis of muscle glycogen in man after glucose and fructose infusion. *Acta Medica Scandinavica*, 182: 93–107.

Bikle, D.D. (2004) Vitamin D and skin cancer. *Journal of Nutrition*, 134: S3472–S3478.

Billat, V., Lepretre, P.M., Heugas, A.M., Laurence, M.H., Salim, D., and Koralsztein, J.P. (2003) Training and bioenergetic characteristics in elite male and female Kenyan runners. *Medicine & Science in Sports & Exercise*, 35: 297–304.

Binzen, C.A., Swan, P.D., and Manore, M.M. (2001) Postexercise oxygen consumption and substrate use after resistance exercise in women. *Medicine & Science in Sports & Exercise*, 33: 932–938.

Bishop, D., Edge, J., Davis, C., and Goodman, C. (2004) Induced metabolic alkalosis affects muscle metabolism and repeated-sprint ability. *Medicine & Science in Sports & Exercise*, 36: 807–813.

Blaak, E.E., van Aggel-Leijssen, D.P.C., Wagenmakers, A.J.M., Saris, W.H.M., and Baak, M.A. (2000) Impaired oxidation of plasma-derived fatty acids in type 2 diabetic subjects during moderate-intensity exercise. *Diabetes*, 49: 2102–2107.

Black, A.E., Prentice, A.M., and Coward, W.A. (1986) Use of food quotients to predict respiratory quotients for the doubly labeled water method of measuring energy expenditure. *Human Nutrition: Clinical Nutrition*, 40: 381–391.

Black, M.M. (2003) Micronutrient deficiencies and cognitive functioning. *Journal of Nutrition*, 133: S3927S–S3931.

Blaxter, K. (1989) *Energy Metabolism in Animal and Man*, Cambridge: Cambridge University Press.

Block, G. (1989) Human dietary assessment: methods and issues. *Preventive Medicine*, 18: 653–660.

Bobkowski, W., Nowak, A., and Durlach, J. (2005) The importance of magnesium status in the pathophysiology of mitral valve prolapse. *Magnesium Research*, 18: 35–52.

Bogardus, C., Thuillez, P., Ravussin, E., Vasquez, B., Narimiga, M., and Azhar, S. (1983) Effect of muscle glycogen depletion on in vivo insulin action in men. *Journal of Clinical Investigation*, 72: 1605–1610.

Boobis, L., William, C., and Wooton, S.A. (1982) Human muscle metabolism during brief maximal exercise. *Journal of Physiology*, 338: 21–22.

Booth, S.L., Broe, K.E., Peterson, J.W., Cheng, D.M., Dawson-Hughes, B., Gundberg, C.M., Cupples, L.A., Wilson, P.W., and Kiel, D.P. (2004) Associations between vitamin K biochemical measures and bone mineral density in men and women. *Journal of Clinical Endocrinology & Metabolism*, 89: 4904–4909.

Borchers, J.R., Clem, K.L., Habash, D.L., Nagaraja, H.N., Stokley, L.M., and Best, T.M. (2009) Metabolic syndromes and insulin resistance in division 1 collegiate football players. *Medicine & Science in Sports & Exercise*, 41: 2105–2110.

Borsheim, E. and Bahr, R. (2003) Effect of exercise intensity, duration, and mode on post-exercise oxygen consumption. *Sports Medicine*, 33: 1037–1060.

Bostick, R.M., Potter, J.D., McKenzie, D.R., Sellers, T.A., Kushi, L.H., Steinmetz, K.A., and Folsom, A.R. (1993) Reduced risk of colon cancer with high intake of vitamin E: the Iowa Women's Health Study. *Cancer Research*, 15: 4230–4237.

Bouchard, C. (2008) Gene–environment interactions in the etiology of obesity: defining the fundamentals. *Obesity*, 16 (Suppl 3): S5–S10.

Bouchard, C., Tremblay, A., Després, J.P., Nadeau, A., Lupien, P.J., Thériault, G., Dussault, J., Moorjani, S., Pinault, S., and Fournier, G. (1990) The response to long-term overfeeding in identical twins. *The New England Journal of Medicine*, 322: 1477–1482.

Bouten, C., Verboeket-van de Venne, W., Westerterp, K., Verduin, M., and Janssen, J. (1996) Daily physical activity assessment: comparison between movement registration and doubly labeled water. *Journal of Applied Physiology*, 81: 1019–1026.

Brage, S., Brage, N., Franks, P.W., Ekelund, U., Wong, M., Anderson, L.B., Froberg, K., and Wareham, N.J. (2003) Branched equation modeling of simultaneous accelerometry and heart rate monitoring improves estimate of directly measured physical activity energy expenditure. *Journal of Applied Physiology*, 96: 343–351.

Braun, B., Clarkson, P.M., Freedson, P.S., and Kohl, R.L. (1991) Effects of coenzyme Q10 supplementation on exercise performance, VO2max, and lipid peroxidation in trained cyclists. *International Journal of Sport Nutrition*, 1: 353–365.

Bravata, D.M., Sanders, L., Huang, J., Krumholz, H.M., Olkin, I., Gardner, C.D., and Bravata, D.M. (2003) Efficacy and safety of low-carbohydrate diets: a systematic review. *Journal of the American Medical Association*, 289: 1837–1850.

Bray, G.A. (1983) The energetics of obesity, *Medicine & Science in Sports & Exercise*, 15: 32–40.

Bray, G.A. and Gray, D.S. (1988) Obesity: part I – pathogenesis. *The Western Journal of Medicine*, 149: 429–441.

Bray, G.A., Whipp, B.J., and Koyal, S.N. (1974) The acute effects of food on energy expenditure during cycle ergometry. *The American Journal of Clinical Nutrition* 27: 254–259.

Bray, G.A., Zachary, B., Dahms, W.T., Atkinson, R.L., and Oddie, T.H. (1978) Eating patterns of the massively obese individual. *Journal of the American Dietetic Association*, 72: 24–27.

Bredle, D.L., Stager, J.M., Brechue, W.F., and Farber, M.O. (1988) Phosphate supplementation, cardiovascular function, and exercise performance in humans. *Journal of Applied Physiology*, 65: 1821–1826.

Brehm, B.J., Seeley, R.J., Daniels, S.R., and D'Alessio, D.A. (2003) A randomized trial comparing a very low carbohydrate diet and a calorie-restricted low fat diet on body weight and cardiovascular risk factors in healthy women. *Journal of Clinical Endocrinology & Metabolism*, 88: 1617–1623.

Bremer, J. (1983) Carnitine: metabolism and functions. *Physiological Reviews*, 63: 1420–1480.

Brodie, D.A. (1988) Techniques of measurement of body composition: part I. *Sports Medicine*, 5: 11–40.

Broeder, C.E., Quindry, J., Brittingham, K., Panton, L., Thomson, J., Appakondu, S., Breuel, K., Byrd, R., Douglas, J., Earnest, C., Mitchell, C., Olson, M., Roy, T., and Yarlagadda, C. (2000) The Andro Project: physiological and hormonal influences of androstenedione supplementation in men 35 to 65 years old participating in a high-intensity resistance training program. *Archives of Internal Medicine*, 160: 3093–3104.

Brooks, G.A., Fahey, T.D., and Baldwin, K.M. (2005) *Exercise Physiology: Human Bioenergetics and Its Applications*, New York: McGraw Hill.

Brown, G.A., Vukovich, M.D., Martini, E.R., Kohut, M.L., Franke, W.D., Jackson, D.A., and King, D.S. (2000) Endocrine responses to chronic androstenedione intake in 30- to 56-year-old men. *Journal of Clinical Endocrinology & Metabolism*, 85: 4074–4080.

Brozek, J., Grande, F., Anderson, J.T., and Keys, A. (1963) Densitometric analysis of body composition; revision of some quantitative assumptions. *Annals of the New York Academy of Sciences*, 110: 113–140.

Bryan, J., Osendarp, S., Hughes, D., Calvaresi, E., Baghurst, K., and van Klinken, J.W. (2004) Nutrients for cognitive development in school-aged children. *Nutrition Reviews*, 62: 295–306.

Buch, I., Hornnes, P.J., and Kuhl, C. (1986) Glucose tolerance in early pregnancy. *Acta Endocrinologica*, 112: 263–266.

Buemann, B. and Tremblay, A. (1996) Effect of exercise training on abdominal obesity and related metabolic complications. *Sports Medicine*, 21: 191–212.

Burke, B.S. (1947) The dietary history as a tool in research. *Journal of the American Dietetic Association*, 23: 1041–1046.

Burke, L.M. and Read, R.S. (1993) Dietary supplements in sport. *Sports Medicine*, 15: 43–65.

Burleson, M.A., O'Bryant, H.S., Stone, M.H., Collins, M.A., and Triplett-McBride, T. (1998) Effect of weight training exercise and treadmill exercise on post-exercise oxygen consumption. *Medicine & Science in Sports & Exercise*, 30: 518–522.

Burstein, R., Epstein, Y., Shapiro, Y., Charuzi, I., and Karnieli, E. (1990) Effect of an acute bout of exercise on glucose disposal in human obesity. *Journal of Applied Physiology*, 69: 299–304.

Cade, R., Conte, M., Zauner, C., Mars, D., Peterson, J., Lunne, D., Hommen, N., and Packer, D. (1984) Effects of phosphate loading on 2,3-diphosphoglycerate and maximal oxygen uptake. *Medicine & Science in Sports & Exercise*, 16: 263–268.

Calvo, M.S., Whiting, S.J and Barton, C.N. (2005) Vitamin D intake: a global perspective of current status. *Journal of Nutrition* 135: 310–316.

Campbell, S.E. and Febbraio, M.A. (2001) Effect of ovarian hormones on mitochondrial enzyme activity in fat oxidation pathway of skeletal muscle. *American Journal of Physiology*, 281: E803–E808.

Cancello, R., Tounian, A., Poitou Ch and Clément, K. (2004) Adiposity signals, genetic and body weight regulation in humans. *Diabetes & Metabolism*, 30: 215–227.

Candow, D.G., Chilibeck, P.D., Burke, D.G., Davison, K.S., and Smith-Palmer, T. (2001) Effect of glutamine supplementation combined with resistance training in young adults. *European Journal of Applied Physiology*, 86: 142–149.

Cannon, B. and Nedergaard, J. (2004) Brown adipose tissue: function and physiological significance. *Physiological Reviews*, 84: 277–359.

Carlson, M.G., Snead, W.L., Hill, J.O., Nurjahan, N., and Campbell, P.J. (1991) Glucose regulation of lipid metabolism in humans. *American Journal of Physiology*, 261: E815–E820.

Carter, S., McKenzie, S., Mourtzakis, M., Mahoney, D.J., and Tarnopolsky, M.A. (2001a) Short-term 17ß-estradiol decreases glucose Ra but not whole body metabolism during endurance exercise. *Journal of Applied Physiology*, 90: 139–146.

Carter, S.L., Rennie, C., and Tarnopolsky, M.A. (2001b) Substrate utilization during endurance exercise in men and women after endurance training. *American Journal of Physiology*, 280: E898–E907.

Caspersen, C.J., Powell, K.E., and Christensen, G.M. (1985) Physical activity, exercise, and physical fitness: definitions and distinctions for health-related research, *Public Health Reports*, 100: 126–131.

Castell, L.M., Poortmans, J.R., and Newsholme, E.A. (1996) Does glutamine have a role in reducing infections in athletes? *European Journal of Applied Physiology and Occupational Physiology*, 73: 488–490.

Cheetham, M.E., Boobis, L.H., Brooks, S., and Williams, C. (1986) Human muscle metabolism during sprint running. *Journal of Applied Physiology*, 61: 54–60.

Chester, N., Reilly, T., and Mottram, D.R. (2003) Physiological, subjective and performance effects of pseudoephedrine and phenylpropanolamine during endurance running exercise. *International Journal of Sports Medicine*, 24: 3–8.

Chin, E.R. and Allen, D.G. (1998) The contribution of pH-dependent mechanisms to fatigue at different intensities in mammalian single muscle fibres. *Journal of Physiology*, 512: 831–840.

Chromiak, J.A. and Antonio, J. (2002) Use of amino acids as growth hormone-releasing agents by athletes. *Nutrition*, 18: 657–661.

Chu, K.S., Doherty, T.J., Parise, G., Milheiro, J.S., and Tarnopolsky, M.A. (2002) A moderate dose of pseudoephedrine does not alter muscle contraction strength or anaerobic power, *Clinical Journal of Sport Medicine*, 12: 387–390.

Clancy, S.P., Clarkson, P.M., DeCheke, M.E., Nosaka, K., Freedson, P.S., Cunningham, J.J., and Valentine, B. (1994) Effects of chromium picolinate supplementation on body composition, strength, and urinary chromium loss in football players. *International Journal of Sport Nutrition*, 4: 142–153.

Clapp, J.F., Wesley, M., and Sleamaker, R.H. (1987) Thermoregulatory and metabolic responses to jogging prior to and during pregnancy. *Medicine & Science in Sports & Exercise*, 19: 124–130.

Cloherty, E.K., Sultzman, L.A., Zottola, R.J., and Carruthers, A. (1995) Net sugar transport is a multistep process: evidence for cytosolic sugar binding sites in erythrocytes, *Biochemistry*, 34: 15395–15406.

Coggan, A.R. and Coyle, E.F. (1991) Carbohydrate ingestion during prolonged exercise: effects on metabolism and performance. *Exercise and Sport Sciences Reviews*, 19: 1–40.

Colberg, S.R., Hagberg, J.M., McCole, S.D., Zumda, J.M., Thompson, P.D., and Kelley, D.E. (1996) Utilization of glycogen but not plasma glucose is reduced in individuals with NIDDM during mild-intensity exercise. *Journal of Applied Physiology*, 81: 2027–2033.

Collins, M.A., Millard-Stafford, M.L., Sparling, P.B., Snow, T.K., Rosskopf, L.B., Webb, S.A., and Omer, J. (1999) Evaluation of the BOD POD for assessing body fat in collegiate football players. *Medicine & Science in Sports & Exercise*, 31: 1350–1356.

Conley, K.E., Jubrias, S.A., and Esselman, P.C. (2000) Oxidative capacity and ageing in human muscle. *Journal of Physiology*, 526: 203–210.

Connor, S.L. and Connor, W.E. (1997) Are fish oils beneficial in the prevention and treatment of coronary artery disease? *The American Journal of Clinical Nutrition*, 66: S1020–S1031.

Costill, D.L., Coyle, E., Dalsky, G., Evens, W., Fink, W., and Hoopes, D. (1977) Effects of elevated plasma FFA and insulin on muscle glycogen usage during exercise. *Journal of Applied Physiology*, 43: 695–699.

Coyle, E.F. (1995) Substrate utilization during exercise in active people. *The American Journal of Clinical Nutrition*, 61: S968–S979.

Coyle, E.F., Hamilton, M.T., Gonzalez-Alonso, J., Montain, S.J., and Ivy, J.L. (1991) Carbohydrate metabolism during intense exercise when hyperglycemic. *Journal of Applied Physiology*, 70: 834–840.

Coyle, E.F., Jeukendrup, A.E., Wagenmakers, A.J., and Saris, W.H. (1997) Fatty acids oxidation is directly regulated by carbohydrate metabolism during exercise. *American Journal of Physiology*, 273: E268–E275.

Craig, W.J., Mangels, A.R., and the American Dietetic Association (2009) Position of the American Dietetic Association: vegetarian diets. *Journal of the American Dietetic Association*, 109: 1266–1282.

Crawford, D., Jeffery, R.W., and French, S.A. (2000) Can anyone successfully control their weight? Findings of a three year community-based study of men and women. *International Journal of Obesity Related Metabolic Disorders*, 9: 1107–1110.

Cunningham-Rundles, S. and McNeeley, D.F. (2005) Mechanisms of nutrient modulation of the immune response. *Journal of Allergy and Clinical Immunology*, 115: 1119–1128.

Cureton, K.J. and Sparling, P.B. (1980) Distance running performance and metabolic responses to running in men and women with excess weight experimentally equated. *Medicine & Science in Sports & Exercise*, 12: 288–294.

Currell, K. and Jeukendrup, A.E. (2008) Superior endurance performance with ingestion of multiple transportable carbohydrates. *Medicine & Science in Sports & Exercise*, 40: 275–281.

Davidson, M. (1979) The effect of aging on carbohydrate metabolism: a review of diabetes mellitus in the elderly. *Metabolism*, 28: 688–705.

Davies, C.T.M., Barnes, C., and Godfrey, S. (1972) Body composition and maximal exercise performance in children. *Human Biology*, 44: 195–214.

De Bandt, J.P., Coudray-Lucas, C., Lioret, N., Lim, S.K., Saizy, R., Giboudeau, J., and Cynober, L. (1998) A randomized controlled trial of the influence of the mode of enteral ornithine alpha-ketoglutarate administration in burn patients. *Journal of Nutrition*, 128: 563–569.

De Glisezinski, I., Harant, I., Crampes, F., Trudeau, F., Felez, A., Cottet-Emard, J.M., Garrigues, M., and Riviere, D. (1998) Effects of carbohydrate ingestion on adipose tissue lipolysis during long-lasting exercise in trained men. *Journal of Applied Physiology*, 84: 1627–1632.

DeFronzo, R.A., Tobin, J.D., and Andres, R. (1979) Glucose clamp technique: a method for quantifying insulin secretion and resistance. *American Journal of Physiology*, 237: E214–E223.

DeFronzo, R.A., Ferrannini, E., Sato, Y., Felig, P., and Wahren, J. (1981) Synergistic interaction between exercise and insulin on peripheral glucose uptake. *Journal of Clinical Investigation*, 68: 1468–1474.

DeFronzo, R., Gunnarsson, R., Bjorkman, D., Olsson, M., and Warren, J. (1985) Effects of insulin on peripheral and splanchnic glucose metabolism in noninsulin-dependent (type II) diabetes mellitus. *Journal of Clinical Investigation*, 76: 149–155.

DeMeersman, R., Gatty, D., and Schaffer, D. (1987) Sympathomimetics and exercise enhancement: all in the mind? *Pharmacology Biochemistry and Behavior*, 28: 361–365.

Deon, T. and Braun, B. (2002) The roles of estrogen and progesterone in regulating carbohydrate and fat utilization at rest and during exercise. *Journal of Women's Health and Gender-Based Medicine*, 11: 225–237.

Devlin, J.T. and Horton, E.S. (1985) Effect of prior high intensity exercise on glucose metabolism in normal and insulin-resistant men. *Diabetes*, 34: 973–979.

Devlin, J.T., Hirshman, M., Horton, E.D., and Horton, E.S. (1987) Enhanced peripheral and splanchnic insulin sensitivity in NIDDM men after single bout of exercise. *Diabetes*, 36: 434–439.

Dietz, W.H. and Bellizzi, M.C. (1999) Introduction: the use of body mass index to assess obesity in children. *The American Journal of Clinical Nutrition*, 70: S123–S125.

Dill, D.B. and Consolazio C.F. (1962) Responses to exercise as related to age and environmental temperature. *Journal of Applied Physiology*, 17: 645–648.

Dohm, G.L. (1986) Protein as a fuel for endurance exercise. *Exercise and Sport Sciences Reviews*, 14: 143–173.

Dohm, G., Kasperek, G.J., Tapscott, E.B., and Barakat, H.A. (1987) Protein degradation during endurance exercise and recovery. *Medicine & Science in Sports & Exercise*, 19: S166–S171.

Dolny, D. and Lemon, P. (1988) Effect of ambient temperature on protein breakdown during prolonged exercise. *Journal of Applied Physiology*, 64: 550–555.

Donati, L., Ziegler, F., Pongelli, G., and Signorini, M.S. (1999) Nutritional and clinical efficacy of ornithine alpha-ketoglutarate in severe burn patients. *Clinical Nutrition*, 18: 307–311.

Donnelly, J.E., Jacobsen, D.J., Snyder-Heelan, K., Seip, R., and Smith, S. (2000) The effects of 18 months of intermittent vs. continuous exercise on aerobic capacity, body weight and composition, and metabolic fitness in previously sedentary, moderately obese females. *International Journal of Obesity*, 24: 566–572.

Duffy, D.J. and Conlee, R.K. (1986) Effects of phosphate loading on leg power and high intensity treadmill exercise. *Medicine & Science in Sports & Exercise*, 18: 674–677.

Dyck, D.J., Putman, C.T., Heigenhauser, G.J.F., Hultman, E., and Spriet, L.L. (1993) Regulation of fat–carbohydrate interaction in skeletal muscle during intense aerobic cycling, *American Journal of Physiology*, 265: E852–E859.

Dyck, D.J., Peters, S.A., Wendling, P.S., Chesley, A., Hultman, E., and Spriet, L.L. (1996) Regulation of muscle glycogen phosphorylase activity during intense aerobic cycling with elevated FFA. *American Journal of Physiology*, 265: E116–E125.

Edwards, H.T., Margaria, R., and Dill, D.B. (1934) Metabolic rate, blood sugar, and the utilization of carbohydrate. *American Journal of Physiology*, 108: 203–209.

Elia, M., Ritz, P., and Stubbs, R.J. (2000) Total energy expenditure in the elderly. *European Journal of Clinical Nutrition*, 54: S92–S103.

Ellis, G.S., Lanza-Jacoby, S., Gow, A., and Kendrick, Z.V. (1994) Effect of estradiol on lipoprotein lipase activity and lipid availability in exercised male rats. *Journal of Applied Physiology*, 77: 209–215.

Emmert, D.H. and Kirchner, J.T. (1999) The role of vitamin E in the prevention of heart disease. *Archives of Family Medicine*, 8: 537–542.

Eriksson, B.O., Karlsson, J., and Saltin, B. (1971) Muscle metabolites during exercise in pubertal boys. *Acta Paediatrica Scandinavica*, 217 (Suppl): 154–157.

Eriksson, B.O., Gollnick, P.D., and Saltin, B. (1973) Muscle metabolism and enzyme activities after training in boys 11–13 years old. *Acta Physiologica Scandinavica*, 87: 485–497.

Esparza, J., Fox, C., Harper, I.T., Bennett, P.H., Schulz, L.O., Valencia, M.E., and Ravussin, E. (2000) Daily energy expenditure in Mexican and USA Pima indians: low physical activity as a possible cause of obesity. *International Journal of Obesity Related Metabolic Disorders*, 24: 55–59.

Essig, D., Costill, D.L., and van Handel, P.J. (1980) Effect of caffeine ingestion on utilization of muscle glycogen and lipid during leg ergometry cycling. *International Journal of Sports Medicine*, 1: 86–90.

Eston, R.G., Rowlands, A.V., and Ingledew, D.K. (1998) Validity of heart rate, pedometry, and accelerometry for predicting the energy cost of children's activities. *Journal of Applied Physiology*, 84: 362–371.

Evans, G.W. (1989) The effect of chromium picolinate on insulin controlled parameters in humans. *International Journal of Biosocial and Medical Research*, 11: 163–180.

Fagard, R.H. and Cornelissen, V.A. (2007) Effect of exercise on blood pressure control in hypertensive patients. *European Journal of Cardiovascular Prevention & Rehabilitation*, 14: 12–17.

Failla, M.L. (2003) Trace elements and host defense: recent advances and continuing challenges. *Journal of Nutrition*, 133: S1443–S1447.

Falk, B., Bar-Or, O., Calvert, R., and MacDougall, J.D. (1992) Sweat gland response to exercise in the heat among pre-, mid-, and late-pubertal boys. *Medicine & Science in Sports & Exercise*, 24: 313–319.

Fawkner, S.G. and Armstrong, N. (2003) Oxygen uptake kinetic response to exercise in children. *Sports Medicine*, 33: 651–669.

Felig, P. and Wahren, J. (1971) Amino acids metabolism in exercising man. *Journal of Clinical Investigation*, 50: 2703–2714.

Ferrando, A.A. and Green, N.R. (1993) The effect of boron supplementation on lean body mass, plasma testosterone levels, and strength in male bodybuilders. *International Journal of Sport Nutrition*, 3: 140–149.

Ferrannini, E. (1988) The theoretical basis of indirect calorimetry: a review. *Metabolism*, 37: 287–301.

Ferrannini, E., Barrett, E.J., Bevilacqua, S., and DeFronzo, R. (1983) Effects of fatty acids on glucose production and utilization in man. *Journal of Clinical Investigation*, 72: 1737–1747.

Ferrara, C.M., Goldberg, A.P., Ortmeyer, H.K., and Ryan, A.S. (2006) Effects of aerobic and resistive exercise training on glucose disposal and skeletal muscle metabolism in older men. *Journal of Gerontology*, 61: 480–487.

Feskanich, D., Rimm, E.B., Giovannucci, E.L., Colditz, G.A., Stampfer, M.J., Litin, L.B., and Willett, W.C. (1993) Reproducibility and validity of food intake measurements from a semiquantitative food frequency questionnaire. *Journal of the American Dietetic Association*, 93: 790–796.

Fielding, R.A. and Parkington, J. (2002) What are the dietary protein requirements of physically active individuals? New evidence on the effects of exercise on protein utilization during post-exercise recovery, *Nutrition in Clinical Care*, 5: 191–196.

Fine, B.J., Kobrick, J.L., Lieberman, H.R., Marlowe, B., Riley, R.H., and Tharion, W.J. (1994) Effects of caffeine or diphenhydramine on visual vigilance. *Psychopharmacology*, 114: 233–238.

Fogelholm, G.M., Näveri, H.K., Kiilavuori, K.T., and Härkönen, M.H. (1993) Low-dose amino acid supplementation: no effects on serum human growth hormone and insulin in male weightlifters. *International Journal of Sport Nutrition*, 3: 290–297.

Foster, G.D., Wyatt, H.R., Hill, J.O., McGuckin, B.G., Brill, C., Mohammed, B.S., Szapary, P.O., Rader, D.J., Edman, J.S., and Klein, S. (2003) A randomized trial of a low-carbohydrate diet for obesity. *The New England Journal of Medicine*, 348: 2082–2090.

Foster-Powell, K., Holt Susanna, H.A., and Brand-Miller, J.C. (2002) International table of glycemic index and glycemic load values. *The American Journal of Clinical Nutrition*, 76: 5–56.

Fraker, P.J., King, L.E., Laakko, T., and Vollmer, T.L. (2000) The dynamic link between the integrity of the immune system and zinc status. *Journal of Nutrition*, 130: S1399S–S1406.

Francis, P.R., Witucki, A.S., and Buono, M.J. (1999) Physiological response to a typical studio cycling session, *ACSM Health & Fitness Journal*, 3: 30–36.

Frayn, K.N. (1983) Calculation of substrate oxidation rates in vivo from gaseous exchange. *Journal of Applied Physiology*, 55: 628–634.

Freedson, P.S. and Miller, K. (2000) Objective monitoring of physical activity using motion sensors and heart rate. *Research Quarterly for Exercise & Sport*, 71: 21–29.

Friedlander, A.L., Casazza, G.A., Horning, M.A., Huie, M.J., Piacentini, M.F., Trimmer, J.K., and Brooks, G.A. (1998) Training-induced alterations of carbohydrate metabolism in women: women respond differently from men, *Journal of Applied Physiology*, 85: 1175–1186.

Friedman, M.I. (1995) Control of energy intake by energy metabolism. *The American Journal of Clinical Nutrition*, 62: S1096–S1100.

Frisancho, A.R. and Flegel, P.N. (1983) Elbow breadth as a measure of frame size for US males and females. *The American Journal of Clinical Nutrition*, 37: 311–314.

Fruin, M.L. and Walberg Rankin, J. (2004) Validity of multi-sensor armband in estimating rest and exercise energy expenditure. *Medicine & Science in Sports & Exercise*, 36: 1063–1069.

Fung, T.T., Hu, F.B., Pereira, M.A., Liu, S., Stampfer, M.J., Colditz, G.A., and Willett, W.C. (2002) Whole-grain intake and the risk of type 2 diabetes: a prospective study in men. *The American Journal of Clinical Nutrition*, 76: 535–540.

Galloway, S.D., Tremblay, M.S., Sexsmith, J.R., and Roberts, C.J. (1996) The effects of acute phosphate supplementation in subjects of different aerobic fitness levels. *European Journal of Applied Physiology and Occupational Physiology*, 72: 224–230.

Gambling, L., Danzeisen, R., Fosset, C., Andersen, H.S., Dunford, S., Srai, S.K., and McArdle, H.J. (2003) Iron and copper interactions in development and the effect on pregnancy outcome. *Journal of Nutrition*, 133: S1554–S1556.

Ganji, S.H., Kamanna, V.S., and Kashyap, M.L. (2003) Niacin and cholesterol: role in cardiovascular disease (review). *Journal of Nutritional Biochemistry*, 14: 298–305.

Gardner, J.W., Kark, J.A., Karnei, K., Sanborn, J.S., Gastaldo, E., Burr, P., and Wenger, C.B. (1996) Risk factors predicting exertional heat illness in male Marine Corps recruits. *Medicine & Science in Sports & Exercise*, 28: 939–944.

Garrow, J.S. (1995) Exercise in the treatment of obesity: a marginal contribution. *International Journal of Obesity*, 19: S126–S129.

Gatalano, P.M., Tyzbir, E.D., and Roman, N.M. (1991) Longitudinal changes in insulin release and insulin resistance in non-obese pregnant women. *American Journal of Obstetrics & Gynecology*, 165: 1667–1672.

Gatalano, P.M., Tyzbir, E.D., Wolfe, R.R., Roman, N.M., Amini, S.B., and Sims, E.A.H. (1992) Longitudinal changes in basal hepatic glucose production and suppression during insulin infusion in normal pregnant women. *American Journal of Obstetrics & Gynecology*, 167: 913–919.

Gauche, E., Lepers, R., Rabita, G., Leveque, J.M., Bishop, D., Brisswalter, J., and Hausswirth, C. (2006) Vitamin and mineral supplementation and neuromuscular recovery after a running race. *Medicine & Science in Sports & Exercise*, 38: 2110–2117.

Giamberardino, M.A., Dragani, L., Valente, R., Di Lisa, F., Saggini, R., and Vecchiet, L. (1996) Effects of prolonged L-carnitine administration on delayed muscle pain and CK release after eccentric effort. *International Journal of Sports Medicine*, 17: 320–324.

Gill, N.D., Shield, A., Blazevich, A.J., Zhou, S., and Weatherby, R.P. (2000) Muscular and cardiorespiratory effects of pseudoephedrine in human athletes. *British Journal of Clinical Pharmacology*, 50: 205–213.

Gillette, C.A., Bullough, R.C., and Melby, C.L. (1994) Post-exercise energy expenditure in response to acute aerobic or resistive exercise. *International Journal of Sports Nutrition*, 4: 347–360.

Gisolfi, C.V. and Cohen, J.S. (1979) Relationships among training, heat acclimation, and heat tolerance in men and women: the controversy revisited. *Medicine & Science in Sports & Exercise*, 11: 56–59.

Gleeson, M. and Bishop, N.C. (2000) Elite athlete immunology: importance of nutrition. *International Journal of Sports Medicine*, 21: S44–S50.

Gollnick, P. (1985) Metabolism of substrates: energy substrate metabolism during exercise and as modified by training. *Federation Proceedings*, 44: 353–356.

Gollnick, P., Piehl, K., and Saltin, B. (1974) Selective glycogen depletion pattern in human muscle fibers after exercise of varying intensity and at varying pedal rates. *Journal of Physiology*, 241: 45–57.

González-Alonso, J., Calbet, J.A., and Nielsen, B. (1999) Metabolic and thermodynamic responses to dehydration-induced reductions in muscle blood flow in exercising humans. *Journal of Physiology*, 520 (Pt 2): 577–589.

Goodpaster, B.H., Kelley, D.E., Wing, R.R., Meier, A., and Thaete, F.L. (1999) Effects of weight loss on regional fat distribution and insulin sensitivity in obesity. *Diabetes*, 48: 839–847.

Goodpaster, B.H., Thaete, F.L., and Kelley, D.E. (2000) Thigh adipose tissue distribution is associated with insulin resistance in obesity and in type 2 diabetes mellitus. *The American Journal of Clinical Nutrition*, 71: 885–892.

Goodpaster, B.H., Wolfe, R.R., and Kelley, D.E. (2002) Effect of obesity on substrate utilization during exercise. *Obesity Research*, 10: 575–584.

Goran, M.I. and Poehlman, E.T. (1992) Total energy expenditure and energy requirements in healthy elderly persons. *Metabolism*, 41: 744–753.

Gortmaker, S.L., Must, A., Sobol, A.M., Peterson, K., Colditz, G.A., and Dietz, W.H. (1996) Television viewing as a cause of increasing obesity among children in the United States, 1986–1990. *Archives of Pediatrics & Adolescent Medicine*, 150: 356–362.

Graham, S., Zielezny, M., Marshall, J., Priore, R., Freudenheim, J., Brasure, J., Haughey, B., Nasca, P., and Zdeb, M. (1992) Diet in the epidemiology of postmenopausal breast cancer in the New York State Cohort. *American Journal of Epidemiology*, 136: 1327–1337.

Graham, T.E. (2001) Caffeine, coffee and ephedrine: impact on exercise performance and metabolism. *Canadian Journal of Applied Physiology*, 26: S103–S119.

Graham, T.E. and MacLean, D.A. (1992) Ammonia and amino acid metabolism in human skeletal muscle during exercise. *Canadian Journal of Physiology and Pharmacology*, 70: 132–141.

Graham, T.E., Helge, J.W., MacLean, D.A., Kiens, B., and Richter, E.A. (2000) Caffeine ingestion does not alter carbohydrate or fat metabolism in human skeletal muscle during exercise. *Journal of Physiology*, 529: 837–847.

Grant, J.P., Custer, P.B and Thurlow, J. (1981) Current techniques of nutritional assessment. *Surgical Clinics of North America*, 61: 437–463.

Green, A.L., Hultman, E., Macdonald, I.A., Sewell, D.A., and Greenhaff, P.L. (1996) Carbohydrate ingestion augments skeletal muscle creatine accumulation during creatine supplementation in humans. *American Journal of Physiology*, 271: E821–E826.

Green, S.M. and Blundell, J.E. (1996) Effect of fat- and sucrose-containing foods on the size of eating episodes and energy intake in lean dietary restrained and unrestrained females: potential for causing overconsumption. *European Journal of Clinical Nutrition*, 50: 625–635.

Greiwe, J.S., Staffey, K.S., Melrose, D.R., Narve, M.D., and Knowlton, R.G. (1998) Effects of dehydration on isometric muscular strength and endurance. *Medicine & Science in Sports & Exercise*, 30: 284–288.

Gross, M.D. (2005) Vitamin D and calcium in the prevention of prostate and colon cancer: new approaches for the identification of needs. *Journal of Nutrition*, 135: 326–331.

Guenther, P.M. (1994) Research needs for dietary assessment and monitoring in the United States. *The American Journal of Clinical Nutrition*, 59 (Suppl.): S168–S170.

Guerrero-Romero, F. and Rodríguez-Morán, M. (2005) Complementary therapies for diabetes: the case for chromium, magnesium, and antioxidants. *Archives of Medical Research*, 36: 250–257.

Hackney, A.C. (1990) Effects of menstrual cycle on resting muscle glycogen content. *Hormone and Metabolic Research*, 22: 647.

Hackney, A.C., McCracken-Compton, M.A., and Ainsworth, B. (1994) Substrate responses to submaximal exercise in the midfollicular and midluteal phases of the menstrual cycle. *International Journal of Sport Nutrition*, 4: 299–308.

Hakim, A.A., Curb, J.D., and Petrovich, H. (1999) Effects of walking on coronary heart disease in elderly men: the Honolulu Heart Study. *Circulation*, 100: 9–13.

Hallmark, M.A., Reynolds, T.H., DeSouza, C.A., Dotson, C.O., Anderson, R.A., and Rogers, M.A. (1996) Effects of chromium and resistive training on muscle strength and body composition. *Medicine & Science in Sports & Exercise*, 28: 139–144.

Halton, R.W., Kraemer, R.R., Sloan, R.A., Hebert, E.P., Frank, K., and Tryniecki, J.L. (1999) Circuit weight training and its effect on excess postexercise oxygen consumption. *Medicine & Science in Sports & Exercise*, 31: 1613–1618.

Hamilton, K.S., Gibbons, F.K., Lacy, D.P., Cherrington, A.D., and Wasserman, D.H. (1996) Effect of prior exercise on the partitioning of an intestinal glucose load between splanchnic bed and skeletal muscle. *Journal of Clinical Investigation*, 98: 125–135.

Hansen, F.M., Fahmy, N., and Nielsen, J.H. (1980) The influence of sexual hormones on lipogenesis and lipolysis in rat cells. *Acta Endocrinologica*, 95: 566–570.

Hargreaves, M. (2006) Skeletal muscle carbohydrate metabolism during exercise. In: *Exercise Metabolism*, Hargreaves, M. and Spriet, L., editors, Champaign, IL: Human Kinetics, pp. 29–44.

Hargreaves, M. and Proietto, J. (1994) Glucose kinetics during exercise in trained men. *Acta Physiologica Scandinavica*, 150: 221–225.

Harris, D.M. and Go, V.L.W. (2004) Vitamin D and colon carcinogenesis. *Journal of Nutrition*, 134: S3463S–S3471.

Haskell, W.L., Yee, M.C., Evans, A., and Irby, P.J. (1993) Simultaneous measurement of heart rate and body motion to quantitate physical activity. *Medicine and Science in Sports*, 25: 109–115.

Hasten, D.L., Rome, E.P., Franks, B.D., and Hegsted, M. (1992) Effects of chromium picolinate on beginning weight training students. *International Journal of Sport Nutrition*, 2: 343–350.

Hatano, Y. (1993) Use of the pedometer for promoting daily walking exercise. *Journal of the International Council for HPER*, 29: 4–8.

Hawley, J.A., Schabort, E.J., Noakes, T.D., and Dennis, S.C. (1997) Carbohydrate-loading and exercise performance: an update. *Sports Medicine*, 24: 73–81.

Hayward and Stolarczyk (1996) *Applied Body Composition Assessment*. Champaign, IL: Human Kinetics.

Hebert, D.N. and Carruthers, A. (1992) Glucose transporter oligomeric structure determines transporter function: reversible redox-dependent interconversions of tetrameric and dimeric GLUT1. *Journal of Biological Chemistry*, 267: 23829–23838.

Heath, B.H. and Carter, J.E. (1967) A modified somatotype method. *American Journal of Physical Anthropology*, 27: 57–74.

Heath, G.W., Gavin, J.R., III, Hinderliter, J.M., Hagberg, J.M., Bloomfield, S.A., and Holloszy, J.O. (1983) Effects of exercise and lack of exercise on glucose tolerance and insulin sensitivity. *Journal of Applied Physiology*, 55: 512–517.

Heath, G.W., Gavin, J.R. III, Hinderliter, J.M., Hagberg, J.M., Bloomfield, S.A., Hebert, D.N., and Carruthers, A. (1992), Glucose transporter oligomeric structure determines transporter function: reversible redox-dependent interconversions of tetrameric and dimeric GLUT1. *Journal of Biological Chemistry*, 267: 23829–23838.

Hebestreit, H., Kriemler, S., Hughson, R.L., and Bar-Or, O. (1998) Kinetics of oxygen uptake at the onset of exercise in boys and men. *Journal of Applied Physiology*, 85: 1833–1841.

Heinonen, O.J. (1996) Carnitine and physical exercise. *Sports Medicine*, 22: 109–132.

Hendelman, D., Miller, K., Baggett, C., Debold, E., and Freedson, P. (2000) Validity of accelerometry for the assessment of moderate intensity physical activity in the field. *Medicine & Science in Sports & Exercise*, 32: S442–S449.

Heyward, V.H. (2002) *Advanced Fitness Assessment and Exercise Prescription*, 4th edn, Champaign, IL: Human Kinetics.

Heyward, V.H. and Stolarczyk, L.M. (1996) *Applied Body Composition Assessment*, Champaign, IL: Human Kinetics.

Hill, R.J. and Davis, P.S. (2002) Energy intake and energy expenditure in elite lightweight female rowers, *Medicine & Science in Sports & Exercise*, 34: 1823–1827.

Holick, M.F. (2004) Sunlight and vitamin D for bone health and prevention of autoimmunine diseases, cancers, and cardiovascular disease. *The American Journal of Clinical Nutrition*, 80 (6): S1678–S1688.

Holloszy, J. (1983) Effects of exercise and lack of exercise on glucose tolerance and insulin sensitivity. *Journal of Applied Physiology*, 55: 512–517.

Holloszy, J. and Coyle, E. (1984) Adaptations of skeletal muscle to endurance exercise and their metabolic consequences. *Journal of Applied Physiology*, 56: 831–838.

Holloszy, J.O., Chen, M., Cartee, G.D., and Young, J.C. (1991) Skeletal muscle atrophy in old rats: differential changes in the three fibers types. *Mechanisms of Ageing and Development*, 60: 199–213.

Holten, M.K., Zacho, M., Gaster, M., Juel, C., Wojtaszewski, J.F., and Dela, F. (2004) Strength training increases insulin-mediated glucose uptake, GLUT4 content, and insulin signaling in skeletal muscle in patients with type 2 diabetes. *Diabetes*, 53: 294–305.

Horowitz, J.F. and Klein, S. (2000) Oxidation of nonplasma fatty acids during exercise is increased in women with abdominal obesity. *Journal of Applied Physiology*, 89: 2276–2282.

Horswill, C.A. (1995) Effects of bicarbonate, citrate, and phosphate loading on performance. *International Journal of Sport Nutrition*, 5: S111–S119.

Hultman, E., Greenhaff, P.L., Ren, J.M., and Söderlund, K. (1991) Energy metabolism and fatigue during intense muscle contraction. *Biochemical Society Transactions*, 19: 347–353.

Hultman, E., Söderlund, K., Timmons, J.A., Cederblad, G., and Greenhaff, P.L. (1996) Muscle creatine loading in men. *Journal of Applied Physiology*, 81: 232–237.

Hunter, G., Blackman, L., Dunnam, L., and Flemming, G. (1988) Bench press metabolic rate as a function of exercise intensity. *Journal of Applied Sport Science Research*, 2: 1–6.

Hunter, G.R., Kekes-Szabo, T., and Schnitzler, A. (1992) Metabolic cost: vertical work ratio during knee extension and knee flexion weight-training exercise. *Journal of Applied Sport Science Research*, 6: 42–48.

Inder, W.J., Swanney, M.P., Donald, R.A., Prickett, T.C., and Hellemans, J. (1998) The effect of glycerol and desmopressin on exercise performance and hydration in triathletes. *Medicine & Science in Sports & Exercise*, 30: 1263–1269.

Institute of Medicine (2005) *Dietary Reference Intakes for Energy, Carbohydrate, Fiber, Fat, Fatty Acids, Cholesterol, Protein, and Amino Acids (Macronutrients)*, Washington, DC: National Academies Press.

Inui, A., Asakawa, A., Bowers, C.Y., Mantovani, G., Laviano, A., Meguid, M.M., and Fujimiya, M. (2004) Ghrelin, appetite, and gastric motility: the emerging role of the stomach as an endocrine organ. *The FASEB Journal*, 18: 439–456.

Isidori, A., Lo Monaco, A., and Cappa, M. (1981) A study of growth hormone release in man after oral administration of amino acids. *Current Medical Research & Opinion*, 7: 475–481.

Issekutz, B. and Paul, P. (1968) Intramuscular energy sources in exercising normal and pancreatecomized dogs. *American Journal of Physiology*, 215: 197–204.

Ivy, J.L., Frishberg, B.A., Farrell, S.W., Miller, W.J., and Sherman, W.M. (1985) Effect of elevated and exercise-induced muscle glycogen levels on insulin sensitivity. *Journal of Applied Physiology*, 59: 154–159.

Ivy, J.L., Katz, A.L., Cutler, C.L., Sherman, W.M., and Coyle, E.F. (1988a) Muscle glycogen synthesis after exercise: effect of time of carbohydrate ingestion. *Journal of Applied Physiology*, 64: 1480–1485.

Ivy, J.L., Lee, M.C., Brozinick, J.T. Jr. and Reed, M.J. (1988b) Muscle glycogen storage after different amounts of carbohydrate ingestion. *Journal of Applied Physiology*, 65: 2018–2023.

Izawa, T., Komabayashi, T., Mochizuki, T., Suda, K., and Tsuboi, M. (1991) Enhanced coupling of adenylate cyclase to lipolysis in permeabilized adipocytes from trained rats. *Journal of Applied Physiology*, 71: 23–29.

Jackson, A.S. and Pollock, M.L. (1978) Generalized equations for predicting body density of men. *British Journal of Nutrition*, 40: 497–504.

Jackson, A.S., Pollock, M.L., and Ward, A. (1980) Generalized equations for predicting body density of women. *Medicine & Science in Sports & Exercise*, 12: 175–182.

Jackson, J.L., Lesho, E., and Peterson, C. (2000) Zinc and the common cold: a meta-analysis revisited. *Journal of Nutrition*, 130: S1512–S1515.

Jakicic, J.M., Wing, R.R., Butler, B.A., and Robertson, R.J. (1995) Prescribing exercise in multiple short bouts versus one continuous bout: effect on adherence, cardiorespiratory fitness, and weight loss in overweight women. *International Journal of Obesity*, 19: 893–901.

Jakicic, J.M., Winters, C., Lang, W., and Wing, R.R. (1999) Effects of intermittent exercise and use of home exercise equipment on adherence, weight loss, and fitness in overweight women: a randomized trial. *Journal of the American Medical Association*, 282: 1554–1560.

Jakicic, J.M., Clark, K., Coleman, E., Donnelly, J.E., Foreyt, J., Melanson, E., Volek, J., and Volpe, S.L. (2001) American College of Sports Medicine position stand: appropriate intervention strategies for weight loss and prevention of weight regain for adults. *Medicine & Science in Sports & Exercise*, 33: 2145–2156.

Jakicic, J.M., Marcus, M., Gallagher, K.I., Randall, C., Thomas, E., Goss, F.L., and Robertson, R.J. (2004) Evaluation of SenseWear Pro ArmbandTM to access energy expenditure during exercise, *Medicine & Science in Sports & Exercise*, 36: 897–904.

Jakubowicz, D., Beer, N., and Rengifo, R. (1995) Effect of dehydroepiandrosterone on cyclic-guanosine monophosphate in men of advancing age. *Annals of the New York Academy of Sciences*, 774: 312–315.

Jang, K.T., Flynn, M.G., Costill, D.L., Kirwan, J.P., Houmard, J.A., Mitchell, J.B., and D'Acquisto, L.J. (1987) Energy balance in competitive swimmers and runners. *Journal of Swimming Research*, 3: 19–24.

Jansson, P.A., Smith, U., and Lonnroth, P. (1990) Interstitial glycerol concentration measured by microdialysis in two subcutaneous regions in humans. *American Journal of Physiology*, 258: E918–E922.

Janz, K.F., Witt, J., and Mahoney, L.T. (1995) The stability of children's physical activity as measured by accelerometry and self-report. *Medicine & Science in Sports & Exercise*, 27: 1326–1332.

Jenkins, A.B., Chisholm, D.J., James, D.E., Ho, K.Y., and Kraegen, E.W. (1985) Exercise induced hepatic glucose output is precisely sensitive to the rate of systemic glucose supply, *Metabolism*, 34: 431–441.

Jenkins, A.B., Furler, S.M., Chisholm, D.J., and Kraegen, E.W. (1986) Regulation of hepatic glucose output during exercise by circulating glucose and insulin in humans. *American Journal of Physiology*, 250: R411–R417.

Jeukendrup, A. and Gleeson, M. (2004) *Sports Nutrition: An Introduction to Energy Production and Performance*, Champaign, IL: Human Kinetics.

Jeukendrup, A. and Jentjens, R. (2000) Oxidation of carbohydrate feedings during prolonged exercise: current thoughts, guidelines and directions for future research. *Sports Medicine*, 29: 407–424.

Johansson, L., Solvoll, K., Bjørneboe, G.E., and Drevon, C.A. (1998) Under- and overreporting of energy intake related to weight status and lifestyle in a nationwide sample. *The American Journal of Clinical Nutrition*, 68 (2): 266–274.

Kaciuba-Uscilko, H. and Grucza, R. (2001) Gender differences in thermoregulation. *Current Opinion in Clinical Nutrition & Metabolic Care*, 4: 533–536.

Kang, J. (2008) *Bioenergetics Primer for Exercise Science*. Champaign, IL: Human Kinetics.

Kang, J., Robertson, R.J., Hagberg, J.M., Kelley, D.E., Goss, F.L., DaSilva, S.G., Suminski, R.R., Utter, A.C. (1996) Effect of exercise intensity on glucose and insulin metabolism in obese individuals and obese NIDDM patients, *Diabetes Care*, 19: 341–349.

Kang, J., Kelley, D.E., Roberston, R.J., Goss, F.L., Suminski, R.R., Utter, A.C., and Dasilva, S.G. (1999) Substrate utilization and glucose turnover during exercise of varying intensities in individuals with NIDDM. *Medicine & Science in Sports & Exercise*, 31: 82–89.

Kang, J., Edward, E.C., Mastrangelo, M.A., Hoffman, J.R., Ratamess, N.A., and O'Connor, E. (2005a) Metabolic and perceptual responses during Spinning® cycle exercise. *Medicine & Science in Sports & Exercise*, 37: 53–59.

Kang, J., Hoffman, J.R., Im, J., Spiering, B.A., Ratamess, N.A., Rundell, K.W., Nioka, S., Cooper, J., and Chance, B. (2005b) Evaluation of physiological responses during recovery following three resistance exercise programs. *Journal of Strength & Conditioning Research*, 19: 305–309.

Kang, J., Hoffman, J.R., Ratamess, N.A., Faigenbaum, A.D., Falvo, M., and Wendell, M. (2007) Effect of exercise intensity on fat utilization in males and females. *Research in Sports Medicine*, 15: 175–188.

Kaplan, G.B., Greenblatt, D.J., Ehrenberg, B.L., Goddard, J.E., Cotreau, M.M., Harmatz, J.S., and Shader, R.I. (1997) Dose-dependent pharmacokinetics and psychomotor effects of caffeine in humans. *Journal of Clinical Pharmacology*, 37: 693–703.

Katan, M.B., Zock, P.L., and Mensink, R.P. (1995) Dietary oils, serum lipoproteins, and coronary heart disease. *The American Journal of Clinical Nutrition*, 61: S1368–S1373.

Katch, V.L., Katch, F.I., Moffatt, R., and Gittleson, M. (1980) Muscular development and lean body weight in body builders and weight lifters. *Medicine & Science in Sports & Exercise*, 12: 340–344.

Katch, V.L., Freedson, P.S., Katch, F.I., Smith, L. (1982) Body frame size: validity of self-appraisal. *The American Journal of Clinical Nutrition*, 36: 676–679.

Kelley, D.E. (2005) Skeletal muscle fat oxidation: timing and flexibility are everything. *Journal of Clinical Investigation*, 115: 1699–1702.

Kelley, D.E., Mokan, M., and Mandarino, L.J. (1992) Intracellular defects in glucose metabolism in obese patients with NIDDM. *Diabetes*, 41: 698–706.

Kendrick, Z.V., Steffen, C., Rumsey, W., and Goldberg, D. (1987) Effect of estradiol on tissue glycogen metabolism in exercise oophorectomized rats. *Journal of Applied Physiology*, 63: 492–496.

Kennedy, C. and Sokoloff, L. (1957) An adaptation of the nitrous oxide methods to the study of the cerebral circulation of children: normal values for cerebral blood flow and cerebral metabolic rate in childhood. *Journal of Clinical Investigation*, 36: 1130–1137.

Kenney, W.L. and Chiu, P. (2001) Influence of age on thirst and fluid intake. *Medicine & Science in Sports & Exercise*, 33: 1524–1532.

Kent, M. (2003) *Food and Fitness: A Dictionary of Diet and Exercise*, New York: Oxford University Press.

Keys, A. and Brozek, J. (1953) Body fat in adult man. *Physiological Reviews*, 33: 245–325.

King, D.S., Dalsky, G.P., Staten, M.A., Clutter, W.E., van Houten, D.R., and Holloszy, J.O. (1987) Insulin action and secretion in endurance-trained and untrained humans. *Journal of Applied Physiology*, 63: 2247–2252.

King, D.S., Sharp, R.L., Vukovich, M.D., Brown, G.A., Reifenrath, T.A., Uhl, N.L., and Parsons, K.A. (1999) Effect of oral androstenedione on serum testosterone and adaptations to resistance training in young men: a randomized controlled trial. *Journal of the American Medical Association*, 281: 2020–2028.

Kjær, M. (1995) Hepatic fuel metabolism during exercise, In: *Exercise Metabolism*, Hargreaves, M., editor, Champaign, IL: Human Kinetics, pp. 73–97.

Kjær, M., Kiens, B., Hargreaves, M., and Richter, E.A. (1991) Influence of active muscle mass on glucose homeostasis during exercise in humans. *Journal of Applied Physiology*, 71: 552–557.

Klein, S. (2004) Clinical trial experience with fat-restricted vs. carbohydrate-restricted weight-loss diets. *Obesity Research*, 12: S141S–S144.

Kline, K., Yu, W., and Sanders, B.G. (2004) Vitamin E and breast cancer. *Journal of Nutrition*, 134: S3458–S3462.

Knuttgen, H.G. and Emerson, K., Jr. (1974) Physiological responses to pregnancy at rest and during exercise. *Journal of Applied Physiology*, 36: 549–553.

Koranyi, L.I., Bourey, R.E., Slentz, C.A., and Holloszy, J.O. (1991) Coordinate reduction of rat pancreatic islet glucokinase and proinsulin mRNA by exercise training. *Diabetes*, 40: 401–404.

Kraemer, W.J., Volek, J.S., Clark, K.L., Gordon, S.E., Puhl, S.M., Koziris, L.P., McBride, J.M.,

Triplett-McBride, N.T., Putukian, M., Newton, R.U., Häkkinen, K., Bush, J.A., and Sebastianelli, W.J. (1999) Influence of exercise training on physiological and performance changes with weight loss in men. *Medicine & Science in Sports & Exercise*, 31: 1320–1329.

Krauss, R.M. (2004) Lipids and lipoproteins in patients with type 2 diabetes. *Diabetes Care*, 27: 1496–1504.

Krauss, R.M., Deckelbaum, R.J., Ernst, N., Fisher, E., Howard, B.V., Knopp, R.H., Kotchen, T., Lichtenstein, A.H., McGill, H.C., Pearson, T.A., Prewitt, T.E., Stone, N.J., Horn, L.V., and Weinberg, R. (1996) Dietary guidelines for healthy American adults: a statement for health professionals from the Nutrition Committee, American Heart Association. *Circulation*, 94: 1795–1800.

Kreider, R.B., Miller, G.W., Williams, M.H., Somma, C.T., Nasser, T.A. (1990) Effects of phosphate loading on oxygen uptake, ventilatory anaerobic threshold, and run performance. *Medicine & Science in Sports & Exercise*, 22: 250–256.

Kreider, R.B., Miller, G.W., Schenck, D., Cortes, C.W., Miriel, V., Somma, C.T., Rowland, P., Turner, C., and Hill, D. (1992) Effects of phosphate loading on metabolic and myocardial responses to maximal and endurance exercise. *International Journal of Sport Nutrition*, 2: 20–47.

Kreider, R.B., Ferreira, M., Wilson, M., and Almada, A.L. (1999) Effects of calcium beta-hydroxy-beta-methylbutyrate (HMB) supplementation during resistance-training on markers of catabolism, body composition and strength. *International Journal of Sports Medicine*, 20: 503–509.

Kushi, L.H., Meyer, K.A., and Jacobs, D.R., Jr. (1999) Cereals, legumes, and chronic disease risk reduction: evidence from epidemiologic studies. *The American Journal of Clinical Nutrition*, 70: S451–S458.

Lamb, D.R. and Shehata, A. (1999) Benefits and limitations to prehydration. *Sports Science Exchange*, 12: 1–6.

Lambert, M.I., Hefer, J.A., Millar, R.P., and Macfarlane, P.W. (1993) Failure of commercial oral amino acid supplements to increase serum growth hormone concentrations in male bodybuilders. *International Journal of Sport Nutrition*, 3: 298–305.

Lancaster, G.I., Khan, Q., Drysdale, P.T., Wallace, F., Jeukendrup, A.E., Drayson, M.T., and Gleeson, M. (2005) Effect of prolonged exercise and carbohydrate ingestion on type 1 and type 2 T lymphocyte distribution and intracellular cytokine production in humans. *Journal of Applied Physiology*, 98: 565–571.

Lang, R. and Jebb, S.A. (2003) Who consumes whole grains, and how much? *Proceedings of the Nutrition Society*, 62: 123–127.

Latzka, W.A., Sawka, M.N., Montain, S.J., Skrinar, G.S., Fielding, R.A., Matott, R.P., and Pandolf, K.B. (1997) Hyperhydration: thermoregulatory effects during compensable exercise-heat stress. *Journal of Applied Physiology*, 83: 860–866.

Latzka, W.A., Sawka, M.N., Montain, S.J., Skrinar, G.S., Fielding, R.A., Matott, R.P., and Pandolf, K.B. (1998) Hyperhydration: tolerance and cardiovascular effects during uncompensable exercise-heat stress. *Journal of Applied Physiology*, 84: 1858–1864.

LeBlanc, J., Diamond, P., Cote, J., and Labrie, A. (1984) Hormonal factors in reduced postprandial heat production of exercise-trained subjects. *Journal of Applied Physiology*, 56: 772–776.

Leder, B.Z., Longcope, C., Catlin, D.H., Ahrens, B., Schoenfeld, D.A., and Finkelstein, J.S. (2000) Oral androstenedione administration and serum testosterone concentrations in young men. *Journal of the American Medical Association*, 283: 779–782.

Lee, J.S., Bruce, C.R., Tunstall, R.J., Cameron-Smith, D., Hugel, H., and Hawley, J.A. (2002) Interaction of exercise and diet on GLUT-4 protein and gene expression in type I and type II rat skeletal muscle. *Acta Physiologica Scandinavica*, 175: 37–44.

Lee, M.F. and Krasinski, S.D. (1998) Human adult-onset lactase decline: an update. *Nutrition Reviews*, 56: 1–8.

Lee-Han, H., McGuire, V., and Boyd, N.F. (1989) A review of the methods used by studies of dietary measurement. *Journal of Clinical Epidemiology*, 42: 269–279.

Leibel, R.L., Rosenbaum, M., and Hirsch, J. (1995) Changes in energy expenditure resulting from altered body weight. *The New England Journal of Medicine*, 332: 621–628.

Lemon, P. and Mullin, J. (1980) Effect of initial muscle glycogen levels on protein catabolism during exercise. *Journal of Applied Physiology*, 48: 624–629.

Lemon, P.W.R., Tarnopolsky, M.A., MacDougall, J.D., and Atkinson, S.A. (1992) Protein requirements and muscle mass/strength changes during intensive training in novice bodybuilders. *Journal of Applied Physiology*, 73: 767–775.

Lieberman, H.R., Wurtman, R.J., Emde, G.G., and Coviella, I.L. (1987) The effects of caffeine and aspirin on mood and performance. *Journal of Clinical Psychopharmacology*, 7: 315–320.

Liu, S., Manson, J.E., Stampfer, M.J., Hu, F.B., Giovannucci, E., Colditz, G.A., Hennekens, C.H., and Willett, W.C. (2000) A prospective study of whole-grain intake and risk of type 2 diabetes mellitus in US women. *American Journal of Public Health*, 90: 1409–1415.

Lockner, D.W., Heyward, V.H., Baumgartner, R.N., and Jenkins, K.A. (2000) Comparison of air-displacement plethysmography, hydrodensitometry, and dual X-ray absorptiometry for assessing body composition of children 10 to 18 years of age. *Annals of the New York Academy of Sciences*, 904: 72–78.

Lohman, T.G. (1988) Anthropometry and body composition. In *Anthropometric Standardization Reference Manual*, Lohman, T.G., Roche, A.F., and Martorell, R., editors, Champaign, IL: Human Kinetics.

Lohman, T.G. and Going, S.B. (1993) Multicomponent models in body composition research: opportunities and pitfalls. *Basic Life Sciences*, 60: 53–58.

Lohman, T.G., Houlkooper, L., and Going, S.B. (1997), Body fat measurement goes high tech: not all are created equal. *ACSM's Health and Fitness Journal*, 1: 32.

Lönnqvist, F., Nyberg, B., Wahrenberg, H., and Arner, P. (1990) Catecholamine-induced lipolysis in adipose tissue of the elderly. *Journal of Clinical Investigation*, 85: 1614–1621.

Lovejoy, J.C., Bray, G.A., Lefevre, M., Smith, S.R., Most, M.M., Denkins, Y.M., Volaufova, J., Rood, J.C., Eldridge, A.L., and Peters, J.C. (2003) Consumption of a controlled low-fat diet containing olestra for 9 months improves health risk factors in conjunction with weight loss in obese men: the Ole' Study. *International Journal of Obesity Related Metabolic Disorders*, 27: 1242–1249.

Lowell, B.B. and Spiegelman, B.M. (2000) Towards a molecular understanding of adaptive thermogenesis. *Nature*, 404: 652–660.

Lukaski, H.C. (1999) Chromium as a supplement. *Annual Review of Nutrition*, 19: 279–302.

Lukaski, H.C., Bolonchuk, W.W., Siders, W.A., and Milne, D.B. (1996) Chromium supplementation and resistance training: effects on body composition, strength, and trace element status of men. *The American Journal of Clinical Nutrition*, 63: 954–965.

Luke, A., Maki, K.C., Barkey, N., Cooper, R., and McGee, D. (1997) Simultaneous monitoring of heart rate and motion to assess energy expenditure. *Medicine & Science in Sports & Exercise*, 29: 144–148.

Lusk, G. (1924) Animal calorimetry-analysis of the oxidation of mixtures of carbohydrate and fat: a correction. *Journal of Biological Chemistry*, 59: 41–42.

Lutz, P.L. (2002) *The Rise of Experimental Biology: An Illustrated History*, Totowa, NJ: Humana Press.

Lyons, T.P., Riedesel, M.L., Meuli, L.E., and Chick, T.W. (1990) Effects of glycerol-induced hyperhydration prior to exercise in the heat on sweating and core temperature. *Medicine & Science in Sports & Exercise*, 22: 477–483.

Ma, Y., Olendzki, B., Chiriboga, D., Hebert, J.R., Li, Y., Li, W., Campbell, M., Gendreau, K., and Ockene, I.S. (2005) Association between dietary carbohydrates and body weight. *American Journal of Epidemiology*, 161: 359–367.

McArdle, W.D., Katch, F.I., and Katch, V.L. (2001) *Exercise Physiology: Energy, Nutrition, and Human Performance*, 5th edn., Baltimore, MD: Lippincott Williams & Wilkins.

McArdle, W.D., Katch, F.I., and Katch, V.L. (2005) *Sports and Exercise Nutrition*, 2nd edn., Baltimore, MD: Lippincott Williams and Wilkins.

McArdle, W.D., Katch, F.I., and Katch, V.L. (2009) *Sports and Exercise Nutrition*, 3rd edn., Baltimore, MD: Lipincott Williams and Wilkins.

McCartney, N., Spriet, L.L., Heigenhauser, G.I.F., Kowalchuk, J.M., Sutton, J.R., and Jones, N.L. (1986) Muscle power and metabolism in maximal intermittent exercise. *Journal of Applied Physiology*, 60: 1164–1169.

McConell, G., Kloot, K., and Hargreaves, M. (1996) Effect of timing of carbohydrate ingestion on endurance exercise performance. *Medicine & Science in Sports & Exercise*, 28: 1300–1304.

McCormack, J. and Denton, R. (1994) Signal transduction by intra-mitochondrial calcium in mammalian energy metabolism. *News in Physiological Sciences*, 9: 71–76.

McCrory, M.A., Gomez, T.D., Bernauer, E.M., and Molé, P.A. (1995) Evaluation of a new air displacement plethysmograph for measuring human body composition. *Medicine & Science in Sports & Exercise*, 27: 1686–1691.

Macdiarmid, J.I. and Blundell, J.E. (1997) Dietary under-reporting: what people say about recording their food intake. *European Journal of Clinical Nutrition*, 51: 199–200.

Macek, M., Vavra, J., and Novosadova, J. (1976) Prolonged exercise in pre-pubertal boys II: changes in plasma volume and in some blood constituents. *European Journal of Applied Physiology*, 35: 299–303.

McManus, T.J. (2000) Helicobacter pylori: an emerging infectious disease. *The Nurse Practitioner*, 25: 40–48.

McNamara, J.P. and Valdez, F. (2005) Adipose tissue metabolism and production responses to calcium propionate and chromium propionate. *Journal of Dairy Science*, 88: 2498–2507.

McNaughton, L., Dalton, B., and Palmer, G. (1999) Sodium bicarbonate can be used as an ergogenic aid in high-intensity, competitive cycle ergometry of 1 h duration. *European Journal of Applied Physiology and Occupational Physiology*, 80: 64–69.

Madsen, K.L. (2001) The use of probiotics in gastrointestinal disease. *The Canadian Journal of Gastroenterology*, 15: 817–822.

Maehlum, S., Hostmark, A.T., and Hermansen, L. (1977) Synthesis of muscle glycogen during recovery after prolonged, severe exercise in diabetic and nondiabetic subjects. *Scandinavian Journal of Clinical and Laboratory Investigation*, 37: 309–316.

Malavolti, M., Pietrobelli, A., Dugoni, M., Poli, M., Romagnoli, E., DeCristofaro, P., and Battistini, N.C. (2007) A new device for measuring resting energy expenditure (REE) in healthy subjects. *Nutrition, Metabolism & Cardiovascular Diseases*, 17: 338–343.

Mannix, E.T., Stager, J.M., Harris, A., and Farber, M.O. (1990) Oxygen delivery and cardiac output during exercise following oral phosphate-glucose. *Medicine & Science in Sports & Exercise*, 22: 341–347.

Manson, J.E., Stampfer, M.J., Hennekens, C.H., and Willett, W.C. (1987) Body weight and longevity: a reassessment. *Journal of the American Medical Association*, 16: 257: 353–358.

Margaret-Mary, G.W. and Morley, J.E. (2003) Physiology of aging invited review: aging and energy balance. *Journal of Applied Physiology*, 95: 1728–1736.

Marker, J.C., Hirsch, I.B., Smith, L.J., Parvin, C.A., Holloszy, J.O., and Cryer, P.E. (1991) Catecholamines in prevention of hyperglycemia during exercise in humans. *American Journal of Physiology*, 260: E705–E712.

Marschall, H.U. and Einarsson, C. (2007) Gallstone disease. *Journal of Internal Medicine*, 261: 529–542.

Martin, A.D., Ross, W.D., Drinkwater, D.T., and Clarys, J.P. (1985) Prediction of body fat by skinfold caliper: assumptions and cadaver evidence. *International Journal of Obesity*, 9 (Suppl 1): 31–39.

Martin, I.K., Katz, A., and Wahren, J. (1995) Splanchnic and muscle metabolism during exercise in NIDDM patients. *American Journal of Physiology*, 269: E583–E590.

Martinez, L.R. and Haymes, E.M. (1992) Substrate utilization during treadmill running in prepubertal girls and women. *Medicine & Science in Sports & Exercise*, 24: 975–983.

Maughan, R.J. (1991) Fluid and electrolyte loss and replacement in exercise. *Journal of Sports Sciences*, 9: 117–142.

Maxwell, N.S., Gardner, F., and Nimmo, M.A. (1999) Intermittent running: muscle metabolism in the heat and effect of hypohydration. *Medicine & Science in Sports & Exercise*, 31: 675–683.

Melanson, E.L., Sharp, T.A., Seagle, H.M., Donahoo, W.T., Grunwald, G.K., Peters, J.C., Hamilton, J.T., and Hill, J.O. (2002) Resistance and aerobic exercise have similar effects on 24-h nutrient oxidation. *Medicine & Science in Sports & Exercise*, 34: 1793–1800.

Melby, C., Scholl, C., Edwards, G., and Bullough, R. (1993) Effect of acute resistance exercise on postexercise energy expenditure and resting metabolic rate. *Journal of Applied Physiology*, 75: 1847–1853.

Merrill, A.L. and Watt, B.K. (1973) *Energy Value of Foods-Basis and Derivation: Agriculture Handbook No. 74*, Washington, DC: US Department of Agriculture. www.nal.usda.gov/fnic/food-comp/Data/Classics/ah74.pdf.

Meyer, K.A., Kushi, L.H., Jacobs, D.R., Jr., Slavin, J., Sellers, T.A., and Folsom, A.R. (2000) Carbohydrates, dietary fiber, and incident type 2 diabetes in older women. *The American Journal of Clinical Nutrition*, 71: 921–930.

Mikines, K.J., Sonne, B., Farrell, P.A., Tronier, B., and Galbo, H. (1988) Effect of physical exercise on sensitivity and responsiveness to insulin in humans. *American Journal of Physiology*, 254: E248–E259.

Mikines, K.J., Sonne, B., Farrell, P.A., Tronier, B., and Galbo, H. (1989a) Effect of training on the dose–response relationship for insulin action in men. *Journal of Applied Physiology*, 66: 695–703.

Mikines, K.J., Sonne, B., Farrell, P.A., Tronier, B., and Galbo, H. (1989b) Effects of training and detraining on dose–response relationship between glucose and insulin secretion. *American Journal of Physiology*, 256: E588–E596.

Miller, W.C., Koceja, D.M., and Hamilton, E.J. (1997) A meta-analysis of the past 25 years of weight loss research using diet, exercise or diet plus exercise intervention. *International Journal of Obesity Related Metabolic Disorders*, 21: 941–947.

Modlesky, C.M., Cureton, K.J., Lewis, R.D., Prior, B.M., Sloniger, M.A., and Rowe, D.A. (1996) Density of the fat-free mass and estimates of body composition in male weight trainers. *Journal of Applied Physiology*, 80: 2085–2096.

Mokdad, A.H., Bowman, B.A., Ford, E.S., Vinicor, F., Marks, J.S., and Koplan, J.P. (2001) The continuing epidemics of obesity and diabetes in the United States. *Journal of the American Medical Association*, 286: 1195–1200.

Montain, S.J., Sawka, M.N., Latzka, W.A., Valeri, C.R. (1998) Thermal and cardiovascular strain from hypohydration: influence of exercise intensity. *International Journal of Sports Medicine*, 19: 87–91.

Mudambo, K.S., Scrimgeour, C.M., and Rennie, M.J. (1997) Adequacy of food ratings in soldiers during exercise in hot, day-time, conditions assessed by doubly labeled water and energy balance methods. *European Journal of Applied Physiology and Occupational Physiology*, 76: 346–351.

Müller-Lissner, S.A., Kamm, M.A., Scarpignato, C., and Wald, A. (2005) Myths and misconceptions about chronic constipation. *The American Journal of Gastroenterology*, 100: 232–242.

Must, A., Spadano, J., Coakley, E.H., Field, A.E., Colditz, G., and Dietz, W.H. (1999) The disease burden associated with overweight and obesity. *Journal of the American Medical Association*, 282: 1523–1529.

Nadel, E. (1979) Temperature regulation. In *Sports Medicine and Physiology*, Strauss, R., editor, Philadelphia, PA: W.B. Saunders.

Nagy, T.R., Goran, M.I., Weinsier, R.L., Toth, M.J., Schutz, Y., and Poehlman, E.T. (1996) Determinations of basal fat oxidation in healthy Caucasians. *Journal of Applied Physiology*, 80: 1743–1748.

National Heart, Lung, and Blood Institute (1998) Clinical guidelines on the identification, evaluation, and treatment of overweight and obesity in adults: the evidence report. *Obesity Research*, 6:S51–S210.

National Institutes of Health: Obesity Education Initiative (2000) *Practical Guide: Identification, Evaluation, and Treatment of Overweight and Obesity in Adults*, US Department of Health and Human Services. www.nhlbi.nih.gov/guidelines/obesity/prctgd_c.pdf.

National Task Force on the Prevention and Treatment of Obesity (2000) Overweight, obesity, and health risk. *Archives of Internal Medicine*, 160: 898–904.

Nielsen, F.H., Hunt, C.D., Mullen, L.M., and Hunt, J.R. (1987) Effect of dietary boron on mineral, estrogen, and testosterone metabolism in postmenopausal women. *The FASEB Journal*, 1: 394–397.

Nissen, S., Sharp, R., Ray, M., Rathmacher, J.A., Rice, D., Fuller, J.C., Jr., Connelly, A.S., and Abumrad, N. (1996) Effect of leucine metabolite beta-hydroxy-beta-methylbutyrate on muscle metabolism during resistance-exercise training. *Journal of Applied Physiology*, 81: 2095–2104.

Norman, R.A., Thompson, D.B., Foroud, T., Garvey, W.T., Bennett, P.H., Bogardus, C., and Ravussin, E. (1997) Genomewide search for genes influencing percent body fat in Pima Indians: suggestive linkage at chromosome 11q21-q22 – Pima Diabetes Gene Group. *The American Journal of Human Genetics*, 60: 166–173.

Olds, T.S. and Abernethy, P.J. (1993) Post-exercise oxygen consumption following heavy and light resistance exercise. *Journal of Strength & Conditioning Research*, 7: 147–152.

Oppliger, R.A., Case, H.S., Horswill, C.A., Landry, G.L., and Shelter, A.C. (1996) American College of Sports Medicine position stand: weight loss in wrestlers. *Medicine & Science in Sports & Exercise*, 28: ix–xii.

Ortiz, O., Russell, M., Daley, T.L., Baumgartner, R.N., Waki, M., Lichtman, S., Wang, J., Pierson, R.N., Jr. and Heymsfield, S.B. (1992) Differences in skeletal muscle and bone mineral mass between black and white females and their relevance to estimates of body composition. *The American Journal of Clinical Nutrition*, 55: 8–13.

Otis, C.L., Drinkwater, B., Johnson, M., Loucks, A., and Wilmore, J. (1997) American College of Sports Medicine position stand: the female athlete triad. *Medicine & Science in Sports & Exercise*, 29: i–ix.

Paddon-Jones, D., Keech, A., and Jenkins, D. (2001) Short-term beta-hydroxy-beta-methylbutyrate supplementation does not reduce symptoms of eccentric muscle damage. *International Journal of Sport Nutrition and Exercise Metabolism*, 11: 442–450.

Page, T.G., Southern, L.L., Ward, T.L., and Thompson, D.L., Jr. (1993) Effect of chromium picolinate on growth and serum and carcass traits of growing-finishing pigs. *Journal of Animal Science*, 71: 656–662.

Pandolf, K.B., Cadarette, B.S., Sawka, M.N., Young, A.J., Francesconi, R.P., and Gonzalez, R.R. (1988) Thermoregulatory responses of middle-aged and young men during dry-heat acclimation. *Journal of Applied Physiology*, 65: 65–71.

Papa, S. (1996) Mitochondrial oxidative phosphorylation changes in the life span: molecular aspects and pathophysiological implications. *Biochimica et Biophysica Acta*, 1276: 87–105.

Pasquali, R., Cesari, M.P., Besteghi, L., Melchionda, N., and Balestra, V. (1987a) Thermogenic agents in the treatment of human obesity: preliminary results. *International Journal of Obesity*, 11: 23–26.

Pasquali, R., Cesari, M.P., Melchionda, N., Stefanini, C., Raitano, A., and Labo, G. (1987b) Does ephedrine promote weight loss in low-energy-adapted obese women? *International Journal of Obesity*, 11: 163–168.

Pate, R.R., Barnes, C., and Miller, W. (1985) A physiological comparison of performance-matched female and male distance runners. *Research Quarterly for Exercise & Sport*, 16: 606–613.

Paul, G.L. (1989) Dietary protein requirements of physically active individuals. *Sports Medicine*, 8: 154–176.

Pencek, R.R., James, F.D., Lacy, D.B., Jabbour, K., Williams, P.E., Fueger, P.T., and Wasserman, D.H. (2003) Interaction of insulin and prior exercise in control of hepatic metabolism of a glucose load. *Diabetes*, 52: 1897–1903.

Penetar, D., McCann, U., Thorne, D., Kamimori, G., Galinski, C., Sing, H., Thomas, M., and Belenky, G. (1993) Caffeine reversal of sleep deprivation effects on alertness and mood. *Psychopharmacology* 112: 359–365.

Percheron, G., Hogrel, J.Y., Denot-Ledunois, S., Fayet, G., Forette, F., Baulieu, E.E., Fardeau, M., and Marini, J.F. (2003) Effect of 1-year oral administration of dehydroepiandrosterone to 60- to 80-year-old individuals on muscle function and cross-sectional area: a double-blind placebo-controlled trial. *Archives of Internal Medicine*, 163: 720–727.

Phillips, W.T. and Ziuraitis, J.R. (2003) Energy cost of the ACSM single-set resistance training protocol. *Journal of Strength & Conditioning Research*, 17: 350–355.

Pichan, G., Gauttam, R.K., Tomar, O.S., and Bajaj, A.C. (1988) Effect of primary hypohydration on physical work capacity. *International Journal of Biometeorology*, 32: 176–180.

Pittler, M.H., Stevinson, C., and Ernst, E. (2003) Chromium picolinate for reducing body weight: meta-analysis of randomized trials. *International Journal of Obesity Related Metabolic Disorders*, 27: 522–529.

Poehlman, E.C., Melby, C.L., and Badylak, S.F. (1988) Resting metabolic rate and postprandial thermogenesis in highly trained and untrained males. *The American Journal of Clinical Nutrition*, 47: 793–798.

Poehlman, E.T. and Melby, C. (1998) Resistance training and energy balance. *International Journal of Sports Nutrition*, 8: 143–159.

Pollock, M.L., Gettman, L.R., Jackson, A., Ayres, J., Ward, A., and Linnerud, A.C. (1977) Body composition of elite class distance runners. *Annals of the New York Academy of Sciences*, 301: 361–370.

Poole, D.C. and Richardson, R.S. (1997) Determinants of oxygen uptake. *Sports Medicine*, 24: 308–320.

Position of the American Dietetic Association, Dietitians of Canada, and the American College of Sports Medicine (2000) Nutrition and athletic performance. *Journal of the American Dietetic Association*, 100: 1543–1556.

Potteiger, J.A., Nickel, G.L., Webster, M.J., Haub, M.D., and Palmer, R.J. (1996) Sodium citrate ingestion enhances 30 km cycling performance. *International Journal of Sports Medicine*, 17: 7–11.

Powers, S.K. and Howley, E.T. (2001) *Exercise Physiology: Theory and Application to Fitness and Performance*, New York: McGraw Hill.

Powers, S.K. and Howley, E.T. (2007) *Exercise Physiology: Theory and Application to Fitness and Performance*, 6th edn., New York: McGraw Hill.

Powers, S.K. and Howley, E.T. (2009) *Exercise Physiology: Theory and Application to Fitness and Performance*, 7th edn., New York: McGraw Hill.

Pruett, E.D.R. (1970) Glucose and insulin during prolonged work stress in men living on different diets. *Journal of Applied Physiology*, 28: 199–208.

Pullinen, T., Mero, A., MacDonald, E., Pakarinen, A., and Komi, P.V. (1998) Plasma catecholamine and serum testosterone responses to four units of resistance exercise in young and adult male athletes. *European Journal of Applied Physiology*, 77: 413–420.

Radecki, T.E. (2005) Calcium and vitamin D in preventing fractures: vitamin K supplementation has powerful effect. *British Medical Journal*, 331: 108.

Raguso, C.A., Coggan, A.R., Sidossis, L.S., Gastaldelli, A., and Wolfe, R.R. (1996) Effect of theophylline on substrate metabolism during exercise. *Metabolism*, 45: 1153–1160.

Rakowski, W. and Mor, V. (1992) The association of physical activity with mortality among older adults in the longitudinal study of aging (1984–1988). *Journal of Gerontology*, 47: M122–M129.

Randle, P.J., Garland, P.B., Hales, C.N., and Newsholme, E.A. (1963) The glucose–fatty acid cycle: its role in insulin sensitivity and metabolic disturbances of diabetes mellitus. *The Lancet*, 1: 785–789.

Ransone, J., Neighbors, K., Lefavi, R., and Chromiak, J. (2003) The effect of beta-hydroxy beta-methylbutyrate on muscular strength and body composition in collegiate football players. *Journal of Strength & Conditioning Research*, 17: 34–39.

Ravussin, E. and Rising, R. (1992) Daily energy expenditure in humans: measurement in a respiratory chamber and by doubly labeled water. In: *Energy Metabolism: Tissue Determinants and Cellular Corollaries*, Kinney, J.M. and Tucker, H.N., editors, New York: Raven Press, pp. 81–96.

Reaven, G. and Miller, R. (1968) Study of the relationship between glucose and insulin responses to an oral glucose load in man. *Diabetes*, 17: 560–569.

Rebro, S.M., Patterson, R.E., Kristal, A.R., and Cheney, C.L. (1998) The effect of keeping food records on eating patterns. *Journal of the American Dietetic Association*, 98: 1163–1165.

Ren, J.M., Semenkovich, C.F., Gulve, E.A., Gao, J., and Holloszy, J.O. (1994) Exercise induces rapid increases in GLUT4 expression, glucose transport capacity, and insulin-stimulated glycogen storage in muscle. *Journal of Biological Chemistry*, 269: 14396–14401.

Rennie, K., Rowsell, T., Jebb, S.A., Holburn, D., and Wareham, N.J. (2000) A combined heart rate and movement sensor: proof of concept and preliminary testing study. *European Journal of Clinical Nutrition*, 54: 409–414.

Richelsen, B., Pedersen, S.B., Moller-Pedersen, T., and Bak, J.F. (1991) Regional differences in triglyceride breakdown in human adipose tissue: effects of catecholamines, insulin, and prostaglandin E2. *Metabolism*, 40: 990–996.

Richter, E.A. (1996) Glucose utilization, In *Handbook of Physiology*, Rowell, L.B and Shepherd, J.T., editors, New York: Oxford University Press, pp. 912–951.

Richter, E.A., Galbo, H., and Christensen, N.J. (1981) Control of exercise-induced muscular glycogeneolysis by adrenal medullary hormones in rats. *Journal of Applied Physiology*, 50: 21–26.

Richter, E.A., Garetto, L.P., Goodman, M., and Ruderman, N.B. (1982) Muscle glucose metabolism following exercise in the rat: increase sensitivity to insulin. *Journal of Clinical Investigation*, 69: 785–793.

Richter, E.A., Garetto, L.P., Goodman, M., and Ruderman, N.B. (1984) Enhanced glucose metabolism after exercise: modulation by local factors. *American Journal of Physiology*, 246: E476–E482.

Richter, E.A., Mikines, K.J., Galbo, H., and Kiens, B. (1989) Effect of exercise on insulin action in human skeletal muscle. *Journal of Applied Physiology*, 66: 876–885.

Richter, E.A., Ploug, T., and Galbo, H. (1985) Increased muscle glucose uptake after exercise. No need for insulin during exercise. *Diabetes*, 34: 1041–1048.

Roberts, S.B., Fuss, P., Dallal, G.E., Atkinson, A., Evans, W.J., Joseph, L., Fiatarone, M.A., Greenberg, A.S., and Young, V.R. (1996) Effect of age on energy expenditure and substrate oxidation during experimental overfeeding in healthy men. *Journal of Gerontology*, 51: B148–B157.

Robinson, S. (1938) Experimental studies of physical fitness in relation to age. *Arbeitsphysiologie*, 10: 251–323.

Romijn, J.A., Coyle, E.F., Sidossis, L.S., Gastaldelli, A., Horowitz, J.F., Endert, E., and Wolfe, R.R. (1993) Regulation of endogenous fat and carbohydrate metabolism in relation to exercise intensity and duration. *American Journal of Physiology*, 265: E380–E391.

Romijn, J.A., Coyle, E.F., Sidossis, L.S., Rosenblatt, J., Wolfe, R.R. (2000) Substrate metabolism during different exercise intensities in endurance-trained women. *Journal of Applied Physiology*, 88: 1707–1714.

Rooyackers, O.E. and Nair, K.S. (1997) Hormonal regulation of human muscle protein metabolism. *Annual Review of Nutrition*, 17: 457–485.

Rosenzweig, P.H. and Volpe, S.L. (1999) Iron, thermoregulation, and metabolic rate. *Critical Reviews in Food Science and Nutrition*, 39: 131–148.

Ross, R., Dagnone, D., Jones, P.J., Smith, H., Paddags, A., Hudson, R., and Janssen, I. (2000) Reduction in obesity and related comorbid conditions after diet-induced weight loss or exercise-induced weight loss in men: a randomized, controlled trial. *Annals of Internal Medicine*, 133: 92–103.

Rothwell, N.J. and Stock, M.J. (1983) Diet-induced thermogenesis. *Advances in Food & Nutrition Research*, 5: 201–220.

Roubenoff, R., Kehayias, J.J., Dawson-Hughes, B., and Heymsfield, S.B. (1993) Use of dual-energy x-ray absorptiometry in body composition studies: not yet a "good standard". *The American Journal of Clinical Nutrition*, 58: 589–591.

Rowland, T.W. and Rimany, T.A. (1995) Physiological responses to prolonged exercise in premenarcheal and adult females. *Pediatric Exercise Science*, 7: 183–191.

Rowland, T.W., Auchinachie, J.A., Keenan, T.J., and Green, G.M. (1987) Physiological responses to treadmill running in adult and pre-pubertal males, *International Journal of Sports Medicine*, 8: 292–297.

Ruby, B.C., Robergs, R.A., Waters, D.L., Burge, M., Mermier, C., and Stolarczyk, L. (1997) Effects of estradiol on substrate turnover during exercise in amenorrheic females. *Medicine & Science in Sports & Exercise*, 29: 1160–1169.

Sady, S. (1981) Transient oxygen uptake and heart rate responses at the onset of relative endurance exercise in pre-pubertal boys and adult men. *International Journal of Sports Medicine*, 2: 240–244.

Sahlin, K., Tonkonogi, M., and Söderlund, K. (1998) Energy supply and muscle fatigue in humans. *Acta Physiologica Scandinavica*, 162: 261–266.

Sallis, J.F., Buono, M.J., Roby, J.J., Carlson, D., and Nelson, J.A. (1990) The Caltrac accelerompeter as a physical activity monitor for school-age children. *Medicine & Science in Sports & Exercise*, 22: 698–703.

Sallis, J.F., Buono, M.J., and Freedson, P.S. (1991) Bias in estimating caloric expenditure from physical activity in children: implications for epidemiological studies. *Sports Medicine*, 11: 203–209.

Sallis, J.F. and Saelens, B.E. (2000) Assessment of physical activity by self-report: status, limitations, and future directions. *Research Quarterly for Exercise & Sports*, 71: 1–14.

Saltin, B., Houston, M., Nygaard, E., Graham, T., and Wahren, J. (1979) Muscle fiber characteristics in healthy men and patients with juvenile diabetes. *Diabetes*, 28 (1): 93–99.

Saris, W.H.M. (1993) The role of exercise in the dietary treatment of obesity. *International Journal of Obesity*, 17: S17–S21.

Sasaki, N., Kusano, E., Takahashi, H., Ando, Y., Yano, K., Tsuda, E., and Asano, Y. (2005) Vitamin K2 inhibits glucocorticoid-induced bone loss partly by preventing the reduction of osteoprotegerin (OPG). *Journal of Bone and Mineral Metabolism*, 23: 41–47.

Sawaya, A.L., Tucker, K., Tsay, R., Willett, W., Saltzman, E., Dallal, G.E., and Roberts, S.B. (1996) Evaluation of four methods for determining energy intake in young and older women: comparison with doubly labeled water measurements of total energy expenditure. *The American Journal of Clinical Nutrition*, 63: 491–499.

Sawka, M.N. and Pandolf, K.B. (1990) Effect of body water loss on physiological function and exercise performance: fluid homeostasis during exercise. In *Perspectives in Exercise Science and Sports Medicine*, vol. 3, Gisolfi, C.V., and Lamb, D.R., editors, Carmel, IN: Benchmark Press, pp. 1–38.

Sawka, M.N., Young, A.J., Francesconi, R.P., Muza, S.R., and Pandolf, K.B. (1985) Thermoregulatory and blood responses during exercise at graded hypohydration levels. *Journal of Applied Physiology*, 59: 1394–1401.

Sawka, M.N., Young, A.J., Latzka, W.A., Neufer, P.D., Quigley, M.D., and Pandolf, K.B. (1992) Human tolerance to heat strain during exercise: influence of hydration. *Journal of Applied Physiology*, 73: 368–375.

Sawka, M.N., Montain, S.J., and Latzka, W.A. (2001) Hydration effects on thermoregulation and performance in the heat. *Comparative Biochemistry and Physiology Part A: Molecular & Integrative Physiology*, 128: 679–690.

Sawka, M.N., Burke, L.M., Eichner, E.R., Maughan, R.J., Montain, S.J., Stachenfeld, N.S., and the American College of Sports Medicine (2007) American College of Sports Medicine position stand on exercise and fluid replacement. *Medicine & Science in Sports & Exercise*, 39: 377–390.

Schoeller, D.A. and van Santen, E. (1982) Measurement of energy expenditure in human by doubly labeled water method. *Journal of Applied Physiology*, 53: 955–959.

Schoffstall, J.E., Branch, J.D., Leutholtz, B.C., and Swain, D.E. (2001) Effects of dehydration and rehydration on the one-repetition maximum bench press of weight-trained males. *Journal of Strength & Conditioning Research*, 15: 102–108.

Schuenke, M.D., Mikat, R.P., and McBride, J.M. (2002) Effect of an acute period of resistance exercise on excess post-exercise oxygen consumption: implication for body mass management. *European Journal of Applied Physiology*, 86: 411–417.

Schutte, J.E., Townsend, E.J., Hugg, J., Shoup, R.F., Malina, R.M., and Blomqvist, C.G. (1984) Density of lean body mass is greater in blacks than in whites. *Journal of Applied Physiology*, 56: 1647–1649.

Schwartz, M.W., Baskin, D.G., Kaiyala, K.J., and Woods, S.C. (1999) Model for the regulation of energy balance and adiposity by the central nervous system. *The American Journal of Clinical Nutrition*, 69: 584–596.

Seagle, H.M., Strain, G.W., Makris, A., Reeves, R.S., and the American Dietetic Association (2009) Position of the American Dietetic Association: weight management. *Journal of the American Dietetic Association*, 109: 330–346.

Seals, D.R., Hagberg, J.M., Allen, W.K., Hurley, B.F., Dalsky, G.P., Ehsani, A.A., and Holloszy, J.O. (1984) Glucose tolerance in young and older athletes and sedentary men. *Journal of Applied Physiology*, 56: 1521–1525.

Segal, K.R. and Gutin, B. (1983a) Thermic effects of food and exercise in lean and obese women, *Metabolism*, 32: 531–589.

Segal, K.R. and Gutin, B. (1983b) Exercise efficiency in lean and obese women. *Medicine & Science in Sports & Exercise*, 15: 106–107.

Segal, R.S., Presta, E., and Gutin, B. (1984) Thermic effect of food during graded exercise in normal weight and obese men. *The American Journal of Clinical Nutrition*, 40: 995–1000.

Seidell, J.C., Muller, D.C., Sorkin, J.D., and Andres, R. (1992) Fasting respiratory exchange ratio and resting metabolic rate as predictors of weight gain: the Baltimore Longitudinal Study on Aging. *International Journal of Obesity Related Metabolic Disorders*, 16: 667–674.

Seip, R.L. and Weltman, A. (1991) Validity of skinfold and girth based regression equations for the prediction of body composition in obese adults. *American Journal of Human Biology*, 3: 91–95.

Shah, M. and Garg, A. (1996) High-fat and high-carbohydrate diets and energy balance. *Diabetes Care*, 19: 1142–1152.

Shephard, R.J. (2000) Exercise and training in women, part II: influence of menstrual cycle and pregnancy on exercise responses. *Canadian Journal of Applied Physiology*, 25: 35–54.

Sherman, W.M., Costill, D.L., Fink, W.J., and Miller, J.M. (1981) Effect of exercise-diet manipulation on muscle glycogen and its subsequent utilization during performance. *International Journal of Sports Medicine*, 2: 114–118.

Sherman, W.M., Morris, D.M., Kirby, T.E., Petosa, R.A., Smith, B.A., and Frid, D.J. (1998) Evaluation of a commercial accelerometer (Tritrac-R3D) to measure energy expenditure during ambulation. *International Journal of Sports Medicine*, 19: 43–47.

Shier, D., Jackie, B., and Lewis, R. (1999) Chemical basis of life. In *Human Anatomy and Physiology*, New York: WCB/McGraw Hill, pp. 36–58.

Shier, D., Butler, J., and Lewis, R. (2010) *Hole's Human Anatomy and Physiology*, 12th edn. New York: McGraw Hill.

Shikany, J.M. and White, G.L., Jr. (2000) Dietary guidelines for chronic disease prevention. *Southern Medical Journal*, 93: 1138–1151.

Shirreffs, S.M. and Maughan, R.J. (2000) Rehydration and recovery of fluid balance after exercise. *Exercise and Sport Sciences Reviews*, 28: 27–32.

Sial, S., Coggan, A.R., Hickney, R.T., and Klein, S. (1998) Training-induced alterations in fat and carbohydrate metabolism during exercise in elderly subjects. *American Journal of Physiology*, 274: E785–E790.

Siri, W.E. (1961) Body composition from fluid space and density. In: *Techniques for Measuring Body Composition*, Brozek, J. and Henschel, A., editors, Washington, DC: National Academy of Sciences.

Sivan, E., Chen, X., Homko, C.J., Reece, E.A., and Boden, G. (1997) Longitudinal study of carbohydrate metabolism in healthy obese pregnant women. *Diabetes Care*, 20: 1470–1475.

Smith, D.J. and Norris, S.R. (2000) Changes in glutamine and glutamate concentrations for tracking training tolerance. *Medicine & Science in Sports & Exercise*, 32: 684–689.

Smolin, L. and Grovenor, M. (2003) *Nutrition Science and Appplication*, 4th edn., Hoboken, NJ: Wiley & Sons.

Snider, I.P., Bazzarre, T.L., Murdoch, S.D., and Goldfarb, A. (1992) Effects of coenzyme athletic performance system as an ergogenic aid on endurance performance to exhaustion. *International Journal of Sport Nutrition*, 2: 272–286.

Sonne, B., Mikines, K.J., Richter, E.A., Christensen, N.J., Galbo, H. (1985) Role of liver nerves and adrenal medulla in glucose turnover of running rats. *Journal of Applied Physiology*, 59: 1640–1646.

Stager, J.M., Cordain, L., and Becker, T.J. (1984) Relationship of body composition to swimming performance in female swimmers. *Journal of Swimming Research*, 1: 21–26.

Standl, E., Lotz, N., Dexel, T., Janka, H.U., and Kolb, J.K. (1980) Muscle triglycerides in diabetic subjects. *Diabetologia*, 18: 463–469.

Starling, R.D., Trappe, T.A., Short, K.R., Sheffield-Moore, M., Jozsi, A.C., Fink, W.J., Costill, D.L. (1996) Effect of inosine supplementation on aerobic and anaerobic cycling performance. *Medicine & Science in Sports & Exercise*, 28: 1193–1198.

Starritt, E.C., Howlett, R.A., Heigenhauser, G.J., and Spriet, L.L. (2000) Sensitivity of CPT I to malonyl-CoA in trained and untrained human skeletal muscle. *American Journal of Physiology*, 278: E462–E468.

Stearns, D.M., Wise, J.P., Sr., Patierno, S.R., and Wetterhahn, K.E. (1995) Chromium (III) picolinate produces chromosome damage in Chinese hamster ovary cells. *The FASEB Journal*, 9: 1643–1648.

Steen, S.N. and Brownell, K.D. (1990) Patterns of weight loss and regain in wrestlers: has the tradition changed? *Medicine & Science in Sports & Exercise*, 22: 762–768.

Steiner, M. (1999) Vitamin E, a modifier of platelet function: rationale and use in cardiovascular and cerebrovascular disease. *Nutrition Reviews*, 57: 306–309.

Stensrud, T., Ingjer, F., Holm, H., and Stromme, S.B. (1992) L-tryptophan supplementation does not improve endurance performance. *International Journal of Sports Medicine*, 13: 481–485.

Stewart, I., McNaughton, L., Davies, P., Tristram, S. (1990) Phosphate loading and the effects on VO2max in trained cyclists. *Research Quarterly for Exercise & Sport*, 61: 80–84.

Stolarczyk, L.M., Heyward, V.H., Goodman, J.A., Grant, D.J., Kessler, K.L., Kocina, P.S., and Wilmerding, V. (1995) Predictive accuracy of bioimpedance equations in estimating fat-free mass of Hispanic women. *Medicine & Science in Sports & Exercise*, 27: 1450–1456.

Stone, N.J. (1997) Fish consumption, fish oil, lipids, and coronary heart disease. *The American Journal of Clinical Nutrition*, 65: 1083–1086.

Strath, S.J., Bassett, D.R., Jr., Swartz, A.M., and Thompson, D.L. (2001a) Simultaneous heart rate–motion sensor technique to estimate energy expenditure. *Medicine & Science in Sports & Exercise*, 33: 2118–2123.

Strath, S.J., Bassett, D.R., Jr., Thompson, D.L., and Swartz, A.M. (2001b) Validity of the simultaneous heart rate–motion sensor technique for measuring energy expenditure. *Medicine & Science in Sports & Exercise*, 34: 888–894.

Street, D., Nielsen, J.J., Bangsbo, J., and Juel, C. (2005) Metabolic alkalosis reduces exercise-induced acidosis and potassium accumulation in human skeletal muscle interstitium. *Journal of Physiology*, 566: 481–489.

Stroud, M.A., Ritz, P., Coward, W.A., Sawyer, M.B., Constantin-Teodosiu, D., Greenhaff, P.L., and Macdonald, I.A. (1997) Energy expenditure using isotope-labeled water (2H18O), exercise performance, skeletal muscle enzyme activities and plasma biochemical parameters in humans during 95 days of endurance exercise with inadequate energy intake. *European Journal of Applied Physiology*, 76: 243–252.

Stryer, L. (1988) *Biochemistry*, 3rd edn., New York: W.H. Freeman and Company.

Stumvoll, M., Goldstein, B.J., and van Haeften, T.W. (2005) Type 2 diabetes: principles of pathogenesis and therapy. *The Lancet*, 365: 1333–1346.

Stunkard, A.J. and Kaplan, D. (1977) Eating in public places: a review of reports of the direct observation of eating behavior. *International Journal of Obesity*, 1: 89–101.

Stunkard, A.J. and Waxman, M. (1981) Accuracy of self-reports of food intake. *Journal of the American Dietetic Association*, 79: 547–551.

Stunkard, A.J., Foch, T.T., and Hrubec, Z. (1986) A twin study of human obesity. *Journal of the American Medical Association*, 256: 51–54.

Stunkard, A.J., Harris, J.R., Pedersen, N.L., and McClearn, G.E. (1990) The body-mass index of twins who have been reared apart. *The New England Journal of Medicine*, 322: 1483–1487.

Suminski, R.R., Robertson, R.J., Goss, F.L., Arslanian, S., Kang, J., DaSilva, S., Utter, A.C., and Metz, K.F. (1997) Acute effect of amino acid ingestion and resistance exercise on plasma growth hormone concentration in young men. *International Journal of Sport Nutrition*, 7: 48–60.

Svensson, M., Malm, C., Tonkonogi, M., Ekblom, B., Sjödin, B., and Sahlin, K. (1999) Effect of Q10 supplementation on tissue Q10 levels and adenine nucleotide catabolism during high-intensity exercise. *International Journal of Sport Nutrition*, 9: 166–180.

Swain, R.A., Harsha, D.M., and Baenziger, J. (1997) Do pseudoephedrine or phenylpropanolamine improve maximum oxygen uptake and time to exhaustion? *Clinical Journal of Sport Medicine*, 7: 168–173.

Tarnopolsky, L.J., MacDougall, J.D., Atkinson, S.A., Tarnopolsky, M.A., and Sutton, J.R. (1990) Gender differences in substrate for endurance exercise. *Journal of Applied Physiology*, 68: 302–308.

Tarnopolsky, M.A., Atkinson, S.A., Phillips, S.M., and MacDougall, J.D. (1995) Carbohydrate loading and metabolism during exercise in men and women, *Journal of Applied Physiology*, 78: 1360–1368.

Tarnopolsky, M.A., Bosman, M., MacDonald, J.R., Vandeputte, D., Martin, J., Roy, B.D. (1997) Post-exercise protein–carbohydrate and carbohydrate supplements increase muscle glycogen in men and women. *Journal of Applied Physiology*, 83: 1877–1883.

Tepperman, J. and Tepperman, H.M. (1987) *Metabolic and Endocrine Physiology*, 5th edn., Chicago, IL: Year Book Medical Publishers.

Terada, S., Yokozeki, T., Kawanaka, K., Ogawa, K., Higuchi, M., Ezaki, O., Tabata, I. (2001) Effects of high-intensity swimming training on GLUT-4 and glucose transport activity in rat skeletal muscle. *Journal of Applied Physiology*, 90: 2019–2024.

Thomas, C.M., Pierzga, J.M., and Kenney, W.L. (1999) Aerobic training and cutaneous vasodilation in young and older men. *Journal of Applied Physiology*, 86: 1676–1686.

Thornton, M.K. and Potteiger, J.A. (2002) Effect of resistance exercise bouts of different intensities but equal work on EPOC. *Medicine & Science in Sports & Exercise*, 34: 715–722.

Toth, M.J., Arciero, P.J., Gardner, A.W., Calles-Escandon, J., and Poehlman, E.T. (1996) Rates of free fatty acid appearance and fat oxidation in healthy younger and older men. *Journal of Applied Physiology*, 80: 506–511.

Tran, Z.V. and Weltman, A. (1988) Predicting body composition of men from girth measurements. *Human Biology*, 60: 167–175.

Trappe, S.W., Costill, D.L., Goodpaster, B., Vukovich, M.D., and Fink, W.J. (1994) The effects of L-carnitine supplementation on performance during interval swimming. *International Journal of Sports Medicine*, 15: 181–185.

Tremblay, A., Coe, J., and LeBlanc, J. (1983) Diminished dietary thermogenesis in exercise-trained human subjects. *European Journal of Applied Physiology*, 52: 1–4.

Tremblay, A., Fontaine, E., and Nadeau, A. (1985) Contribution of postexercise increment in glucose storage to variations in glucose-induced thermogenesis in endurance athletes. *Canadian Journal of Physiology and Pharmacology*, 63: 1165–1169.

Tremblay, A., Després, J.P., Leblanc, C., Craig, C.L., Ferris, B., Stephens, T., and Bouchard, C. (1990) Effect of intensities of physical activity on body fatness and fat distribution. *The American Journal of Clinical Nutrition*, 51: 153–157.

Tremblay, A., Simoneau, J.A., and Bouchard, C. (1994) Impact of exercise intensity on body fatness and skeletal muscle metabolism. *Metabolism*, 43: 814–818.

Treuth, M.S., Hunter, G.R., Weinsier, R., and Kell, S. (1995) Energy expenditure and substrate utilization in older women after strength training: 24 hour calorimeter results. *Journal of Applied Physiology*, 78: 2140–2146.

Trumbo, P., Schlicker, S., Yates, A.A., Poos, M., the Food and Nutrition Board of the Institute of Medicine and the National Academies (2002) Dietary reference intakes for energy, carbohydrate, fiber, fat, fatty acids, cholesterol, protein and amino acids. *Journal of the American Dietetic Association*, 102: 1621–1630.   .

Tso, P. and Liu, M. (2004) Ingested fat and satiety. *Physiology & Behavior*, 81: 275–287.

Ukropcova, B., McNeil, M., Sereda, O., de Jonge, L., Xie, H., Bray, G.A., and Smith, S.R. (2005) Dynamic changes in fat oxidation in human primary myocytes mirror metabolic characteristics of the donor. *Journal of Clinical Investigation*, 115: 1934–1941.

Van Etten, L.M., Westerterp, K.R., and Verstappen, F.T.J. (1995) Effect of weight-training on energy expenditure and substrate utilization during sleep. *Medicine & Science in Sports & Exercise*, 27: 188–193.

van Staveren, W.A., de Boer, J.O., and Burema, J. (1985) Validity and reproducibility of a dietary history method estimating the usual food intake during one month. *The American Journal of Clinical Nutrition*, 42: 554–559.

Vander, A.J., Sherman, J.H., and Luciano, D.S. (2001) *Human Physiology: The Mechanisms of Body Function*, 7th edn., New York: McGraw Hill.

Varnier, M., Leese, G.P., Thompson, J., and Rennie, M.J. (1995) Stimulatory effect of glutamine on glycogen accumulation in human skeletal muscle. *American Journal of Physiology*, 269: E309–E315.

Vasankari, T.J., Kujala, U.M., Vasankari, T.M., Vuorimaa, T., and Ahotupa, M. (1997) Increased serum and low-density-lipoprotein antioxidant potential after antioxidant supplementation in endurance athletes. *The American Journal of Clinical Nutrition*, 65: 1052–1056.

Vescovi, J.D., Zimmerman, S.L., Miller, W.C., Hildebrandt, L., Hammer, R.L., and Fernhall, B. (2001) Evaluation of the BOD POD for estimating percentage body fat in a heterogeneous group of adult humans. *European Journal of Applied Physiology*, 85: 326–332.

Villareal, D.T. and Holloszy, J.O. (2006) DHEA enhances effects of weight training on muscle mass and strength in elderly women and men. *American Journal of Physiology: Endocrinology and Metabolism*, 291: E1003–E1008.

Vincent, J.B. (2003) The potential value and toxicity of chromium picolinate as a nutritional supplement, weight loss agent and muscle development agent. *Sports Medicine*, 33: 213–230.

Visser, M., Launer, L.J., Deurenberg, P., and Deeg, D.J.H. (1997) Total and sports activity in older men and women: relation with body fat distribution. *American Journal of Epidemiology*, 145: 752–761.

Vukovich, M.D., Costill, D.L., Hickey, M.S., Trappe, S.W., Cole, K.J., and Fink, W.J. (1993) Effect of fat emulsion infusion and fat feeding on muscle glycogen utilization during cycle exercise. *Journal of Applied Physiology*, 75: 1513–1518.

Vukovich, M.D., Costill, D.L., and Fink, W.J. (1994) Carnitine supplementation: effect on muscle carnitine and glycogen content during exercise. *Medicine & Science in Sports & Exercise*, 26: 1122–1129.

Wagenmakers, A.J.M. (1999) Nutritional supplements: effects on exercise performance and metabolism. In *Perspectives in Exercise Science and Sports Medicine*. vol. 12, Lamb, D.R. and Murry, R., editors, Carmel, IN: Cooper Publishing Group, pp. 2007–2259.

Wagner, D.R. (1999) Hyperhydrating with glycerol: implications for athletic performance. *Journal of the American Dietetic Association*, 99: 207–212.

Wagner, D.R., Heyward, V.H., and Gibson, A.L. (2000) Validation of air displacement plethysmography for assessing body composition. *Medicine & Science in Sports & Exercise*, 32: 1339–1344.

Wahren, J., Hagenfeldt, L., and Felig, P. (1975) Splanchnic and leg exchange of glucose, amino acids, and free fatty acids during exercise in diabetes mellitus. *Journal of Clinical Investigation*, 55: 1303–1314.

Wahren, J., Felig, P., and Hagenfeldt, L. (1978) Physical exercise and fuel homeostasis in diabetes mellitus. *Diabetologia*, 14: 213–222.

Wahren, J., Sato, Y., Ostman, J., Hagenfeldt, L., and Felig, P. (1984) Turnover and splanchnic metabolism of free fatty acids and ketones in insulin-dependent diabetics at rest and in response to exercise. *Journal of Clinical Investigation*, 73: 1367–1376.

Wahrenberg, H., Bolinder, J., and Arner, P. (1991) Adrenergic regulation of lipolysis in human fat cells during exercise. *European Journal of Clinical Investigation*, 21: 534–541.

Wallace, J.P. (1997) Obesity. In: *ACSM's Exercise Management for Persons with Chronic Diseases and Disabilities*, Champaign, IL: Human Kinetics, pp. 106–111.

Wallace, M.B., Lim, J., Cutler, A., and Bucci, L. (1999) Effects of dehydroepiandrosterone vs androstenedione supplementation in men. *Medicine & Science in Sports & Exercise*, 31: 1788–1792.

Walsh, J. (2004) Vitamin D and breast cancer: insights from animal models. *The American Journal of Clinical Nutrition*, 80: S1721–S1724.

Wang, J., Thornton, J.C., Kolesnik, S., and Pierson, R.N., Jr. (2000) Anthropometry in body composition: an overview. *Annals of the New York Academy of Sciences*, 904: 317–326.

Wardlaw, G. and Smith, A. (2006) *Contemporary Nutrition*, 7th edn., New York: McGraw Hill.

Washburn, R.C., Cook, T.C., and LaPorte, R.E. (1989) The objective assessment of physical activity in an occupationally active group. *Journal of Sports Medicine and Physical Fitness*, 29: 279–284.

Wasserman, D.H., Lickley, H.L.A., and Vranic, M. (1984) Interactions between glucagon and other counter-regulatory hormones during normoglycemic and hypoglycemic exercise in dogs. *Journal of Clinical Investigation*, 74: 1404–1413.

Wasserman, D.H., Williams, P.E., Lacy, D.B., Green, D.R., and Cherryington, A.D. (1988) Importance of intrahepatic metabolisms to gluconeogenesis from alanine during exercise and recovery. *American Journal of Physiology*, 254: E518–E525.

Wasserman, D.H., Spalding, J.A., Lacy, D.B., Colburn, C.A., Goldstein, R.E., and Cherrington, A.D. (1989) Glucagon is a primary controller of hepatic glycogenolysis and gluconeogenesis during muscular work. *American Journal of Physiology*, 257: E108–E117.

Webster, M.J., Scheett, T.P., Doyle, M.R., and Branz, M. (1997) The effect of a thiamin derivative on exercise performance. *European Journal of Applied Physiology and Occupational Physiology*, 75: 520–524.

Weglicki, W., Quamme, G., Tucker, K., Haigney, M., and Resnick, L. (2005) Potassium, magnesium, and electrolyte imbalance and complications in disease management. *Clinical and Experimental Hypertension*, 27: 95–112.

Weinsier, R.L., Nelson, K.M., Hendsrud, D.D., Darnell, B.E., Hunter, G.R., and Schutz, Y. (1995) Metabolic predictors of obesity: contribution of resting energy expenditure, thermic effect of food, and fuel utilization to four year weight gain of post-obese and never-obese women. *Journal of Clinical Investigation*, 95: 980–985.

Weinsier, R.L., Nagy, T.R., Hunter, G.R., Darnell, B.E., Hensrud, D.D., and Weiss, H.L. (2000) Do adaptive changes in metabolic rate favor weight regain in weight-reduced individuals? An examination of the set-point theory. *The American Journal of Clinical Nutrition*, 72: 1088–1094.

Welk, G.J. and Corbin, C.B. (1995) The validity of the Tritrac-R3D activity monitor for assessment of physical activity in children. *Research Quarterly for Exercise & Sport*, 66: 202–209.

Weltman, A., Seip, R.L., and Tran, Z.V. (1987) Practical assessment of body composition in adult obese males. *Human Biology*, 59: 523–555.

Weltman, A., Levine, S., Seip, R.L., and Tran, Z.V. (1988) Accurate assessment of body composition in obese females. *The American Journal of Clinical Nutrition*, 48: 1179–1183.

Whitney, E. and Rolfes, S. (2005) *Understanding Nutrition*, 10th edn., Belmont, CA: Thomson Wadsworth.

Willett, W.C., Sampson, L., Stampfer, M.J., Rosner, B., Bain, C., Witschi, J., Hennekens, C.H., and Speizer, F.E. (1985) Reproducibility and validity of a semiquantitative food frequency questionnaire. *American Journal of Epidemiology*, 122: 51–65.

Willett, W.C., Stampfer, M., Manson, J., and VanItallie, T. (1991) New weight guidelines for Americans: justified or injudicious? *The American Journal of Clinical Nutrition*, 53: 1102–1103.

Williams, M.H. (2005) *Nutrition for Health, Fitness, and Sport*, 7th edn., New York: McGraw Hill.

468

Williams, M.H., Kreider, R.B., Hunter, D.W., Somma, C.T., Shall, L.M., Woodhouse, M.L., and Rokitski, L. (1990) Effect of inosine supplementation on 3-mile treadmill run performance and VO2 peak. *Medicine & Science in Sports & Exercise*, 22: 517–522.

Willoughby, D.S., Chilek, D.R., Schiller, D.A., and Coast, J.R. (1991) The metabolic effects of three different free weight parallel squatting intensities. *Journal of Human Movement Studies*, 21: 51–67.

Wilmore, J.H. (1995) Variations in physical activity habits and body composition. *International Journal of Obesity*, 19: S107–S112.

Wilmore, J.H. (1996) Increasing physical activity: alterations in body mass and composition. *The American Journal of Clinical Nutrition*, 63: S456–S460.

Wilmore, J.H. and Brown, C.H. (1974) Physiological profiles of women distance runners. *Medicine & Science in Sports & Exercise*, 6: 178–181.

Wilmore, J.H. and Costill, D.L. (2004) Sex difference in sports and exercise. In: *Physiology of Sports and Exercise*, 3rd edn., Champaign, IL: Human Kinetics, pp. 566–602.

Wilmore, J.H., Parr, R.B., Ward, P., Vodak, P.A., Barstow, T.J., Pipes, T.V., Grimditch, G., and Leslie, P. (1978) Energy cost of circuit weight training. *Medicine & Science in Sports & Exercise*, 10: 75–78.

Wing, R.R. and Phelan, S. (2005) Long-term weight loss maintenance. *The American Journal of Clinical Nutrition*, 82: S222–S225.

Wolfe, R.R., Peters, E.J., Klein, S., Holland, O.B., Rosenblatt, J., Gary, H., Jr. (1987) Effect of short-term fasting on lipolytic responsiveness in normal and obese human subjects. *American Journal of Physiology*, 252: E189–E196.

Woods, S.C., Seeley, R.J., Porte, D., Jr. and Schwartz, M.W. (1998) Signals that regulate food intake and energy homeostasis. *Science*, 280: 1378–1383.

Yamamoto, J.B., Yamamotoa, B.E., Yamamotoa, P.P., and Yamamoto, L.G. (2008) Epidemiology of college athlete sizes, 1950s to current. *Research in Sports Medicine*, 16: 111–127.

Yanagimoto, S., Aoki, K., Horikawa, N., Shibasaki, M., Inoue, Y., Nishiyasu, T., and Kondo, N. (2002) Sweating response in physically trained men to sustained handgrip exercise in mildly hyperthermic conditions. *Acta Physiologica Scandinavica*, 174: 31–39.

Young, L.R. and Nestle, M. (2002) The contribution of expanding portion sizes to the US obesity epidemic. *American Journal of Public Health*, 92: 246–249.

Zachwieja, J.J., Witt, T.L., and Yarasheski, K.E. (2000) Intravenous glutamine does not stimulate mixed muscle protein synthesis in healthy young men and women. *Metabolism*, 49: 1555–1560.

Zanconato, S., Buchthal, S., Barstow, T.J., and Cooper, D.M. (1993) 31P-magnetic resonance spectroscopy of leg muscle metabolism during exercise in children and adults. *Journal of Applied Physiology*, 74: 2214–2218.

Zinman, B., Zuniga-Guajardo, S., and Kelly, D. (1984) Comparison of the acute and long-term effects of physical training on glucose control in type 1 diabetes. *Diabetes Care*, 7: 515–519.

Zuliani, U., Bonetti, A., Campana, M., Cerioli, G., Solito, F., and Novarini, A. (1989) The influence of ubiquinone (Co Q10) on the metabolic response to work. *Journal of Sports Medicine and Physical Fitness*, 29: 57–62.

Index

Page numbers in *italics* denote tables, those in **bold** denote figures.